Politics, Statistics and Weather Forecasting, 1840–1910

Weather forecasting is the most visible branch of meteorology and has its modern roots in the nineteenth century when scientists redefined meteorology in the way weather forecasts were made, developing maps of isobars, or lines of equal atmospheric pressure, as the main forecasting tool. This book is the history of how weather forecasting was moulded and modelled by the processes of nation-state building and statistics in the Western world.

Aitor Anduaga is an Ikerbasque Research Professor (Ikerbasque: Basque Foundation for Science) at the Basque Museum of the History of Medicine and Science, University of the Basque Country.

Routledge Studies in the History of Science, Technology and Medicine

For the full list of titles in the series, please visit: https://www.routledge.com/
Routledge-Studies-in-the-History-of-Science-Technology-and-Medicine/
book-series/HISTSCI

Politics, Statistics and Weather Forecasting, 1840–1910

Taming the Weather

Aitor Anduaga

Routledge
Taylor & Francis Group

LONDON AND NEW YORK

First published 2020
by Routledge
2 Park Square, Milton Park, Abingdon, Oxon OX14 4RN

and by Routledge
605 Third Avenue, New York, NY 10017

First issued in paperback 2021

Routledge is an imprint of the Taylor & Francis Group, an informa business

British Library Cataloguing-in-Publication Data
A catalogue record for this book is available from the British Library

Library of Congress Cataloging-in-Publication Data
Names: Anduaga, Aitor, author.
Title: Politics, statistics and weather forecasting, 1840–1910 : taming the weather / Aitor Anduaga.
Description: Abingdon, Oxon ; New York : Routledge, 2020. | Includes bibliographical references and index.
Identifiers: LCCN 2019016481 (print) | LCCN 2019018332 (ebook)
Subjects: LCSH: Weather forecasting—Methodology—History—19th century. |Weather forecasting—Mathematical models—History—19th century. | Weather forecasting—Political aspects—History—19th century. | Science and state—History—19th century.
Classification: LCC QC995 (ebook) | LCC QC995.A52 2020 (print) | DDC 551.6309/034—dc23
LC record available at https://lccn.loc.gov/2019016481

ISBN 13: 978-0-367-78550-5 (pbk)
ISBN 13: 978-0-367-24459-0 (hbk)

Typeset in Times New Roman
by codeMantra

In Memoriam

Javier Egaña
(1936–2018)

Ez da, ez, lan makala izango,
zuk, gehienetan isilean,
hitzekin baino ekintzekin,
emandako irakaspenak jarraitzea:
zintzotasuna, apaltasuna
eta umiltasuna.

Contents

List of figures

List of tables

Acknowledgements

For assistance with reference to archive materials, I am grateful to Jocelyne Carminati, at the Archives départementales de la Côte-d'Or, Dijon, France; the staff at the Bibliothèque de l'Observatoire de Paris; Olivier Damme, at the Archives de l'Académie Royale des Sciences, des Lettres et des Beaux-Arts de Belgique, Brussels; Martín María Morales SJ, at the Archivio Storico, Pontificia Università Gregoriana, Rome; and the staff at the National Archives, Kew, England. I also appreciate the assistance of Amaia Guerrero, at the Library of the University of the Basque Country, Bilbao; and Elena Morato, Maribel Castillo and Elena Rioja, at the Library of the Agencia Estatal de Meteorología, Madrid.

There are also a myriad of individuals who, with their assistance and guidance, have helped in the construction of this work in various ways and at diverse levels. I would especially like to thank Josep Batlló, from the University of Lisbon; Pasqual Bernat, from the Centre d'Història de la Ciència, Barcelona; Ileana Chinnici, from the INAF-Osservatorio Astronomico di Palermo, Italy; Luigi Iafrate, from the Unità di Ricerca per la Climatologia e la Meteorologia applicate all'Agricoltura, CREA-CMA, Rome; Sylvie Jourdain from the Direction de la Climatologie et des Service Climatiques, Météo-France, Toulouse; and Pierre Paillot from the Bibliothèque Météo-France, Paris. In addition, Amy Rodgers and Daisy Lopez provided useful suggestions in the editing stage.

At the University of California, San Diego, the Scripps Institution of Oceanography honoured me with the Ritter Memorial Fellowship in 2019, a fellowship that gave me the opportunity to deliver some lectures and exchange views with colleagues, enhancing my perceptions of what was important in this historical account and how to present it. Art Miller, at Scripps, and Tal Golan, at the Department of History of the mentioned university, provided invaluable assistance during my stay. Kara Henry and Amy Butros also provided priceless assistance with archival and library sources. At Routledge publishers, an imprint of Taylor and Francis Group, I am indebted to

Robert Langham, my editor, for the confidence placed in me, and to Dana Moss, editorial assistant, and Assunta Petrone, project manager.

Finally, I am grateful to the Spanish Ministry of Economy and Competitiveness and the European Regional Development Fund, for the help provided from 2016 to 2018, the research period of the present work (project HAR2015–67831-P [MINECO/FEDER)].

Abbreviations

ABIF	Archives de la Bibliothèque de l'Institut de France
ADCO	Archives départementales de la Côte-d'Or
ANF	Archives Nationales, France
AOP	Archives de l'Observatoire de Paris
APUG	Archivio della Pontificia Università Gregoriana, Rome, "Correspondencia Angelo Secchi, S.J."
ARCSO	*Annual Report of the Chief Signal Officer made to the Secretary of War for the Year...*
ARSB	Archives of the Académie Royale des Sciences de Belgique, Brussels
BAAS	British Association for the Advancement of Science
BOP	Bibliothèque de l'Observatoire de Paris
CR	*Comptes Rendus Hebdomadaires des Séances de l'Académie des Sciences*, Paris
NAUK	National Archives of England, Wales and the United Kingdom, Kew
PP	Parliamentary Papers, UK
RS	HS Royal Society of London, Archives, Herschel Papers

Introduction

Science deals with methods and principles. At different times and in different and varied disciplines, scientists have developed inductive and deductive modes of reasoning, field- and laboratory-based methodologies, and theoretical and experimental approaches to the problems of the natural world. The diversity of scientific methodology has persisted through the ages, in spite of the attempts often made to establish and maintain the hegemony of particular approaches. Science, however, also deals with the organization that renders all these applications possible. As important as methods and principles is the organizational environment within which the scientific profession is conducted. The observatory sciences are a paradigmatic case in this regard, in which cognitive aspects and organizational aspects are interrelated and mutually reinforcing.

Interestingly, historians of science demonstrated through case studies that scientific principles are affected by social forces.[1] History is a witness of ideologies and currents of thought that shaped scientific knowledge. It is also a witness of how industry and commerce can do the same.[2] However, if it has been proven that principles are permeable and vulnerable to social environments, what about scientific methods? Might they be affected by such environments? If so, how and to what extent? These questions lead to a central problem of the history and philosophy of science: if methodological approaches are conditioned by social forces, what does this say about the nature of scientific method? And if the scientific method is fragile

---

1 A widely cited example is Paul Forman's (1971) argument that the cultural values prevalent in Weimar Germany in the 1920s, through their emphasis on acausality and individuality, powerfully contributed to the acceptance of quantum mechanics. Forman argues that post-war intellectual revolts against rationalist stances advocating Newtonian causality, determinism, and materialism exerted a direct influence on scientists' conceptions in quantum mechanics. For contemporary perspectives on the Forman thesis, see Carson, Kojevnikov, and Trischler (2011).
2 For the emergence of realist attitudes among interwar geophysicists towards the entities of the upper atmosphere and the Earth's crust as a result of the influence of industrial and commercial environments on the scientific community, see Anduaga (2016).

and socially vulnerable, how can we justify its use in social and political decision-making?

Meteorology can serve here as a testing ground. When we think about modern meteorology we often envision weather maps with fronts, depressions and anticyclones (i.e. low- and high-pressure systems), and lines connecting places of equal atmospheric pressure, called isobars. These are characterizing features that we take for granted, as the usual way of representing weather changes. Yet isobaric lines have not always enjoyed this role, and in fact their adoption as a standard mode of representation is the result of a relatively recent development. The first serious attempts to study the patterns of meteorological elements by graphic techniques happened during the consolidation of the modern Western states in the eighteenth century. However, it is especially when the nation-state building and its consequent centralism imposed themselves in Europe and America, in the second half of the nineteenth century, and after the revolutions of 1848 (when there were republican revolts against European monarchies), that weather cartography became accepted practice and attracted the legitimate interest of state rulers and meteorologists alike.

In retrospect, this relationship between weather maps and state building may seem surprising to some people, although many would understand it. We live in a society in which we see weather representations on a daily basis—representations that allow us to assimilate a wealth of information at a glance, once their graphic codes are learnt. Maps inform us of what the weather will be in nearby and distant places, help create an imaginary national space, and reinforce our notions of common belonging. In the broad sense of the word, maps are, as geographer David N. Livingstone suggests, 'repositories of trust' that are regarded by the public as 'an accurate representation of the world under scrutiny'. Owing to the impression of objectivity, exactitude, and precision that they convey, maps have been used and viewed as 'an analogy for scientific theory itself'.[3]

And yet, as Livingstone himself holds and many scholars have shown, the idea that maps are 'a straightforward representation of reality is deceptively simple'.[4] Rather than accurate and credible repositories of knowledge, maps are controlled fictions or, at least, projections of what is intended to be represented. As projections, they are 'selective representations of reality' that have been traced to emphasize particular spatial relationships while distorting the very reality in one way or another.[5] Every weather map is a two-dimensional projection of a three-dimensional atmospheric reality. Weather maps must necessarily simplify the complexity of natural phenomena in order to be informative and comprehensible for the public. Or, to

3 Livingstone (2003, p. 154).
4 Ibid.
5 For the idea of maps as 'selective representations of reality' in the context of their application to politics, see: Black (1997, p. 11).

put it simply, we should show, as map historian Mark Monmonier suggests, a 'healthy degree of scepticism' about the reality they claim to represent.[6]

Weather maps may be viewed as the systematic publication of the meteorological knowledge collected by government-sponsored institutions. But they also play a more subtle and less visible role as a symbol of power, a symbol that is associated with the maintenance of order and the reinforcement of status quo and security. 'Maps are preeminently a language of power, not of protest'—J. Brian Harley wrote in his essay on maps, knowledge, and power.[7] The means of weather map production, whether official or commercial (as was the case in the United States), were and are still controlled by holders of power.[8] The weather map serves as a symbol of national meteorological institutes, reflecting not only the knowledge produced but also the ruling power and the public service rendered. For all these reasons, the maps that are viewed as an impersonal and unproblematic knowledge 'desocialise' the reality they represent. They transmit the notion of a 'socially empty space'. Paraphrasing Harley, we can say that any meteorological history that ignores the social and political significance of weather mapping relegates itself to being 'ahistorical'.[9]

In addition to its close link with nation-state building, the nineteenth-century development of meteorology—and weather mapping, in particular—parallels the rise of statistics and its related historical processes, which historians describe through the *etiquettes* of industrialization, metropolization, globalization, and the formation of an information society. Because meteorology deals with large amounts of data, it might seem logical to us to represent these complex natural processes by graphic methods; some scholars have even gone further in highlighting the role played by statistical science, in its attempt to summarize, through mathematical tools (such as probability, mean values, etc.), a diversity that is presumably unmanageable. Yet the reason why we envisage weather forecasts the way we do—that is, as expressed by isobaric maps—is, in turn, the result of the rise of statistics to a position of influence and hegemony—a rise of statistics not only as a mode of graphic representation but also as a politico-administrative tool in nation-state building.

In this regard, it is necessary to clarify here how I use the term 'statistics' as it has had at least two meanings throughout its turbulent history. In its modern sense, the term 'statistics' concerns, on one hand, the organization of data collecting and processing whether by the state or by private agencies, and, on the other hand, the science for the mathematical treatment of

---

6 Monmonier (1996, p. 1).
7 Harley (1988, p. 301).
8 According to Harley (1988, pp. 289–94), the undeclared means of domination can be found in the 'hidden rules' of cartographic imagery, rules related to map geometry, omissions in the content of maps, and hierarchical relations in representation.
9 Harley (1988, p. 303).

quantitative data. This double meaning actually symbolizes the convergence of several and historically distinct traditions in the nineteenth century: that of administrative records and surveys about social issues, fostered by the Western nation-states, within the framework of the German 'statistics' or the English 'political arithmetic'; and that of the practice of measurements, typical of the eighteenth-century natural sciences, like astronomy.[10] In this book I incorporate the two senses—the politico-administrative and the cognitive. So when I hold that statistics contributed to the shaping of weather forecasting, I refer to both its contribution as a 'mode of representation' to the imagining of a national meteorological space, and the role its methods and principles played in the treatment of meteorological data.[11]

While scholars have come to analyze the use of statistical methods and principles in the treatment of weather and climatological data, too little attention has been paid to the influence of organizational and institutional statistics on the practice of meteorology—especially in the context of nation-state building.[12] The situation is slightly different with respect to weather maps, whose perceived precision and exactitude were often viewed as an analogy for scientific knowledge about the atmosphere. Most historians simply accepted their objectivity without reservation or hesitation, so that surface maps of isobars, for example, were taken as true representations of storm motion in a given place and time.[13] Arthur H. Robinson, for example, carefully examined the historical genealogy of the isolines used in meteorology and thematic cartography in general, showing how some common symbols evolved and even migrated from one area to another, and what forces were involved in their origin and spread.[14] It is worth noting, however, that these genealogies have often expressed in almost teleological terms of attribution of a function or natural end to concrete processes, and of acceptance of an objective cartographic standard. Although they may occasionally mention particular social contexts, they tend to convey the idea that weather patterns could be objectively and scientifically mapped and that the knowledge generated was impersonal and socially 'neutral'. Only recently have historians such as Katharine Anderson begun to question the

10 There is obviously a wide body of literature on this convergence of traditions in the history of statistics. Some classic works would include Lazarsfeld (1970), Porter (1986), and Stigler (1986). I also use additional sources on this subject throughout this research work.

11 I use the term 'representation' in the sense suggested by Roger Chartier (1988, pp. 13–14), i.e. as to its capability to order and classify the social. For this French historian, a member of the Annales school, representations cannot be separated from the social practices and institutions whereby those modes of classification and ordering become socially accepted.

12 For a survey of the history of the statistical method in meteorology, see Sheynin (1984). On the application of the method of least squares to the treatment and analysis of meteorological data in the 1850s, especially by Dutch polymath Christophorus H.D. Buys Ballot, see Boumans (2015, pp. 68–76).

13 See, for example, Robinson and Wallis (1967, 1982), Cox (2002).

14 Robinson (1971, 1982), Robinson and Wallis (1967).

appropriateness of these approaches, as part of an attempt to problematize the dominant narratives and practices in meteorological mapping.[15] Yet even these attempts to show the practice of weather mapping as a means of elucidating the relationship between local and global knowledge are generally treated as something that is neutral and unconnected to any process of nation-state building.[16]

However, I will show that the use of isobaric maps and statistics to analyze weather patterns was not an intrinsically valuable means of representation, nor was the believability of these means unanimously assumed. Rather, their application was the result of a choice between several approaches. Before they were used to create a national meteorological space, a series of state interests related to order, security, and surveillance came into play. Meteorologists had to make clear decisions on what exactly should be measured and represented and how such observations were to be organized—that is, how, in the end, was one to proceed to transform an abstract and physically ambiguous concept such as the weather into a discernible and predictable object?[17] Even the very notion of synoptic chart—which was first defined by Robert FitzRoy in 1863 as the graphic representation of 'consecutive simultaneous states of the atmosphere'—was not entirely unproblematic.[18] Moreover, the methodological and institutional mechanisms of data gathering and recording also played an important role in establishing what meteorological elements were chosen as a workable way of quantifying the weather, socially consolidating the scientific authority of the devices adopted. To clarify this idea in a broader sense, *Politics, Statistics and Weather Forecasting* shows how the process of learning to both see and foresee the weather was produced, shaped, and without doubt constrained by the various actors and state interests that enabled it.

The problem of weather forecasting

The story of weather forecasting illustrates how social factors and environments influenced the choice of scientific methods. In the second half of the nineteenth century, a number of scientists—most notably Hippolyte Marié-Davy, under Urbain Le Verrier's direction—proposed that the advance of *depressions* (or low-pressure areas) could be forecast by maps of isobars. From the 1870s on, meteorologists and forecasters from different national

---

15 Anderson (2006, pp. 69–71).
16 One of the few exceptions to this rule is the work by Deborah Coen (2006), who has examined the resurgence of weather science in Vienna in the 1920s and its relationship with non-scientific narratives on the Austrian nation and the more amorphous entity known as Central Europe. Even so, her study falls outside the nineteenth-century processes of nation-state building.
17 On the transformation of abstract qualities into numerical values, see Gould (1981).
18 FitzRoy (1863, p. 103).

weather services embraced this method and assumed that the horizontal distribution of atmospheric pressure (i.e. the 'baric field') determined weather changes. Yet fifty years later, the basic central idea behind this method—that the weather travelled through maps, being determined by the baric field—began to be uniformly rejected, not simply as unproved, but as wrong and physically absurd. This is how Vilhelm Bjerknes, the Bergen school's leader and precursor of the analysis of fronts in weather forecasting, remembered this rejection: 'During fifty years meteorologists all over the world looked at weather maps without discovering their most important causes. I only gave the right kind of maps to the right young men'.[19] Here the question arises: why did distinguished scientists almost universally accept as valid a method then adamantly rejected as wrong, unfounded, unsustainable?

The most common argument which may be found in many meteorology textbooks and some works on the history of science is that weather forecasting during its early era (1860–1920) was *too empirical*. Because it was largely based on techniques of extrapolation, rules of thumb, and intuition, scholars have concluded that it was not *sufficiently* scientific. But this is a clear example of a rational reconstruction in the history of science. Because physical and theoretical principles played little role, scholars have tended to assume that if forecasting was practised by earlier scientists, then those scientists should be viewed as empiricists, forecasters, or prophets, more than anything else. After all, as they like to say, 'forecasting was more an art than a science'.[20] This interpretation assumes that there is one standard scientific method, and early forecasting methods do not meet those standards. The reasons for this are various: in part, because of the minor role played by theoretical physics; in part, because of the scarcity of observations, especially for the upper air and over the oceans; and because of the fair amount of guesswork based on intuition and personal judgement. In other words, early methods were clearly inadmissible by today's standards.

However, the inadmissibility of early forecasting methods does not explain by itself why the isobaric mapping method was accepted almost universally. It might explain why later meteorologists rejected it as inappropriate, but it says nothing about the reasons why earlier scientists adopted it so convincingly. Fortunately, historians of science have recently begun to identify the values and factors that favoured the development of synoptic forecasting techniques during the second half of the nineteenth century. Fabien Locher attributes this development to the emergence of a certain 'intellectual and material culture of dynamic meteorology from 1850 onwards'.[21] This culture was, in his view, inspired by important technical innovations, especially in

19 Quoted by Reed (1977, p. 393), extracted from *In Memory of Vilhelm Bjerknes on the 100th Anniversary of His Birth*. Oslo: Universitetsforlaget, 1962, p. 18.
20 Lynch (2006, p. 4).
21 Locher (2007, p. 492). For a cultural history of observatory techniques, see Aubin, Bigg, and Sibum (2010, esp. introduction).

the field of graphic and cartographic representations. This culture appears so clearly defined to Locher that he does not hesitate to state that it was the Paris Observatory which set the pace, with the introduction by Le Verrier of electromechanical devices and a division of labour based on industrial production techniques.[22]

Yet, the thesis that there was one intellectual and material culture is false. In the 1850s and 1860s, the most important identifying feature of meteorology was the confrontation of *two* rival conceptions of how air motion should be studied: one, the Eulerian approach, which was based on the synoptic representation of isobars and the determining role of pressure distribution in weather conditions; the other, the Lagrangian approach, which emphasized the air trajectories and streamlines while distrusting the weather interpretations in merely isobaric terms. This book shows how the former was first adopted by French and Italian weather scientists and then extended to other countries while the latter was well regarded in small scientific circles in Britain and Germany.[23] By the 1880s, different techniques (such as the search for statistical regularities, weather types, etc.) had been proposed for weather forecasting, all of them leading to the main feature that would define meteorology until the 1910s: the hegemonic position of the Eulerian approach.

Confrontation between the two approaches defined the conceptual framework of the nineteenth-century weather forecasting. Most importantly, it brought with it a social component that shaped how meteorologists conceptualized weather forecasting and envisaged observational systems and their respective world views. For while the defenders of the Lagrangian approach advocated autonomy, local power, and the freedom of individual judgement, those who championed the Eulerian approach defended the centralism, authority, and state control that would end up triumphing in most national weather services.

So why was the Eulerian approach adopted? My argument is that the Paris Observatory's scientists (led by Le Verrier) altered their usual scientific methods and adopted the Eulerian approach to storm warning not because

---

22 In his well-documented and thorough work *Le savant et la tempête. Étudier l'atmosphère et prévoir le temps au XIXe siècle*, Locher (2008) places the rise of synoptic forecasting techniques in the context of a society transformed by industrialization, the expansion of press and telegraph, and the globalization of exchanges—that is, as a 'complex social construction, sometimes precarious and controversial, sometimes victorious'.

23 Besides Locher (2008), for an outline of the institutionalization of meteorology in France see Davis (1984) and Fierro (1991); and for the case of Italy, Iafrate (2008) and Eredia (1914). On the history of nineteenth-century meteorology in Britain, see in particular Burton (1986, 1988), Halford (2004), Anderson (2005), Walker (2012); and for the case of Germany, see the articles in the collective volume edited by Tetzlaff, Lüdecke, and Behr (2008) and Wege (2002). Some classic works on the history of meteorology in general in the nineteenth and twentieth centuries include Brunt (1951), Khrgian (1970), Schneider-Carius (1975), Kutzbach (1979), Nebeker (1995), Lynch (2006), and Harper (2008).

there were substantial theoretical grounds (there was none), but because they were impelled by the political opportunism and the needs for military and maritime security arisen in the months following the 1854 Black Sea storm.[24] Practical aims and needs, rather than physical principles, were the reason for adopting such an approach. For Le Verrier and his followers, the Eulerian approach became an effective tool for achieving his goals of order, control, and centralization in the organization of scientific activities. It was a means to both scientific and sociopolitical ends.

In this book, I argue that *scientific methods are not only affected by social forces*, but these forces can likewise have a *selective effect* on such methods. This selection between two approaches was not so much a process of state-backed experts who wished to impose a particular scientific method from the highest echelons of power, but rather a delicate intersection of wide-ranging state interests that transformed weather phenomena into seemingly precise graphic forms that were perceived as scientifically predictable.

Two approaches: Eulerian and Lagrangian

Among possible examples of the influence of social factors and environments on scientific methods, the case of weather forecasting deserves special attention. It is well known that science is inseparable from the sociocultural traditions from which it emerges. But perhaps this is all the more true in our case. While meteorology is an institutionalized branch of the science, weather forecasting is more than this. Weather forecasting is both a public service and a professional occupation—it has, therefore, a clear social dimension. Science is only one aspect, as far as the forecaster is concerned, and, as a prestigious meteorologist states, 'perhaps not always the most important; equally important is the organisational environment within which the profession is conducted', along with a wise judgement.[25]

Weather forecasting is concerned with air motions or fluid dynamics. There has been, and there are, two ways to describe fluid flows. In the first, known as the Eulerian approach, fluid flow properties are expressed as 'fields', and each field is expressed as a function of space and time. Here the observer is concerned about the pressure, velocity, etc., of a fluid flow at a given place, rather than about the location or velocity of any particular particle. This approach produces isobars and wind streamlines represented

---

24 The idea that the Black Sea storm (also known as the Great Storm of 14–15 November 1854 in Crimea) was the starting point of modern meteorology is a common theme in the history of science. See, e.g., Danjon (1946), Landsberg (1954), Cox (2002, pp. 85–90), Locher (2008), Barboza (2012), and Lequeux (2013, pp. 280–84). However, none of them has related this historical episode to the adoption of an Eulerian approach to the predictive study of storms.

25 As British meteorologist Reginald C. Sutcliffe (1952, p. 291) held in his article 'Principles of synoptic weather forecasting'.

on a map. By contrast, in the Lagrangian approach individual fluid particles are tracked, much like the tracking of billiard balls in a physics experiment. The particles or parcels of air are 'marked', their velocities, positions, etc., being expressed as a function of time. Here the observer is concerned about following a particular air parcel as it moves through space and time. Instead of maps, this approach produces pathways and trajectory analyses.[26]

However, it is misleading to interpret the adoption of any of these approaches as the mere application of Euler's equations (or their Lagrangian equivalent) to the patterns of air motion, as deduced from the textbooks on fluid mechanics.[27] According to these textbooks, one can mathematically formulate the conservation laws involving fluid flow by following either an Eulerian or a Lagrangian approach.[28] The two approaches described in this book and which so antagonistically confronted each other in the 1850s and the 1860s brought with them different observing systems and world views. Depending on the approach adopted, the groups of scientists discussed here often held diametrically opposing views about the organization of observations and the graphic representation of physical properties, and without necessarily implying the use of any mathematical equation.

The Eulerian and Lagrangian specifications of the flow field deal not only with equations but also with organizational aspects and observational frames of reference. The Eulerian insight is a way of looking at fluid motion from fixed observation points in which the fluid flows as time passes. It is like sitting on the bank of a river and watching the water pass. One places stationary instruments at some fixed point near the ground (usually in a station), and measures weather variables over time. Usual instruments are the barometer, anemometer, and thermometer. In contrast, the Lagrangian insight is a way of looking at fluid motion from a moving point in which the observer follows the air parcel trajectories. It is like sitting in a boat and drifting down a river. Here usual systems include cloud observation, smoke puffs, kites, and balloons equipped with recording instruments.

Why are these two approaches especially influenceable by social factors? Throughout the 1850s and the 1860s and for both technical and economic reasons, scientists were concerned primarily with surface meteorological observations and rarely with measurements of upper-air properties. The characteristics of the Eulerian scheme—with its spatially fixed coordinates, geared towards routine surface observations, and appropriate for spatially coherent structures such as cyclones—made it well-suited for centralized and highly standardized observing systems. This kind of system was traditionally welcomed by scientific elites. The reason was deliberate: scientific

26 For an overview of the Eulerian and Lagrangian fluid descriptions, and the differences between the two methods, see Young (2011, pp. 103–07).

27 For a description of the Eulerian differential equations and the Lagrangian integral of fluid motion, see Tokaty (2013, pp. 73–80), first ed. in 1972; and Darrigol (2008, pp. 23–30).

28 See, for example, Granger (1995); Bennett (2006, pp. 1–24); James (2015, pp. 448–51).

elites and rulers were interested in the control of observational data and their immediate transmission and application, not in improvization or unmanageability. Viewing science as a rational, state enterprise, scientists such as Le Verrier were especially reluctant to the observing systems which were not easily controllable or manageable, and did not guarantee order and national security. They were suspicious of a scheme, like the Lagrangian, that was geared towards asynoptic non-routine measurements, despite being physically more meaningful.

The study of methods also sheds much light on the social aspects of observing systems and organization: the Eulerian approach appears to imply order, control, and highly centralized observing systems, as will be shown in this book. By its very nature and definition, in the Eulerian approach each field of fluid flow needs to be expressed as a function of *both* space and time, hence the use of spatially fixed coordinates (observing stations) and routine surface observations. On the contrary, the Lagrangian approach appears to imply freedom, imagination, and autonomous observing systems. The fields are expressed as a function of time—not of space. While the use of Lagrangian approach entails unmanageability and verticality, the use of the Eulerian approach brings with itself territoriality and horizontality. This fact, which has gone unnoticed by historians and sociologists of science, allows us to study interesting transitions between the scientific and the social worlds.

The very nature of the forecasting practice in part explains this social dependence. The establishment of observational systems required some sort of balancing between several factors: the challenge of new printing technologies, map production, telegraph communications, the personnel required to chart and analyze data, the techniques of observing all considered on the basis not only of costs but also of the paramount time factor. In weather forecasting, it could be better to accept some observations from inaccurate measurements than to demand high quality standards and be unable to defray the costs, it could be more useful to gather a limited amount of information immediately than to receive large amounts of accurate data some time later, and it could be more practical to make an acceptable approximate analysis of the weather quickly than to perform a detailed study of the situation subsequently.

Certainly, the observational system is not the only place in which scientific and social interests have intertwined historically to influence the choice of method. From the beginning, the growing need for creating small-scale schematic pictures of the weather led to the synoptic representation of the physical properties of the atmosphere. Impersonal, precise, and easy to visualize and understand, weather maps quickly conveyed to viewers a sense of spatial specificity between the categories of 'local', 'national', and 'global'. In this race, the Eulerian approach started out with an advantage. Given its stationary coordinate system and the fact that Eulerian equations can be solved easily for certain flows, this approach was easier to handle mathematically and graphically than the Lagrangian approach. So, the Eulerian approach became synoptic. The main isolines were the mean sea-level isobars, a somewhat abstract notion of an imaginary territory often confused with national space, which provided a continuous field representation of atmospheric pressure. Isobars were thus both very practical and easily intelligible, projecting a sense of predictability about the depressions they described.[29]

Historians of science have often examined the social and cultural contexts in their reconstructions of the development of weather forecasting in the nineteenth century and discussed the conditions of the synoptic meteorology at its early stages. Christina Barboza has defined the establishment of Le Verrier's telegraphic meteorological network in the 1850s as an attempt to politically control the Paris Observatory's scientific activities.[30] Fabien Locher has applied Peter Galison and Lorraine Daston's discourse of mechanical objectivity to the process of production of synoptic maps, associating this process with the formation of an 'intellectual and material culture' in the field of dynamic meteorology.[31] Other historians have emphasized the deep-rooted presence of a material culture in nineteenth-century observatories, including Adolphe Quetelet's role in this formation.[32] The importance that debates about the credibility and authority of scientists' assessments had for the social legitimization of meteorology has been also underlined in studies that have stressed how public expectations often shaped the practice of weather forecasting.[33] In the early years of the British Meteorological Office, this controversial science actually became, as Katharine Anderson has observed, a 'public spectacle', with its forecasts open to daily scrutiny in the press, and often confronted with the predominant values of Victorian culture.[34]

29 For interesting reflections on the characteristics of the synoptic weather forecasting and the advantages and limitations of the synoptic method as seen from the perspective of the 1950s, see Sutcliffe (1952).
30 Barboza (2005, 2012).
31 Locher (2008, pp. 139–45). Locher has also analyzed how Marié-Davy devised the concept of atmospheric depression as the common representation of the weather from the identification and tracking of storms in the 1860s.
32 Aubin, Bigg, and Sibum (2010, pp. 1–32). For Quetelet's contribution to the formation of 'observatory culture', see, e.g., Aubin (2005, pp. 287–91). For an urban culture of circulation and observation at the Paris Observatory, see Aubin (2003).
33 See, for example, Burton (1986, 1988), Jankovic (1998, 2001), Dry (2009), for the case of Britain; Pelosse (1990), Locher (2008), Barboza (2012), for France; and Gross (1972), Fleming (1990, 2000), and Bergman (2016), for the United States.
34 Among the many virtues of Anderson's book, *Predicting the Weather*, one of the most remarkable ones is her contribution to our understanding of Victorian science-in-culture, showing carefully the fraught relations between the social acceptance of meteorologists' expertise through professionalization and institutionalization, on one hand, and the popular wisdom about natural knowledge, on the other hand.

But if historians of science have thus often noticed that the knowledge of synoptic meteorology was of a social and statistical nature, and have written extensively on the cultural history of weather forecasting, nonetheless they have not yet identified the two approaches—Eulerian and Lagrangian—that defined the nineteenth-century forecasting practice and the study of storms.[35] Nor have they written about how these approaches contributed to shape the image of what the weather was about, or how specific interests related to nation-state building had a 'selective effect' on the choice of one approach over another. This work will explore these issues and will show how these approaches competed against each other in the 1860s before the ascendency of the Eulerian approach and its hegemony until the 1910s. Thus, I contend that the Western states, through national weather services, not only performed a work of social legitimation of this approach and its outcome, the isobaric map, but also contributed to the shaping, the 'taming' if I may, of the weather, that is of the very entity that they supposedly aimed to forecast.

The state and meteorological cartography

At the same time, this book is also a study on the processes of nation-state building—for as is well known, the establishment of weather forecasting services did not occur in a vacuum. Rather, they emerged against a broader background of increasing national consciousness in Europe and America throughout the second half of the nineteenth century. As this consciousness became directly related to economics and politics, it tended to focus on issues that were perceived as reasons and interests of state such as centralism, security, order, and surveillance. In this context, political authorities and professional meteorologists (though not all of them) often turned to weather maps and forecasts as a means for achieving the objectives set. Thus weather maps became the tools in a building work driven by a variety of agents and institutions, contributing to creating the image of a national meteorological space for their states. By the 1870s, weather maps of isobars had moved away from the early representations of local conditions and individual details towards a more globalist, internationally based variant that viewed the weather changes through nationalist lenses.

Historians have often described these developments as part of a wider revolution in the visual presentation of scientific information that affected fields such as printing technology and the telegraph.[36] This evolution was often depicted in the standard genealogies written by meteorologists

---

35 The lone exception to this oversight has been historian James Rodger Fleming (2016, p. 16), who briefly describes the Eulerian and Lagrangian approaches to the study of fluid dynamics in a book on the foundations of twentieth-century meteorology.

36 Literature on meteorological cartography include Harrington (1895), Robinson and Wallis (1967), Robinson (1971), Kington (1980), Monmonier (1999), Anderson (2006).

themselves in the 1890s and subsequently by historians of weather cartography, in which the conventions of mapping and the various nomenclatures and techniques of isolines seem to have followed logical processes, or, at least, rational patterns.[37] More recent scholarship, however, has questioned these accounts, seeking instead the explanation in the relationship between local and global conditions, between the professional forecaster and the observer.[38]

Without underestimating these proposals, the present book raises a new argument, suggesting that this evolution was above all the result of a *struggle* of forces expressed at three levels: local, national, global. Rather than a relation of balance, the evidence gathered here seems to show a confrontational relationship between popular and expertise knowledge, in which the global was often confused with a voluntarist internationalism that was subordinate to national interests. I argue that specific interests related to national security, order, and surveillance shaped the way weather forecasting could or should be practised, giving preference to the application of methods based on isoline mapping. From the 1870s on, the state support to national weather services reinforced the isolinear character of weather charts. The states themselves, through these institutions, soon realized that they had in their hands a prognostic—rather than just diagnostic—tool with which to measure changes in the atmospheric pressure distribution in a given region, allowing them to predict weather changes in the short and medium term. The motion of low-pressure areas could be tracked on synoptic charts (often in the form of storm tracks), while new 'weather types' could be identified with certain forms of isobars. No longer would their storm warnings and forecasts have to depend on the 'obscure' methods of intuitive empiricists like Robert FitzRoy, whose zeal for autonomy and freedom of individual judgement was always looked down upon with disdain by London's scientific elite and the state. Now forecasts could be made mechanically in state-controlled centralized institutions, thereby guaranteeing the public the daily mass production of weather synoptic maps.

Of course, as weather forecasting became more and more a routine task of state administrations, the states could direct such services to safeguard their most immediate interests—especially with regard to the exposure of commercial, agricultural, and military interests to uncontrollable natural threats. Here meteorological cartography could fulfil this national protective role through a variety of modes and forms. Synoptic maps could be

---

Some examples relevant to the history of mapping include Robinson and Pentchenik (1976), Konvitz (1987), Camerini (1993), Palsky (1996).

37 A widely cited example of standard genealogy is Gustav Hellmann's work on *Meteorologische Karten* (Berlin, 1897), in which the German meteorologist outlines the main milestones of this field through a classic sequence of synoptic maps.

38 The most notable recent study on how synoptic weather mapping showed in the same way the relationship between local and global knowledge is Anderson (2006).

used to foretell the approach of storms and the incidence of floods (as in the case of the large rivers of the United States); or they could be harnessed to show the geographic distribution of monthly and annual average values in order to know national climate (as the *Atlas météorologique* of France). They could also be configured to describe the spatial arrangement of the national territory, either through the establishment of decentralized systems (such as the U.S. district forecast centres),[39] or the articulation of regional committees (such as the *commissions météorologiques départementales* in France).[40]

The invention or configuration of a national meteorological space, therefore, was not only an emblematic part of the nation-state building but also constitutive of the same. It created government-sponsored bureaucratized services in which the production of synoptic maps and weather forecasts was a routine part of the functions of the state. Indeed, as will be discussed later, these established patterns of seeing and building the nation differed little, in their forms and expressions, from the knowledge about the nation generated by statistics and cartography. Uncovering the power of weather maps as instruments of order and control thus forms a crucial part of any effort to understand the shaping of weather forecasting by the states over the course of the second half of the nineteenth century.

Local and global knowledge

The developments I discuss in this book also took place against the broader backdrop of the relations between local and global knowledge, and, in a broad sense, they show the quest for national security and social order in the Western world over the second half of the long nineteenth century (1840–1910). This book questions the traditional narratives regarding the use of statistical sciences and cartography in environmental sciences, showing how the authority of devices associated with weather forecasting, such as maps and empirical regularities, was not the result of any balance between the local and the global, but rather was built up by nation-states alongside a diverse variety of local and supranational entities. However, contrary to what one might think, the idea of nation-states imposing scientific knowledge 'from above' on the basis of specific interests and social needs, quite a different idea, is defended in this book, namely the *struggle* between some nation-states trying to pursue national security and order through scientific prediction and local and supranational entities whose interests and organization frequently escaped from the control of the nation-states themselves.[41] The actions of the various agents at stake are interpreted, therefore, as a

---

39 For the decentralization of weather forecasting in the United States from the 1890s onward, see Whitnah (1961, pp. 64–67).
40 Fierro (1991, pp. 206–08).
41 For a recent collection of case studies on the role played by nation-states in coping with global environmental problems and issues, such as the exploitation of rivers, migratory

struggle of forces that is expressed, in weather forecasting and the science of meteorology in general, at three levels: local, national, and global.[42]

Over the course of the past two decades, historians of environmental sciences have highlighted the double-dimension (both local and global) of this field. Where scientific developments in different national settings were surveyed, the primary objective was often to tie the universal to the intimate, link global knowledge to local phenomena.[43] Global knowledge was thus envisaged as a set, fundamental, and unproblematic category. Although few can doubt that such developments constantly moved back and forth between popular and expert knowledge, between local and global world views, I think however that this mode of analyzing the issue is not entirely convincing. In this work, global knowledge is used in two ways: first, as a form of knowledge that moves between and beyond national boundaries and connects individuals and institutions distributed around the world; and, second, as the understanding of natural phenomena and patterns presented on a trans-continental scale.[44] Stressing the double sense of global knowledge is not incompatible with the plurality of forces proposed in this book; quite the contrary, the multiplicity of global knowledge can only be understood in connection with the struggle of forces expressed at three levels, rather than to an allegeable balance between the local and the global.[45]

Likewise, over the last several years, historians have come to recognize the important role played by cultural factors and values in helping create what sociologist Gary Alan Fine has referred to as the 'culture of prediction' in a group (i.e. the set of shared knowledge, values, practices, and rules around the profession of weather forecasting).[46] Here culture is often identified as a 'mangle' in which the practices and meaning of a scientific community are

---

wildlife, control of natural resources, and hurricane prediction, see the contributions to Bsumek, Kinkela, and Lawrence (2013).

42 This struggle of forces seems to evoke early modern globalizations, as might be inferred from historian Christopher Alan Bayly's words, who holds the existence of 'older networks and dominances created by geographical expansion of ideas and social forces from the local and regional level to the inter-regional and inter-continental level'. See Bayly (2004, pp. 41–42).

43 See, for example, the contributions to the work edited by James Rodger Fleming, Vladimir Jankovic, and Deborah R. Coen, *Intimate Universality: Local and Global Themes in the History of Weather and Climate* (2006). See also Andrea Westermann's (2011) study on earthquake observation in Switzerland and Germany at the turn of the nineteenth century.

44 In this point I follow Diarmid A. Finnegan, Jonathan Jeffrey Wright, 'Introduction: Placing Global Knowledge in the Nineteenth Century', in Finnegan and Wright (2016, p. 5).

45 For an updated discussion of the issue of global kinds of knowledge, see Hulme (2010). The volume edited by Crawford, Shinn, and Sörlin (1993) is still a useful work to understand the nationalization and subsequent denationalization of the sciences, especially after the growth of transnational science and the international scientific practice in the twentieth century.

46 Fine (2007) introduces this term to describe the culture of weather forecasting in the US National Weather Service's Chicago office in the late twentieth century.

embedded,[47] and whose members shared methods, values, and expectations that contribute to a recognition of belonging.[48] Cultural narratives instinctively tend to divide the knowledge produced into categories of 'local' and 'global', establishing clear lines of demarcation between monolithic blocks. They create idealized worlds in which each block is a staple, homogeneous, and unproblematic category. For this reason scholars such as Katharine Anderson have highlighted their important role in weather forecasting, arguing how meteorologists 'sought to develop a balance' between 'particular observations and general dynamical laws', between 'meteorology as a public science' and 'meteorology as a research science'.[49]

While this book acknowledges the importance of cultural values and practices in the construction of knowledge about the weather, it deviates to large extent from the considerable body of literature on the cultural and ethno-sociological studies on scientific prediction inspired by the work of Bruno Latour in the 1980s[50] (or the 'narrative culture' of the workplace as Karin Knorr-Cetina appropriately titled these accounts in her essay on *Epistemic Cultures: How the Sciences Make Knowledge*).[51] This is not a study of knowledge culture or construction. Rather, it shows how the tensions between the expectations about order and security which the Eulerian and Lagrangian approaches were supposed to produce and the garnered results hindered the immediate implementation of the same, thereby opening the door to an intense rivalry between them in the 1860s. Cartographic and statistical knowledge about the weather and climate was in effect a battleground, as scientists and rulers from the same country (e.g. the French Navy and Paris Observatory, or the British seamen and the Royal Society) began to interpret facts often in diametrically opposed terms. Hence the resultant weather forecast was the result, not so much of any imposition of power from above through maps and statistics, but rather of tensions between the diverse actors involved that would be finally resolved in favour of the Eulerian approach. These tensions are only explicable by reference to a struggle—not a balance—of forces expressed at three levels: local, national, and global.[52]

---

47 Pickering (1993).

48 For the emergence and development of 'cultures of prediction' in atmospheric and climate science, and how these cultures were shaped by computer modelling and simulation in the twentieth century, see the several case studies in the collective volume edited by Heymann, Gramelsberger, and Mahony (2017). For a cultural history of prediction in the American gilded age (from the 1870s to about 1900), see Pietruska (2009).

49 Anderson (2006, p. 86).

50 In these cultural and ethno-sociological studies on scientific prediction, weather bureaus are described as to what Bruno Latour (1987) has defined as 'centres of calculation'.

51 Knorr-Cetina (1999, p. 203).

52 On the notion of a struggle of forces expressed at three levels in an imperial context (i.e. cyclone prediction by Spanish Jesuits in Cuba and the Philippines in the second half of the nineteenth century), see Anduaga (2017, pp. xviii–xxiii, 166–67).

Structure of the study

The book is organized into eight chapters, each of which deals with a specific issue of how both state interests and statistics tamed and shaped weather forecasting.

Chapter 1 examines the theories of storms elaborated in Europe and the United States in the first half of the nineteenth century, and their relation to the gradual distancing between meteorology and climatology. I account for the reception of three theories based on mechanical and thermal principles; on one hand, Heinrich W. Dove's opposing current theory, and on the other hand, the two antagonistic models that gave rise to the so-called 'storm controversy' in America—William C. Redfield's whirlwind theory and James P. Espy's convective theory. I then look at the two wavelike solutions proposed to this problem, Elias Loomis's cold waves in America, and John Herschel and William Birt's atmospheric waves in Britain. Although the theoretical debates generated by these scientists are familiar to scholars, they had been analyzed mainly in the context of storm controversy. My contention is that they well reflect the transition from a climatological to a 'weatherological' view in the study of meteorological phenomena.

In Chapter 2, I move on to examine what I call the application of statistical methods to meteorology during the pre-synoptic era—i.e. before the first maps of weather patterns in the mid-1850s. This application centred on the attempts made in Europe to determine the mean state of the atmosphere, and especially those which paved the way for studies on the temporal evolution of mean values. A closer examination of this application shows the close connection between statistical and cartographic techniques. I emphasize the graphical method for representing atmospheric waves devised by Adolphe Quetelet. His method became the umbilical cord between pre-synoptic techniques and modern weather maps, and provided a basis to analyze the progress of storms.

Chapter 3 deals with a key episode in the history of predictive meteorology, the adoption of an Eulerian approach to weather forecasting for reasons of state interest. Focusing on the political context of France under Napoleon III, the chapter shows how the 1854 Black Sea storm was the catalyst that led Urbain Le Verrier from the Paris Observatory to adopt such an approach. This adoption, I contend, resulted from an accommodation of scientific thought to the state's interests. Developed by Emmanuel Liais and Hippolyte Marié-Davy in France and Angelo Secchi in Italy, the new weather maps of isobars were not only able to reveal weather patterns but also served as an instrument to achieve order, security, and centralism in those states.

In Chapter 4 I investigate the peculiarities and repercussions of an approach that acted as a counterweight of the Eulerian approach and that was characterized by the quest for autonomy and individual judgement in weather forecasting—just the opposite of what its rival approach entailed.

I show how the first to choose this (Lagrangian) approach—British Robert FitzRoy, George Jinman, and Francis Galton (though the latter indeed opted for a semi-Lagrangian frame)—proposed storm flow-models and iconic maps of weather. In particular, this chapter discusses the observational system and outcomes of FitzRoy's attempts to forecast the weather by combining empirical rules and instrumental readings. FitzRoy's contributions and his stormy life are well known by historians, but have been examined primarily in the context of confrontation with the values of Victorian culture. My contention is that he championed a Lagrangian approach diametrically opposed in forms and aims to the methods used in continental Europe.

The convergence between statistical science and the synoptic method of weather representation during the 1850s and 1860s is the subject of Chapter 5. This convergence, I argue, was largely driven by Christophorus Buys Ballot's studies on deviations from mean values (including his known 'baric wind law') and the analytical method of weather charts devised by Alexander Buchan. This chapter shows how meteorological science took on certain statistical forms that affected not only conceptual and procedural questions but also organizational aspects related to the standardization and coordination of observational systems. Thus, a new synoptic meteorology was projected in the first international meteorological meetings in Leipzig and Vienna in 1872–73: an official, state-sponsored science that combined the languages of statistical quantification and cartography.

Chapter 6 deals with the hegemony of the Eulerian approach and the beginning of its end in the period 1870–1910, considering both the potential and limitations of the synoptic maps of isobars as a tool for weather forecasting. I show how this approach reached its peak with the search for 'weather types' and isobaric forms in the 1880s and 1890s, and how it began to wane progressively in the 1900s with the first attempts to seek for forecasting methods based on the principles of physical theory. I trace this development in two parts. On one hand, I chart the search for 'objective' criteria that could be used to establish a methodical forecasting, especially empirical rules and statistical regularities in weather patterns. I analyze the seven forms of isobars proposed by Ralph Abercromby and his attempts to identify them with certain weather types. His isobaric geometry, I contend, was the epitome of the use of the Eulerian frame in weather forecasting. On the other hand, I examine the search for 'scientific' forecasting methods on the basis of the theories of hydrodynamics and thermodynamics among prominent European physicists such as Napier Shaw, Max Margules, Felix Exner, and Vilhelm Bjerknes. I note the shift from the Eulerian idea—that the pressure field (through its isobaric forms) determined weather changes from the 1870s till the 1890s—to the Lagrangian concept of path (or trajectory of parcels of air), which was viewed as the tool par excellence for predicting the weather scientifically.

Chapter 7 focuses on the public institutions that massively produced the weather maps of isobars that were published and disseminated by the national weather services in the period 1850–1910. I examine in particular the allocation of public funds for weather forecasting by different governments in France, Britain, and the United States, and their impact on meteorological science. I show how all this state support led to a public service-oriented meteorology rather than a research-oriented meteorology. Government patronage, I argue, was extremely important because it contributed to shape not only synoptic forecasting by financing the processes of data collection and image-production but also the character and direction of the research itself. As a result, meteorological research always remained subordinate to public service.

Finally, Chapter 8 charts the development of more than six decades of meteorological cartography, showing how the synoptic weather map went from being a tool for *diagnosis* of atmospheric conditions to a tool for *prognosis* or the weather forecast. This evolution reflects well the coexistence of two opposing approaches (the Eulerian and the Lagrangian) in the 1850s and 1860s. Yet by the 1870s meteorologists had shifted the focus from observation points to isolines. By then isobars had become the dominant feature of synoptic maps. The primacy of isolines, I contend, was both stimulated by the expansion of statistical cartography and nurtured by national weather services through public funds. The chapter ends with some considerations of the relation between local observation, national expertise, and global knowledge which strongly conditioned the development of weather maps. That the relation between the local and the global, far from being balanced—as has been defended by many scholars—ended up becoming a *struggle* of forces expressed at three levels (local, national, global), as I will show, is an issue that merits special reflection in the history of science.

References

Anderson, Katharine, 2005. *Predicting the Weather: Victorians and the Science of Meteorology*. Chicago: The University of Chicago Press.

Anderson, Katharine, 2006. 'Mapping Meteorology'. In James Rodger Fleming, Vladimir Jankovic, and Deborah R. Coen eds., *Intimate Universality. Local and Global Themes in the History of Weather and Climate*. Sagamore Beach, MA: Watson Publishing International, 69–91.

Anduaga, Aitor, 2016. *Geophysics, Realism and Industry: How Commercial Interests Shaped Geophysical Conceptions, 1900–1960*. Oxford: Oxford University Press.

Anduaga, Aitor, 2017. *Cyclones and Earthquakes: The Jesuits, Prediction, Trade and Spanish Dominion in Cuba and the Philippines, 1850–1898*. Quezon City: Ateneo de Manila University Press.

Aubin, David, 2003. 'The Fading Star of the Paris Observatory in the Nineteenth Century: Astronomers' Urban Culture of Circulation and Observation'. *Osiris*, 18, 79–100.

Aubin, David, 2005. 'Observatory Mathematics in the Nineteenth Century'. In Eleanor Robson, and Jacqueline Stedall eds., *The Oxford Handbook of the History of Mathematics*. Oxford: Oxford University Press, 273–98.

Aubin, David, Bigg, Charlotte, and Sibum, H. Otto eds. 2010. *Observatories and Astronomy in Nineteenth-Century Science and Culture*. Durham, NC: Duke University Press.

Barboza, Christina Helena da Motta, 2005. 'Nice Weather, Meteors at the End of the Day'. In Stefan Emeis, and Cornelia Lüdecke eds., *From Beaufort to Bjerknes and Beyond: Critical Perspectives on Observing, Analyzing, and Predicting Weather and Climate*. Augsburg: Rauner, 157–68.

Barboza, Christina Helena da Motta, 2012. *As viagens do tempo. Uma história da meteorologia em meados do século XIX*. Rio de Janeiro: E-papers Serviços Editoriais Ltda.

Bayly, Christopher Alan, 2004. *The Birth of the Modern World, 1780–1914: Global Connections and Comparisons*. Oxford: Blackwell.

Bennett, Andrew, 2006. *Lagrangian Fluid Dynamics*. Cambridge: Cambridge University Press.

Bergman, James, 2016. 'Knowing Their Place: The Blue Hill Observatory and the Value of Local Knowledge in an Era of Synoptic Weather Forecasting, 1884–1894'. *Science in Context*, 29 (3), 305–46.

Black, Jeremy, 1997. *Maps and Politics*. Chicago: University of Chicago Press.

Boumans, Marcel, 2015. *Science outside the Laboratory: Measurement in Field Science and Economics*. Oxford: Oxford University Press.

Brunt, David, 1951. 'A Hundred Years of Meteorology'. *The Advancement of Science*, 30, 114–24.

Bsumek, Erika Marie, Kinkela, David, and Lawrence, Mark Atwood eds., 2013. *New Approaches to International Environmental History*. New York: Oxford University Press.

Burton, James, 1986. 'Robert FitzRoy and the Early History of the Meteorological Office'. *British Journal for the History of Science*, 19, 147–76.

Burton, James, 1988. 'The History of the British Meteorological Office to 1905'. The Open University, PhD diss.

Camerini, Jane R., 1993. 'The Physical Atlas of Heinrich Berghaus: Distribution Maps as Scientific Knowledge'. In Renato G. Mazzolini ed., *Non-Verbal Communication in Science Prior to 1900*. Florence: Leo S. Olschki, 479–512.

Carson, Cathryn, Kojevnikov, Alexei, and Trischler, Helmuth eds., 2011. *Weimar Culture and Quantum Mechanics. Selected Papers by Paul Forman and Contemporary Perspectives on the Forman Thesis*. London: Imperial College Press.

Chartier, Roger, 1988. *Cultural History: Between Practices and Representations*. Cambridge: Polity Press.

Coen, Deborah R., 2006. 'Scaling Down: Mapping the Austrian Climate between Empire and Republic'. In James Rodger Fleming, Vladimir Jankovic, and Deborah R. Coen eds., *Intimate Universality: Local and Global Themes in the History of Weather and Climate*. Sagamore Beach, MA: Science History Publications, 115–40.

Cox, John D., 2002. *Storm Watchers: The Turbulent History of Weather Prediction from Franklin's Kite to El Niño*. Hoboken, NJ: John Wiley & Sons, Inc.

Crawford, Elisabeth, Shinn, Terry, and Sörlin Sverker eds., 1993. *Denationalizing Science: The Contexts of International Scientific Practice*. Dordrecht: Kluwer.

Danjon, André, 1946. 'Le Verrier, créateur de la météorologie'. *La Météorologie*, 1 (1046), 863–82.

Darrigol, Olivier, 2008. *Worlds of Flow. A History of Hydrodynamics from the Bernoullis to Prandtl*. Oxford: Oxford University Press.

Davis, John L., 1984. 'Weather Forecasting and the Development of Meteorological Theory at the Paris Observatory, 1853–1878'. *Annals of Science*, 41 (4), 359–82.

Dry, Sarah, 2009. 'Safety Networks: Fishery Barometers and the Outsourcing of Judgement at the Early Meteorological Department'. *The British Journal for the History of Science*, 42 (1), 35–56.

Eredia, Filipo, 1914. 'L'organizzazione del servizio dei presagi del tempo in Italia'. *Rivista Meteorico-Agraria*, 35, 1001–48.

Fierro, Alfred, 1991. *Histoire de la météorologie*. Paris: Éditions Denoël.

Fine, Gary Alen, 2007. *Authors of the Storm: Meteorologists and the Culture of Prediction*. Chicago: University of Chicago Press.

Finnegan, Diarmid A., and Wright, Jonathan Jeffrey eds., 2016. *Spaces of Global Knowledge: Exhibition, Encounter and Exchange in an Age of Empire*. London: Routledge.

FitzRoy, Robert, 1863. *The Weather-Book: A Manual of Practical Meteorology*. London: Longman, Green, Longman, Roberts, & Green.

Fleming, James Rodger, 1990. *Meteorology in America, 1800–1870*. Baltimore, MD: Johns Hopkins University Press.

Fleming, James Rodger, 2000. 'Storms, Strikes, and Surveillance: The U.S. Army Signal Office, 1861–1891'. *Historical Studies in the Physical and Biological Sciences*, 30, 315–32.

Fleming, James Rodger, 2016. *Inventing Atmospheric Science. Bjerknes, Rossby, Wexler, and the Foundations of Modern Meteorology*. Cambridge, MA: The MIT Press.

Forman, Paul, 1971. 'Weimar Culture, Causality, and Quantum Theory, 1918–1927: Adaptation by German Physicists and Mathematicians to a Hostile Intellectual Environment'. *Historical Studies in the Physical Sciences*, 3, 1–115.

Gould, Stephen Jay, 1981. *The Mismeasure of Man*. New York: W.W. Norton.

Granger, Robert A., 1995. *Fluid Mechanics*. New York: Dover Publications, Inc.

Gross, Walter E., 1972. 'The American Philosophical Society and the Growth of Meteorology in the United States: 1835–1850'. *Annals of Science*, 29, 321–38.

Halford, Pauline, 2004. *Storm Warning. The Origins of the Weather Forecast*. Stroud: Sutton Publishing.

Harley, John Brian, 1988. 'Maps, Knowledge and Power'. In Denis Cosgrove, and Stephen Daniels eds., *The Iconography of Landscape. Essays on the Symbolic Representation, Design and Use of Past Environments*. Cambridge: Cambridge University Press, 277–312.

Harper, Kristine C., 2008. *Weather by the Numbers. The Genesis of Modern Meteorology*. Cambridge, MA: The MIT Press.

Harrington, Mark W., 1895. 'History of the Weather Map'. In Oliver L. Fassig ed., *Report of the International Meteorological Congress, Chicago, Ill., August 21–24, 1893. Part II*. Washington, DC: Weather Bureau, 327–35.

Hellmann, Gustav, 1897. *Meteorologische Karten 1688, 1817, 1846, 1863, 1864. Sechs tafeln in lichtdruck mit einer einleitung*. Berlin: Asher.

Heymann, Matthias, Gramelsberger, Gabriele, and Mahony, Martin eds., 2017. *Cultures of Prediction in Atmospheric and Climate Science: Epistemic and Cultural Shifts in Computer-based Modelling and Simulation*. New York: Routledge.

Hulme, Mike, 2010. 'Problems with Making and Governing Global Kinds of Knowledge'. *Global Environment Change*, 20 (4), 558–64.

Iafrate, Luigi, 2008. *Fede e scienza: un incontro proficuo. Origini e sviluppo della meteorologia fino agli inizi del '900*. Roma: Ateneo Pontificio Regina Apostolorum.

James, Richard D., 2015. 'Continuum Mechanics'. In Nicholas H. Higham ed., *The Princeton Companion to Applied Mathematics*. Princeton: Princeton University Press, 446–57.

Jankovic, Vladimir, 1998. 'Ideological Crests versus Empirical Troughs: John Herschel's and William Radcliffe Birt's Research on Atmospheric Waves, 1843–50'. *British Journal for the History of Science*, 31, 21–40.

Jankovic, Vladimir, 2001. *Reading the Skies: A Cultural History of English Weather 1650–1820*. Chicago: University of Chicago Press.

Khrgian, Aleksandr Khristoforovich, 1970. *Meteorology: A Historical Survey*, 2nd ed. Jerusalem: Israel Program for Scientific Translations.

Kington, John A., 1980. 'Daily Weather Mapping from 1781: A Detailed Synoptic Examination of Weather and Climate during the Decade leading up to the French Revolution'. *Climatic Change*, 3, 7–36.

Knorr-Cetina, Karin, 1999. *Epistemic Cultures: How the Sciences Make Knowledge*. Cambridge, MA: Harvard University Press.

Konvitz, Joseph W., 1987. *Cartography in France, 1660–1848. Science, Engineering and Statecraft*. Chicago: Chicago University Press.

Kutzbach, Gisela, 1979. *The Thermal Theory of Cyclones: A History of Meteorological Thought in the Nineteenth Century*. Boston, MA: American Meteorological Society.

Landsberg, Helmut, 1954. 'Storm of Balaklava and the Daily Weather Forecast'. *The Scientific Monthly*, 79 (6), 347–52.

Latour, Bruno, 1987. *Science in Action: How to Follow Scientists and Engineers through Society*. Milton Keynes: Open University Press.

Lazarsfeld, Paul F., 1970. 'Notes sur l'histoire de la quantification en sociologie: les sources, les tendances, les grands problèmes'. In Paul F. Lazarsfeld ed., *Philosophie des sciences sociales*. Paris: Gallimard, 75–162.

Lequeux, James, 2013. *Le Verrier–Magnificent and Detestable Astronomer*. New York: Springer. Edited and with an introduction by William Sheehan. Translated By Bernard Sheehan.

Livingstone, David N., 2003. *Putting Science in Its Place: Geographies of Scientific Knowledge*. Chicago: University of Chicago Press.

Locher, Fabien, 2007. 'The Observatory, the Land-based Ship and the Crusades: Earth Sciences in European Context, 1830–50'. *The British Journal for the History of Science*, 40 (4), 491–504.

Locher, Fabien, 2008. *Le savant et la tempête. Étudier l'atmosphère et prévoir le temps au XIX siècle*. Rennes: Presses Universitaires de Rennes.

Lynch, Peter, 2006. *The Emergence of Numerical Weather Prediction: Richardson's Dream*. Cambridge: Cambridge University Press.

Monmonier, Mark S., 1996. *How to Lie with Maps*. Chicago: The University of Chicago Press.

Monmonier, Mark, 1999. *Air Apparent. How Meteorologists Learned to Map, Predict, and Dramatize Weather*. Chicago: The University of Chicago Press.

Nebeker, Frederik, 1995. *Calculating the Weather. Meteorology in the 20th Century.* New York: Academic Press.

Palsky, Gilles, 1996. *Des chiffres et des cartes: Naissance et développement de la cartographie quantitative française au XIXe siècle.* Paris: Comité des Travaux Historiques et Scientifiques.

Pelosse, Valentin, 1990. 'Observation météorologique et sociétés savantes de province, ou la désignation du bon objet scientifique, 1821–1878'. *Études Rurales*, 118–119, 69–82.

Pickering, Andrew, 1993. 'The Mangle of Practice: Agency and Emergence in the Sociology of Science'. *American Journal of Sociology*, 99, 559–89.

Pietruska, Jamie L., 2009. 'Propheteering: A Cultural History of Prediction in the Gilded Age'. Massachusetts Institute of Technology. PhD. Diss., Program in Science, Technology and Society.

Porter, Theodore M., 1986. *The Rise of Statistical Thinking 1820–1900.* Princeton: Princeton University Press.

Reed, Richard J., 1977. 'The Development and Status of Modern Weather Prediction'. *Bulletin of the American Meteorological Society*, 58, 390–99.

Robinson, Arthur H., 1971. 'The Genealogy of the Isopleth'. *Cartographical Innovations*, 8, 49–53.

Robinson, Arthur H., 1982. *Early Thematic Mapping in the History of Cartography.* Chicago: University of Chicago Press.

Robinson, Arthur H., and Pentchenik, Barbara Bartz, 1976. *The Nature of Maps: Essays towards Understanding Maps and Mapping.* Chicago: Chicago University Press.

Robinson, Arthur H., and Wallis, Helen, 1967. 'Humboldt's Map of Isothermal Lines: A Milestone in the History of Thematic Cartography'. *Cartographic Journal*, 2, 119–23.

Schneider-Carius, Karl, 1975. *Weather Science, Weather Research: History of Their Problems and Findings from Documents during Three Thousand Years.* New Delhi: Indian National Scientific Documentation Centre; originally published as *Wetterkunde, Wetterforschung: Geschichte ihrer Probleme und Erkenntnisse in Dokumenten aus drei Jahrtausenden.* Freiburg: Verlag Karl Albert, 1955.

Sheynin, Oscar Boris, 1984. 'On the History of the Statistical Method in Meteorology'. *Archive for History of Exact Sciences*, 31 (1), 53–95.

Stigler, Stephen M., 1986. *The History of Statistics. The Measurement of Uncertainty before 1900.* Cambridge, MA: The Belknap Press of Harvard University Press.

Sutcliffe, Reginald C., 1952. 'Principles of Synoptic Weather Forecasting'. *Quarterly Journal of the Royal Meteorological Society*, 78 (337), 291–320.

Tetzlaff, Gerd, Lüdecke, Cornelia, and Behr, Hein Dieter eds., 2008. *125 Jahre Deutsche Meteorologische Gesellschaft. Festveranstaltung am 7. November 2008 in Hamburg.* Offenbach am Main: Deutschen Wetterdienstes.

Tokaty, Grigori Alexandrovich, 2013. *A History and Philosophy of Fluid Mechanics.* New York: Dover Publications, Inc. First ed. in 1972.

Walker, J. Malcolm, 2012. *History of the Meteorological Office.* Cambridge: Cambridge University Press.

Wege, Klaus, 2002. *Die Entwicklung der meteorologischen Dienste in Deutschland.* Offenbach am Main: Selbstverlag des Deutschen Wetterdienstes

Westermann, Andrea, 2011. 'Disciplining the Earth: Earthquake Observation in Switzerland and Germany at the Turn of the Nineteenth Century'. *Environment and History*, 17, 53–77.

Whitnah, Donald R., 1961. *A History of the United States Weather Bureau*. Urbana: University of the Illinois Press.

Young, Donald F., 2011. *A Brief Introduction to Fluid Mechanics*, 5th ed. Hoboken, NJ: Wiley.

1 Pre-1850 conceptualizations of storms

Introduction

The transition from a climatological analysis of meteorology to a genuinely predictive analysis of the weather occurred gradually in the middle decades of the nineteenth century through reversible analogies between different fields of observational sciences. Not only did graphical representation and cartography change, but the theory of storms, atmospheric waves, wind statistics, telegraph techniques, and even systems of systematic observations changed as well. The modern use of such synoptic predictive techniques implied a great deal of implicit knowledge that only acquired its full meaning when compared with older practices.

In Europe and America, this transition was by no means free of obstacles. A tension persisted between the climatological and 'weatherological' perspectives in the study of meteorological phenomena: while statistics-oriented scientists leaned mostly towards the former, empirically minded natural philosophers and seamen held to the latter.[1] For climatologists, the study of the weather was not sufficiently precise; for natural philosophers and seamen, it was not only insufficiently empirical, but also should be an object of inquiry per se and as such—not as an appendage of climate.[2]

The study of storms was a key piece of this transition. At the end of the 1840s, different views of the origin and maintenance of storms competed with each other just before meteorologists and engineers succeeded in establishing the first telegraphic systems of meteorological observations.

---

1 The frontier between naval officers and men of science at that time was more blurred than what one might think today; on their relationship in maritime meteorology and hydrography, see Achbari (2015, pp. 257–61, 2017, pp. 21–68).

2 The term 'weatherology' referring to the study of the weather was used in the popular jargon in mid-nineteenth century almanacs and journals. Lord Byron encapsulated the essence of 'weatherology' in his poem of canto X III, Don Juan: 'The London winter's ended in July– / Sometimes a little later. I don't err / In this: whatever other blunders lie / Upon my shoulders, here I must aver / My Muse a glass of weatherology; / For parliament is our barometer: / Let radicals its other acts attack, / Its sessions form our only almanack'. Lord Byron (1825, p. 147).

The nature of storm motion became a major bone of contention in the 1820s, and talented men in Prussia, such as Heinrich W. Dove, professor of physics at the University of Berlin, championed their theories with determination. For Dove, midlatitude storms arose from the conflict of two opposite currents, while tropical storms were of vortical nature. Again, in America William C. Redfield and James P. Espy were elaborating in the 1830s two antagonistic theories that gave rise to the so-called 'storm controversy': the whirlwind theory versus the convective theory. Later, two wavelike solutions to this controversy were proposed: Elias Loomis's cold waves in America, and John Herschel and William Birt's atmospheric waves in Britain. These theories form the pre-1850 conceptualizations of storms.

Meteorology qua climatology

Towards the mid-1840s European climatologists, like the scholars of the atmosphere, continued to value with satisfaction the application of statistical methods to the science in which they had been engaged for two decades. They, like many astronomers familiar with statistical techniques, believed that behind the irregular variations of weather elements there was a regularity that was analogous, or at least comparable, to the regularity of the motions of the planets. The climatologist examining his tables and graphs was convinced that the essence of climatology rested on averages and normals, and on seeking numerical expressions for regular patterns. Mean values, or mean states, 'show us the permanent in the change',[3] Alexander von Humboldt noted in his book *Kosmos* in 1845, so that the study of all natural phenomena both periodical and irregular is subject to the determination of mean values. An example of this was the called isotherms—or lines that joined places having the same average temperature—that he introduced in 1817 in order to study the distribution of heat over the earth's surface.[4] His compatriot Heinrich Wilhelm Dove shared this statistical view, and established a triple desideratum for meteorology, ranging from the determination of mean values, through the determination of the laws of periodical variations, to specify the rules for irregular phenomena.[5] And all these desiderata seemed assured even many years later as Joseph Henry from the Smithsonian Institution proclaimed in 1855 that the irregular weather variations could be well analyzed in terms of the regularity of climate.[6] Who, seeing

---

3 Humboldt (1845, Bd. 1, p. 82).
4 Unlike astronomer Tobias Mayer, who in 1755 assumed a model of temperature distribution—a geometric function of latitudes—over the surface of the earth based on the solar heating, Humboldt calculated mean temperatures from original observations. Humboldt (1817, p. 462); Körber (1959, pp. 289–335).
5 Dove (1838, p. 285).
6 Henry (1855, p. 58)—quoted by Thornthwaite and Leighly (1943, p. 458): 'uniting the results of observations with those of experiments in the laboratory, and mathematical

all these expressions of satisfaction and confidence, could have foreseen that with the advent of telegraph and synoptic techniques in mid-century, meteorology would alienate itself from climatology and would become merely *weatherology?*[7]

It is not therefore surprising that the heirs of this largely Baconian data-gathering tradition all thoroughly agreed that meteorological science was conceptually and objectively inherent to climatology. There were plenty of displays in this regard. Thus, drawing upon atmospheric changes observed in different places throughout the world, in 1847 Russian meteorologist Mikhail F. Spasskii described certain constant elements in them in his book *On the climate of Moscow.* These elements defining the concept of climate consisted in either a succession of known atmospheric phenomena, a 'prevalence of some specific type of weather', or 'limits between which atmospheric changes of a given type complete their cycles'.[8] By then, his European colleagues did not hesitate to assert that it was not only the average state of meteorological phenomena that typified a given climate but also the weather variability and its deviations from the mean states. In fact, so close was the identification of meteorology with knowledge of weather and of climate that any factor that could affect this relationship was viewed as a threat. Only thus can Humboldt's letter to Élie de Beaumont be understood, when in the heat of the Paris Observatory's ambitious plans he warned him: 'telegraphic meteorology will create even more confusion than telegraphic diplomacy'.[9]

The natural philosopher or the seaman, however, emphasizing the importance of natural laws as objects of inquiry and the desirability of studying the origin, size, and progress of individual storms, openly voiced dissatisfaction with the constraints of statistical methods. As Swedish scientist Frih Ehrenheim noted as early as 1824, the arithmetic treatment in meteorology was not as successful as in astronomy: 'On the one hand, [it] is the most regular order to be found in Nature'; on the other hand, 'the most confused variability, so that we see no way in which any common periodicity may be found'.[10] Ironically, the arithmetic mean, which proved to be

---

deductions from astronomical and other data, we are enabled, not only to refer the periodic changes to established laws, but also to trace to their source, various perturbing influences which produce the variations from the mean, and thus arrive, at least, at an approximate explanation of the meteorological phenomena which are constantly presented to us'.

7 On the adoption of statistical methods in climatology, especially mean values by Humboldt and the deviation from arithmetic means by Dove, see Sheynin (1984, pp. 68–70, 72–73).

8 Quoted by Khrgian (1970, p. 307).

9 Quoted by Alfred Dove (1873, p. 374). In his letter, printed in 1855 in the *Comptes Rendus de l'Académie des Sciences*, 'Sur les Sociétés de Météorologie et les observations météorologiques', Humboldt also steadfastly supported the separation of meteorological sections from astronomical observatories.

10 Ehrenheim (1824, p. 206)—quoted by Thornthwaite and Leighly (1943, p. 458).

so unsatisfactory for many natural philosophers as a representative value, was the most prominent constituent of the summaries of observations of weather made by climatologists. These summaries included probabilities of occurrence of meteorological phenomena at specific periods of daily and annual cycles. Whereas for the latter the arithmetic mean was a useful tool to eliminate chance errors in physical and astronomical measurements, for the former it was incapable to reveal any law from the flux of atmospheric phenomena. Certainly there was no reason to think that Friedrich von Schiller's famous poem—'the tranquilizing pole in the procession of phantoms' that arithmetic mean could well have meant for climatologists—alleviated natural philosophers'disquiet.[11]

While it would be extensive to depict the discrepancies between them, one can at least point to some brief passages alluding to new ways of understanding meteorology which reflect what was in some natural philosophers' mind at that time. In 1830, John Frederick William Herschel, one of the leading figures in science at the time, described the goals and objects of meteorology, and physical geography in general (of which it was part), in his book *Preliminary Discourse on the Study of Natural Philosophy*. His essay was a song to the progress of sciences, understood as an evolution from natural history to natural philosophy, from nomenclature and classification to the search for laws and causes. In his first paragraphs he stated that 'it is principles, not phenomena—laws, not insulated independent phenomena—which are the objects of inquiry to the natural philosopher'.[12] Likewise, he conceded that meteorology, 'one of the most complicated and difficult' sciences, was subject to physical research, and as such, scientists should aim for the disclosure of the relations and laws governing the atmosphere.[13] Understanding meteorology implied, therefore, discovering general laws, or ascertaining how 'the physico-chemical fluid' behaved on a great scale and in all conditions—of temperature, pressure, etc. To this end, he previously defined an overall system of universal induction in which large numbers of global observations over long periods of time were required, if one wished to discover the universal physical laws.[14] In subsequent works, he emphasized that meteorology was a science of induction, not a deductive science.[15] His ideas were definitely at variance with the paladins of deductive reasoning

---

11 'Den ruhenden Pol in der Erscheinungen Flucht', as quoted by Leighly (1949, p. 658).

12 Herschel (1830, pp. 13–14).

13 Herschel (1830, pp. 328–29).

14 On a first stage of induction, Herschel (1830, pp. 144–220) included the discovery of proximate causes and the laws of the lowest degree of generality; on a subsequent stage (of inductive generalization), the formation and verification of theories. For a discussion on Herschel theoretical and organizational involvement in meteorology, particularly his new approach to meteorology and to terrestrial physics in general, see Good (2006, pp. 37–41).

15 Herschel's most developed expression of a general view of meteorology can be found in his book *Meteorology* (Edinburgh, 1861), which was originally published in the

who appropriated meteorology; the correspondence between the professors of natural philosophy, William Whewell and James Forbes,[16] makes clear this aspect: to the allusion that observation was the maiden of theoretical inquiry, Forbes agreed—'meteorology most of all needed its Newton'.[17]

Divergent as it may have been the views on meteorology of climatologists and natural philosophers, or even within the latter between inductionists and deductionists, there were other aspects to be considered. For all of them climate embodied the Hippocratic tradition of the study of airs, waters, and places, an historical inquiry on nature and place. They would all agree that the study of climate represented a key issue of inquiries and debates about navigation, sanitation, public health, imperial policy, and even national security and order. However, their fields of application and interest were often very different. While ideas about climate were developed by climatologists and usually emerged within accounts of medicine, expeditions, geology, and paleontology, theories about the dynamics of the atmosphere were formulated by natural philosophers and navigators and often emerged within scholars' memoirs, marine diaries, and logbooks. Furthermore, whereas the former flowed within the tradition of climate and locality, the latter reached wider banks, fuelling controversies over determinism and induction, the move from local to global, from particular event to general laws.[18]

When the development of telegraphy, which seemed merely experimental in early 1840s, became operational with the opening of the first commercial telegraphic network in Washington 1845, the scientists and the navigators found themselves in possession of a tool with tremendous potential for data transmission.[19] Their valuation of the science of meteorology and its

---

Encyclopaedia Britannica, 1857, 8th ed., 14, 636–90. In this article, he emphasized that meteorology was a dynamic but not strictly deductive science.

16 On the need for the disciplinary and institutional emancipation of meteorology, as claimed by natural philosophers like James Forbes, see Garber (1976, pp. 52–53).

17 As quoted by Anderson (2005, p. 7). For the letter from William Whewell to James David Forbes, see Todhunter (1876, vol. 2, pp. 165–67). Before we accept as an isolated fact the mention of Newton as 'the spiritual saviour of meteorology', we should hear other testimonies on the need for deterministic laws such as Newton's law of universal gravitation and Kepler's laws of planetary motion. In 1867, British meteorologist Georges Symons (1867, p. 34) noted: 'when this Newton of Meteorology is to arise, we know not'. And Achbari and Lunteren (2016, p. 8), bringing up some words expressed by the director of the Magnetic and Meteorological Observatory at Batavia, Johannes Paulus van der Stok (1899, p. 65), state: 'Theoretical foundations had not been developed or, to put it differently, the field still awaited its Newton'.

18 On the historical relations between climate and locality, see for example Feldman (1993), Fleming (1998), Jankovic (2001), and Ospovat (1977). For the relations between climate and health, see for example Kevan (1993), Sargent (1982), and Thomson (1979).

19 The first commercial telegraph line was established by Samuel Morse and Alfred Vail, and went from Baltimore, Maryland to Washington, D.C. On this company, see Prime (1875, pp. 512–15). For the telegraph's commercial and communicational potential in the United States, see Thompson (1947, pp. 21–34).

prospects created were substantially altered. That was, largely, their perception of the new scenario. Although there were other factors that came into play as well (e.g. the irruption of synoptic maps), this effect can be inferred from the optimistic tone of some talks and articles by meteorologists and navigators addressed to scientific and academic audiences in the late 1840s. While these and other testimonies will be shown in detail in the following chapters, here we will just mention a few passages conveying confidence in the prospects and opportunities created for the study of weather and storm forecasting. Thus, in September 1846 William C. Redfield, the first president of the American Association for the Advancement of Science, who authored a theory of hurricanes, described the tracks of three hurricanes in the American seas and their relation to the northers.[20] It is no coincidence that at the end of his extensive study (of 116 pages), he dealt with the telegraphic warning for navigators, and proceed to extol the prospects opened up: 'In the Atlantic ports of the United States the approach of a gale', he said, 'may be made known by means of the electric telegraph'. This 'will enable the merchant to avoid exposing his vessel to a furious gale soon after leaving her port'.[21] In the same vein, in 1848 Irish glaciologist and politician John Ball read the paper 'On rendering the electric telegraph subservient to meteorological research' before the British Association for the Advancement of Science (BAAS) in Swansea. Ball expressed his conviction on the intrinsic possibilities of telegraphy for prediction, and concluded his speech with this hope: 'I do not pretend to say' that 'we should be enabled to predict changes in the weather with absolute certainty'. But 'there is no reason to doubt that in a short time the determinations thus arrived at would possess a high degree of probability'. He even dared to mention the calculation of the state of the weather 'for twenty-four hours in advance'.[22]

Indeed, these 'calls for telegraphic-meteorology', these 'weatherological' currents which emerged after the opening of the first commercial telegraphic lines in Britain and the United States seemed so compelling that on August 1848 an English newspaper, the *Daily News*, first published a telegraphic daily weather report.[23] Only a few months earlier, Joseph Henry proposed to the Smithsonian's Board of Regents to organize a system of meteorological observations in North America. On this occasion the aim was not the study of climate, but rather the position and progress of storms: 'the extended lines of the telegraph', he wrote to the Board on 8 December 1847, 'will furnish a ready means of warning the more northern and eastern observers to be on

20 Known chiefly in the United States, a norther is a sudden cold gale coming from the north.
21 Redfield (1846, p. 334).
22 Ball (1848, p. 12).
23 Marriott (1903, pp. 123–26).

the watch for the first appearance of an advancing storm'.[24] For both Henry and the men behind the weather reports and maps in England, the telegraph enabled them not only bring together almost instantly observations made simultaneously in different places but also to determine the position and properties of individual storms on successive days.

Therefore, the result of this first approach to the challenge of proving the selective effect that certain national interests had on methods of weather prediction is unequivocal: the science of meteorology, theretofore conceived as climatology, began to focus on the weather. By 1850, the focus in meteorology began to shift: whereas from the 1820s to the 1840s its object of investigation was climate, by the 1850s it gradually became the weather and individual storms. By then, many meteorologists realized that the weather—and especially storms of all sorts—was an object of inquiry that deserved to be studied per se and from itself, rather than from other frameworks. In the next section and before broaching the mentioned challenge, I shall briefly outline the three storm models that became especially well known in the first half of the nineteenth century. For the purposes of this book, it is important to describe the state of storm theory and the stated causes for storm formation before the 1850s. An overview of those conceptualizations will help to determine in what sense and to what extent specific interests linked to national defence, security, and order may be regarded as responsible for the choice of an Eulerian approach in weather forecasting.

Three storm models

One of the unresolved problems of the first half of nineteenth-century meteorology was the formation of storms. Throughout this period, meteorologists interest focused on the nature of the wind field associated with midlatitude storms, both for their practical importance as safe solutions for navigation and maritime trade, and for their scientific importance as responses to the physical processes involved in the general circulation of the atmosphere. In the 1830s and 1840s causal theories of storm formation became a primary focus of scientific concern both in Europe and the United States. While Dove in Prussia and Redfield in America proposed theories exclusively based on mechanical principles, James Pollard Espy held that thermal convection played a decisive role in atmospheric motions on these scales. These three men held very different views of storm formation.

24 Joseph Henry, 'Report of the Secretary of the Smithsonian Institution to the Board of Regents, December 8, 1847'; reprinted in *Eighth Annual Report of the Board of Regents of the Smithsonian Institution, up to January 1, 1854, and the Proceedings of the Board up to July 8, 1854*. Washington, DC: 1854, House of Representatives, Misc. Doc., 33d Cong., 1st sess., no. 97, 119–39, on p. 139—quoted by Monmonier (1999, p. 39).

Dove's opposing current theory

Heinrich Wilhelm Dove was a professor of physics at the universities of Königsberg and Berlin who dedicated his career mainly to experimental physics, as well as climatology—as director of the Prussian Meteorological Institute from 1849 to 1879.[25] In 1827, he derived his 'law of gyration of the wind' (known as *Drehungsgesetz*) from the statistics of changes of local wind direction, from which he later articulated the theory of conflicting linear currents as the physical mechanism of storm formation.[26] (His theories were expounded in German in the journal *Annalen der Physik* from 1828 to 1841, and condensed in his book *The Law of Storms*, translated into English by Robert H. Scott in 1862.)[27] Dove likened the process of storm formation to the counteraction of two opposite currents and suggested that weather changes were closely linked to changes in the direction of the prevailing winds. The shape of wind-roses, he proclaimed, allowed us to deduce the struggle of these currents in midlatitudes.[28]

According to Dove's theory, the formation of midlatitude storms was explained in terms of 'air-currents' of equatorial and polar origin.[29] In his global wind scheme, there were two air currents blowing throughout the whole atmosphere. One was the equatorial currents (warm and humid) that could be seen as the continuation of the southwesterly winds above the *trade winds*.[30] The other was the polar currents (cold and dry) or northeasterly surface winds that changed into the trade winds on their way to the equator. Interestingly, both storms and weather changes in midlatitudes followed this global wind scheme. Storms were the result of 'the conflict between two currents which alternately displaced each other at the place of observation'.

---

25 Dove was a prolific writer who published more than 330 papers, among them more than one hundred on topics of experimental physics. Climatology featured prominently in his bibliography: sixty-one out of eighty-four articles signed by Dove in the Proceedings of the Prussian Meteorological Institute from 1837 to 1876 dealt with this science.

26 Dove (1827, pp. 545–90).

27 In this regard, relevant reference works are Dove (1828a, 1828b, 1841a). Dove's book was originally published as *Über das Gesetz der Stürme*. Berlin: D. Reimer, 1857 (1861, 2nd ed.). It was translated into French by Alexandre LeGras, *La loi des tempêtes considérée dans ses rapports avec les mouvements de l'atmosphère*. Paris: Dupont, 1864.

28 For biographical notes, Neumann (1925), Kutzbach (1979, pp. 230–31), and Pedersen (1971, pp. 174–75). For Dove's theory, Bernhardt (2008, pp. 61–100), Carramolino (1994, pp. 103–19), Khrgian (1970, pp. 168–71), Kutzbach (1979, pp. 11–18, 21, 25, 36), and Schneider-Carius (1975, pp. 188–91, 217–20).

29 For an exposition of his theory in English, see his article entitled 'On the Influence of the Rotation of the Earth on the Currents of its Atmosphere; Being Outlines of a General Theory of the Winds', Dove (1837, esp. pp. 361–63).

30 The trade winds are the dominant pattern of easterly winds found in tropics. These usually blow from the northeast in the Northern Hemisphere and from the southeast in the Southern Hemisphere. The importance of these winds for navigation was recognized by Portuguese traders as early as the fifteen century.

Weather change was the result of the 'sole predominance of one of these currents'.[31]

The continual replacement of the equatorial current by the polar current and vice versa explained a number of puzzling atmospheric disturbances such as the absolute extremes of weather changes and the formation of mid-latitude storms and gales.[32] However, the theory disproved the widespread belief that all storms were necessarily of vortical nature: at midlatitudes, storms stemmed from the conflict of two opposite currents, within which the changes of barometric pressure were accompanied by variations of temperature; by contrast, in tropical regions storms were rotary and there was no change of temperature as pressure fluctuated.[33] What underlay this difference in storm's movement was his law of gyration. This law defined the veering of the wind when the equatorial current replaced the polar one (and vice versa). According to Dove's law, in this case winds veered from south over southwest to northwest (vice versa in the opposite scenario). The effects of Dove's law of gyration, therefore, could not be confounded with those of rotary storms.

This struggle between radically different air currents was represented by Dove by means of abstract diagrams in his 1828 article, in which he described the turning winds and pressure for a strong storm that hit Europe on 24 December 1821.[34] As shown in Figure 1.1, one of these diagrams is a grid of lines that depict wind and pressure. A large arrow points out the storm's passage from the southeast, as the polar current is gradually displaced by the equatorial current. The lines set the pace of pressure: while line c–d represents how the line of minimum pressure moves forward and therefore the storm's area, line a–b shows the way followed by the lowest pressure along the depression.[35] No less important, Dove's diagram followed the ideas of other precursors in at least two respects: first, directional arrows representing winds; and second, his midlatitude wind system, fitting in with the global circulation of the atmosphere. Edmond Halley had introduced directional arrows in the map of trade winds and monsoons that this astronomer published in 1886. He had sought the cause of the trade wind circulation

---

31 Dove (1841b, p. 7)—quoted by Kutzbach (1979, p. 14).
32 On Dove's flow-model of midlatitude storms, see Ludlam (1967, pp. 23–24).
33 Dove's view on storm motion was not always consistent. In 1828, he considered *all* storms as whirling winds, and held the idea of a circular flow turning around the minimum of barometric pressure, as advocated by Redfield. Although he formulated his law of gyration drawing on the data provided by Brandes (1827), he was against the view of his teacher in meteorology, Brandes, who asserted that all storms were centripetal in nature, with a radial flow towards the centre of barometric diminution. Brandes's view was subsequently embraced by Espy.
34 Dove (1828a, fig. 8, on plate 7).
35 Only two of his nine diagrams can be regarded as maps. For Dove's 1828 maps, see Monmonier (1999, pp. 29–30).

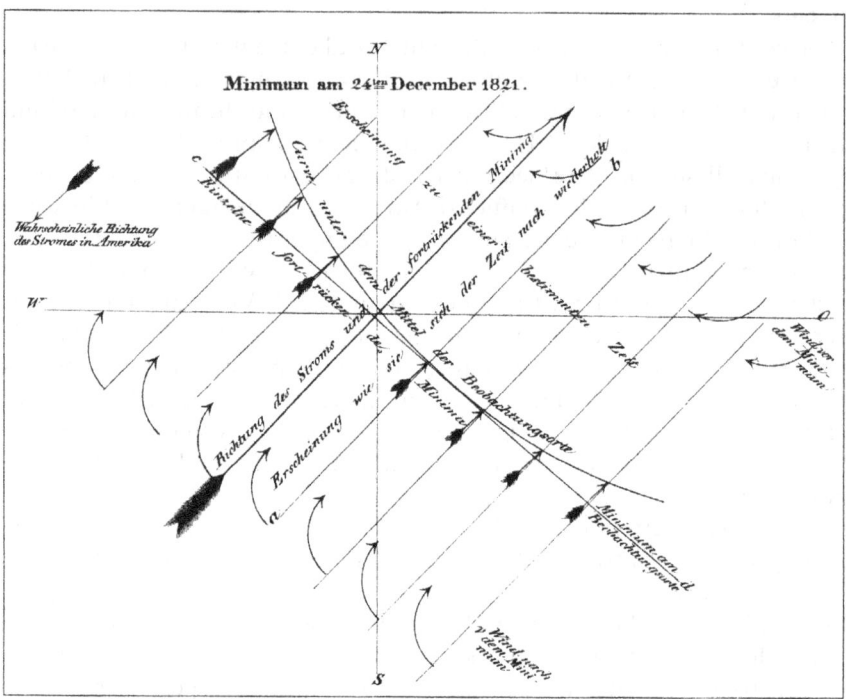

Figure 1.1 Dove's chart showing wind and pressure patterns associated with the
storm of 24 December 1821.

Source: Dove (1828a, fig. 8, on plate 7).

in thermal convection, rather than the earth's rotation.[36] This interpretation had been modified by George Hadley in 1735, when he suggested the
existence of twin belts of descending air balancing rising air in equatorial
regions. He also had explained the deflecting force of the earth's rotation,
even if only its east-west component.[37] Dove's scheme amalgamated in part
all these ideas: the global circulation was the effect of both the earth's rotation and solar heating.[38]

Dove's opposing current theory was a unifying account of atmospheric
dynamics based on mechanical principles that explained both the storms'
formation and the weather changes in midlatitudes. In historical perspective,

36 According to Halley (1686), rising air caused by solar heating in equatorial regions produced a vacuum at low altitudes, which the cooler air that would form the trade winds
   tended to fill.
37 Hadley (1735, pp. 58–62) showed how this force turned winds to the right in the Northern
   Hemisphere and to the left in the opposite hemisphere.
38 On the early ideas about the general circulation of the atmosphere, especially those of
   Halley and Hadley, see Burstyn (1966, pp. 171–72, 181–86) and Lorenz (1983, p. 731).

however, his work was the object of contradictory assessments. In a letter dated on 27 December 1848, Humboldt wrote: 'You are, my dear Dove, the founder of modern meteorology as a science, as Leo[pold] von Buch is the founder of modern geology'.[39] Others, however, such as Austrian meteorologist Julius von Hann, who continued using Dove's law of gyration as late as 1873, declared with no qualm a few years later: 'it hardly can be denied that Dove's quite unphysical theories have retarded the progress of meteorology for a long time'.[40] Analyzing the emergence of the convective or thermal theory of cyclones, historian Gisela Kutzbach took a more balanced view on his work:

> At a time when the baric wind relationship, describing the interdependence of the pressure field and wind field, had not yet been established and when no synoptic charts were available that could have provided a grasp of the spatial distribution of weather phenomena at a given instance in time, Dove's scheme had the definite advantage of providing a temporal framework into which series of local observations could be fitted. For this reason Dove's law of gyration, first expounded in 1827, was commonly used in the interpretation of weather changes in Europe up to the 1860's.[41]

Dove's opposing current theory was thus the culmination both of one man's Humboldtian climatological training and of a long-standing tradition in European science bound to the dynamic study of tropical weather. But while storm theory in Europe was viewed as a continuation of Halley's and Hadley's trade winds theories, the scenario in the United States was different. There, William Redfield developed an entirely different theory of atmospheric dynamics in which small-scale storms (especially hurricanes and tornados) will stand out as objects of study.

Redfield's whirlwind theory

A son of a seafarer, William C. Redfield initially developed a career linked to crafts and business, first as a saddle—and harness—maker in Connecticut and then as a marine engineer and promoter in New York. He worked as superintendent of the Steam Navigation Company, and was a promoter of railroad transportation. An avid reader, he was self-taught in meteorology.[42]

---

39 Quoted by Bernhardt (2008, p. 64). Early in his career, Dove had adopted the Humboldtian model of studying the distribution of the heat in the earth's system, as evidenced by the title of his postdoctoral work in Königsberg, *De distributione caloris per tellurem* (1826).

40 Hann (1885, p. 395).

41 Kutzbach (1979, p. 15).

42 For Redfield's life and early activities, Olmsted (1857), Burstyn (1981, pp. 340–41), Kutzbach (1979, pp. 241–42).

In 1821, at the age of 23, he observed the orientation of the fallen trees caused by a hurricane in New England, a fact that would be the source of his scientific inspiration. Ten years later, in the *American Journal of Science*, he depicted the origin of the 1821 storm, traced its movement, and concluded that it was developed 'in the form of a great whirlwind'.[43] In this paper, he first formulated a theory of cyclonic whirlwinds moving forward from southwest to northwest along a parabolic path. Thereafter, he devoted much of his time to the study of storms.[44] He would rather confute to his critics and order his ideas from the reflexive environment of his home office, spreading them in articles in scientific and marine journals, than to personally defend them at academic meetings or lecture before public audiences.[45]

In Redfield's view, hurricanes were large circular vortexes, where winds blew counterclockwise around a centre advancing in the direction of the prevailing winds. John Farrar from Harvard University had suggested in 1819 that a hurricane was a 'moving vortex and not the rushing forward of a great body of the atmosphere'.[46] Redfield noted that winds circulated counterclockwise in the Northern Hemisphere, and predicted that they would circulate clockwise south of the equator. As regards their direction, he concluded that it was 'in all cases, compounded of both the rotative and progressive velocities of the storm'.[47] The existence of a low barometric pressure Redfield attributed to 'the centrifugal tendency, or action', which pertained to 'all revolving or rotary movements'. He illustrated said action with this example: just as in bringing water in a container into rotary motion 'we shall find that the surface of the fluid immediately becomes depressed by the centrifugal action', so, too, rotary motion produces an atmospheric depression at the storm's centre induced by the centrifugal action, causing the fall of barometric pressure.[48]

Redfield's interests were both synthetic and cartographic, and from 1831 to 1856 he afforded further evidence for the revolving character of storms, including tropical hurricanes and tornados. Perhaps his best-known work was his analysis of a storm of 1839, which he read before the American

---

43 Redfield (1831, p. 333). This paper brought him immediate recognition in both Britain and America: his theory was disseminated through E. and G.W. Blunt's *American Coast Pilot*, as well as through Captain Henry Piddington's 1848 book, *The Sailor's Hornbook of Storms in All Parts of the World*.
44 For Redfield's whirlwind theory, Cox (2002, pp. 27–32), Davis (1895, pp. 306–09), Fleming (1990, pp. 23–73), Khrgian (1970, pp. 165–67), Kutzbach (1979, pp. 16–18, 22, 25), and Schneider-Carius (1975, pp. 220–202).
45 Ludlum (1969, p. 229).
46 Farrar had studied the Boston hurricane of 23 September 1819, describing the extent of damage and the veering of the winds.
47 Redfield (1831, p. 29).
48 Ibid., p. 46.

Philosophical Society in 1841.[49] This work introduced two elements worth mentioning: he included a map of winds and he distanced himself from his early notion of pure circular rotation in storms. Examining forty-eight observations compiled from sources as diverse as ships' logs, newspapers, and personal correspondence, he noticed that revolving storms presented a slight vortical movement. He noted this effect—to be precise, 'an inward inclination of about six degrees'—by drawing a map in which wind direction was represented by arrows and observations were identified by numbers (Figure 1.2). He added concentric lines, 'drawn at intervals of thirty miles', 'not as precisely indicating the course of the wind, but to afford better means of comparison for the several observations'.[50] Interestingly, Dove had published a similar map, a circular vortex over Western Europe, just two years before him.[51]

Unlike Dove's work, focusing on large-scale midlatitude cyclones, Redfield's theory provided explanations of small-scale storms (usually, hurricanes and tornados).[52] In many aspects, they arrived at completely opposed results, in part because of their different objects of study, but also because of their different modus operandi: while Dove largely drew on observations made by the so-called local method—i.e. scrutiny of the temporal sequence of phenomena at a fixed station—Redfield often based on the testimonies of the casual observer, more given to emphasize the storm's revolving nature than anything else. Notwithstanding these differences, they shared one common method: they first examined the facts and then attempted to construct a theory for them—they were inductionists, not deductionists.[53]

Because Dove and Redfield primarily dealt with the kinematic description of the wind field, their storm theories came to be labelled as mechanical theories. Hence, although clearly a type of mechanistic theory, resting as it did on the premise of centrifugal action and devoid of sophisticated physical concepts, Redfield's work was never perceived by his American colleagues as standing in opposition to the conflicting current theory articulated by Dove. But still, Redfield's whirlwind theory did not become widely accepted in the United States. In some ways, 'centrifugal' theory was unpersuasive, because it left the door open for any centripetal theory that could be based

49 Redfield (1843). For the role played by this society in the progress of meteorology in the United States between 1835 and 1850, see Gross (1972, esp. pp. 322–26).

50 Redfield (1843, p. 77). For a fuller account of Redfield's contribution to cartographic meteorology, see Monmonier (1999, p. 31–33).

51 Dove (1841b, p. 194, Fig. 5) published a map of the weather conditions in Europe on 24 December 1821, in the fourth volume on meteorology in *Repertorium der Physik*, printed in Berlin in 1841. For further details, see Bernhardt (2008, p. 77).

52 Kutzbach (1979, p. 16) has emphasized the 'loose terminology' used at the time to designate storms of various scale size and origin, which considerably contributed to misunderstandings among meteorologists.

53 This point was highlighted by Bôcher (1888, p. 5). For a contemporary evaluation of Redfield's theory, see Davis (1844, pp. 336–41).

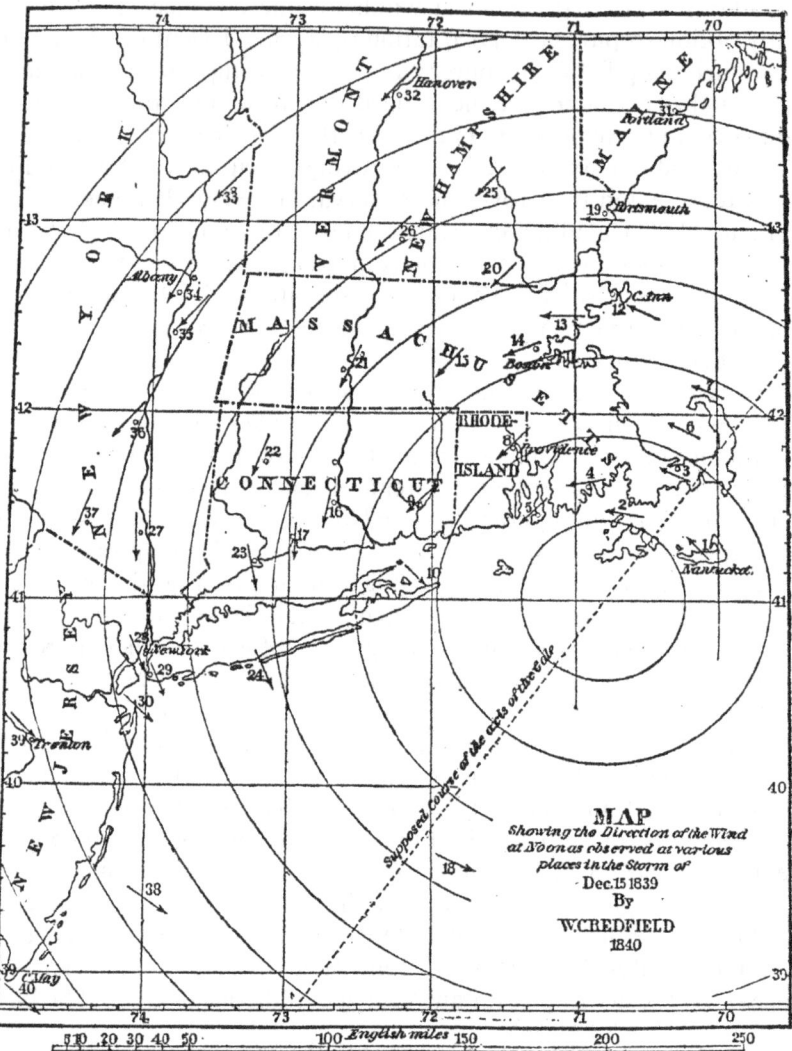

Figure 1.2 A whirlwind storm of vertical nature, as traced by Redfield. While arrows
show wind direction, the numbers beside arrows indicate observation
stations.

Source: Redfield (1843, p. 80).

on the assumption of a radial movement towards the storm centre. Did
winds blow in circles or directly towards the centre? To address this question Redfield had to contend with, and to a great extent refute, the work of
his foremost adversary, James Pollard Espy, the precursor of the thermal
theory of cyclones.

Espy's convective theory

James Pollard Espy, a teacher at the Franklin Institute in Philadelphia in the 1820s, was the first official 'American meteorologist', a post created by the U.S. government for him in 1842 that he kept it until his retirement in 1857.[54] In 1836, he founded a meteorological network in Pennsylvania which subsequently merged into Joseph Henry's corps of observers of the Smithsonian Institution. Espy emphatically rejected Redfield's 'centrifugal' theory as unsupported by experimental evidence. Indeed, he was greatly impressed by John Dalton's work on the laws of vapours and investigated their meteorological implications, i.e. the relationship between rain and the quantity of vapour in the atmosphere. Drawing on laboratory experiments on the expansion and compression of dry and moist air, Espy identified physical processes that eventually led him to propose a centripetal theory of storms: a thermal theory stemming from the linkage between the concepts of adiabatic change and thermal convection.[55]

To begin with, Espy sought an explanation of the large-scale formation of clouds—and, why not, of precipitation—that could be based on the expansion and cooling of rising air. He sought to replace earlier interpretations of this process with his own. According to his theory, when water vapour was condensed into cloud its air became heavier than it was before. Espy invented a device (the nepheloscope) that simulated the behaviour of clouds, allowing him to measure the adiabatic cooling rates. Drawing on his experiments, he argued that during cloud formation air did not become heavier but lighter than it was before.[56] Condensation of water vapour, therefore, favoured (not inhibited) the rising of air, namely, thermal convection.[57] The expansion of air took place in response to the latent heat released by condensation of water vapour. It was, therefore, the latent heat (not the centrifugal forces) that provided the energy for precipitation and ascending movement of clouds.[58]

Espy suggested that the nature of storms could be elucidated by his hypothesis of the formation of clouds. Thus, processes such as thermal convection and adiabatic change contributed to the movement of storms in a kind of positive-feedback loop that resulted in a centripetal inflow of air

---

54 The post was the professorship of meteorology at the War Department. For biographical notes on Espy, see Bache (1860, pp. 108–11), Fleming (1990, pp. 23–28, 43–59, 67–81), Kutzbach (1979, pp. 231–32), and Reingold (1971, pp. 410–11).

55 For Espy's convective theory of storms, see Cox (2002, pp. 33–40), Davis (1895, pp. 309–15), Garber (1976, pp. 53–55), Khrgian (1970, pp. 165–67), Kutzbach (1979, pp. 3–4, 21–27), Ludlum (1969, pp. 224–27), McDonald (1963, pp. 634–41), and Schneider-Carius (1975, pp. 222–28).

56 This process was explained in the first of a series of three essays published by Espy in the *Journal of the Franklin Institute* in 1836. See Espy (1836a, p. 241).

57 Middleton (1966, pp. 155–60).

58 For Espy's investigations of adiabatic change see Kutzbach (1979, pp. 22–24) and McDonald (1963, pp. 634–41).

in storms. As water vapour was condensed, the latent heat released expanded air; the expansion of this air caused a rise in temperature in the ascending air; and this permitted more vapour to be condensed into cloud—for as this air rose, other air rushed in to take its place, and this, in turn, rose. Eventually, the vapour would become so dense that a vortex would be formed, near whose centre would be rain, and by some mechanism linked to barometric pressure, the wind would radially flow towards this centre.[59] Thus, the centripetal inflow of air in storms was causally linked to the adiabatic expansion of rising air, and any formation of storms was the result of the latent heat released during condensation of water vapour.

The cartographical representation of the weather was an additional motivation for the Franklin Institute and the U.S. government to trust Espy's studies on storms. Typical of his representations was what was regarded as 'one of the first birds'-eye-view weather maps of a specific weather phenomenon' published in America, 'if not *the* first',[60] a winds map depicting the storm that hit Silver Lake, New York, on 20 June 1836.[61] In this map, Espy's arrows of wind showed an inward inflow towards a centre of minimum pressure (just the opposite of Redfield's maps, whose arrows described a circular pattern).[62] This centre represented the point of greatest rain. Over the ensuing decades Espy constructed many maps for the government. His later maps were innovative for meteorological cartography: they included advancing lines of high and low pressure, traced by thick black and red lines, respectively (see Figure 1.3). Like in Brandes's map, the wind field was represented as an inward flow of air towards a centre of low pressure—an effect barely noticeable on Redfield's map.[63]

In the United States—where there was a heated controversy between Espy and Redfield that lasted for twenty years—neither theory was fully accepted.[64] A tension persisted between the convectional and mechanistic perspectives, in which mathematically oriented physicists opted mostly for the former, and empiricists and inductionists leaned towards the latter. For Espy's supporters, who at the time interpreted adiabatic processes

---

59 Espy (1836b, pp. 105–08).
60 According to Spitz (1944, p. 281). Italics in original.
61 Originally published by Espy (1837, p. 19) in the *Journal of the Franklin Institute* to account for a report of the Joint Committee on Meteorology, formed by the Franklin Institute and the American Philosophical Society in Philadelphia in 1834, the map was reproduced in Espy's well-known work, *The Philosophy of Storms,* in 1841, on p. 105.
62 On this map and Espy's contribution to meteorological cartography in general, see Monmonier (1999, pp. 32–35).
63 Espy's advancing lines can be found, for example, in his *First Report on Meteorology to the Surgeon General of the United States Army* that dates from 1845, and includes twenty-one maps illustrating the weather in 1843. Espy (1843).
64 The American storm controversy, which began with the third paper published by Espy (1836b) in the *Journal of the Franklin Institute,* involved a good number of scientists and theories. For a thorough account, see Fleming (1990, pp. 23–73). On the storm controversy as viewed in Britain by Herschel and William Reid, see Jankovic (1998, pp. 26–28) and Halford (2004, pp. 33–45).

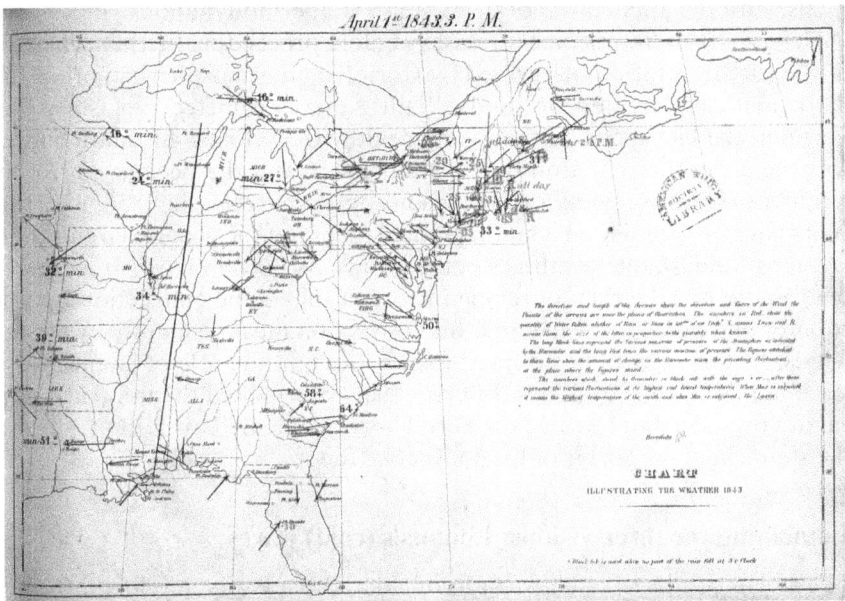

Figure 1.3 Espy's chart illustrating the weather on 1 April 1843. The arrows show
the direction of the winds, and the length of arrows shows its force. The
points of the arrow are near the points of observation. The numbers in
red represent the quantity of water fallen in hundredths of an inch—R
means rain, S snow. The thick black lines represent the various maxima
of barometric pressure, and the red lines represent the various minima
of pressure. The figures attached to them show the amount of change in
the barometer since the preceding reading. The numbers with the sign +
or − after them indicate the fluctuations of the thermometer at its highest
and lowest temperatures.

Source: Espy (1851, p. 1).

according to the caloric theory of heat, the motive power of storms was the
release of latent heat. In fact, the spirit of Espy's theory was convectionist—
with the heat causing air to rise and winds to flow radially towards the centre. In contrast, Redfield's theory was explicitly mechanistic. He believed
that gravity and the earth's rotation were the driving force behind the winds,
which, in turn, explained differences in temperature, pressure, and moisture. Unfortunately for Espy, he was unaware of Coriolis's work on the
earth's rotational force,[65] and, perhaps worst of all, the conversion processes
between heat and work could not be quantitatively calculated by the caloric

65 As U.S. Weather Bureau meteorologist Eric R. Miller wrote years later, Espy 'could not
admit the rotation observed by Redfield because he knew of no force capable of producing it' —the *Coriolis effect* was announced in a seminal paper published in 1835, once
Espy had formulated his storm theory. Quoted in Cox (2002, pp. 36–37).

theory—indeed, such a calculation only became possible in the 1850s and 1860s, with the application of the first law of thermodynamics.

Here, a slight digression is necessary before proceeding further. The effect of the earth's rotation, described by Coriolis, on atmospheric motions was mathematically treated by the brilliant American meteorologist William Ferrel in the late 1850s. Although he dealt with cyclones, his main concern was the general circulation of the atmosphere and the oceans. Ferrel first wrote a descriptive essay on winds and ocean currents in 1856, and three years later he formulated a mathematical theory of the general circulation.[66] His most remarkable result is what is known as Ferrel's law.[67] It is interesting to note how he then developed the idea that some convection between equator and pole must exist and this convection must be mainly by westerly winds. In this regard, he adopted Espy's ideas on the role of latent heat release in midlatitude cyclones. In brief, Espy overlooked the influence of the earth's rotation, but he saw how Ferrel—who remedied his failing—became the staunchest supporter of his convective theory.[68]

Combining the three visions: Loomis's (cold) waves

The storm controversy between Redfield and Espy was developed from evidence on hurricanes, tornados, and other tropical storms, but meteorologists soon analyzed extratropical storms. Among these was Elias Loomis, a professor of mathematics and natural philosophy at Western Reserve College, Ohio, and from 1844 to 1860, at the University of the City of New York.[69] From 1836 to 1837, Loomis studied in Paris, attending the lectures of François Arago, Jean-Baptiste Biot, and Siméon-Denis Poisson, among others. Loomis was one of the first who attempted to settle the controversial question of the surface wind pattern in storms by studying midlatitude cyclones. In a series of papers published from 1841 to 1846, Loomis attempted to clarify the centrifugal-centripetal character of winds and soon realized neither of these models faithfully represented winter storms. Indeed, he discerned opposing winds rushing towards one another, as Dove and Espy had supposed, and defended thermal convection. However, in so doing, he introduced a new concept: his synoptic maps showed that a *wave of cold air* 'advanced more rapidly than the center of the storm', and was propagated by

---

66 Ferrel (1856, 1859–1860).
67 Ferrel (1858, p. 99): 'if a body is moving in any direction, there is a force, arising from the earth's rotation, which always deflects it to the right in the northern hemisphere, and to the left in the southern'.
68 On Ferrel, see Abbe (1895) and Burstyn (1971). For his contribution to meteorology and especially the thermal theory of cyclones see Davis (1891), Kutzbach (1979, pp. 35–41, 110–117), and Cox (2002, pp. 65–74).
69 For Loomis's biographical notes, cf. Newton (1895, esp. pp. 215–18, 227–36), Cox (2002, pp. 41–49), Kutzbach (1973, p. 487), and Daniels (1994, pp. 87–101, 216–17).

mechanical laws long after its cause ceased to act.[70] This concept embodies the embryonic idea of what today is known as a cold front.

The assumption of storm rotation was premised on Redfield's arguments on hurricanes and Espy's laboratory experiments and evidence on tropical cyclones. Therefore storm rotation had to be evaluated in the context of storm formation in temperate latitudes. Persuaded that meteorology should be promoted, not so much by theoretical deductions or by taking the mean of long continued observations, as by examining the formation of particular storms, Loomis resolved to 'select some single storm of strongly marked characteristics, and trace its progress as extensively and minutely as possible'.[71] The chance arose in 1835 when, under the auspices of the BAAS, Herschel invited him to participate in a systematic collection of hourly meteorological observations.[72] Whether winds blew in circular whirls or towards a centre, Loomis would know it by studying a large-scale storm in the winter of 1836.[73]

As a first point in his analysis Loomis passed mechanical and thermal considerations through the filter of meteorological evidence. Did meteorological data support the centrifugal theory or the centripetal theory? Neither—or better, both in certain aspects. A thorough examination of the most important weather elements, Loomis argued, was unable to confirm neither of the theories; the final verdict should be postponed to future study. Yet, how could the passage of a storm be accounted for over a region as vast as the United States? The answer lay with Loomis's *waves*. 'A bare inspection of the meteorological registers', he wrote, 'is sufficient to show' that there was 'a sudden depression of the barometer, immediately succeeded by an equally sudden rise; that the barometric minimum' passed 'like a *wave* over the entire country, from west to east'.[74] Barometric waves bore an analogy to the waves of the ocean agitated by a tempest: 'just as an elevation or depression of the surface of the ocean which arrives successively at distant stations, we call one wave', even if it 'may be formed by entirely different particles of water', so too 'the same atmospheric wave pass[es] over

---

70 Loomis (1845, p. 176).

71 Loomis (1841, p. 125).

72 Herschel's plan, addressed to meteorologists from anywhere in the world to promote hourly observations of temperature, barometric pressure, etc., during the equinoxes and solstices, provided Loomis with probably the most comprehensive information anyone had ever gathered for any storm. He had at hand barometric data from twenty-seven stations, hourly readings from eight eastern stations from Quebec to Baltimore, as well as additional data from much of the country east. For further details about this plan, see Herschel (1836). This plan fit into Herschel's system of inductive universalism that he had defined in his book *Preliminary Discourse on the Study of Natural Philosophy* (1830). On Herschel's ideas on inductive science, cf. Carter (2004, esp. pp. 512–15).

73 The results of this study were read before the American Philosophical Society in Philadelphia in 1840, and published in the *Transactions* of this society in 1841.

74 Italics added. Loomis (1841, p. 128).

the continent'.[75] However, how could the rate of progress of these waves be known? A synoptic map was required to ascertain this progress: first, 'we must fix upon some particular phase of the wave'; this could be done by comparing either the barometric maximum or minimum, though the latter was preferable; finally, a map of the 'lines connecting all those places where the barometer attains its minimum at the same instant' should be drawn.[76] Once the lines were traced, Loomis concluded, their progress on the map would show how the storm had travelled. Moreover, arrows added would indicate the movement of the air.[77]

This graphical procedure led Loomis back to the question that had motivated Redfield and Espy and virtually all meteorologists interested in the features of storm movement: How do winds or air currents move? Examining precipitation, wind, and temperature patterns, Loomis inferred that cold air currents would flow under opposing warm ones. A warm southeast current, pushed up off the earth's surface, would blow towards the centre of the barometric depression. Concurrently, a cold current from the northeast would displace the warm current by flowing under it. The result would be intense cold outbreaks advancing over the continent. In fact, Loomis illustrated his thesis in a chart that symbolizes the first cross section of what the modern reader would call a cold front (Figure 1.4).[78] Although Loomis does not mention him, this same reader might easily identify these currents with Dove's equatorial and polar currents.[79] Interestingly, this very mechanism could explain the cause of the rain along the line of minimum pressure. The inflow of cold air from the one side of this line would lead warm humid air to be lift on the other side. On being raised to higher altitudes, the pressure of this warmer air would decrease, and, consequently, be cooled, forming clouds and causing their precipitation.

Although Loomis credited his ideas only to Espy's and Redfield's insights, his vision of storms was a blend of the Dove and Espy models. Dove's model assumed the struggle of two opposing currents on both global and local scales, but Loomis held a scheme of two currents dissociated from the general circulation and from any whirling effect. Again, there is little doubt that many of his ideas about rain and cloud formation were in line with the process of convection described by Espy. Although he did not embrace

75 Loomis (1841, pp. 128–29).This analogy was more emphatically expressed by Loomis in his 1845 paper: 'The undulations thus excited in the atmosphere bear considerable analogy to the waves of the ocean agitated by a tempest, and which are propagated by mechanical laws long after the first exciting cause has ceased to act'. Loomis (1845, p. 176).

76 Loomis (1841, p. 129).

77 For Loomis's early investigations on storms, see Cox (2002, pp. 41–49), Davis (1844), Khrgian (1970, pp. 166–67), Kutzbach (1979, pp. 27–35), Newton (1895, pp. 227–36), Schneider-Carius (1975, pp. 228–34), and Waldo (1895, pp. 322–25).

78 Kutzbach (1979, p. 30).

79 Loomis did not mention Dove's work, although he was probably a major source of inspiration. Cf. Kutzbach (1979, p. 30).

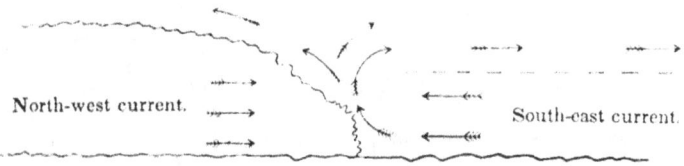

North-west current.

South-east current.

Figure 1.4 Loomis's representation of the confluence of warm southeast and cold
northwest currents of air.
Source: Loomis (1841, p. 159).

Espy's theory of storms, Loomis adopted his arguments on the formation
of rain by ascent and consequent cooling of moist air.[80] This blended view
of storms implied that the mechanisms for the formation and progress of
storms were of both a mechanical and thermal nature. Yet this left Loomis
with the unresolved question of what the effect of the earth's rotation was,
as well as the stages in storm formation. Loomis addressed these issues in a
study on two great storms read before the American Philosophical Society
in May 1843. The outcome was an extraordinarily detailed account of the
weather over the United States during the passage of two storms. He summarized in tables hundreds of observations, and then reproduced these data
on charts. In this paper the most remarkable innovation was the construction of lines of equal barometric deviation as opposed to minimum pressure: this distinction earned him recognition as the initiator of the synoptic
map—I shall address this issue in the next chapter.[81]

A first major difference arose from the cartographic representation of
barometric pressure and temperature. While Redfield represented the wind
field by lines and arrows and later Espy mapped the lines of minimum pressure, Loomis obtained more comprehensive weather charts by tracing deviations from normal pressure and lines of equal temperature deviation.[82]
When the wind field was analyzed, Redfield's diagrams showed circling
winds and Espy's charts showed radial winds. In the 1845 paper, Loomis
looked more closely to the behaviour of winds. Unlike the 1836 storm, the
centres of the 1842 winter storms that swept across the United States were
within the area of observation stations. Loomis plotted data from 131 observers on thirteen synoptic weather charts and discovered that neither of

80 The following sentence by Loomis (1845, p. 176) could be attributable to Espy: 'air sud-
   denly transported into elevated regions' must 'be allowed to be by far the most efficient
   of all causes of rain'.
81 This recognition with reference to the significance of Loomis's weather maps can be
   found in work of scholars such as Hellmann (1897a, pp. 6–7; 1897b), and Scultetus (1943,
   pp. 419–22).
82 Monmonier (1999, pp. 35–38) summarizes Loomis's cartographic system and its features.
   See also Newton (1895, pp. 230).

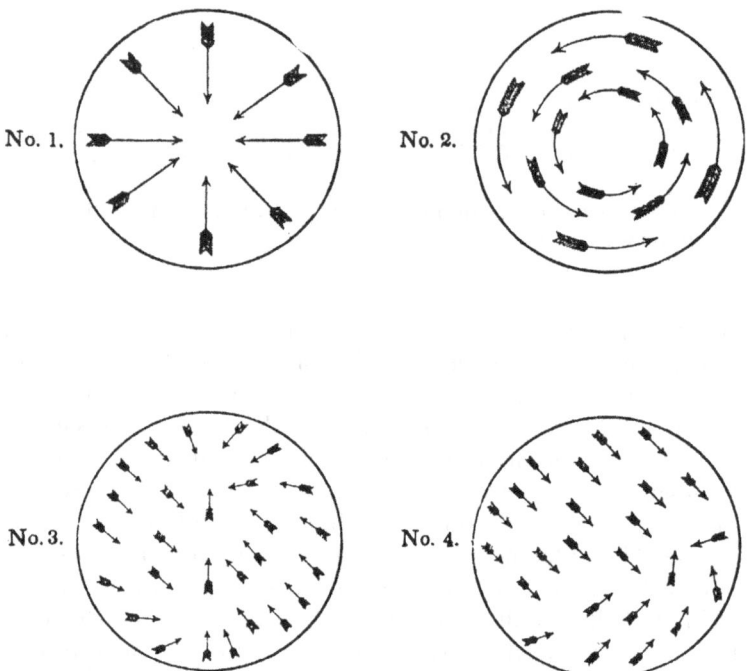

Figure 1.5 Wind diagrams in cyclones according to three models: (1) Espy's radial
            winds; (2) Redfield's circling winds; (3 and 4) Loomis's confluence of two
            currents.
Source: Loomis (1845, p. 180).

the two foregoing diagrams faithfully represented the storms investigated:
indeed, 'a combination of the two is frequently seen', he wrote.[83] When
Loomis pursued the question, he found that a factor had been omitted: the
deflecting force of the earth's rotation provoked a deviation of the motion of
a body, as he had learned from Poisson, Arago, and Biot during his stay in
Paris.[84] It was this force, he claimed, that was responsible for the spiral inflow towards the centre of minimum pressure. Loomis showed in a diagram
that this movement to the centre was 'not exactly in radii, but circuitously'
(Figure 1.5).[85]

The idea of a cold air wave was also essential for Loomis's version of
the origin and maintenance of storms, and he presented his weather charts
and Espy's convective arguments on storms as confirmatory evidence for his

83 Loomis (1845, p. 180).
84 In this point, Loomis seems to follow George Hadley's theory, who considered only the
    east-west component of this deflecting force, as suggested by Kutzbach (1979, p. 34).
85 Loomis (1845, p. 175).

views. As regards the origin of storms, 'an abnormal current at the earth's surface' would interfere with the dominant westerly current. Next, southerly and easterly currents would give rise to cloud formation and rain in just the manner described by Espy in his convective theory: 'the heat liberated' in cloud formation, he wrote, 'raises the thermometer, causing a more decided tendency of the air inward towards the region of condensation'. This elevation of temperature increases the velocity of the inward current of air. 'Thus the storm gains violence by its own action'.[86] As to the storm's demise, his reasoning here was very specific, and pointed to the action of the cold wave:

> [The] cold north-west wave advanced more rapidly than the centre of the storm. It was beginning to counteract the increase of temperature arising from the condensed vapour, and after that the violence of the storm must rapidly decline.[87]

The wave motion of cold air mass was already well established in Loomis's 1841 paper, which showed how cold wave was far in advance of the centre of the barometric depression.[88] But Loomis also extended this undulant form to large-scale meteorological elements, including the most conspicuous of all—the oscillations of the barometer, as we saw earlier. Although he did not define it as such, one key empirical innovation was that the velocity and form of the barometric wave could be calculated by analyzing synoptic charts. In fact, by drawing the barometric curves of a good number of stations and plotting the lines of greatest depression for every six hours on a map of the United States, Loomis was able to determine its velocity. He found that it was not uniform: it was higher in the north of the country (varying from seventeen to thirty-seven miles per hour) than in the south (from seventeen to twenty-nine miles per hour).[89] Building apparently on Loomis's ideas, from 1843 to 1849 two astronomers in Britain developed a theory of barometric waves, based on the analysis of graphs on pressure variation rather than on synoptic maps.

---

86 Ibid., p. 174.
87 Ibid., p. 176.
88 For a modern reader, it is obvious that Loomis's treatment of cold waves was basically descriptive. Indeed, he could not state much more else, since at that time the laws of thermodynamics remained to be developed. Even so, his proposals were not on the wrong track. In 1888, A. W. Greely, forecaster and Chief Signal Officer, wrote in his book *American Weather* that 'the cold wave occurs not with the centre of the anticyclone, but on its outskirts, far in advance of the centre' (p. 213). He previously stated that 'the term *cold wave* is a technical one devised by the United States Signal Service, not to represent the intensity of the cold, except within certain limits, but... to show very decided falls in temperature within a limited time' (p. 211). Quoted by Kutzbach (1979, pp. 30–31).
89 Loomis (1841, p. 129).

Herschel and Birt's barometric waves

In 1843, Sir John Herschel, perhaps the highest authority in science in Britain, read a report on the reduction of meteorological observations before the BAAS. He coined the term *barometric wave* to announce a new, original solution to the famous American storm controversy. With the help of his assistant William Birt, and like his predecessors, he sought to lay down the empirical laws of the atmosphere rather than predicting storms. Although he never credited his ideas to Loomis's insight, his theory had a clear Loomisian imprint: with graphic rather than cartographic methods, it suggested storms were the result of the crossing of two large atmospheric waves moving in different directions.[90]

Since eighteenth-century meteorology and astronomy, the idea of air disturbances propagating over the earth's atmosphere in vast waves vaguely floated on scientists' minds.[91] Herschel himself conjectured about the winds and 'great fluctuations, of the nature of waves' prevailing in the atmosphere as early as 1833, in his book *Treatise on astronomy*.[92] Later his participation in the simultaneous meteorological observation campaign, within the great voyage to the Cape of Good Hope, gave substantive form to this idea. In a letter addressed to James Prinsep, from Calcutta, he posited the existence of an 'alternate, annual transfer' of air between the two hemispheres, a fact that would cause a seasonal variation of atmospheric pressure across the equator. Far from being a local phenomenon, this would 'communicate motion to immense masses of air'.[93] Nevertheless, this idea did not carry the notion of barometric wave with it yet.[94]

However, this idea did carry with it—or better, did reflect—two aspects that played a prominent role in the genesis of said notion. On one hand, the 1830s and 1840s witnessed a profound renewal in the earth sciences, as well as the first attempts to globally analyze all fields of the terrestrial environment. Topics of practical interest were the intensity and direction of the geomagnetic field, but also meteorological readings at solstices and equinoxes. Herschel's campaign covered all these subjects, and was promoted by the

---

90   Herschel (1844, p. 98). For Herschel and Birt's atmospheric waves, cf. Jankovic (1998) and Good (2006, pp. 46–54). See also Walker (2012, pp. 16–17).

91   For example, in his 1832 report on the 'Recent progress and present state of meteorology', read at the Oxford meeting of the BAAS, James Forbes alluded to the accidental variations of atmospheric pressure as influenced by latitude, and considered these variations as resulting from 'great atmospheric tidal waves which are perpetually traversing oceans and continents'. Forbes (1833, p. 235).

92   Herschel (1833, p. 26), quoted by Good (2006, p. 51). I follow this source for the rest of the paragraph.

93   Herschel to Prinsep, 2 December 1834, RS: HS, 25.54.4, quoted by Good (2006, p. 51). This letter was written during Herschel's voyage to the Cape.

94   In 1835, Herschel speculated about an 'annual transfer of atmospheric pressure' between the two hemispheres. Herschel to Friedrich Wilhelm Bessel, 16 May 1835, RS:HS, 25.4.10, quoted by Good (2006, p. 52).

first circle linked to the BAAS.[95] On his return to London, his involvement in both the British 'magnetic crusade', and his works on reduction, tabulation, and plotting of barometric data, together with Birt in the early 1840s, were two sides of the same coin.[96] On the other hand, Herschel championed the idea of universal induction in science. Unlike Humboldt, he believed in a system of inductive universalism demanding large sets of global observations for extended periods in order to determine general physical laws that accounted for the behaviour of particular phenomena. Barometric, geomagnetic, and tidal oscillations were all appropriate tests for his philosophical system.[97]

After his stay in South Africa, Herschel found stimulation for his theoretical inspirations in Birt, on whom he had drawn to reduce barometric readings, as well as in the Committee on Physics and Meteorology for the Royal Society of London, which he chaired at that time. Although in the correspondence with Birt we find explicit mentions of 'barometric curves' in 1841 and even of 'atmospheric curves' in June 1842, it was not until July 1843 that Herschel spoke to Birt of the 'barometric wave' as a defined concept.[98] It is worthwhile reproducing this letter extensively because it shows the order of ideas and possible influences:

> I have already come to the conclusion that our conception of a barometric wave can be easily enlarged (i.e. if we set out with my notion of their analogy to sea-waves) so as to take in vast (?) flattened (blanket-like if I may be allowed such a homely comparison) heavings more like the tide waves of the ocean [...] And if we consider (which is already well-known of the phenomena of the barometer) that we have often a rising glass for whole month and vice versa – and sometimes a general high or low level ... for one or two months, it will be clear that our atmosphere is a vehicle for a wavelike movements which may embrace in their single swell & fall a whole quadrant of a globe.
>
> My view of the matter to which I have alluded is that every barometric movement is essentially and intimately connected with an appropriate and corresponding wind – as ... every wavelike movement in a fluid consists of two distinct things – an advancing form and a molecular movement – ... the advancing form is indicated by the barometer – the

95 This circle of researchers included Edward Sabine and Humphrey Lloyd. For the renewal of the earth sciences, including the British and German geomagnetic crusades, see Locher (2007b, esp. pp. 492–501). For the shift of leadership in this field from France to Britain and Germany, see Cawood (1977, pp. 583–87).

96 On the British magnetic crusade in the 1830s and its aims, see Cawood (1977, pp. 584–87; 1979).

97 For Herschel's system of universal induction, cf. Carter (2004, esp. pp. 512–15). For his use of hypotheses as a means to develop generalizations, cf. Good (2006, pp. 47–48).

98 These mentions can be found in Herschel to Birt, 16 March 1841, RS:HS 19.88; 8 June 1842, RS:HS 19.95; and 28 July 1843, RS:HS 19.106, respectively. See Good (2006, p. 52).

molecular movement by the wind, and between these two phenomena there exists of necessity a close and purely dynamical connection.
[Knowing barometrical fluctuations one would be able to predict wind]
for this could lead to prediction of great storms.[99]

At this point, we must above all ask what came to pass in those months that
Herschel and Birt opted for the conceptual shift from barometric curves to
barometric waves. First and foremost was the issue of storm controversy:
since Redfield's and Espy's theories, the paramount problem of atmospheric
physics was the formation of storms and their centrifugal or radial motion.
Espy had just published barometric graphs with curves.[100] In March 1840,
Loomis announced his ideas on waves of cold air and barometric waves as
a solution to the American storm controversy[101]; and on 26 May 1843—i.e.
two months before Herschel's letter—he read before the American Philosophical Society his study on the two 1842 storms, including his analogy
with oceanic waves.[102] That Herschel, a man who followed very closely
and in-depth the controversy, was not knowledgeable about Loomis's ideas
seems almost inconceivable.[103] Second, Herschel's suggestion fell within a

99 Herschel to Birt, 28 July 1843, RS:HS, 19.106, as reproduced by Jankovic (1998, pp. 28–29).
100 See, for example, Espy's seminal book *The Philosophy of Storms*, 1841, pp. 549–50. The
    graphical representation of barometric oscillations was common practice in the meteor-
    ological campaigns at the Bossekop Observatory and on board the ship *La Recherche*,
    where French Navy's researchers Auguste Bravais and Victor Lottin plotted smoothed
    curves and bar graphs in 1838. In this campaign, the instructions given by Herschel were
    followed. Locher (2007, pp. 495–99).
101 These ideas, read before the American Philosophical Society on 20 March 1840, were
    published in the *Transactions* of this Society in 1841. Loomis (1841, plates 1 and 2) also
    included barometric charts with curves.
102 The study was published in 1845. See Loomis (1845, p. 176) for the analogy with oceanic
    waves.
103 The diffusion and repercussion of the American storm controversy in Britain was almost
    immediate. Thus, for example, at the 10th Meeting of the BAAS, held in Glasgow in
    August 1840, in which Herschel and Espy participated, allusions to the debate were nu-
    merous. Espy presented a paper 'On Storms', and an 'Extract of a letter from Mr. Redfield
    to Sir J. F. W. Herschel' was read after Espy's paper (although Redfield was invited, he
    was unable to attend the meeting). Revealing for its testimony to the extent of interest
    which American storm theories received in Britain—and for aspects that would portrait
    Herschel—is the 'Supplementary report on meteorology', read by James D. Forbes at the
    same meeting, which stated the following: 'Major Sabine has referred me to a clear and
    able analysis of the effects of the storm of the 20th December, 1836, by Mr. Loomis, in the
    first part of the 7th vol. of the American Transactions (which had not reached Scotland
    when this report was written), who appears to infer, in that particular case, a rotation
    round a horizontal axis like that imagined by Mitchell. The paper is highly worthy of
    consultation'. It also said: 'Espy maintains a peculiar physical theory of storms, which,
    since this report was written, has been brought in several forms before the British public'
    (p. 111). In fact, in this meeting Herschel himself extolled the success in America of his
    campaign of simultaneous meteorological observations—of which Loomis was an active
    member: 'thanks to American zeal and industry, the group including the United States

wider context: the search for hypothesis in science and as a response to the contemporary rhetoric of the immaturity of meteorology. He aimed to find diurnal and monthly barometric regularities on which he could later build and formulate the general laws of the atmosphere.[104] And finally, there was a factor linked to national self-esteem. Herschel regarded as worrisome the secondary role played by British scientists in the storm controversy.[105] The BAAS strove for ameliorating the situation, and provided financial support for Herschel and Birt's research project on barometric waves. The hope was that these waves could lead Britain at the forefront of meteorology.[106]

Herschel published his ideas on barometric waves in the *Report* of the meeting of the BAAS, held in Cork in 1843. He criticized Redfield and Reid's notion of 'funnel-shaped revolving depressions', and added the crucial caveat that two barometric waves traversing in different direction could intersect each other and form hurricanes or tornados. Or, as he succinctly expressed it, storms could be produced by 'the crossing of two large long waves moving in different directions'.[107] Herschel propounded it as a hypothesis, and soon entrusted Birt with the task of corroborating their effect by graphically representing barometric data. In an article on storm-paths printed in *Philosophical Magazine* in 1846, Birt warned of the need to accompany all the wind observations with barometric readings, and noted that these readings 'must not be confined to the mere period of the passing of the gale', but must include 'the entire system of observations appertaining to both waves, not only in time, but also in space'.[108] In this regard, he added, worthy of note are Loomis's 'interesting charts' regarding the passage of two

---

promises to be the most prominent'. Herschel (1840, p. 434). Presumably Herschel would have regarded the notion of barometric wave which Loomis propounded in that paper as the scientific materialization (at least in part) of his own early speculations.

104 Jankovic (1998, p. 28).

105 By 1838, the only British involved in the storm controversy was the lieutenant-colonel of the Royal Engineers, William Reid, who was at that time in charge of rebuilding the government edifices damaged by storms on the island of Barbados. At the 1839 meeting of the British Association, held in Newcastle, Reid inaugurated Herschel's section with a paper on Redfield's law of storms—Reid (1839, pp. 21–25). For the collaboration and lifelong friendship forged by Reid and Redfield, see Cox (2002, pp. 31–32).

106 Although Jankovic (1998) admits that Herschel's conceptualization of barometric waves was geared particularly to facilitating the solution of the storm controversy, he places greater emphasis on analyzing the model of 'credibility-investment' proposed by Bruno Latour and Steve Woolgar in 1979, regarding the reasons which led Herschel and Birt to undertake the research on atmospheric waves. This fact might explain why Jankovic overlooks Loomis's work, as well as the persistence—as will be shown in next chapters—of the concept of barometric wave in meteorology in the 1850s and 1860s. Good (2006, pp. 46–54) also fails to notice Loomis's work, but he highlights and shows Herschel's unyielding defence of atmospheric waves in those decades—see, e.g., his six-pages section on atmospheric waves and tides written in his article on meteorology for the *Encyclopaedia Britannica* (1857, vol. 14, pp. 636–90) and Herschel (1861).

107 Herschel (1844, p. 100).

108 Birt (1846a, p. 382).

storms over the United States in February 1842—*nota bene*, announced two months before Herschel's crucial letter to Birt. Significantly for our present purposes, Loomis's 'lines of equal pressure' indicate, according to Birt, 'that two waves passed over the United States' and 'that in the point of intersection the storm raged'.[109]

In both 1843 *Report* and 1843 letter to Birt, Herschel differentiated between form and movement in a wave. While the barometer indicates the advancing form, the wind shows the molecular movement. There should be, he said, 'a close and purely dynamical connexion'.[110] As was shown, Loomis also distinguished between barometric waves and cold air waves. In his analysis of the 'Great November Wave' of 1842, also published in the journal *Philosophical Magazine* in 1846, Birt found graphic evidence that the molecular movement was likely connected dynamically with the wind: in fact, it seemed to be strongest in the wave troughs and geared towards said troughs from each side.[111] Nevertheless, Birt was much more cautious two years later; in an essay on 'Barometric undulations' written for Edward J. Lowe's *Treatise on Atmospheric Phenomena*, he suggested that while the advancing form could be represented by waves, it was 'not so when we treat of the molecular motions or wind'.[112] For Birt, Loomisian waves of air were not reflected in his graphs based on anemometric records.

Indeed, the reason behind Birt's caution about the reality of waves was the lack of consistence between empirical data and hypotheses. From 1845 to 1847, Birt found difficulties in the observation of waves, as well as numerous anomalies (some waves were erratic, others seemed to be superimposed). The fact was not inconsequential: a substantial shift in the ontological meaning of waves took place for reasons of empirical inconsistence, as a result of the barometric studies undertook by Birt. Where in 1844 Herschel spoke of 'realities with forms' analogous to astronomical elliptic orbits and 'of a higher order' than those hitherto conceived in meteorology,[113] three years later Birt had to admit that the barometric curve 'does not represent the form of any reality in nature'.[114] Barometric waves became, therefore, merely instrumental concepts with no ontological attribute of reality.

109 Ibid.
110 Herschel (1844, p. 99).
111 Birt (1846b, p. 358).
112 Birt (1846c, p. 361).
113 Herschel to Birt, 1 September 1844, RS:HS 19.114—quoted by Jankovic (1998, p. 30).
114 Birt (1847, p. 491). Birt not only admitted the limitations of atmospheric waves, but seems to have welcomed an idealist stance regarding those waves in a spirit of resignation and abnegation: 'the term atmospheric wave has been used to designate *ideal individuality* which the mind attributes to the process which it observes of the successive change of place which the barometric maxima and minima undergo'. Birt (1847, p. 485). For Jankovic (1998, pp. 34–35), Birt's 'ideal individuality was certainly a far cry from Herschel's Borlaseian naïvety'. [William Borlase was 'perhaps the most conscientious weather observer during the eighteenth century'.]

Or, to put it in another way, where in 1844 Herschel and Birt conceived atmospheric waves as entities formulable in terms of wave mechanics, later Birt had to admit they were not reconcilable with hydrodynamics—in fact, they could not explain the observations of the direction and force of winds.[115] This, however, was expectable to a certain extent: inasmuch as these waves referred, by definition, to a wavelike form of the curve reflecting changes in barometric pressure, they were by nature graphical rather than dynamical.[116] This instrumental nature will continue to characterize the atmospheric waves used for weather forecasting in coming decades, as shall be seen later.

Birt's criticism denoted disaffection rather than denial. In fact, Birt (and even more so Herschel) continued to discuss these waves in subsequent works.[117] The two pursued the laws of the atmosphere, rather than prediction; the former connoted order and precision—just the opposite of prediction.[118] Herschel and Birt walked a hierarchical pathway, passing from the knowledge of periodical regularities to that of non-periodical phenomena (storms), from the general to the local. Atmospheric waves were conceived in the framework of the quest for laws of storms and therefore equating the search for laws with that for weather prediction was not supported by the facts. I will come back to this issue later.[119]

---

115 This limitation was explicitly depicted by Birt in a letter addressed to Herschel:

> The principal objection to this definition of an atmospheric wave consists I apprehend in its non-agreement with hydrodynamical principles. So far as the term *wave* has been employed *by myself* in immediate reference to these particular phenomena I most readily admit it was *ill-chosen*.

Birt to Herschel, 1 April 1848, RS:HS 19.207—quoted by Jankovic (1998, p. 37).

116 This point has been emphasized by Jankovic (1998, p. 37).

117 It would be more accurate to say that, although in private Birt was highly sceptical about these waves—in 1847 he wrote in a letter that 'he failed in contemplating any single wave in its totality'[quoted in Jankovic (1998, p. 37)]—in public his stance was more temperate. In his book *The Hurricane Guide*, published in 1850, he stated that the object of the work was to exhibit the connection between the terrific revolving storms and the 'more extensive and occult but not less important phenomena, atmospheric waves' (p. 7), so that seamen were able to avoid the centres of storms. In the same vein, his book *Handbook of the Law of Storms*, printed in 1856, devotes the last chapter to 'atmospheric waves', even if now they seem to refer only to the wind field: 'each of [the] sets of parallel currents of air blowing in opposite directions constitute "an atmospheric wave"' (p. 144). For Herschel's views on atmospheric waves published in the 1850s and 1860s, see Good (2006, pp. 52–54).

118 Except in Herschel's 1843 letter to Birt—partially reproduced in Herschel (1844, p. 99)—in which the 'prediction of great storms' appears as a mere wish and speculation, none of the aforementioned Herschel's and Birt's papers and books allude to prediction or the anticipation of weather changes as a working hypotheses, let alone as an object of study. The emphasis lay on the discovery of the laws of the general movement of the atmosphere and, in particular, of storms.

119 Analyzing Jankovic's historical study on the atmospheric waves in the 1840s, Barboza (2012, pp. 188–89) warned against the tendency (widespread among historians of meteorology, including Jankovic) to confuse the attempts to seek the laws of the general movement of the atmosphere with the laws of weather behaviour (and weather prediction, in general). I find this consideration relevant.

Two views, three (or more) theories

By the end of the first half of the nineteenth century, meteorology remained part of climatology and its scholars frequently sought to ascertain the laws behind periodical regularities. Nevertheless, European and American meteorologists soon found themselves holding incompatible views of storm origin. From the same object of study—the formation and progress of storms—two different views emerged. In the mechanistic view, a storm was an air flow whose motion was subject to mechanical principles. Its study was based on the kinematic description of the wind field. Here two theories stand out: storms could result from the counteraction of two opposite currents (Dove), or could be seen as depressions in rotary or whirling motion (Redfield). While Dove's theory received great attention in Europe, especially in Central Europe until the 1870s, Redfield's whirlwind theory was born in America, and then spread across Western Europe. In the thermal view, a storm was an air flow whose formation was the result of the expansion and cooling of rising air. Its motion was explained in terms of adiabatic change and thermal convection. Here one theory gained relevance in America: storms could be seen as centripetal inflows of air around a centre of minimum pressure (Espy).

The three theories differently weighted the available data and tools. Dove emphasized the large-scale storms in temperate latitudes, and the local method of the temporal sequence of phenomena at a fixed station. Redfield emphasized the small-scale tropical storms (or hurricanes), and the circular rotation of storms in wind charts. And Espy underscored the evidence on tropical cyclones and laboratory experiments, as well as the weather maps with arrows showing wind direction and with lines of minimum pressure.

If Redfield and Espy opened the debate on the centrifugal-centripetal nature of storms, cold waves and barometric waves certainly emerged as solutions to the controversy. By examining precipitation, wind, and temperature patterns in a synoptic map, Loomis first called into question the inadequacy of the centrifugal-centripetal quandary. Cold air waves overcame Espy's convective view by incorporating the scheme of the two opposing currents, while barometric waves overcame Dove's mechanical view by incorporating thermal processes. By assimilating Espy's convective theory, Loomis redefined the issue of the origin and maintenance of storms. By assimilating Dove's opposing current theory, he re-opened the question of the formation and motion of midlatitude cyclones. But unlike them, he brought synoptic map to its highest expression, tracing deviations from normal pressure and lines of equal temperature deviation.

In Great Britain—which had an entrenched tradition in tidal waves studies—mechanical theories gained greater acceptance than thermal ones. Here too, barometric waves were propounded—this time by Herschel and Birt—as a solution to the American storm controversy, probably following

Loomis's wake. But the tone of Loomis's theory was multicolour—with its blend of mechanical and thermal causes, and its distinction between cold air waves and barometric waves. In contrast, Herschel's theory was exclusively mechanistic. He believed that storms were the result of the crossing of two atmospheric waves advancing in different directions. Moreover, unlike Loomis, his empirical method to detect such waves was graphic, not cartographic. Birt traced the curves representing barometric oscillations in times of storms, and the coveted atmospheric waves proved to be erratic and superimposed. For Birt, barometric waves were elusive per se, but for Herschel the curves reflected *'realities* with forms of a higher order'.[120]

Thus, by the end of the 1840s, two different views of storm motion vied to gain ground: mechanists embraced the vision of an essentially revolving storm—with the wind field as a salient feature—whereas convectionists embraced the vision of a much more complex storm—with both the wind and pressure fields as salient features.

References

Abbe, Cleveland, 1895. 'Memoir of William Ferrel, 1817–1891'. In *Biographical Memoirs of the National Academy of Sciences*. Washington, DC: National Academy of Sciences, 265–309. https://www.worldcat.org/title/memoir-of-william-ferrel-1817-1891/oclc/1026689339&referer=brief_results

Achbari, Azadeh, 2015. 'Building Networks for Science: Conflict and Cooperation in Nineteenth-Century Global Marine Studies'. *Isis*, 106 (2), 257–82.

Achbari, Azadeh, 2017. *Rulers of the Winds. How Academics Came to Dominate the Science of the Weather, 1830–1870*. Amsterdam: Vrije Universiteit.

Achbari, Azadeh, and Lunteren, Frans van, 2016. 'Dutch Skies, Global Laws: The British Creation of "Buys Ballot's Law"'. *Historical Studies in the Natural Sciences*, 46 (1), 1–43.

Anderson, Katharine, 2005. *Predicting the Weather: Victorians and the Science of Meteorology*. Chicago: The University of Chicago Press.

Bache, Alexander Dallas, 1860. *Smithsonian Institution Annual Report for 1859*. Washington, DC: Smithsonian Institution, 108–11.

Ball, John, 1848. 'On Rendering the Electric Telegraph Subservient to Meteorological Research'. *Report of the Annual Meeting/British Association for the Advancement of Science*. London, 12–13.

Barboza, Christina Helena da Motta, 2012. *As viagens do tempo*. Uma história da meteorologia em meados do século XIX. Rio de Janeiro: E-papers Serviços Editoriais Ltda.

Bernhardt, Karl-Heinz, 2008. 'Heinrich Wilhelm Dove (1803–1879) und seine Stellung in der Geschichte der Berliner Meteorologie'. *Dahlemer Archivgespräche*, 14, 61–100.

Birt, William Radcliffe, 1846a, 'On the Storm-Paths of the Eastern Portion of the North American Continent'. *Philosophical Magazine*, 28, 380–82.

---

120 Herschel to Birt, 1 September 1844, RS:HS 19.114—quoted by Jankovic (1998, p. 30).

Birt, William Radcliffe, 1846b, 'On the Great Symmetrical Barometric Wave of November'. *Philosophical Magazine*, 29, 356–59.

Birt, William Radcliffe, 1846c, 'On Barometric Undulations'. In Edward J. Lowe ed., *A Treatise on Atmospheric Phenomena*. London: Longman, Brown, Green, and Longmans, 356–61.

Birt, William Radcliffe, 1847, 'On Certain Atmospheric or Barometric Waves Which Traversed Europe during November 1842'. *Philosophical Magazine*, 30, 482–502.

Bôcher, Maxime, 1888. 'The Meteorological Labors of Dove, Redfield, and Espy'. *American Meteorological Journal*, 5, 1–13.

Brandes, Heinrich Wilhelm, 1827. 'Einige meteorologische Untersuchungen über den Wind'. *Annalen der Physik und Chemie*, 87, 545–90.

Burstyn, Harold L., 1966. 'Early Explanations of the Role of the Earth's Rotation in the Circulation of the Atmosphere and the Ocean'. *Isis*, 57 (2), 167–87.

Burstyn, Harold L., 1971. 'Ferrel, William'. In Charles Coulston Gillispie ed., *Dictionary of Scientific Biography*. New York: Charles Scribner's Sons, 4, 590–593.

Burstyn, Harold L., 1981. 'Redfield, William C.'. In Charles Coulston Gillispie ed., *Dictionary of Scientific Biography*. New York: Charles Scribner's Sons, 11, 340–41.

Carramolino, David, 1994. 'La "Ley de Giro" (Drehungsgesetz) de Dove (1827) y el nacimiento de la dinámica atmosférica en Alemania'. *Éndoxa: Series Filosóficas*, 3, 95–119.

Carter, Christopher, 2004. 'Herschel, Humboldt, and Imperial Science'. In Raymond Erickson, Mauricio A. Font, and Brian Schwartz eds., *Alexander von Humboldt. From the Americas to the Cosmos: An International Interdisciplinary Conference, October 14–16, 2004*. New York: The City University of New York, Bildner Center for Western Hemisphere Studies, 509–18.

Cawood, John, 1977. 'Terrestrial Magnetism and the Development of International Collaboration in the Early Nineteenth Century'. *Annals of Science*, 34, 551–87.

Cox, John D., 2002. *Storm Watchers: The Turbulent History of Weather Prediction from Franklin's Kite to El Niño*. Hoboken, NJ: John Wiley & Sons, Inc.

Daniels, George H., 1994. *American Science in the Age of Jackson*. Tuscaloosa: University of Alabama Press.

Davis, Charles H., 1844. 'Redfield, Reid, Espy and Loomis on the Theory of Storms'. *North American Review*, 58, 335–71.

Davis, William Morris, 1891. 'Ferrel's Contributions to Meteorology'. *American Meteorological Journal*, 8, 348–59.

Davis, William Morris, 1895. 'The Redfield and Espy Period'. In Oliver L. Fassig ed., *Report of the International Meteorological Congress, Chicago, Ill., August 21–24, 1893*. Washington, DC: Weather Bureau, Part 2, 305–17.

Dove, Heinrich Wilhelm, 1827. 'Einige meteorologische Untersuchungen über den Wind'. *Annalen der Physik*, 11, 545–90.

Dove, Heinrich Wilhelm, 1828a. 'Über barometrische Minima'. *Annalen der Physik*, 13, 596–613.

Dove, Heinrich Wilhelm, 1828b. 'Ueber das Gewitter'. *Annalen der Physik und Chemie*, 89 (7), 419–33.

Dove, Heinrich Wilhelm, 1837. 'On the Influence of the Rotation of the Earth on the Currents of Its Atmosphere; Being Outlines of a General Theory of the Winds'. *The London and Edinburgh Philosophical Magazine and Journal of Science*, 11, 227–39, 353–63.

Dove, Heinrich Wilhelm, 1838. 'Über die nicht periodischen Änderungen der Temperaturverteilung auf der Oberfläche der Erde'. *Abhandlungen der Königlich Preussische Akademie der Wissenschaften zu Berlin, A. Physikalische Abhandlungen,* 285–415.

Dove, Heinrich Wilhelm, 1841a. 'Über das Gesetz der Stürme'. *Annalen der Physik,* 52, 1–41.

Dove, Heinrich Wilhelm, 1841b. *Repertorium der Physik: enthaltend eine vollständige Zusammenstellung der neuern Fortschritte dieser Wissenschaft. 4. Band, Meteorologie, specifische Wärme, strahlende Wärme.* Berlin: Verlag von Veit & Comp.

Dove, Alfred, 1873. 'Alexander von Humboldt. The Meridian and Decline of Life; Berlin, 1827–1859'. In Julius Löwenberg, Robert Avé-Lallemant, and Alfred Dove eds., *Life of Alexander Von Humboldt: Compiled in Commemoration of the Centenary.* London: Longmans, Green, and Co., 75–416.

Ehrenheim, Frih. 1824. *Om climaternes rörlighet.* Stockholm: Tryckt hos P.A. Norstedt & Söner.

Espy, James Pollard, 1836a. 'Essays on Meteorology. No 1. Theory of Hail'. *Journal of the Franklin Institute,* 17, 240–46.

Espy, James Pollard, 1836b. 'Essays on Meteorology. No III. Examination of Hutton's Redfield's and Olmstead's Theories'. *Journal of the Franklin Institute,* 18, 100–08.

Espy, James Pollard, 1837. 'Third Report of the Joint Committee on Meteorology, of the American Philosophical Society and the Franklin Institute of the State of Pennsylvania, for the Promotion of the Mechanic Arts'. *Journal of the Franklin Institute,* 19, 17–21.

Espy, James Pollard, 1841. *The Philosophy of Storms.* Boston, MA: Charles C. Little and James Brown.

Espy, James Pollard, 1843. *First Report on Meteorology to the Surgeon General of the United States Army.* Washington, DC.

Feldman, Theodore, 1993. 'The Ancient Climate in the 18th and Early 19th Century'. In M. Shortland ed., *Science and Nature: Essays in the History of the Environmental Sciences.* Stanford in the Vale: British Society for the History of Science, 23–40.

Ferrel, William, 1856. 'Essay on the Winds and Currents of the Ocean'. *Nashville Journal of Medicine and Surgery,* 6, 287–301, 375–89. It was reprinted as *Professional Papers of the U.S. Signal Service, no. 12.* Washington, DC. 1882.

Ferrel, William, 1858. 'The Influence of the Earth's Rotation upon the Relative Motion of Bodies Near Its Surface'. *Astronomical Journal,* 5, 97–100.

Ferrel, William, 1859–1860. 'The Motions of Fluids and Solids Relative to the Earth's Surface'. *Mathematical Monthly,* 1 (1859), 140–48, 210–16, 300–07, 366–73, 379–406; 2 (1859–1860), 89–97, 339–46, 374–90. It was republished with notes by Waldo as *Professional Papers of the U.S. Signal Service, no. 8.* Washington, DC. 1882.

Fleming, James Rodger, 1990. *Meteorology in America, 1800–1870.* Baltimore, MD: Johns Hopkins University Press.

Fleming, James Rodger, ed., 1998. *Historical Perspectives on Climate Change.* New York: Oxford University Press.

Forbes, James D., 1833. 'Report upon the Recent Progress and Present State of Meteorology'. In *Report of the First and Second Meeting of the British Association for the Advancement of Science, 1831 and 1832; at York in 1831, and at Oxford* in 1832: Including Its Proceedings, Recommendations and Transactions. London: John Murray, 196–258.

Garber, Elizabeth, 1976. 'Thermodynamics and Meteorology (1850–1900)'. *Annals of Science*, 33 (1), 51–65.

Good, Gregory A., 2006. 'A Shift of View: Meteorology in John Herschel's Terrestrial Physics'. In James Rodger Fleming, Vladimir Jankovic, and Deborah R. Coen eds., *Intimate Universality. Local and Global Themes in the History of Weather and Climate*. Sagamore Beach, MA: Watson Publishing International, 35–67.

Gross, Walter E., 1972. 'The American Philosophical Society and the Growth of Meteorology in the United States: 1835–1850'. *Annals of Science*, 29, 321–38.

Hadley, George, 1735. 'Concerning the Cause of the General Trade–Winds'. *Philosophical Transactions*, 39, 58–62.

Halford, Pauline, 2004. *Storm Warning. The Origins of the Weather Forecast*. Stroud: Sutton Publishing.

Halley, Edmond, 1686. 'An Historical Account of the Trade Winds, and Monsoons, Observable in the Seas between and Near the Tropicks, with an Attempt to Assign the Physical Cause of the Said Winds'. *Philosophical Transactions*, 16, 153–68.

Hann, Julius, 1885. 'Einige Bemerkungen zur Entwickelungs-Geschichte der Ansichten über den Ursprung des Föhn'. *Meteorologische Zeitschrift*, 2, 393–99.

Hellmann, Gustav, 1897a. *Meteorologische Karten 1688, 1817, 1846, 1863, 1864. Sechs tafeln in lichtdruck mit einer einleitung*. Berlin: Asher.

Hellmann, Gustav, 1897b. *Neudrucke von Schriften und Karten über Meteorologie und Erdmagnetismus. N. 8. Meteorologische Karten*. Berlin: A. Asher & Co.

Henry, Joseph, 1854. 'Report of the Secretary of the Smithsonian Institution to the Board of Regents, December 8, 1847'. Reprinted in *Eight Annual Report of the Board of Regents of the Smithsonian Institution, up to January 1, 1854, and the Proceedings of the Board up to July 8, 1854*. Washington, DC: A.O.P. Nicholson, 119–39.

Henry, Joseph, 1855. 'Meteorology in Its Connection with Agriculture'. *Report of the Commissioner of Patents for the Year 1855*. Washington, DC.

Herschel, John Frederick William, 1830. *A Preliminary Discourse on the Study of Natural Philosophy*. London: Longman.

Herschel, John Frederick William, 1833. *A Treatise on Astronomy*. London: Longman.

Herschel, John Frederick William, 1835. *Instructions for Making and Registering Meteorological Observations in Southern Africa and Other Countries in the South Seas, as Also at Sea*. London: Bradbury and Evans.

Herschel, John Frederick William, 1836. 'Instructions for Making and Registering Meteorological Observations at Various Stations in Southern Africa and Other Countries in the South Seas, as Also at Sea'. *Edinburgh New Philosophical Journal*, 21, 135–49.

Herschel, John Frederick William, 1841. 'Report on the Reduction of Meteorological Observations Made at the Equinoxes and Solstices, on the Part of a Committee Appointed by the British Association at Newcastle'. *Report of the Tenth Meeting of the British Association for the Advancement of Science, Held at Glasgow in August 1840*. London: John Murray, 10, 432–34.

Herschel, John Frederick William, 1844. 'Report on the Reduction of Meteorological Observations'. In *Report of the British Association for the Advancement of Science, Held at Cork in August 1843*. London: John Murray, 13, 60–100.

Herschel, John Frederick William, 1861. *Meteorology*. Edinburgh: Adam and Charles Black. First published in *Encyclopaedia Britannica*, 1857, 8th ed., 14, 636–90.

Humboldt, Alexandre de, 1817. 'Des lignes isothermes et de la distribution de la chaleur sur le globe'. *Mémoires de physique et de chimie, de la Société d'Arcueil*, 3, 462–602.

Humboldt, Alexander, 1845–1858. *Kosmos: Entwurf einer physischen Weltbeschreibung*. Stuttgart, Augsburg: J. G. Cotta, 4 vols.

Jankovic, Vladimir, 1998. 'Ideological Crests versus Empirical Troughs: John Herschel's and William Radcliffe Birt's Research on Atmospheric Waves, 1843–50'. *British Journal for the History of Science*, 31, 21–40.

Jankovic, Vladimir, 2001. *Reading the Skies: A Cultural History of English Weather 1650–1820*. Chicago: University of Chicago Press.

Kevan, Simon M., 1993. 'Quests for Cures: A History of Tourism for Climate and Health'. *International Journal of Biometeorology*, 37, 113–24.

Khrgian, Aleksandr Khristoforovich, 1970. *Meteorology: A Historical Survey*, 2nd ed. Jerusalem: Israel Program for Scientific Translations.

Körber, Hans Günther, 1959. *Über Humboldts Arbeiten zur Meteorologie und Klimatologie*. Berlin: Akademie-Verlag.

Kutzbach, Gisela, 1973. 'Loomis, Elias'. In Charles Coulston Gillispie ed., *Dictionary of Scientific Biography*. New York: Charles Scribner's Sons, 8, 487.

Kutzbach, Gisela, 1979. *The Thermal Theory of Cyclones: A History of Meteorological Thought in the Nineteenth Century*. Boston, MA: American Meteorological Society.

Leighly, John, 1949. 'Climatology since the Year 1800'. *Transactions, American Geophysical Union*, 30 (5), 658–72.

Locher, Fabien, 2007. 'The Observatory, the Land-based Ship and the Crusades: Earth Sciences in European Context, 1830–50'. *The British Journal for the History of Science*, 40 (4), 491–504.

Loomis, Elias, 1841. 'On the Storm Which Was Experienced throughout the United States about the 20th of December, 1836'. *Transactions of the American Philosophical Society*, 7 (7), 125–63.

Loomis, Elias, 1845, 'On Two Storms Which Were Experienced throughout the United States in the Month of February, 1842'. *Transactions of the American Philosophical Society*, 9 (2), 161–84.

Lord Byron, George Gordon, 1825. *The Complete Works of Lord Byron with a Biographical and Critical Notice by J.W. Lake, Esq. Vol. 3, Don Juan, etc. etc*. Paris: Baudry and Amyot.

Lorenz, Edward N., 1983. 'A History of Prevailing Ideas about the General Circulation of the Atmosphere'. *Bulletin of the American Meteorological Society*, 64 (7), 730–69.

Ludlam, Frank Henry, 1967. *The Cyclone Problem: A History of Models of the Cyclonic Storm*. London: Imperial College of Science and Technology.

Ludlum, David M., 1969. 'The Espy-Redfield Dispute'. *Weatherwise*, 22, 224–29, 261–62.

Marriott, William, 1903. 'The Earliest Telegraphic Daily Meteorological Reports and Weather Maps'. *Quarterly Journal of the Royal Meteorological Society*, 29, 123–31.

McDonald, James E., 1963. 'James Espy and the Beginnings of Cloud Thermodynamics'. *Bulletin of the American Meteorological Society*, 44, 634–41.

Middleton, William Edgar Knowles, 1966. *A History of the Theories of Rain and Other Forms of Precipitation*. New York: Franklin Watts.

Monmonier, Mark, 1999. *Air Apparent. How Meteorologists Learned to Map, Predict, and Dramatize Weather*. Chicago, IL: The University of Chicago Press.

Neumann, Hans, 1925. *Heinrich Wilhelm Dove, eine Naturforscher Biographie*. Liegnitz: H. Krumbhaar.

Newton, Hubert Anson, 1895. 'Biographical Memoir of Elias Loomis, 1811–1889'. *Biographical Memoirs, National Academy of Sciences*, 3, 213–52.

Olmsted, Denison, 1857. 'Biographical Memoir of William C. Redfield'. *American Journal of Science*, 24, 355–73.

Ospovat, Don, 1977. 'Lyell's Theory of Climate'. *Journal of the History of Biology*, 10, 317–39.

Pedersen, Kurt Møller, 1971. 'Dove, Heinrich Wilhelm'. In Charles Coulston Gillispie ed., *Dictionary of Scientific Biography*. New York: Charles Scribner's Sons, 4, 174–75.

Prime, Samuel Iranaeus, 1875. *The Life of Samuel F. B. Morse, LL. D.: Inventor of the Electro-Magnetic Recording Telegraph*. New York: Appleton.

Redfield, William Charles, 1831. 'Remarks on the Prevailing Storms of the Atlantic Coast of the North American States'. *American Journal of Science*, 20, 17–51.

Redfield, William Charles, 1843. 'Observations on the Storm of December 15, 1839'. *Transactions of the American Philosophical Society*, 8, 77–80.

Redfield, William Charles, 1846, 'On Three Several Hurricanes of the American Seas, and their Relation to the Northers so Called of the Gulf of Mexico and the Bay of Honduras, with Charts Illustrating the Same'. *The American Journal of Science* (new series), 1, 1–16, 153–69, 333–69; 2, 162–87, 311–34.

Reid, William, 1839. 'A Statement of the Progress Made towards Developing the Law of Storms; and of What Seems Further Desirable to be Done, to Advance Our knowledge of the Subject'. *Report of the Eighth Meeting of the British Association for the Advancement of Science*. London: John Murray, 7, 21–25.

Reingold, Nathan, 1971. 'Espy, James Pollard'. In Charles Coulston Gillispie ed., *Dictionary of Scientific Biography*. New York: Charles Scribner's Sons, 4, 410–11.

Sargent, Frederick, 1982. *Hippocratic Heritage: A History of Ideas about Weather and Human Health*. New York: Pergamon Press.

Schneider-Carius, Karl, 1975. *Weather Science, Weather Research: History of Their Problems and Findings from Documents during Three Thousand Years*. New Delhi: Indian National Scientific Documentation Centre; originally published as *Wetterkunde, Wetterforschung: Geschichte ihrer Probleme und Erkenntnisse in Dokumenten aus drei Jahrtausenden*. Freiburg: Verlag Karl Albert, 1955.

Scultetus, Hans Robert, 1943. 'Dove und Loomis als Wegbereiter der Synopsis'. *Meteorologische Zeitschrift*, 60, 419–22.

Sheynin, Oscar Boris, 1984. 'On the History of the Statistical Method in Meteorology'. *Archive for History of Exact Sciences*, 31 (1), 53–95.

Spitz, Armand N., 1944. 'Meteorology in the Franklin Institute'. *Journal of the Franklin Institute*, 237, 271–87, 331–57.

Symons, George J., 1867. 'Reviews: Sunshine and Showers: Their Influence throughout Creation: A Compendium of Popular Meteorology, By Andrew Steinmetz'. *Symons's Monthly Meteorological Magazine*, 2, 33–34.

Thompson, Robert Luther, 1947. *Wiring a Continent: The History of the Telegraph Industry in the United States, 1832–1866*. Princeton: Princeton University Press.

Thomson, William, 1979. *A Change of Air: Climate and Health*. London: Black.

Thornthwaite, Charles Warren, and Leighly, John, 1943. 'Status and Prospects of Climatology'. *The Scientific Monthly*, 57 (5), 457–65.

Todhunter, Isaac, 1876. *William Whewell, D.D., Master of Trinity College, Cambridge: An Account of His Literary and Scientific Correspondence*. London: Macmillan & Co.

Van der Stok, Johannes Paulus, 1899. 'Levensbericht C.H.D. Buys Ballot'. In Koninklijke Akademie van Wetenschappen ed., *Jaarboek der Koninklijke Nederlandsche Akademie van Wetenschappen*. Amsterdam: C.G. van der Post, 59–100.

Waldo, Frank, 1895. 'Some Remarks on Theoretical Meteorology in the United States, 1855 to 1890'. In Oliver L. Fassig ed., *Report of the International Meteorological Congress, Chicago, Ill., August 21–24, 1893*. Washington, DC: Weather Bureau, Part 2, 318–25.

Walker, J. Malcolm, 2012. *History of the Meteorological Office*. Cambridge: Cambridge University Press.

2 Meteorology and statistics before 1854

Introduction

Dove's, Redfield's, and Espy's schemes dominated storm theories by the middle of the century. Yet the physicists' conceptions of the weather went far beyond storms. This diversity can only be understood by taking into account the contemporary studies on climate and weather.

A driving force of the meteorology of the pre-synoptic era was the application of statistical methods. The foregoing theories of atmospheric dynamics provided a frame that explained the temporal and spatial evolution of meteorological elements during storms, all kinds of statistical considerations going unnoticed. However, during the same period, there were systematic research programs in Europe that intended to determine, with statistical tools, the mean state of the atmosphere. Some of Germany's most gifted physicists and geographers contributed to this enterprise. Most importantly, some of these investigations paved the way for studies on the mean evolution of atmospheric phenomena. This chapter recounts the emergence, development, and embodiment of this shift of focus until 1854.

The result of this widening of meteorological science in the pre-synoptic era was an efflorescence of ideas and methods, both statistical and cartographic. The first sections of this chapter show how successful and innovative these investigators were at determining the mean state of the atmosphere. The second part deals with an outstanding yet little known contribution: Adolphe Quetelet's graphical method for representing atmospheric waves. With his method, Quetelet wanted to ascertain the regularities that lie behind non-period changes, to exemplify leadership in this subject, and to stimulate international cooperation. He found some success with his work, even though Birt had by then questioned the existence of these waves.

Two episodes of this chapter are worth mentioning upfront. First, a detailed discussion is given of Quetelet's atmospheric waves and their graphic representation. Nearly all the historians of science have ignored this work, presumably because it seemed obsolete after Birt's questioning, and unconnected to weather forecasting.[1] However, it was crucial in bridging pre-synoptic techniques and synoptic weather mapping; it laid the groundwork for the process of dynamization of the atmospheric space; and, as will be shown in Chapter 3, it provided a technical basis to study storm progress. Another particularity of this chapter is the special emphasis on Quetelet's kaleidoscopic view on statistics, observatory sciences, and nation-building. This is certainly justified by his extraordinary character as a builder of intellectual institutions, but also by the legacy of leadership in science-building and state-building—as two sides of the same coin—he left to his followers in Europe.[2]

Although the statistical approach became characteristic of the empirical studies on the atmosphere, its development can hardly be explained unless the context of internationalization in observatory sciences is taken into consideration; a part of this is described in the first section.

Observatory sciences and internationalization

The two decades immediately before the 1854 Crimean War witnessed a blossoming in the organization of global meteorological projects as savants from across Europe ceased to act individually and engaged in large-scale data collection and international collaboration. In 1829 Humboldt visited the Russian Tsar Nicholas I as a result of which the Russian government allowed a series of magnetic and meteorological stations to be established across northern Asia.[3] In 1833, Humboldt himself participated in German physicists Carl Friedrich Gauss and Wilhelm Weber's venture, the *Magnetischer Verein*. In this venture over twenty observatories from different parts of the world conducted simultaneous observations of the earth's magnetic field, with Gottingen as the coordination centre.[4] With the foundation of the British Association for the Advancement of Science (BAAS), the desire for emulating this plan received strong impetus in Britain, resulting in the organization of Edward Sabine's *Magnetic Crusade* (1838) and the establishment of a global network of geomagnetic observatories,[5] as well as John Herschel's program of simultaneous meteorological observations.[6] All these

---

1 Not all. Barboza (2012, pp. 176–78) and especially Locher (2008, pp. 24–27) have analyzed Quetelet's work on atmospheric waves.
2 However, these aspects of Quetelet's personality and legacy have not been studied by Locher (2008, pp. 24–27) and Barboza (2012, pp. 176–78)—see previous footnote.
3 Sachs (2007, pp. 81–85).
4 Wolfschmidt (2005, esp. p. 127 et seq.). For a historical review of the measurements of terrestrial magnetism at Gottingen Observatory, see Jungnickel and McCormmach (1990, vol. 1, pp. 63–77).
5 For an examination of the development of the British magnetic crusade, including the geomagnetic expedition to the southern polar regions, see Cawood (1977, 1979, esp. pp. 551–52); Morrell and Thackray (1981, pp. 353–70).
6 For Herschel's efforts to organize global research in meteorology, see Good (2006, pp. 54–58).

initiatives were not isolated facts, but were part of a global interest in the study of the distribution of scientific objects and variables on the earth's surface. This interest encompassed the geographical and temporal distribution not only of meteorological, magnetic, geodetic, and oceanic phenomena, but also of animals, plants, and minerals.

The surging enthusiasm for large-scale international projects was linked to the European expansionism and the politics of the 1840s—a time when continental Europe undertook a transition from the old regime state to the incipient nation-state. In this regard, projects like that promoted by Gauss and Weber in part fulfilled a political and propagandistic purpose, seeking for unity out of the chaos of the fractured Prusso-German political reality. The standardization of the torsion balance promoted by them for terrestrial magnetism implied not only the imposition of a set of procedures but also the coherence of group practices—i.e. the sacrifice of local praxis for unified and global standards.[7] In the British context, the ability to complement and extend a system of colonial observatories became especially useful in a time of rapid change in the structure of British science, providing her government with sound reasons to enhance the British science and extol national glory.[8] The Magnetic Crusade enabled the Admiralty and the War Office to apply the values of geomagnetic variations to the field of navigation. The establishment of physical observatories could also serve to reinforce the sense and position of leadership of the metropolis over colonies. Herschel's *Instructions for making observations* (1835), for example, depicted 'a concerted plan of contemporaneous observations' encompassing all areas of terrestrial physics, from which a synoptic view of global phenomena could only be achieved by a colonial network led by Britain.[9]

However, there is a paradox in this picture of the rise of the scientific internationalization by state-nations and their imperial surrogates. While projects of international campaigns, observations, and coordination increased significantly, nation-states vigorously endeavoured to control and direct them to their own ends. Even when these projects claimed to represent universal principles, such as standardization and the pursuit of synoptic views of global phenomena, they were structured by the participation of nation-states under a single leadership. Observatory science's internationalization linked to global-scale standardization fostered the exclusion of the weakest countries. In Europe, the emergence of several new nation-states (Belgium, Germany, and later Italy), and the humiliation of others, like France and German Austria after the Napoleonic Wars, were reflected in this process of internationalization. As a consequence of the rise of nationalist discourses, France and the Paris Observatory lose their leadership in the field of terrestrial magnetism.[10]

7 Dörries (1994, pp. 121–22).
8 Cawood (1979, pp. 517–18).
9 Herschel (1835, p. 1).
10 Analyzing the standardization of the torsion balance in the European projects on terrestrial magnetism, Dörries (1994, pp. 144–47) argues that the loss of French leadership

Humboldt's global view of a natural world in which a general equilibrium of forces reigned played a key role in promoting the internationalization and globalization of observatory sciences. This view distinguished itself from the mechanic and self-regulating world system suggested by Pierre-Simon Laplace some years before in which the balance of forces was mathematically demonstrable.[11] Humboldtian natural order emerged from complexity and observational simultaneity rather than from mathematical mechanism. This idea led him to move from the sporadic measurements made in his various travels in the 1810s to the pursuit of empirical precision by comparing the observations simultaneously made at fixed observatories in the 1830s. The greater the number of forces observed, and the greater the accuracy of such measurements and the number and extension of the observatories involved, he thought, the more clearly would natural order emerge.[12]

Another fundamental question was the need to develop a holistic view of observatory sciences, so that observers would be able to find relations between the possible sources of error, accurate measurements, and mathematical laws in a wide variety of phenomena. This view was holistic as it included sciences like meteorology, terrestrial magnetism, geodesy, atmospheric electricity, and the tides (and even botany and geology in the broadest sense of Humboldtian science), but only insofar as these were capable of being quantified, arranged, and compared. This conception had a profound effect on the perception of the data that observers transmitted, moving from an individual sense of the record of observations and phenomena to one that was spatially and thematically coordinated. Instructions for making observations, like Herschel's ones, therefore, would include various sciences, and only then could one discover the order of nature and its physical laws.[13]

---

cannot be merely explained by isolation or by a fervid nationalism. Indeed, Britain bore its share of responsibility for this exclusion. According to him, it was not until 1838 that the nationalist discourse in the project of magnetic crusade appeared in abundance— until then the issue of cooperation had dominated discourses. In 1838, the BAAS presidential discourses began to distinguish for the first time the British methods from the continental ones: 'The resources of other societies are employed on their publications or collections; and it is one of the rules of the German scientific Reunions, that they shall possess no property. Our objects were more extensive than theirs, and therefore, our plan was different'. Or a little earlier, 'to ask nothing of Government but what belongs to the national interests or honour to effect, and what cannot be effected but by national means'. The two extracts are taken from the *Address by Mr. Murchison* (1838), BAAS, 1839, xxxi–xliv, on p. xxxvi and xxxii, respectively—quoted by Dörries (1994, p. 146).

11 For the term 'Humboldtian science' and its meaning, Cannon (1978, esp. pp. 73–110). See also Dettelbach (1996, pp. 287–91).

12 The debate over the different methods for measurements on terrestrial magnetism clearly reflects the tensions between the parties involved in the internationalization of observatory sciences. While held an empirical and global view of the natural world, Gauss focused on those aspects that were subject to the application of mathematical methods. On this debate, see Dörries (1994, pp. 126–29).

13 Herschel's presidential speech to the BAAS in 1845 is a particularly good example of the ways in which the holistic views of nature favoured large-scale data collection and

Paralleling this increasing influence of Humboldt's holistic vision of nature on earth scientists was a third major development that helped make observatory sciences more globalizable: the technical and material culture of precision.[14] Here the mobilization of techniques being specific of astronomical observatories was an important factor, enabling the extraction of numbers from precision instrumentation that were then tabulated and used as basis for establishing comparisons and theories.[15] These techniques paved the way to the preparation of geomagnetic surveys such as the aforementioned efforts by Humboldt. In 1828, he established a magnetic observatory in Berlin, from where he launched a program of observations at different places at prearranged times. In fact, it was the astronomical techniques that enabled him to accurately determine the time and geographical location of each participant.[16] Yet from a training standpoint, the observatory also played a key role in fostering the disciplination of observers, leading the effort to publish monthly and annual observations which then served as the school manuals for trainees.[17]

Like the history of Western industrialization, the evolution of state-sponsored observatories was also marked by a steady movement from the individual action and effort towards the introduction of factory-like methods and increasing professionalization. The first impulses towards introducing factory-like methods of management and production were heavily tied to military and maritime needs, first for the rating and maintenance of chronometers for navigation and then later to facilitate astronomical and

---

international collaboration. Herschel tended to regard meteorology as part of a larger enterprise of terrestrial physics—a vision that justified the need to expand observations to all continents. 'The gigantic problems of meteorology, magnetism, and oceanic movements', Herschel (1846, p. xxxiv) stated, 'can only be resolved by a far more extensive geographical distribution of observing stations, and by a steady, persevering, systematic attack, to which every civilized nation [...] ought to feel bound to contribute its contingent'—quoted by Good (2006, pp. 57–58).

14 On Humboldt's inspirational (rather than practical) influence on Herschel and the Magnetic Crusade, see Carter (2004, p. 512).

15 For the role of astronomical techniques in laboratory physics, see Aubin (2005, pp. 31–36). See also Aubin, Bigg and Sibum (2010, pp. 8–11); and Locher (2007, p. 492).

16 One of the most famous applications of astronomical techniques to geomagnetic surveys was carried out by Gauss, as director of the Gottingen Observatory. Drawing on his knowledge and expertise in instrumental precision, Gauss adapted a telescope to Gambey's dip magnetometer, which enabled him to observe the needle at a great distance and thus elude any possible disturbance. In a letter addressed to German physician and astronomer H.W.M. Olbers, he was exultant with this devise as 'the horizontal part of the earth's magnetic field can now be observed as precisely as the stars in the sky'. Gauss to Olbers, 2 September 1837, in: *Briefwechsel zwischen C. F, Gauss und H. W. M. Olbers.* Hildesheim: Georg Olms, 1976, 2, p. 649—quoted by Aubin (2005, p. 35).

17 At the 1845 BAAS meeting at Cambridge, Herschel (1846, p. xxix) acknowledged the value of the training role: 'Every astronomical observatory which publishes its observations becomes a nucleus for the formation around it of a school of exact practice'.

magnetic tasks.[18] This development was pioneered by the Royal Observatory at Greenwich under the direction of astronomer George B. Airy in the late 1830s. Airy was attracted by the ability of first engineers and industrialists to transform both the manufacturing methods and mechanization.[19] By the 1840s the strict hierarchic organization of labour and the rational and efficient discipline were so deeply rooted at Greenwich that Airy was soon admired and emulated by other directors of European observatories. In France, this came largely from the Paris Observatory's director, Urbain Le Verrier (as will be discussed later), while in Belgium the Brussels Observatory's director, Adolphe Quetelet, was unstinting in his praise of Airy's merits.[20] By the middle of the century, Greenwich became a central element in international programs of observatories, not only for its role in both studies on terrestrial magnetism and the calibration of instruments but also for its factory-like mentality.

Organized more efficiently and with a highly disciplined staff, Greenwich began to quickly publish its scientific production, whether through tables and monographs, or other types of formats, like monthly and annual reports. This production on a large scale served as the reference point for coming worldwide meteorological projects. Data collected about observatory sciences could be broken down by branches and topics and quickly transmitted to other fellow members for further consideration and comparison. Here Airy opened the way by introducing greater simplification and standardization at Greenwich.[21]

The result of all these mentioned developments was the construction of extensive meteorological and geomagnetic networks in which national observatories were key nodes. Local data gathered by these observatories were collected in standardized and simplified ways, and centralized at a few coordinated, carefully chosen centres. However, meteorological enterprises depended on extensive data-gathering networks that were open to statistical treatments. The suggestive nature of early theories of errors like Gauss's and Laplace's was replaced with a new statistical orthodoxy, in which meteorology resorted to the mathematics of probability to discover order rather than to measure chance. Here Quetelet was the path breaker in showcasing the affinity and power of cooperation between meteorology and statistics—a partnership that made Quetelet Europe's torchbearer in meteorological cartography by the middle of the nineteenth century.

---

18 Chapman (1985, pp. 321–22).
19 Schaffer (1988) and Smith (1991).
20 For Le Verrier's inspiration into Greenwich methods in the 1850s, see Locher (2008, pp. 53–56).
21 Chapman (1985, p. 323); Smith (1991, pp. 14–15).

Probability and natural law

The history of measurement has been often viewed as a process of reaching objectivity by progressively replacing personal judgement by theory and calculi of observational errors. Techniques of probability calculations, such as those developed by Gauss and Pierre-Simon Laplace, aimed to control measurement errors in astronomy and therefore to reach an objective calculus of observations. These techniques included the Gaussian (later called normal) distribution representing random variables, arithmetical mean as the most probable value of the measured parameter, and the method of least squares as the criterion of optimization.[22] All these tools were intended to reach the ideal of complete objectivity. This ideal, however, was never achieved, and there always remained a residual of personal judgement in measurement.[23]

The method of least squares or the error law, firmly established in astronomy by the late 1810s, epitomizes the pursuit of this ideal. This method was devised as a mechanical procedure for evaluating accuracy of observations. According to this rule, when astronomers handle a large number of observations of a star, its most probable position is that for which the sum of the squares of the difference between the observed value and the fitted value is a minimum. This formulation was appropriate for a science like astronomy, where the physical quantities measured (i.e. stellar positions) apparently did not vary and where accuracy became a quality criterion and rigour. Since then, 'most probable' was usually interpreted as 'most accurate' in astronomical language.

These conditions, however, were not met in other observatory sciences like meteorology. First, the two disciplines were not equally viable with regard to their capacity for reducing the complexity of phenomena to simple laws. While in astronomy a single cause (gravitational force) and one theory (Newton's laws) guided the calculi of observational errors and the pursuit of objectivity, in meteorological science a host of possible causes and the lack of deterministic laws placed the observer before more intricate and broader problems. Moreover, nobody in the 1820s could hold that measurements in meteorology were equally reliable as in astronomy, or that they involved physical quantities that were not subject to any significant variation. In the light of all this, the need to think and act in terms of probability would soon become patent.

The great scholars of climate of the first half of the nineteenth century were less concerned with the calculi of observational errors than with data gathering and analysis. Available knowledge in this field was mostly empirical.

---

22 Sheynin (1984a).
23 On the concept of mechanical objectivity, see Porter (1995, p. 4); Daston and Galison (2007, p. 121). On how measurement implies a combination of mechanical objectivity and trained judgment, see Daston and Galison (2007).

Adolphe Quetelet, a young Belgian mathematician and astronomer study-ing the theory of probabilities in Paris, was a noteworthy exception. From his first works in the early 1820s to his popular books on probability, he sought to apply the error law to any field or subject of study that could be measured and counted.[24] These fields could range from climate and plants to studies on mortality and incidence of crime, and in all of them he worked to search regularities. He performed the first application of the Gaussian distribution used in astronomy to other measurements such as those of the height of conscripts in a regiment.[25]

The greatest novelty of Quetelet's work was his study of what he con-ceptualized as the 'average man' (*l'homme moyen*)—or what historians re-garded as the birth of 'social physics' or even 'social meteorology'.[26] His first successful analogy in this regard was to liken the forms of the mentioned two distributions and, in this way, define an ideal of human measurement from which all actual individual would deviate. In the same way that the real position of a star was the 'cause' explaining the Gaussian distribution of error measurements, and whose most probable position was the mean of the positions observed, it too made full sense to propose an average man that would be the 'cause' for the distribution of heights, and whose most probable height would be the mean of the heights observed. This analogy, he argued, opened the possibility of making a 'mécanique sociale' just as there is a 'mécanique céleste'.[27]

Quetelet's interest in social mechanics derived from the studies on celes-tial mechanics performed by Laplace decades earlier.[28] In his seminal book *Essai philosophique sur les probabilités,* the French astronomer had showed how the accuracy attained by the error law could reveal the laws of celestial mechanics.[29] In 1835, Quetelet published a memoir entitled *Sur l'homme,* whose cover reproduced as an epigraph the passage from Laplace's *Es-sai:* 'Let us apply to the political and moral sciences the method founded upon observation and calculation, which has served us so well in the natu-

---

24 For Quetelet's contribution to error law and to probability theory and mathematical sta-tistics in general, see Sheynin (1986, pp. 306–20).

25 Desrosières (2002, p. 4).

26 Stigler (1997, p. 53): 'We can see most of Quetelet's "social physics" [...] as a sort of "social meteorology", where the propensity for crime takes the place of barometric pressure'.

27 Quetelet to Ministre de l'Intérieur de Belgique, Sylvain Van de Weyer, 22 August 1834. General State Archives Brussels, *Papiers Van de Weyer,* 247—partly reproduced by Del-mas (2004, pp. 57–58): 'The great problems of population motion will be as solvable as those concerning the motion of celestial bodies; and what is most remarkable is the as-tonishing analogy that exists between the formulas that are used for the computations. I think I have partly realized what I have been saying for a long time about the possibility of making a social mechanics, just as we have a celestial mechanics'—according to the English translation by Aubin (2009, p. 289).

28 See, e.g., Courgeau (2012, pp. 204–05); Daston (1988, pp. 107–08); Porter (1986, p. 101); Stigler (1986, p. 162, esp. ftn. 4).

29 Laplace (1814, pp. 51–60).

ral sciences'.[30] Quetelet not only echoed Laplace's call, extending the error law to the study of society, but he also sought to elucidate the natural laws producing the regularities regarding to climate, population, and crime, just as Laplace did in astronomy. Most importantly for our purposes, he took the idea that law was the mere regularity of fact conforming to the theory of probabilities to the extreme.[31] As shown below, this conception of law played a paradigmatic role in defining a dynamical cartography of meteorology.

This probabilistic approach of scientific law, both in observatory sciences and in social and moral sciences, corresponded to a particular view of mathematical rigour and a high degree of engagement with astronomical questions. Quetelet's way of injecting statistical reasoning into observational analysis was familiar to Laplace and Gauss, who were less involved in meteorological studies and more versed in astronomy. Yet many meteorologists will judge Quetelet's approach to be too statistical and too idealized. Causal factors and the mode of action of the phenomena involved went unnoticed, and at times even were disdained. Acutely aware of these insufficiencies, Quetelet developed innovative methods that combined the demands of error calculations and practical usefulness.

Meteorology and statistical methods

The mathematical study of probability described earlier was but one side of the larger quest to quantitatively measure natural regularities; for any statistical inquiry implied the use of mathematical techniques, not only probabilistic. Without the application of graphical methods of statistics and techniques such as the calculation of mean values and arithmetical means, no effort of theorizing would have had the authority necessary to make the meteorologists' claims about the weather and its forecasting credible. In fact, the meteorologist often claimed his authority through the idea that such knowledge had been obtained by a long process of rational and coordinated observation—a process that demanded the gathering, tabulation, and manipulation of data in order to make them universally comparable. In the present context of examining weather forecasting, however, this presented a big problem since it did not seem feasible for a meteorologist to apply statistical methods to non-periodic or variable phenomena like storms. An observational system was therefore required that could accurately collect all the relevant data possible from key locations around the world. As was shown, scientists like Humboldt soon discovered that system in the form of observatory networks, which had the qualities needed to conduct large-scale meteorological investigations by applying statistical techniques. Furthermore, over the course of the 1830s and 1840s observatory scientists

---

30 Quetelet (1835, vol. 1).
31 Lottin (1912, p. 111); Dufour (1948, p. 63).

would endeavour to mathematize the social sciences and foster the migrations of techniques from meteorology to these sciences in order to interpret macro-social regularities. By the middle of the nineteenth century, meteorologists would realize that the origin of the temporal and spatial variations of the weather and storms could only be investigated by worldwide permanent and simultaneous observations from global and coordinated networks of observatories.

This dual relationship between meteorology and statistics would have puzzled a savant from the eighteenth century. At that time, the interface between the two disciplines could be regarded as sporadic and restricted to a few facets. National observatories did not usually promote international cooperation programs and the application of mathematics to phenomena with complex and variable causes was viewed with suspicion and distrust, especially those related to the weather and climate.[32] The Societas Meteorologica Palatina, for example, was a private international network of observers founded in 1780 that attained a standardization of their readings theretofore without equal.[33] Still, it remained anchored to pure empiricism and practical applications, leaving aside the study of climate and its evolution—the object pursued by experimental climatology.[34] It was not until 1820 that, drawing on records of the Societas, H.W. Brandes from the University of Breslau investigated the relation between air pressure and wind and is said to have plotted a map of the storm on 6 March 1783.[35] The French centralized bureaucratic administration under Napoleon I also compiled their own statistics on elements related to climate and its effects (such as agricultural and medical consequences), though this information was basically part of auxiliary devices of control and surveillance.[36] These surveys were based on a climatologic program devised by naturalist Jean-Baptiste Lamarck who, with the support of the minister of Interior Antoine-Claude Chaptal, proposed from the use of tables and records of simultaneous and comparable observations, tabulated in standard ways, to the establishment of a *bureau central* and secondary stations, which would allow local specificities to be eliminated.[37] Yet, his reluctance to accept the validity of mathematics to solve complex phenomena was a handicap to further progress—a constraint

---

32 For a review on the early history of international meteorological observations, Landsberg (1980).

33 Cassidy (1985, pp. 8–9); Davies (2000, pp. 964–72); Lüdecke (2014, pp. 123–25).

34 Feldman (1990, pp. 168–72).

35 The possible drawing of this map, which has been a matter of debate, is discussed in Chapter 8. See also Monmonier (1999, pp. 18–22); Robinson (1982, pp. 73–74).

36 Bourguet (1988, pp. 68–106).

37 Lamarck put forward these proposals in the pages of his *Annuaires météorologiques* (1799–1810). For further details on Lamarck's program on the organization of observations, see Lamy (2007, Chapter 5) and Delange (1997, pp. 123–36); on the role of Chaptal in promoting a plan of departmental statistics in France, see Desrosières (1994, pp. 47–48) and Bourguet (1988, p. 68).

that would prevent any critical statistical analysis (such as those by Quetelet some decades later) from coming to fruition.[38]

The Napoleonic wars and France's capability to centralize the collection and distribution of records regarding the climate, geography, and population prompted a change in the neighbouring states' attitude towards statistical science. In Germany, initiatives were implemented to emulate the mobilizing action of the French Bureau of Statistics (established in 1800), resulting in the establishment of professional statistical bureaus in Prussia (1805), Bavaria (1808), and other places after the wars. In this new context of collaboration between the state and demographers and scholars in general, whose investigations could be valuable in producing practical knowledge for the state, Humboldt introduced statistical techniques into meteorology. A key member of the Academies of Science in Berlin and Paris who possessed a massive amount of data as a result of his scientific expeditions to America, the issue of how to ascertain the numerical laws of meteorology was of great interest to Humboldt.[39] Fascinated by the way in which global order emerged from local peculiarities in Laplace's classic theory of celestial equilibrium and very familiar with the magnetic lines employed by Edmund Halley and others in their isogonic maps, Humboldt proposed a new and subtle yet critical application of the isometric lines—or lines on a map along which some value is constant.[40] Like the planetary system that follows a cycle with almost no change in the angle of the ecliptic (according to Laplace's *système du monde*), the distribution of heat in the atmosphere would oscillate around a mean state, from which it would never deviate 'by more than very slight quantities'[41] (Humboldt's *physique du monde*). With the aim of showing the points receiving an equal quantity of heat, he was the first to devise the isothermal lines—i.e. the annual isotherms linking points of same

38 According to Lamarck, the application of mathematics to such situations was of zero value. For more on Lamarck and his statistical thinking, see Sheynin (1984b, pp. 65–66, 81–89). Lamarck introduced and frequently used the term *Météorologie-statistique*, though he formulated different—and at times contradictory—definitions, often identifying it with *statistique atmosphérique*. Here is an example of Lamarck's definitions: 'The object of statistical meteorology is to know better, not only the general nature of the climate in the country where [this science] will be established, but also in each of the most important points of the country the order of occurrence of winds [...wind forces, meteors...], in brief, the influence of these meteors on animals, plants and the ground itself in each of the places'. Lamarck, *Annuaire météorologique*, 1803, 4, pp. 153–54)—quoted by Sheynin (1984b, p. 85).
39 Humboldt visited America from 1799 to 1804.
40 Robinson and Wallis (1967, 1982, p. 157, 221, 228); Konvitz (1987, p. 134); Wise (2010, pp. 88–90).
41 According to Humboldt's manuscript note, quoted by Dettelbach (1996, p. 295). For the similarity between Humboldt's isotherms (1817) and Halley's isogones (1701), see Robinson and Wallis (1967, p. 121).

mean annual temperature.[42] Here, the isotherm was envisioned as the embodiment of global order, emerging amidst local averages.[43] Humboldt described it in almost physical-geographical terms, defining it as the graphical
representation of a 'physical continuity', in which the mean value was 'the
final object, the expression of physical laws', showing 'the constant amidst
the flux and flight of phenomena', while isotherm was the outcome of averaging and interpolation.[44] Rather than mathematical expressions, rather
than geometrical manifestations of some entity or value, isotherms encapsulated local, physical quantities, and as such, Humboldt argued, were the
embodiment of nature's order.[45]

Humboldt's method, advanced first in an essay on isotherms but whose
map appeared in an abstract published in another journal—the *Annales de
chimie et de physique* in 1817—was in many ways groundbreaking.[46] On one
hand, he was essentially the precursor of a fecund line of European meteorological cartographers that ran from Ludwig F. Kämtz and Heinrich
Berghaus to Wilhelm Mahlmann and that was predicated on the application
of the isarithmic method to phenomena such as barometric pressure, annual
precipitation, and thunderstorms.[47] As historians of scientific cartography
A.H. Robinson and H.M. Wallis noted in 1967, Humboldt's 1817 map of
isothermal lines meant 'the catalyst for many similar uses of the isarithm in
the thematic cartography'.[48] In fact, it was frequently copied and emulated
far and wide: it was reproduced in G.W. Muncke's *Gehler's Physikalischen
Worterbuch* in 1827[49]; and it was given pride of place in Berghaus's monumental 1838 work, *Physikalischer Atlas*, by applying the isarithmic method
to a variety of meteorological phenomena.[50]

42 Humboldt was probably the first to attach the prefix 'iso-' to a cartographic term in
his 1804 map of isodynamic zones, based on his geomagnetic measurements. Robinson
(1982, p. 216).

43 In an article on the distribution of vegetation (1816), Humboldt first defined the term
'isothermal parallel' as a 'curve drawn through the points on the globe which receive an
equal quantity of heat'. Robinson and Wallis (1967, p. 121). See also Horn (1959).

44 Humboldt—quoted by Dettelbach (1996, p. 298).

45 Humboldt even wrote a work on 'Lines' for the *Dictionnaire des sciences naturelles* to
show the analytical power of cartography to terrestrial physics. See Humboldt (1823).

46 Humboldt (1817a, 1817b).

47 The isarithmic technique involves drawing thematic maps that represent a continuous
field by lines or symbols in order to connect points of similar value.

48 Robinson and Wallis (1967, p. 119). On his contribution to thematic cartography see Palsky (1998, pp. 45–47); Zogner (1983, pp. 393–406).

49 Muncke (1827, vol. 3).

50 France was an exception. Although his activities had considerable impact there (in
fact, he stayed in Paris from 1804 to 1827, where he read his essay on isothermal lines),
Humboldt never left a mark or attracted cartographers in France. See Konvitz (1987,
p. 134). For the importance of Berghaus's *Physikalischer Atlas* in the expansion of thematic cartography, see Camerini (1993, pp. 479–512).

Humboldt's isometric technique moved away from the focus on the individual as an object of analysis, endeavouring instead to search for appropriate mean values (mean states) and their deviations. Or to put it another way, Humboldt was more interested in deriving the physical laws associated to the thermal distribution than in offering a compendium of temperatures.[51] As he noted in his 1818 work on the influence of the solar declination on equatorial rains, the meteorologist should ascertain the average movements of the atmosphere in order to distinguish a 'certain type in the sequence of phenomena' from which to derive the empirical physical laws.[52]

Humboldt's methodology was also significant from a statistical viewpoint for it expressed the need to know not only the mean state of the atmosphere but also the 'constant kinds of its variations' before examining the causes of local disturbances.[53] In this regard, the most fundamental was the idea that deviations themselves from mean states clearly became an object of inquiry—a feature that will be essential in mid-century meteorology. More will be said about these questions in coming sections on meteorological cartography; here it will be sufficient to note that Humboldt's technique transformed isometric line into a fundamental analytical and communicative tool for meteorological cartography.[54]

Humboldt's technical innovation found widespread appeal among engineers and demographers, to the extent that isotherms marked the shift from a topographic technique to a technique of thematic cartography in these fields. In effect, Humboldt's innovation concerned one of the two big classes of symbols related to thematic mapping: the isometric lines; the other class was the isopleths. While the former portrayed numerical distributions belonging to basic geographical facts such as air temperature, atmospheric pressure, elevation, or depth, the latter portrayed distributions regarding much more complex geographical concepts including functions of two or more variables such as density, spacing, correlation, or ratios. According to German geographer Max Eckert, isometric lines represented absolute

---

51 Schneider-Carius (1975, pp. 164–79).
52 Humboldt (1818, p. 179).
53 Ibid., p. 190.
54 Part of the explanation for the success of Humboldt's graphic method was the complete simplicity of his map and its seemingly great power to project statistical values. However, there were both aesthetic and methodological reasons for this that attracted its attention and interest beyond meteorological circles. To begin with, Humboldt thought that a map, 'overcharged with signs, becomes confused, and loses its principal advantage, the power of conveying at once a great number of relations'—quoted by Robinson and Wallis (1967, p. 120). Moreover, his earlier statistical studies on physical and plant geography (developed since the early 1790s) were helpful in this sense, as the concern with the relation between isothermal lines and the geographical latitude that led him to plot his 1817 map attracted special attention among geographers and naturalists.

values because they could de facto exist at points on the earth, while isopleths denoted relative values because these were derived by functions.[55]

If Humboldt's isotherms provided the basis for graphical representation of absolute values in meteorology, then the work of French engineer Léon Lalanne and his efforts to define isopleths meant the isometric application to thematic cartography.[56] As a graduate at the École Polytechnique de Paris who worked in the division of Ponts et Chaussées (bridges and roads) from 1831, he pushed the focus of quantitative cartography from meteorology to transportation and demography.[57] At the heart of Lalanne's work lay the idea of a new cartographical method—the isopleths—that could show the relations between population distribution and channels of communication, especially the large railway lines. Its background can be found in a long appendix, entitled 'Sur la représentation graphique des tableaux météorologiques et des lois naturelles en général', that he wrote in 1843 for Kämtz's well-known textbook *Cours complet de météorologie*.[58] In that work, he provides graphs of forty-two numerical tables, in some of which he faces with a crucial problem: how can z dimensions be shown on a xy plane? Or, in other words, how can a flat representation be obtained in a three dimension space? Then Lalanne devises a method of contours with tables of double entry. As an example of application, he plots the variables 'time of day' and 'month of year' on the axes of x and y, respectively, while he draws the variable 'temperature' on the z-axis. By doing so, he obtains contours or curves of equal temperature projected on the xy plane that form a topographic surface. An example of this is his law of variation of average temperature per hour in different months in a year in Halle (Figure 2.1).[59]

Lalanne's next step, advanced in a paper read before the Academy of Science in Paris in 1845, in response to a suggestion by a certain Morlet regarding curves of population *d'égale excentricité*, was revolutionary.[60] Thus, he essentially broke with a long tradition of isometric cartography that ran through Marcellin Du Carla, Halley, and Humboldt, replacing one or some

---

55 For more on the conceptual distinction between isometric lines and isopleths, as well as on Max Eckert's views, see Robinson (1971, pp. 49–50).
56 The most comprehensive study on Lalanne's cartographical work and his technique of isopleths is Palsky (1996, pp. 102–13). See also Robinson (1971, pp. 51–53; 1982, pp. 216–17); Konvitz (1987, pp. 151–53); Funkhouser (1937, pp. 301–02).
57 On Lalanne's early life and training, see Palsky (1996, pp. 102–03).
58 Lalanne (1843), in Kämtz (1843, appendix 1, pp. 1–35).
59 Lalanne's contour diagram enables the observer to deduce key data such as maximum, minimum, relative, and absolute temperatures. For the figure on temperature in Halle, see Kämtz (1843, plate 1, fig. 1).
60 Lalanne (1845). Concerned about questions of political economy, Morlet sought to reduce administrative costs through studies on population distribution. He devised a geometrical method implying concentric lines of force around a centre, the distance between which being equivalent to a constant value for each variable considered. In Morlet's lines (or *courbes d'égale excentricité*), the sum of the distances from these lines to the elements representing a population was constant.

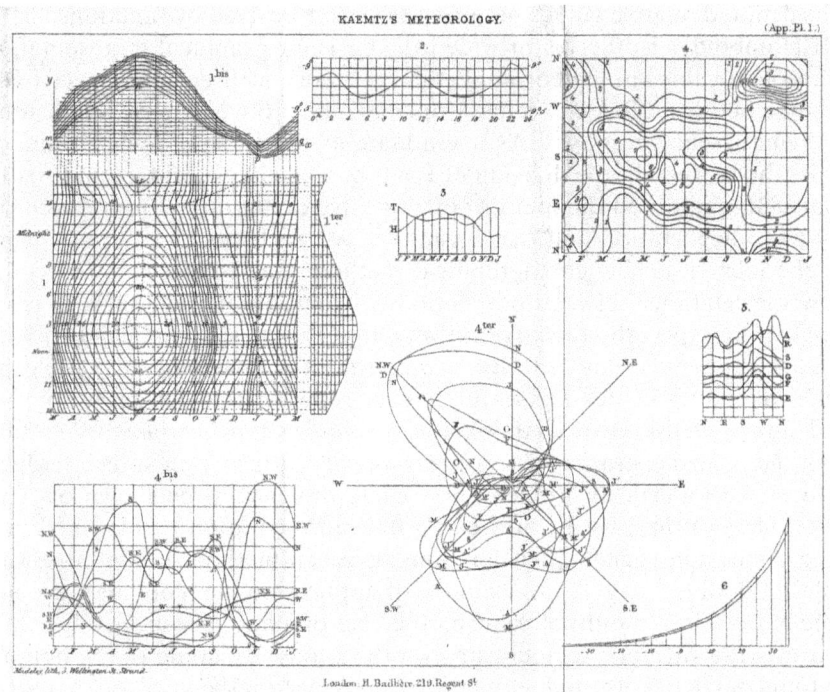

Figure 2.1 Lalanne's contour diagrams and curves of equal temperature showing
the variation of average temperature per hour for the different months
of the year in Halle.
Source: Kämtz (1845, plate 1, fig. 1).

of the variables (i.e. absolute values) by a function of the variable (relative
values). Or, in other words, he substituted the geometrical space (of natural
laws) for the geographical space (of population), by replacing algebraic scales
by logarithmic ones. As he noted in his 1845 article, Humboldt's example in
the construction of isothermic curves 'is not the only application possible,
or the most important', regarding the study of population. In effect,

> Suppose [...] that one partitions the territory of a country into a large
> number of sufficiently small parts so that it would provide a division as
> extensive as the communes of France; that at the centre of each of these
> divisions one raises a vertical, proportional to the specific population,
> or in other words, to the number of inhabitants per square kilometre in
> the territory of the commune in question; that one joins the extremities
> of all these verticals with a continuously curved surface; and finally one
> projects on a map, at a convenient scale, the contours traced on that

surface which correspond to equidistant integral elevations: One will thus have lines of equal specific population.[61]

Lalanne's cartographical proposal moved beyond the idea of the isometric lines as a symbol of the geometrical space, transcending it instead by contours that dissected and illustrated the elements of population. Like a topographic map, his lines of equal specific population would show 'undulations, steep summits, craters, passes and valleys'. Thus, he said, 'our populous valleys would appear as chains of mountains' while our deserted summits 'would look like deep funnels'.[62] While Lalanne never mapped the contours of population above visualized, his concept of isopleths as an abstract representation of variables achieved popularity a decade later and became one of the pillars of thematic cartography.[63]

Thematic cartography was not the sole beneficiary of Humboldt's isometric innovation. Indeed, Humboldt was not alone in his belief that the graphical representation of mean values could generate a global view of climate. The decades after his seminal 1817 isothermic chart witnessed an irruption of climatologic maps as scientists from Central Europe began to transform the values obtained by averaging long series of observations of meteorological elements into visual images. In 1832, the professor of physics at the University of Halle Ludwig F. Kämtz published the second volume of his *Lehrbuch der Meteorologie*, which included annual isotherm charts for the northern hemisphere.[64] He also included a partial world map of isobarometric lines—or lines of equal barometric variation—covering the North Atlantic and four continents.[65] The first world chart of annual isotherms appeared in Prussia in 1841, and was prepared by Wilhelm Mahlmann using data from 305 stations.[66] With the renewed attempts to build a German state in the 1840s, the interest in climatologic charts received fresh impetus, which resulted not only in the search for mean values but also in efforts to study the deviations from mean values—as was shown, both searches had been suggested by Humboldt.

---

61 Lalanne (1845, p. 440), as translated by Robinson (1982, p. 217).

62 Ibid.

63 The first graphical application of Lalanne's concept took place in 1857, when Danish officer Nils F. Ravn published isopleth maps for the Statistical Bureau in Copenhagen; they were population density maps for the years 1845 and 1855. See Robinson (1971, pp. 52–53). According to Funkhouser (1937, p. 302), although statisticians usually avoided contour maps, Francis Galton drew on them in his early studies on correlation and regression whereby he defined the normal correlation surface; he even plotted contours representing lines of equal birth rate (called 'isogens').

64 Kämtz (1832, fig. 1).

65 Ibid., fig. 3.

66 Mahlmann (1841). Mahlmann became the first director of *Preußisches Meteorologisches Institut* in 1847.

The most remarkable of these efforts precisely occurred during this period in Prussian lands, where the new director of the Prussian Institute of Meteorology H.W. Dove reached the culmination of two decades of studies on climatology. During years, he understood climatology essentially in the same way as Humboldt did it, as a descriptive science based on measurements and observations.[67] He compiled available observations on temperature conditions and then analyzed them comprehensively in order to ascertain the laws of periodic changes. The results were published in voluminous tables rather than in striking graphs.[68]

However, as a Humboldt's follower and thus very familiar with the issue on 'the distribution of heat on the Earth'—the title of his doctoral thesis (1826)—the issue of how to study non-periodic changes soon became of great interest to Dove. Troubled by the fact that mean temperatures (as well as average pressures and humidity) were not directly apparent, but rather were masked by the periodic and irregular changes, Dove proposed a subtle yet crucial distinction of the matter. The task set for meteorology, he stated in 1838, should be threefold: 'to determine the average, to ascertain the laws of periodic changes and to state the rules for the irregular ones'. While no one of these three problems 'is near a complete solution', he added, the first two have been treated with greater zeal by investigators, at the cost of the latter—the non-periodic changes.[69]

Dove's search for the rules for non-periodic changes, developed in a series of articles published at the Academy of Science in Berlin from 1841 to 1848, was particularly revolutionary. On one hand, he overcame Humboldt's approach of yearly isotherms that was predicated on the search for laws of periodical changes rather than of non-periodical ones. Instead of limiting himself to plot lines of equal mean yearly temperatures (which was essentially a visual description of mean values), he would introduce monthly mean values and the term 'non-periodic changes'—an expression for the deviations from mean values.[70] On the other hand, he defined a new concept, 'normal', as the mean value derived from a long series of observational years.[71] Here, 'normal' was conceived as a reference category, an

67   Dove's dedication to climatology is beyond question: sixty-one out of eighty-four papers on meteorology published by Dove in the Proceedings of the Prussian Academy of Science between 1837 and 1876 deal with climatological issues. See Harnack (1900, pp. 55–61) and Bernhardt (2008, pp. 70–71).

68   Although traditionally regarded as reluctant to synoptic method, Dove graphically represented qualitative wind observations as early as 1828. For a defence of Dove's synoptic facet, see Scultetus (1943).

69   Dove (1838, p. 285), as translated from original German by Schneider-Carius (1975, p. 188).

70   Dove (1848a).

71   Landsberg (1955, p. 7) states that the word 'normal' appeared for the first time in meteorological literature in the first monograph of Dove's series on the temperature variations over the globe, even if it was used rather vaguely. Dove (1840, p. 4).

adjective qualifying any meteorological element. Dove spoke of it in almost statistical-moral terms, regarding as 'normal temperature' the place where temperature is the average of its geographical latitude. If a place has a lower temperature than the normal, that place is relatively cold; if it has a higher temperature, it is relatively hot: the spatial-temporal distribution of temperature, Dove argued, should be determined by normal values.[72]

Indeed, Dove's procedure represented the most developed application of statistical thought to the problem of spatial-temporal distribution of air temperature, leading him, as John Leighly has noted, to the introduction of 'a welcome kinetic element into discussion'.[73] Although the construction of Dove's 1848 isothermal charts intended to show the monthly distribution of temperature on the globe, at the heart of his methodology was also an attempt to reflect the spatial scatter of the deviations of temperature.[74] With this aim, he read a paper before the 17th BAAS meeting held at Oxford in 1847, where he suggested that normal values of temperature could be associated to latitude. In fact, he calculated some normal values by averaging observations around a latitudinal circle, and then drew a series of charts showing deviations against these latitudinal averages.[75]

While the impetus to represent the spatial-temporal distribution of air temperature met with the goals of those meteorologists who saw non-periodic variations as a key factor for understanding the real climate, practical concerns demanded a further conceptual refinement before Dove's approach could produce effective results. He developed a concept that he believed would accurately and spatially show non-periodic changes. Thus, employing temperature records from over 1,500 stations, he first determined the latitudinal mean values (in spaces of 10°), then obtained the deviations of the mean temperature from these values, and finally drew maps of monthly isotherms. With this material, he traced lines delineating areas with the same values of specific heat anomalies—which he termed 'temperature isanomals' (*isanomalen*).[76] This greatly facilitated efforts to understand non-periodic long-time temperature changes over extended areas.[77]

Dove's approach is especially interesting for our purposes for the way in which he introduced motion into the distribution of temperature over the Earth's surface. Earlier methods of isothermal charts denoted to a large extent fixity and motionlessness, since the archaic technique used could hardly

---

72 Dove (1848a, p. 393).
73 Leighly (1949, p. 661).
74 The importance of spatial distribution was underscored by Dove (1838, p. 401). For this point as well as for Dove's ideas about deviations from mean values, see Sheynin (1984b, p. 73).
75 Dove (1848b, pp. 373–76).
76 Dove (1852). The maps can be found between pp. 22 and 23, and pp. 24 and 25. It also includes five charts of thermal anomalies.
77 For an earlier formulation of thermal anomalies, see Dove (1851, pp. 619–26).

reflect how isolines of mean values changed over time and space. Even in the subsequent discussions of temperature maps by Kämtz and Berghaus the active verb for 'isotherms' was usually the verb 'to be'. Dove addressed this question in two ways. First, by adopting a language aimed at depicting deviations from mean values, he projected their ability to 'move'. Isotherms and isanomals would henceforth be independent entities that were constantly in motion. As proven by his 1852 discussion on the maps of mean monthly temperature, isotherms *sich bewegen, wandern*, and *warden steiler*, while areas of thermal anomalies expanded, contracted, and vanished.[78] Second, by plotting all these lines on the same maps, he could easily monitor their variations. Monthly maps would clearly indicate the extremes of the annual pendulation of these lines; and by printing as many of maps as possible on the same sheet, he could follow isothermal changes from one map to the next.[79]

Despite all that progress, however, problems remained that undermined the credibility of the application of statistical methods to meteorology. To begin with, temporal deviations of meteorological observations from the mean values did not possess the usual properties of statistical science. These deviations tended to be considered as time series, and as such their statistical meaning was never an object of inquiry by Humboldt or Dove. Moreover, given that atmospheric temperature and pressure varied significantly, it was by no means easy to obtain reliable average results.[80] The use of arithmetic means and probable errors was too arbitrary, open to too many variables in the choice of the criteria of measurement. Most meteorologists took for granted that observational errors were consistent with a normal distribution.[81]

78 This point was underlined by Leighly (1949, p. 665). Leighly set up Dove's kinetic view against the static view of climatology adopted at the end of the nineteenth century, and instantiated by John G. Bartholomew and Andrew J. Herbertson's discussion of temperature maps in their *Atlas of Meteorology* (1899). It would be more accurate to say that although Dove's isothermal maps clearly showed for the first time the differences between continental and maritime climates, as well as the influence of the air currents from the sea, he tended to explain these processes in terms of conflict between southward-moving cold air and northward-moving warm air. In this regard, Dove has been regarded as a precursor of dynamical climatology. See Chromow (1930, p. 312).

79 Leighly (1949, p. 661) expressed the kinetic potentiality of Dove's isothermal charts very forcefully: 'The isotherms that Dove thus followed in their annual pendulations represented, it is true, only monthly means; but it requires only a slight further effort of the imagination to extend the same concept to the movements of isotherms from day to day, or even from hour to hour'.

80 This problem was noted early by German astronomer and director of the Royal Observatory of Bogenhausen, Johann von Lamont. See Lamont (1839, p. 263).

81 Lamont noted this fact in 1867 in an interesting critical study in which he introduced probability calculations in meteorology. See Lamont (1867, p. 243). Sheynin (1984b, p. 55) points out that Lamont was the only scholar of mid-nineteenth century to consider

In a more specific aspect, statistical treatment in meteorology also inherited some of the problems of the methods used to deal with measurement errors in astronomy and geodesy. Most astronomers and geodesists agreed that the properties of a large number of observations of stable patterns, like planetary orbits, could be known by applying the method of least squares.[82] However, the application of this method to meteorology was more problematic. Meteorologist was not familiar with the repetition of measurements of physical quantities experiencing minor variations, but rather with periodic and non-periodic changes in elements as varied as temperature, sunshine, or humidity. Moreover, the method of least squares was commonly used to assess the measurements performed by precise instruments. But in meteorology each measurement was performed by a different instrument of which its reliability was often not known. The outcome of all this is that, while meteorologists borrowed from statistics the methods originally devised by astronomers/geodesists to deal with errors, they used them to deal with variation in changing values like monthly temperature. Here the observer in charge of measuring these values presupposed that errors could be ignored, most likely due to a lack of statistical training. Meteorologist interpreted those variations as deviations from *normal values*, in the sense of Dove's term.[83]

For all these reasons—the use of unreliable instruments collected from different places, a lack of certainty derived from statistical ignorance, and the arbitrariness to accept errors or not—meteorologists themselves tended to view their works as approximations rather than conclusive assertions on the evolution of atmospheric phenomena. We shall see in Chapter 5 how C.H.D. Buys Ballot, the founder of the Royal Dutch Meteorological Institute, held in 1851 that deviations from normals were more reliable than direct measurements.[84] In fact, a method based on this principle would lead him to enunciate his celebrated law relating wind direction with the horizontal pressure distribution in 1857. However, Adolphe Quetelet—a follower of Humboldt who had received instruction in the theory of probabilities with Laplace in Paris—preferred to think of maps representing not only the variations of barometric pressure but also the motion of atmospheric waves.[85] His approach would pave the way to storm warnings.

---

the fundamental issue of the use of probability calculation and arithmetic mean in meteorology.

82 For the introduction of statistical methods in astronomy and geodesy, see Stigler (1986, pp. 9–158); Sheynin (1984a, 1993).

83 For the properties of measurement in meteorology and its differences with those in astronomy and geodesy, see Boumans (2015, pp. 68–69).

84 Buys Ballot (1850, pp. 42–49).

85 For the relationship between Humboldt and Quetelet, see Avé-Lallemant (1873, pp. 57–58). Quetelet even wrote as a posthumous homage, *Notice sur le Baron F.A.H. de Humboldt* (1860).

Quetelet

From periodic to non-periodic variations

Meteorological elements, like temperature and pressure, are subject to two kinds of causes: constant and variable. Among the former, mid-century meteorologists identified latitude, longitude, nearness to the sea, etc.; and among the latter, clouds, storms, solar radiation, etc. At the same time, the very variable causes, according to their recurrent or non-recurrent nature, could produce periodic or non-periodic variations in elements. These variations were assumed to be the result of natural causes that were often inexplicable for meteorologists. Mainly thanks to Dove's efforts, these variations were interpreted as deviations from normal values. By the late 1840s, periodical variations (such as daily, monthly, seasonal or yearly changes) were rather well known, while non-periodic variations attracted far less attention. By this time too, meteorologists began to realize that the climate was largely characterized by normal values, which they obtained by averaging long series of observations to eliminate fortuitous irregularities. However, they saw, the weather was characterized by an entirely different factor—i.e. by day-to-day fluctuations of elements, caused by atmospheric disturbances. The change of focus from periodic to non-periodic variations contributed to this realization—and in the near future, to the alienation of meteorology from climatology.

This shift of focus and the emphasis on statistical methods in meteorology can be seen in the work of Adolphe Quetelet, whose career was marked by the founding of Belgium as a nation-state after the revolution in 1830 and capabilities of the observatory he directed. Initially trained as a mathematician—his doctoral thesis dealt with geometry, the first thesis read at the University of Ghent—Quetelet had lectured in these subjects and physics at Brussels Museum of Sciences before the revolution. Enterprising and energetic, he had been sent to Paris in 1823 to learn the praxis and instruments of astronomy while receiving instruction in the theory of probability under the direction of Laplace. This field would henceforth occupy a preferential place in his thought.[86] Following the independence of Belgium he was named astronomer of the Royal Observatory at Brussels, where he served as director until his death in 1874.[87]

---

86 It would be more accurate, perhaps, to say that although he did indeed acknowledge the importance of probabilities in scientific researches, the preeminent place was for mathematics in general. For example, after his return from Paris, he gave a course on the history of the sciences at the Museum, the opening of which read: 'The more advanced the sciences have become, the more they have tended to enter the domain of mathematics, which is a sort of center towards which they converge. We can judge of the perfection to which a science has come by the facility, more or less great, with which it may be approached by calculation'—quoted by Hankins (1908, p. 16).

87 Apart from the literature discussed below in connection with his studies on periodicity, the abundant Queteletian bibliography is almost exclusively focused on, and limited to,

Complementing this twofold interest in mathematics and observatory sciences was his increasing involvement in practical activities related to official statistics. His proposal for a complete population census proved to be particularly stimulating as it was ordered by the government in 1828, and this and subsequent nationwide censuses became model in many European countries.[88] He was later named supervisor of statistics in the Belgian administration, where he became the architect of the organization of the *Commission Centrale de Statistique* in 1841. Under his presidency—a position he held until his death—this Commission brought the Belgian official statistics to a standard of thoroughness and professionalism.[89]

Quetelet's meteorological work was built around four main principles, the foremost of which was the systematic application of the calculation of probabilities in meteorology. Here his leitmotif was to find natural laws—understood as regularities of facts in accordance with the theory of probabilities. His efforts were directed at determining not so much the average evolution of individual phenomena, as the mean state of the atmosphere (or the climate). Instead of elucidating the physical causes and the mode of action of phenomena, he endeavoured to gather as many observations as possible of each phenomenon in particular, in order to determine their mean values and, through these values, the empirical rules guiding phenomena. Thus to increase the accuracy of measurements he tried to transform the process of individual observation into one that involved both simultaneity and uniformity of observations. To effect such change he took up the torch left by Herschel regarding his program of simultaneous meteorological observations, replacing this with an extensive scheme of investigation that he implemented from 1840 onward. His plan of simultaneous campaigns not only increased the accuracy of measurements—in 1842 he had up to

---

appraisals and analyses on his pioneering role in the history of statistical thinking and human sciences. See, e.g., Hankins (1908), Lottin (1912), Sheynin (1986), Stigler (1986, pp. 161–220; 1997, pp. 47–61). In particular, the search for his influence on these sciences has largely focused upon Quetelet's 'social physics' and his notion of *l'homme moyen* ('the average man') as the basis of this theory. This issue, which I am not directly concerned with here, has been extensively treated by Halbwachs (1913), Stigler (1986, pp. 169–74; 1999, pp. 51–65), Porter (1986, pp. 42–56), Hacking (1990, pp. 105–11), Callens (1997, pp. 193–252), Desrosières (1993, pp. 94–99; 1997, pp. 179–98; 2002, pp. 3–7). Studies on his contribution to physical sciences are otherwise far less numerous, and have mainly focused on earth sciences; see, e.g., Mailly (1875), Elkhadem (1975, pp. 46–58), Anderson (2005, pp. 83, 153–55), Locher (2008, pp. 14–16, 23–28), Aubin (2009, pp. 287–91). One of the few dealing in part with climatology and meteorology is Donnelly (2017).

88 Quetelet took charge of the 1842 census of Brussels and the national censuses of 1846, 1856, and 1866. According to Lebrun (1974, p. 18), 'the 1846 census was universally considered as a model thanks to the boldness of some of its ideas and the perfection of its execution'. For the expansion of Belgian statistical model to Europe and America, see Prévost and Beaud (2012, pp. 54–61).

89 For an overview on Quetelet as a state statistician, see Prévost and Beaud (2012, pp. 49–61).

forty-one observers throughout Europe—but also enlarged the field of inquiry to which statistical tools could be applied.[90] Unlike his predecessor's plan, largely focused on thermometric and barometric observations, Quetelet implemented a scheme covering meteorology, terrestrial magnetism, and the annual and diurnal habits of plants and animals. Periodical anomalies of temperature, magnetic variations, and even the relations between temperature and the phases of vegetation could all be investigated on a more systematic and uniform basis.[91]

As Quetelet worked to apply probability calculation to meteorology at the Brussels Observatory, he also focused on the pursuit of precision. Here his efforts were directed at a second major principle—drawing on 'observatory material culture', or making observatory techniques as accurate and adaptable a tool as possible for all needs, whether physical or social.[92] Under his direction, observatory techniques were employed for the extraction of numbers from precision instrumentation, their manipulation and graphical representation, all on the basis of the experience provided by astronomy. For the manipulation of numbers in large quantities, a whole array of techniques would be mobilized, including table construction, averaging, data standardization, and even preparation of graphics.[93] The tabulation of results would also become centralized, with the Brussels Observatory acting as the focal point in the processing of numerical data. In his search for precision Quetelet promoted the distribution of standardized instruments, which quickly mushroomed across Belgium.[94]

Finally, Quetelet's third principle was that meteorological data should be internationally uniform and comparable with each other. For him, as with statistical data, the international uniformity in close connection with the coordination of observations made it possible to find true averages and establish comparability between the individual states. He therefore began to promote the international coordination of both statistical and meteorological data through two concomitant conferences, the first International Maritime Conference (August 1853) and the first International Congress on Statistics (September 1853), as shall be shown below.[95] This was the first of a series of nine congresses (the last in 1876), held in different European cities

---

90 Quetelet abandoned the direction and coordination of this plan in 1844 due to lack of economic resources, after having completed a total of seventeen campaigns. See Locher (2008, pp. 14–24).

91 For Quetelet's plan and instructions for observations of periodic phenomena, see Quetelet (1842, pp. 65–95). See also Hankins (1908, pp. 26–27).

92 The idea is essentially that developed most fully and forcefully by Aubin (2009, pp. 288–90), who holds the importance of observatory techniques in Quetelet's approach to social sciences.

93 For the manipulation of large numbers in the context of observatories, see Aubin, Bigg. and Sibum (2010, pp. 12–15).

94 On Quetelet and the Brussels Observatory, see Mailly (1875, pp. 68–85, 124–29).

95 Desrosières (2002, pp. 4–5).

with the aim of promoting uniformity and of standardizing measurement procedures in order to facilitate comparisons in time and space.[96]

The results of Quetelet's investigations were quite extraordinary, both with regard to climatology and to the study of atmospheric electricity and the life of plants and animals. From 1841, his studies on the annual and diurnal variations of temperature and atmospheric electricity began to regularly be published in local scientific journals and handbooks such as the *Bulletins* and the *Mémoires de l'Académie Royale des Sciences de Bruxelles.* By the mid-1840s, the volume of material and data gathered by the Brussels Observatory was so great (especially that collected in 1840–44 campaigns) that Quetelet could publish his first compilation work on the matter (*Sur le climat de la Belgique*, in two volumes) in which he analyzed variables such as temperature, pressure, rain, solar radiation, and their periodicity.[97]

As regards data manipulation, Quetelet was also one of the first to use statistical tools in climatology. In fact, as early as 1846, he published his book *Letters on the Theory of Probability,* where the laws of distributions and means were illustrated by meteorological examples.[98] For instance, letter 13 included a representation of the mean daily temperatures of July in Brussels from 1833 to 1842 by a table and a histogram while remarking on the symmetry of the values measured.[99] Letter 26, in contrast, placed emphasis on the asymmetry found in a series of daily ranges of temperature and speculated about the physical reasons underlying these skewed distributions.[100] By 1852, Quetelet would decide to apply probability calculation to the case of persistent rain, in the hope of establishing the dependence of weather on previous states.[101]

The search for periodicity through the use of statistical tools was also reflected in the studies on the plant and animal kingdom.[102] Plant observations of differing types (flowering, leafing, and leaf fall) could be conducted on a comparable and systematic basis, with the implementation of Quetelet's 1841 programme across Belgium.[103] The number of observers involved

96 For the ambitions and shortcomings of international statistical congresses, see Randeraad (2011, pp. 50–62).
97 A. Quetelet, *Sur le climat de la Belgique*, t. 1–2. Bruxelles, 1849–1857.
98 Here I will use the English translation by O.G. Downes, dating from 1849.
99 Quetelet (1849, pp. 48–57).
100 Ibid., pp. 122–23.
101 Sheynin (1984b, pp. 78–79).
102 Donnelly (2017, pp. 54–67) shows how, in Quetelet's effort to create a scientific movement in Belgium, the study of periodical phenomena, and what he later called global physics, 'was far better suited' for a politically peripheral and small country like Belgium than a science based on large organizing structures such as astronomy.
103 These observations were regularly published in the *Mémoires* of the Belgian Royal Academy. An overview of the observations on periodical phenomena can be found in Chapter IV of the work *Sur le Climat de la Belgique.* It includes a note on the *Instructions pour l'observation des phénomènes périodiques* regarding plants. See Quetelet (1849, pp. 174–83).

also hugely increased, to the extent that a network of phenological (or plant) observations was established throughout Europe.[104] The program was aimed at observing periodic phenomena and, in particular, at determining the relationship between temperature and the different phases of plant flowering by applying statistical methods. Thus, probability theory enabled him to distinguish statistically relevant variations of temperature from irrelevant ones, and so he could find out if regularity or anomaly defined such phenomena.[105] Insofar as the study of periodical phenomena could contribute to a better understanding of both the nature and characteristics of the Belgian climate—as he argued—Quetelet's view on the subject was climate-oriented and statistical rather than botanical.[106]

By the time of the 1848 Revolution, however, Quetelet's probabilities theory had evolved from an array of techniques used to determine the mean state of an atmosphere whose periodic phenomena could easily be studied by detecting the patterns of order underlying variations, to a powerful tool that would seemingly enable to follow the average evolution of non-periodic phenomena. Two years after his investigations on atmospheric electricity and phenology he turned his attention to the mass of observations of hourly barometric pressures collected in his simultaneous observational campaigns.[107] Sophisticate techniques of graphical representation allowed him to study the succession of variations in pressure, which he identified with the motion of *atmospheric waves* (as defined by Herschel and Birt in 1843, as was shown in Chapter 1). Thus he could visualize such non-periodic phenomena as atmospheric waves on a chart, observing how barometric pressures varied or oscillated.[108] More generally, analyzing the variations of temperature in 1853, Quetelet was able to differentiate periodical causes from non-periodic ones, and urged meteorologists to study the latter.[109] By then, he had already transformed probabilities theory into a tool for tracing regularities in all sorts of phenomena, whether periodic or non-periodic. Simply put, by early 1850s probabilities theory had become the tool for data

---

104 For the networking of Quetelet's phenological programme across Belgium and throughout Europe, see Demarée (2009, pp. 22–23).
105 Quetelet's letter 33 from his work *Letters on the Theory of Probability* tries to explain the variability of phenological observations in probabilistic terms. Quetelet (1849, pp. 161–73).
106 Quetelet's climate-oriented definition of 'periodical phenomena' for the plant kingdom was in sharp contrast with more botany-oriented views such as that of naturalist Charles Morren (the proposer of the term 'phenology'). In 1848, Morren expressed his position in that regard: 'The ensemble of the returning phenomena constitutes [...] the manifestation of life ruled by the time. That ensemble has been named [by Quetelet] "periodical phenomena". The name seems to us too vague because he is too general [...] Our goal must be to restrict this study to the essentials in these relations with the organized living beings'. Quoted by Demarée and Rutishauser (2011, p. 758).
107 Mailly (1875, p. 149); Dufour (1948, p. 69).
108 This issue will be discussed in next section. See also Locher (2008, pp. 24–27).
109 Quetelet (1853a, p. 4).

analysis Quetelet needed in order to discover the regularity and order underlying barometric variations.

Armed with a methodology that statistically reduced barometric variations to a singular and objectively quantifiable entity (the *atmospheric wave*), by 1851 Quetelet began to plot isolines on a geographical map of Eurasia that appeared to really show the motion of certain barometric waves. In essence, he drew his inspiration from Birt's programme on barometric waves (as was shown in Chapter 1, Birt postulated the propagation of pressure waves on the Earth's surface, some of which were at the origin of storms), as well as from Herschel's ideas.[110] Already by 1843, Herschel had acknowledged that Brussels could be a nodal point for the observational system of atmospheric waves: 'Very deep waves, it is true, and very extensive ones, ride over' Brussels; 'but with regard to smaller ones, it may be regarded as a nodal point where irregularities are smoothed down, and oscillatory movement in general is more or less checked'.[111] The correspondence between him and Herschel is proof of this influence.[112] Yet, Quetelet also was stimulated by his own experience, as he evinced in his book *Sur le Climat de la Belgique* (1851), in which he addressed the atmospheric-wave theory, and recognized that his studies of periodic phenomena gave him the confidence he needed for 'this vast scientific crusade'.[113] Stimulated therefore by Herschel's and Birt's investigations, and legitimated by both his international observational campaigns (on which he drew to obtain his data of barometric oscillations) and by the state (through its financial support), he became the authoritative voice in the subject of propagation of atmospheric waves. More will be said about the consequences of his approach for storm warning and forecasting in Chapter 3. First, however, it will be necessary to analyze the strategies of visual communication and nation-building that

---

110 Quetelet (1851, pp. 74–77) explicitly acknowledged Herschel's and Birt's influence.
111 As stated at the BAAS meeting held at Cork in August 1843. See the report by Herschel (1844, p. 80)—quoted by Quetelet (1851, p. 77). In a letter written to Quetelet on 14 September 1843, Birt mentioned Herschel's words in this regard: 'In Europe, Brussels is clearly entitled to be regarded as a point of comparatively gentle barometrical disturbance'. See ARSB, Quetelet Papers, 392, W.R. Birt.
112 For example, Herschel emphasized the importance of examining very carefully barometric fluctuations in a letter addressed to Quetelet as early as 3 December 1841 (ARSB, Quetelet Papers, 1289, J.F.W. Herschel), in which, among other suggestions to Quetelet, he remarked that 'there cannot be a doubt that the longer series offers a much greater chance of following' these fluctuations, 'so as to include the passage of a *complete wave* over the places of observation (a point on which Birt has lately initiated with much force), and [...] one need only look at your projection of the 36 hour series for March 1841'.
113 Quetelet (1851, p. 75). By 1843, Quetelet was fully aware of the huge potentiality of barometric oscillations: 'The study of atmospheric oscillations, undertaken by a large number of observers, [will] give us knowledge of the vastness of atmospheric waves, their average speed of progress, the general sense of their motion, the places where they are formed and disappear, the influence that mountains or certain places can have to change them, [...]' Quetelet (1843, pp. 554–55)—quoted by Quetelet (1851, p. 74).

made it possible to establish the pillars of a dynamical cartography of meteorological phenomena—the subject of the next two sections.

From static to dynamic cartography

In few places were meteorology and statistics as intertwined as in the Netherlands and German states, both methodologically and institutionally. In Western countries, the 1840s represented an opening of opportunities and perspectives for these two sciences due to the growing interest in gathering numerical facts about a state, especially for the benefit of commerce, agriculture, and public health.[114] But also in those places the institutional interweaving between both sciences and Humboldt's long shadow brought with them a mastery of graphic methods that led those countries to set the pace in the study of meteorology by 1850. In fact, the Prussian meteorological institute was established in 1847 as a branch of the state statistical office rather than as an independent entity; and Quetelet simultaneously organized two international conferences—the two being pioneers—on statistics and maritime meteorology in Brussels in 1853.

Moreover, the increasingly influential role nationality began to play in the transformation of European politics—magnified by the 1848 national revolutions in places like Belgium and Germany—led scientists to regard nationality and its identifying elements (climate, population, etc.) as variables worth representing graphically.[115] The best example of the close connection between science and the cartography of the nation was Berghaus's *Physical Atlas* (Gotha, 1845), which contained not only statistical maps on such subjects as weather and oceanography but also ethnographic maps of Europe, with national borders coloured or at times their indoor areas as well.[116] Likewise, the 1846 census on Belgian population, industry, and agriculture, prepared by the Central Commission for Statistics, under the impulse of Quetelet, acted as a lever for promoting statistical cartography and data standardization in early 1850s.[117] This work has been regarded as the first general census of a scientific nature in the history of statistics.

Meteorological cartography noticed these changes at two levels, conceptual and methodological. In the 1830s meteorologists had used isobarometric lines in drawing their maps, but this type of isoline never ceased to be a graphic element of a static nature. For instance, when Kämtz first introduced

---

114 In England, the relation between the weather and the death rate and other vital statistics was the objective for which the statistical office of the Register General requested Greenwich Observatory that any meteorological data be submitted in good and due form.

115 For the flowering in the publication of ethnographic maps in Europe before and after the 1848 revolutions, see Seegel (2012, pp. 134–38).

116 This two-volume work was commissioned by Humboldt as a graphic complement to his colossal work the *Cosmos*.

117 For the importance of this census as a mirror of the development of Belgian agriculture, see Goossens (1993, pp. 48–49).

these lines in 1827, he conceived as lines of equal mean barometric variation describing a regular pattern.[118] In fact, when he mapped these lines of the earth in his *Lehrbuch der Meteorologie* (1832), he envisaged them as parallels of latitude.[119] Or for example, when Berghaus published his *Physikalischer Atlas* in 1839, he drew Kämtz's isolines in the form of straight lines parallel to the equator.[120] Both in Kämtz's works and in Berghaus's *Atlas*, the outcome of their graphic representations was unequivocal: the isobarometric lines were a static image of the mean state of the atmospheric pressure.

In the 1840s, however, the situation changed. On one hand, efforts to define the mean state of the atmosphere were combined with a growing interest in the origin and progress of storms arisen in the late 1830s. On the other hand, the conceptualization of 'atmospheric waves' —proposed by Herschel and Birt in the context of the American storm controversy—introduced a dynamic component in the determination of the average evolution of meteorological phenomena. Both atmospheric waves and their predecessors, Loomis's cold air waves, opened the door to the dynamization of an atmospheric physical space that had theretofore been subject to the Humboldtian pursuit of 'general equilibrium'—i.e. the use of isolines as forms of equilibrium.

The dynamization of atmospheric space became evident, above all, at the methodological level—with Loomis's cartographical innovation as a key step.[121] Unlike Herschel and Birt, who drew on graphic rather than cartographic methods to show atmospheric waves, Loomis made the progress of cold air waves more visible and more easily measurable. In his 1841 paper, he explained that this could be obtained by plotting on a map the lines connecting all those places where the barometric pressure was minimal for every six hours. By measuring spaces and times he was able to determine the velocity and form of the cold wave, which was—as he asserted—far in advance of the centre of the barometric depression, the storm.

Two years later, he took a further step with a comparative study on two storms experienced in the United States in 1842. Now, he was concerned not so much with cold air waves, as with mapping the weather conditions accompanying storms. Yet, unlike Espy who had plotted lines of minimum pressure on his weather charts, Loomis drew lines of equal deviation from normal pressure with the aim of compensating for geographic variations in elevation. He used the term 'mean pressure' rather than 'normal':

I have also determined the barometric and thermometric observations in the following manner: Having determined, as well as I was able, the mean height of the barometer at each station, I compared

118 Kämtz (1827, pp. 168–71).
119 Kämtz (1832, p. 339, and chart III).
120 'Übersicht Der Mittleren Barometerstände Am Meere Und Der Oscillationen Des Luftdrucks', Berghaus (1839).
121 See Chapter 1.

each observation with the mean. I then drew a line passing through all the places, where the barometer stands at its mean height. This line is marked – – 0, and may be called the line of mean pressure. I then drew a line through all the places where the barometer stands two inches above the mean. This line is marked – – – +.2, and so of the others. In like manner [for thermometric observations]. Nearly every circumstance essential to a correct understanding of the phenomena of the storm is thus presented to the eye at a single glance.[122]

Loomis's studies were contemporary with the first investigations conducted by Quetelet at Brussels on barometric periodicity and atmospheric waves. In 1840, Quetelet had seen Herschel's retirement from the program of simultaneous observations as an opportunity to place his Observatory and Belgium within Europe's scientific map. A decade later, when he analyzed all the barometric observations gathered from Europe and Asia from June to August 1841, many questions came to mind: what was the best method for studying their propagation? Should not one represent them cartographically rather than graphically? Indeed, his ideas about the concept and nature of atmospheric waves were not at all original if we pay attention to the sources referred in his book *Sur le Climat de la Belgique* (1851)—he mentioned Herschel, Birt, and Espy as well as Redfield and Dove, though not Loomis. Unlike Loomis, who focused on the cold air waves originated from individual storms, Quetelet placed the emphasis on the progress of the pressure waves propagating permanently on the Earth's surface (as suggested by Birt and Herschel), which were not necessarily at the origin of storms.[123]

Although he introduced no conceptual innovation on the subject, Quetelet did apply a new cartographic method to atmospheric waves.[124] In the maps of Kämtz and Berghaus the isobarometric lines represented the mean states of pressure in the atmosphere. Quetelet reconsidered isobarometric lines as a tool of analysis, redefining their role by identifying them as the crests of atmospheric waves. Or to put this another way, Quetelet was more interested in investigating the propagation of atmospheric waves than in determining the mean state of the atmosphere. Thus in his 1851 seminal book he traced the curves showing the barometric variations in around twenty stations from Brussels to Beijing so that he could determine the date and time of extreme pressure in these places. Then he plotted on a map of Eurasia isolines linking points in which the barometric value was at its maximum for a given data and hour (Figure 2.2).[125] Quetelet's isochronic lines

122 Loomis (1845, p. 164).
123 A testament to the sources from which he drew inspiration is his correspondence with Herschel and Birt; see, e.g., Birt to Quetelet, 14 September 1843, 28 December 1844, 26 October 1849. ARSB, Quetelet Papers, 392, W.R. Birt; and Herschel to Quetelet, 3 December 1841, ARSB, Quetelet Papers, 1289, J.F.W. Herschel.
124 Locher (2007, p. 502; 2008, pp. 24–27).
125 For maps of isochronic pressure lines, see Quetelet (1851, p. 104 *et seq.*)

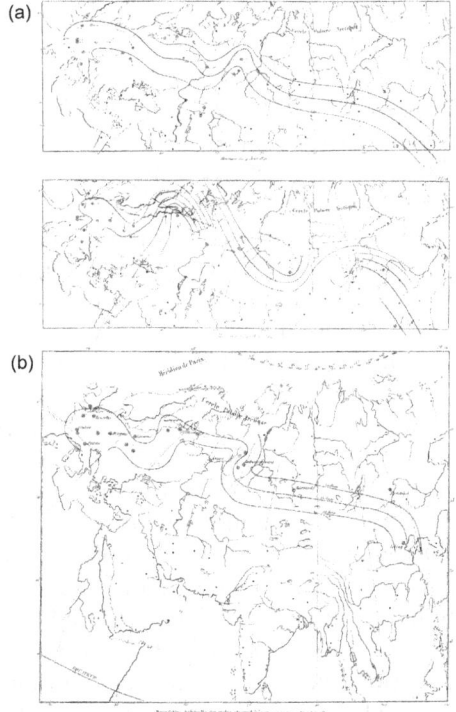

Figure 2.2 Quetelet's isochronic lines of barometric pressure plotted on a map of Eurasia; these isolines link points where the value of barometric pressure is at its maximum for a given day and hour (4 August 1841).
Source: Quetelet (1851, p. 104).

embodied not only the maximum values of pressure but also—and most importantly for our purposes—the crests of atmospheric waves.[126]
Quetelet did not mention Loomis's work on cold air waves, probably because neither of his main sources (Herschel and Birt) mentions it. Nor did

126 Quetelet stresses repeatedly in his *Sur le Climat de la Belgique* (1851, p. 74, 91, 92) that directions of barometric waves are not directly related to wind directions. 'We should not confuse these waves indicated by barometer with the atmospheric currents that exhibit in general wind directions. This distinction is important' (p. 74)—here he mentions Herschel's distinction between 'the advancing form of the fluid' indicated by the barometer and 'the molecular movement' by the wind (Herschel, 1844, p. 99). From a theoretical standpoint, his position is not unexpected. Although Quetelet held Dove in high esteem, he deemed Dove's air currents as a working hypothesis and was studiously vague about the variegated and 'certainly very numerous' causes producing barometric waves: These waves 'would not be only the result of pressures exerted by the air which during the cooling process flows towards the pole due to tropical current; atmospheric oscillations could also have other causes' (p. 90).

he refer to Loomis's lines of equal barometric deviation despite the relative similarity between his and Loomis's ways of displaying the progress of waves through atmospheric space.

In meteorological science, unlike its geomagnetic counterpart, the application of isobarometric maps was conditioned by Humboldtian search for order, resulting in a *static* cartography of climate. Kämtz's and Berghaus's efforts to define the mean state of the atmosphere were limited by Humboldt's ideas about order and the general equilibrium of forces. Again, although precursors in conceptualizing atmospheric waves, Herschel and Birt did not incorporate isoline methods into their investigations. It was Quetelet who introduced isometric technique as a tool of analysis for this type of wave. It is no small irony underlying these two facts. On one hand, Quetelet partly drew on Kämtz's and Berghaus's isobarometric method—in fact, in his charts of pressure lines through Europe and Asia, isochronic lines were practically equidistant and parallel. However, these isochronic lines representing the maximum values of pressure were intended to determine the mean evolution of atmospheric waves rather than the mean state of the atmosphere—which was Kämtz and Berghaus's aim. On the other hand, Quetelet, who had introduced the mathematics of probability not to measure chance but to discover *order* (in the most Humboldtian sense of the word), aimed to detect regularity in nature. These facts led him to show, as a way of visualizing regularities, the *dynamics* of atmospheric waves—an essentially idealist and questionable scientific concept, as defined by Birt as early as 1847.[127] In fact, Quetelet's pressure lines propagated at a constant speed and with no deformation. His idea of natural law as the regularity of fact in accordance with the theory of probabilities only intensified his search.[128]

Science-building and state-building go together

Quetelet showed extraordinary skill to cover fields as varied as meteorology, statistics, and social physics. The theoretical goals that guided his research seem to be well-founded, but it is evident that, at least in the case of observatory sciences, he built a series of scientific institutions that enabled him not only to conduct his investigations but also to drive forward the process of nation-state building in Belgium. This clearly illustrates a pattern that will be repeated later in France, i.e. that the concentration of power, above all in a process of national building, can determine the shape of the objectives and practices of a scientific discipline.

Historian K. Donnelly has analyzed how Quetelet devised a controversial science of man (*physique sociale*) in the 1820 and 1830s.[129] As a part of the

---

127 As is described in Chapter 1. See also Birt (1847, p. 485), Jankovic (1998, pp. 34–35).
128 Locher (2008, p. 27) holds that, with Quetelet, the first steps for a dynamic cartography in meteorology were taken: his 1851 maps 'are as the birth certificate of dynamical meteorology'.
129 Donnelly (2015, esp. pp. 159–65).

United Kingdom of the Netherlands, Belgium lagged behind the European powers, lacked scientific structures typical of a state, like an observatory or a national academy of sciences, and was unable to compete on equal terms in the era of great savants. In this context, Quetelet sought to elevate Belgium to the head of the leading scientific nations by creating a revolutionary discipline (social physics) headed by *l'homme moyen*—a revolution that would have consisted of avoiding the egoistic, 'more egocentric' attitudes of individual genius, which Quetelet regarded as typical of the big countries and of the past, and of stimulating the creation of *average men* in the Queteletian sense of the word (much more collective, committed to data gathering and the networking of observatories). International collaboration and large-scale data collection would redeem Belgium.

Donnelly underlines that part of what drove Quetelet to create *physique sociale* were certain theoretical and practical objectives related to probabilities theory and administrative concerns. But, above all else, what drove him was the standardized data collection that could ensure state patronage and leverage with government policy and practice. It is this practice of data collection and counting what would have enable him to sell numerical data to state bureaucrats—which, in turn, would produce further needs for counting and tabulating, and ultimately more administrators, as a loop. A premise of Donnelly's discussion is that the 'average men' were competent observers and data collectors keeping track the distribution of physical and moral traits, as well as the population. Both state and scientific community were obviously interested in supporting those initiatives that allowed the two to reap the benefits obtained.

The case of observatory sciences in Quetelet seems to have similarities in many respects. The Brussels Observatory and the Royal Academy of Sciences (of which he was responsible for the foundation and reformation, respectively) played a crucial role in his atmospheric research projects, as well as in nation-building. Through these institutions Quetelet conducted his large-scale programs for barometric and phenological observations; at the same time he also envisaged specific strategies to persuade state administrators of the importance of these institutions.[130] Drawing a parallel with

130 The spirit of cooperation developed in Belgium for the search of laws regarding meteorology was in good part a consequence of government's support through the Academy of Sciences and the Brussels Observatory. But not entirely. If in general Belgium did not go so far in the establishment of a national service as those in Holland, Prussia, and France, that was also in part—and paradoxically—because of the lack of government's endorsement. In an appeal entitled *Météorologie et Physique du globe*, published three years after his study on atmospheric waves, Quetelet bemoaned the fact that the first attempt in this regard (the program of simultaneous observations) was abandoned in 1843 due to the lack of governmental support. In the 1840s the Belgian government seems to have supported more another system of observations (that of periodical phenomena regarding plants and animals), made up of a dozen or more Belgian stations, with whom their relations could in some respects be closer (due to their interest in agriculture) than those which they had maintained with the foreign observatories in the above-mentioned program. In the same

Donnelly's analysis, we could state that Quetelet's promotion of large-scale meteorological programs was not only the means whereby he achieved research aims but also proved to be often ends in themselves.[131]

The meteorological programs promoted by Quetelet in the 1840 and 1850s, during the period of nation-building, reflect many traits of Belgium's aims: the pursuit of a mediating and central role among the powers of Europe, investigations directed towards fields that took large-scale collaboration into consideration, promotion of international ventures (such as the 1853 congresses on statistics and maritime meteorology), a conception of intellectual enterprise that valued collectivism rather than individualism (he hoped 'the man of talent' would cease 'to act as an individual'[132]), and a conception of research in which data gathering was deemed more important than theory.

Quetelet's large-scale, state-financed projects found widespread appeal in the field of phenology, to the extent that the way in which he issued instructions and created a network of observers inspired other meteorological services in Europe.[133] In 1842, he first published a set of instruction for periodical phenomena including those related to plants and animal life—a set that would be completed in 1853, with a fifteen-pages leaflet printed by the Belgian Academy.[134] Quetelet's project was developed within the framework of the Royal Academy of Sciences (of which he was the secretary until his death); this symbiosis was crucial for its success, for Quetelet conceptualized plant phenological observations through his instructions at the same time the Academy published his instructions and results and provided a questionnaire to the members of the international phenological network.[135] During the period between 1840 and 1870, this network totalised about eighty stations spread out over Belgium and Europe.[136] In this field Belgium was exemplary, as Quetelet's project found its way into the phenological programs established in stations across Europe, including those from the *Bureau Central Météorologique* in France and the Observatory of Urbino in Italy.[137]

---

appeal, he urged government to establish a centre or regular organization for meteorology and the physics of globe similar to those of the neighbouring countries. See Quetelet (1854, pp. 1–2). For the history of meteorology in Belgium, see Dufour (1951, p. 361).

131 Donnelly (2015, p. 4).

132 In his book *Sciences mathématiques et physiques chez les belges* published in 1866 Quetelet emphasized 'an intellectual innovation of great importance' as a key element for Belgium to have regained its scientific status: 'The man of talent, in certain cases, ceased to act as an individual and became a fraction of the body that attained the most important results'. Quetelet (1866, p. 5), as translated by Donnelly (2015, p. 1).

133 Demarée (2009); Demarée and Chuine (2007, pp. 817–18); Koch et al. (2008, pp. 12–16).

134 Quetelet (1842, 1853b).

135 Phenological observations were published in the *Memoirs* of the Belgian Royal Academy of Sciences and the *Annuaire de l'Observatoire royal de Bruxelles*.

136 Demarée (2009, p. 22).

137 In France, phenological observations started in 1880 under the guidance of Eleuthère Mascart and then Alfred Angot. In Italy, Father Serpieri published plant observations including the dates of flowering for the period 1857 till 1865. See Demarée (2009, p. 23).

Between Quetelet's large-scale programs on meteorological observations and his national initiatives on practical statistics, there was a bridge which seems to have attracted his supporters within government, namely, the idea that leading the international scientific collectivism could elevate Belgium to the ranks of the elite nations. Quetelet was able to materialize this idea in a double encounter that took place in Brussels in 1853, the first General Statistical Congress (organized under his direction),[138] and the first International Maritime Conference (as a host).[139] His participation in the maritime conference was requested by Matthew F. Maury, the US Naval Observatory's director. Maury had been working for several years on mapping sea currents and prevalent winds, and planned a collaborative project aimed at finding the laws governing the sea wave and atmospheric motions for the sake of navigators. To this end, he proposed to collect logbooks from ships of all countries. Every abstract log kept should contain a series of observations.[140] Quetelet supported Maury's project, thus becoming the host of a meeting that will precede its twin sister statistics in the space of one month.[141]

Like Quetelet's kaleidoscopic oeuvre, this double encounter embodied the confluence between statistics as a set of tools emanating from mathematics through probability theory and applying to fields as varied as climate and population, and statistics as a data-gathering enterprise (including ocean and atmospheric data) performed by states.[142] Inasmuch as statistics was a fluctuating and extensive concept that at that time could be equated with good administration, Quetelet acknowledged, without pretence, that the statistical congress was comparable to the maritime conference in aims and tendencies. In the opening address delivered at the Statistical Congress, he pointed out certain similarities:

> Some days ago, Brussels hosted another congress, having the same tendencies, the same object than ours. It also aimed at establishing relations of benevolence between observers from different countries, proposing uniform methods in order to simplify their works and make results comparable. The aim was to study the large currents of the atmosphere and the major seas on the globe; our [aim] is also to study, in another context, the fluctuations in modern societies, as well as their movements and obstacles.[143]

---

138 Randeraad (2011, pp. 53–54).

139 Lewis (1927, pp. 57–59); Williams (1963, pp. 216–22). For the confluence of ideas and differences in the two events, see Donnelly (2017, pp. 68–72).

140 For the use of ships' logbooks in the British and Spanish Navies' maritime tradition, see Naylor (2015) and Anduaga (2017, pp. 254–63), respectively.

141 For a personal account of the Brussels Maritime Conference, see Quetelet (1874). For a comprehensive study on its organization, including the points of conflict and cooperation, see Achbari (2015, pp. 274–80; 2017, pp. 57–64). See also Davis (1984, pp. 362–67).

142 The encounter of two currents in Quetelet's kaleidoscopic work has been noted by Desrosières (2002, pp. 6–7), and Prévost and Beaud (2012, p. 50).

143 *Compte rendu des travaux du congrès général de statistique: Réuni à Bruxelles les 19, 20, 21 et 22 septembre 1853.* Bruxelles: M. Hayez, 1853, p. 23. All French translations are my own unless otherwise noted.

When Quetelet in the same address analyzed the challenges facing statistics in the early 1850s, he underlined three of them: a common language, standardization of measurements, and cooperation between countries. Challenges of statistics, which were identical to atmospheric and marine sciences, thus became ends in themselves. The historical mediating role of Belgium among European powers appears to have prompted Quetelet to offer invited representatives his country's leadership and stability in the pursuit of these challenges. When, two decades later, Quetelet took stock of the progress generated by these meetings, he stated that the first maritime conference succeeded in 'covering the entire globe with a vast network of observers' in such a way that 'the eye of science remains, so to speak, incessantly open on everything that happens on the surface of our planet'.[144] The implementation of the devices required for this achievement—i.e. standardization of procedures, precision of measurements, uniformity in notations, and supranational structures within which these standards could be attained—made the dream of 'the eye of science' come to fruition.[145]

Historians have often highlighted Quetelet's multifaceted profile, as well as his kaleidoscopic view and oeuvre. They also have held that he was an 'actualizer' rather than 'pure creator'.[146] Many of the features of Quetelet's versatility can be identified, as has been discussed here, in his activities during the 1830s, 1840s, and early 1850s: an ability to institutionalize ideas and build intellectual institutions, a leadership to found national and international organizations, an instinct to open prospects and ways in new fields like statistics of society and anthropometry, no inclination to develop original methods or concepts, and a propensity to prioritise data gathering over theory (which, for Quetelet, was often an afterthought).[147] In the case of meteorological science, these features implied not so much the lack of original ideas or concepts, as the opening of new horizons. His meteorological contribution par excellence is the anticipation of important notions that will come to fruition in France and Italy in the very near future: the dynamization of the atmospheric space, his tenacity to apply statistics to the study of meteorology, and his conviction that meteorology-building and state-building went together. Quetelet's greatest legacies were a dynamical cartography and models of transnational cooperation and networking that will serve to articulate projects of storm warnings at the service of nation-building and state leadership.

---

144 Quetelet (1866, p. 23).
145 On the concomitance between the two meetings, see Brian (1989, pp. 122–24), Desrosières (2002, pp. 6–7).
146 Donnelly (2015, 12).
147 Quetelet called 'social physics' to statistics of society. The 'average man' was one of the few original concepts devised by Quetelet.

Conclusion

Although Quetelet's application of statistical methods to meteorology is remarkable, it was not the only achievement prior to the 1850s on this emerging yet thorny theme. For the scientists mentioned in this chapter it was clear that many of the phenomena occurring in the atmosphere had the unpredictable, large-scale whirling character of the air flows or turbulences, and their observations did not follow stable patterns like planetary orbits. Studying the mean state of the atmosphere was the only thing they could do. Humboldt was the first to draw isothermal lines on a map. Next, Kämtz, Berghaus, and Mahlmann applied isarithmic methods to several meteorological phenomena. Dove introduced statistic approaches to non-periodic changes in temperature, based on the concept of the 'normal' and on deviations from mean values. In this framework, Quetelet sought regularity in non-periodic variations by using statistical tools like probabilities theory, with a particular focus on atmospheric waves.

This early history of meteorology provides a good assortment of the cartographic techniques that physicists may develop when tackling the complex evolution of atmospheric phenomena. Most basically, physicists may envisage idealizations and draw isolines in the form of straight lines. This is what Kämtz and Berghaus did with isobarometric lines, obtaining a static image of the mean state of the atmosphere. In the more refined technique devised by Dove, they may try to map monthly isotherms, to delineate areas with the same values of heat anomalies, and to show how temperature patterns changed over extended areas, thus introducing a kinetic element into their cartographic representation. At a still more advanced stage, they may show the dynamics of the waves that propagate through the atmosphere. This is what Loomis and then Quetelet did by identifying isolines with cold air waves and atmospheric waves, respectively.

Following Quetelet's shift of focus from periodic variations to non-periodic ones, we began with his program of simultaneous observations, continued with plant phenological studies, and ended with atmospheric waves. Quetelet jumped from one topic to the next always bearing in mind probability theory and the error law. Most importantly for our purposes, he identified the isochronic lines of atmospheric pressure as the crests of barometric waves, and inaugurated a powerful approach to weather science in which isobarometric geometry would reveal the state of the atmosphere and its impending changes. In this approach, the pressure field became the indicator par excellence of the atmospheric state.

Although the scope of Quetelet's contribution to the science of weather has gone almost unnoticed by historians, its practical importance became apparent during the Crimean War.[148] In the mid-1850s, Quetelet's atmospheric

---

148 Two laudable exceptions to this oversight are Locher (2008, pp. 24–27) and Barboza (2012, pp. 176–78).

waves offered more manageable and controllable solutions to the French and Italian meteorologists interested in storm warnings than mechanistic and convective theories of storms did it. As will be shown in next chapters, Dove's local method to study midlatitude storms in terms of conflict of opposite currents turned out to be inadequate, and Espy's solutions in terms of thermal convection were dismissed. Less known, but no less important, Quetelet's unshakeable belief in international cooperation, scientific leadership, and the orderly and centralized system of observations—in other words, that meteorology-building was closely intertwined with nation-state building—was soon transmitted to Le Verrier in France. Quetelet did not know precisely how to relate his atmospheric waves with physical processes like heat exchange or cloud formation; however, he did provide a scheme in which the field of barometric pressure and fixed coordinates took precedence—that is, an Eulerian framework.

References

Achbari, Azadeh, 2015. 'Building Networks for Science: Conflict and Cooperation in Nineteenth-Century Global Marine Studies'. *Isis*, 106 (2), 257–82.

Achbari, Azadeh, 2017. *Rulers of the Winds. How Academics Came to Dominate the Science of the Weather, 1830–1870*. Amsterdam: Vrije Universiteit.

Anderson, Katharine, 2005. *Predicting the Weather: Victorians and the Science of Meteorology*. Chicago, IL: The University of Chicago Press.

Anduaga, Aitor, 2017. *Cyclones and Earthquakes: The Jesuits, Prediction, Trade and Spanish Dominion in Cuba and the Philippines, 1850–1898*. Quezon City: Ateneo de Manila University Press.

Aubin, David, 2005. 'Astronomical Precision in the Laboratory: The Role of Observatory Techniques in the History of the Physical Sciences'. In Georg Heinrich Borheck, and Klaus Beuermann eds., *Grundsätze über die Anlage neuer Sternwarten mit Beziehung auf die Sternwarte der Universität Göttingen*. Göttingen: Universitätsverlag Göttingen, 31–36.

Aubin, David, 2009. 'Observatory Mathematics in the Nineteenth Century'. In Eleanor Robson, and Jacqueline Stedall eds., *The Oxford Handbook of the History of Mathematics*. Oxford: Oxford University Press, 273–98.

Aubin, David, Bigg, Charlotte, and Sibum, H. Otto eds., 2010. *Observatories and Astronomy in Nineteenth-Century Science and Culture*. Durham, NC: Duke University Press.

Avé-Lallemant, Robert, 1873. 'Alexander von Humboldt. Sojourn in Paris from 1808 to 1826'. In Karl Bruhns ed., *Life of Alexander von Humboldt*. London: Longmans, Green, and Co., 2, 3–74.

Barboza, Christina Helena da Motta, 2012. *As viagens do tempo. Uma história da meteorologia em meados do século XIX*. Rio de Janeiro: E-papers Serviços Editoriais Ltda.

Berghaus, Heinrich, 1839. *Physikalischer Atlas oder Sammlung von Karten*. Gotha: Justus Perthes.

Bernhardt, Karl-Heinz, 2008. 'Heinrich Wilhelm Dove (1803–1879) und seine Stellung in der Geschichte der Berliner Meteorologie'. *Dahlemer Archivgespräche*, 14, 61–100.

Birt, William Radcliffe, 1847. 'On Certain Atmospheric or Barometric Waves which Traversed Europe during November 1842'. *Philosophical Magazine*, 30, 482–502.

Boumans, Marcel, 2015. *Science outside the Laboratory: Measurement in Field Science and Economics*. Oxford: Oxford University Press.

Bourguet, Marie-Noëlle, 1988. *Déchiffrer la France: la statistique départementale à l'époque napoléonienne*. Paris: Éditions des Archives Contemporaines.

Brian, Eric, 1989. 'Observation sur les origines et sur les activités du Congrès international de statistique (1853–1876)'. *Bulletin de l'Institut international de statistique* (47e session), 53, 121–38.

Buys Ballot, Christoph Hendrik Diederik, 1850. 'On the Great Importance of *Deviations* from the Mean State of the Atmosphere for the Science of Meteorology'. *London, Edinburgh and Dublin Philosophical Magazine*, 37 (247), 42–49.

Callens, Stéphane, 1997. *Les Maîtres de l'erreur: mesure et probabilité au XIXe siècle*. Paris: Presses Universitaires de France.

Camerini, Jane R., 1993. 'The Physical Atlas of Heinrich Berghaus: Distribution Maps as Scientific Knowledge'. In Renato G. Mazzolini ed., *Non-Verbal Communication in Science Prior to 1900*. Florence: Leo S. Olschki, 479–512.

Cannon, Susan Faye, 1978. *Science in Culture: The Early Victorian Period*. New York: Science History Publications.

Carter, Christopher, 2004. 'Herschel, Humboldt, and Imperial Science'. In Raymond Erickson, Mauricio A. Font, and Brian Schwartz eds., *Alexander von Humboldt. From the Americas to the Cosmos: An International Interdisciplinary Conference, October 14–16, 2004*. New York: The City University of New York, Bildner Center for Western Hemisphere Studies, 509–18.

Cassidy, David C., 1985. 'Meteorology in Mannheim: The Palatine Meteorological Society, 1780–1795'. *Sudhoffs Archiv*, 69, 8–25.

Cawood, John, 1977. 'Terrestrial Magnetism and the Development of International Collaboration in the Early Nineteenth Century'. *Annals of science*, 34, 551–87.

Cawood, John, 1979. 'The Magnetic Crusade: Science and Politics in Early Victorian England'. *Isis*, 70, 493–518.

Chapman, Allan, 1985. 'Sir George Airy (1801–1892) and the Concept of International Standards in Science, Timekeeping and Navigation'. *Vistas in astronomy*, 28, 321–28.

Chromow, Swoboda P., 1930. '"Dynamische Klimatologie" und Dove'. *Zeitschrift fur angewandte Meteorologie*, 48 (10), 312–14.

Courgeau, Daniel, 2012. *Probability and Social Science: Methodological Relationships between the two Approaches*. New York: Springer.

Daston, Lorraine, 1988. *Classical Probability in the Enlightenment*. Princeton, NJ: Princeton University Press.

Daston, Lorraine, and Galison, Peter, 2007. *Objectivity*. Brooklyn, NY: Zone Books.

Davies, Trevor D., 2000. 'Meteorological Observation Networks in the Eighteenth Century, Especially that of the Societas Meteorologica Palatina to which Toaldo Contributed'. In Luisa Pigatto ed., *Giuseppe Toaldo e il suo tempo nel bicentenario della morte: scienze e lumi tra Veneto e Europa*. Cittadella: Bertoncello Artigrafiche, 964–72.

Davis, John L., 1984. 'Weather Forecasting and the Development of Meteorological Theory at the Paris Observatory, 1853–1878'. *Annals of Science*, 41 (4), 359–82.

Delange, Yves, 1997. 'Les phénomènes de l'atmosphère et la météorologie de Lamarck'. In Goulven Laurent ed., *Jean-Baptiste Lamarck 1744–1829*. Paris: Comité des Travaux Historiques et Scientifiques, 123–36.

Delmas, Bernard, 2004. 'Pierre-François Verhulst et la loi logistique de la population'. *Mathématiques et sciences humaines / Mathematics and Social Sciences*, 42 (167), 27–58.

Demarée, Gaston R., 2009. 'The Phenological Observations and Networking of Adolphe Quetelet at the Royal Observatory of Brussels'. *Italian Journal of Agrometeorology*, 1, 22–24.

Demarée, Gaston R., and Chuine, Isabelle, 2007. 'A concise History of the Phenological Observations at the Royal Meteorological Institute of Belgium'. In *Proceedings of the HAICTA2006 Conference, Volos, Greece, 20–23 September 2006*, 3, 815–24.

Demarée, Gaston R., and Rutishauser, This, 2011. 'From "Periodical Observations" to "Anthrochronology" and "Phenology"—The Scientific Debate between Adolphe Quetelet and Charles Morren on the Origin of the Word "Phenology"'. *International Journal of Biometeorology*, 55, 753–61.

Desrosières, Alain, 1993. *La politique des grands nombres – Histoire de la raison statistique*. Paris: La Découverte. English translation by Camille Naish as: *The Politics of Large Numbers. A History of Statistical Reasoning*. Cambridge, MA: Harvard University Press, 1992.

Desrosières, Alain, 1994. 'Le territoire et la localité. Deux langages statistiques'. *Revue de synthèse*, 7 (25), 46–58.

Desrosières, Alain, 1997. 'Quetelet et la sociologie quantitative: du piédestal à l'oubli'. In Jean-Jacques Droesbeke ed., *Quetelet Actualité et universalité de la pensée scientifique d'Adolphe Quetelet*. Bruxelles: Académie royale de Belgique, 179–98.

Desrosières, Alain, 2002. 'Adolphe Quetelet'. *Courrier des statistiques*, 104, 3–8.

Dettelbach, Michael, 1996. 'Humboldtian Science'. In Nicholas Jardine, James A. Secord, and Emma C. Spary eds., *Cultures of Natural History*. Cambridge: Cambridge University Press, 287–304.

Donnelly, Kevin, 2015. *Adolphe Quetelet, Social Physics and the Average Men of Science, 1796–1874*. London: Routledge.

Donnelly, Kevin, 2017. 'Redeeming Belgian Science: Periodic Phenomena and Global Physics in Brussels, 1825–1853'. *History of Meteorology*, 8, 54–73.

Dörries, Matthias, 1994. 'La standardisation de la balance de torsion dans les projets européens sur le magnétisme terrestre'. In Christine Blondel, and Matthias Dörries eds., *Restaging Coulomb: Usages, controversies et réplications autour de la balance de torsion*. Firenze: Leo S. Olschki, 121–49.

Dove, Heinrich Wilhelm, 1838. 'Über die nicht periodischen Änderungen der Temperaturverteilung auf der Oberfläche der Erde'. *Abhandlungen der Königlich Preussische Akademie der Wissenschaften zu Berlin, A. Physikalische Abhandlungen*, 285–415.

Dove, Heinrich Wilhelm, 1840–47. *Über die nicht periodischen Änderungen der Temperaturverteilung auf der Oberfläche der Erde*. Berlin: Sanders, Part 1 (1840), P. 2 (1841), P. 3 (Reimer, 1844), P. 4 (Reimer, 1847).

Dove, Heinrich Wilhelm, 1848a. 'Über die Isothermen in Januar und July'. *Bericht über die zur Bekanntmachung geeigneten Verhandlungen der Königlich Preussische Akademie der Wissenschaften zu Berlin*, 389–408.

Dove, Heinrich Wilhelm, 1848b. 'Temperature Tables'. In *Report of the Seventeenth Meeting of the British Association for the Advancement of Science; held at Oxford in June 1847*. London: John Murray, 373–465.

Dove, Heinrich Wilhelm, 1851. 'Über thermische Anomalien (Linien gleicher thermischer Abweichung)'. *Monatsberichte der Preußischen Akademie der Wissenschaften* (Berlin), 619–26.

Dove, Heinrich Wilhelm, 1852. *Die Verbreitung der Warme auf der Oberflüche der Erde erlautert durch Isothermen, Thermische Isanomalen und Temperatur curven.* Berlin: Dietrich Reimer.

Dufour, Louis, 1948. 'Quelques considérations sur l'æuvre météorologique de A. Quetelet (1796–1874)'. *Ciel et Terre*, 64, 58–71.

Dufour, Louis, 1951. 'Sketch History of Meteorology in Belgium'. *Weather*, 6 (12), 359–64.

Elkhadem, Hossam, 1975. 'La Correspondance d'Adolphe Quetelet avec sir John Herschel: un exemple de la richesse du fond Quetelet'. In Académie Royale des Sciences, des Lettres et des Beaux-Arts de Belgique eds., *Adolphe Quetelet 1796–1874: Hommages et contributions.* Bruxelles: Académie Royale de Belgique, 46–58.

Feldman, Theodore Sherman, 1990. 'Late Enlightenment Meteorology'. In Tori Frängsmyr, John L. Heilbron, and Robin E. Ryder eds., *The Quantifying Spirit in the Eighteenth Century.* Berkeley: University of California Press, 143–78.

Funkhouser, H. Gray, 1937. 'Historical Development of the Graphical Representation of Statistical Data'. *Osiris*, 3, 269–404.

Good, Gregory A., 2006. 'A Shift of View: Meteorology in John Herschel's Terrestrial Physics'. In James Rodger Fleming, Vladimir Jankovic, and Deborah R. Coen eds., *Intimate Universality. Local and Global Themes in the History of Weather and Climate.* Sagamore Beach, MA: Watson Publishing International, 35–67.

Goossens, Martine, 1993. *The Economic Development of Belgian Agriculture: A Regional Perspective, 1812–1846.* Leuven: Leuven University Press.

Hacking, Ian, 1990. *The Taming of Chance.* Cambridge: Cambridge University Press.

Halbwachs, Maurice, 1913. *La théorie de l'homme moyen. Essai sur Quetelet et la vstatistique morale,* Paris: Alcan.

Hankins, Frank Hamilton, 1908. *Adolphe Quetelet as Statistician.* New York: Columbia University.

Harnack, Adolf, 1900. *Geschichte der Königlich Preußischen Akademie der Wissenschaften zu Berlin, Bd. 3.* Berlin, 55–61.

Herschel, John Frederick William, 1835. *Instructions for Making and Registering Meteorological Observations in Southern Africa and other Countries in the South Seas, as also at Sea.* London: Bradbury and Evans.

Herschel, John Frederick William, 1844. 'Report on the Reduction of Meteorological Observations'. In *Report of the British Association for the Advancement of Science, Held at Cork in August 1843.* London: John Murray, 13, 60–100.

Herschel, John Frederick William, 1846. 'Address'. In *Report of the Fifteenth Meeting of the British Association for the Advancement of Science.* London: John Murray, xxvii–xliv.

Horn, Werner, 1959. 'Die Geschichte der Isarithmenkarten'. *Petermanns Geographische Mitteilungen*, 103, 225–32.

Humboldt, Alexandre de, 1817a. 'Des lignes isothermes et de la distribution de la Chaleur sur le globe'. *Mémoires de physique et de chimie*, 3, 462–602. Full English translation, 'On Isothermal Lines and The Distribution of Heat over the Globe'. *Edinburgh Philosophical Journal*, 1820, 3, 1–19; 256–74; 1820–21, 4, 23–27, 262–81; 1821, 5, 28–39.

Humboldt, Alexandre de, 1817b. 'Sur les lignes isothermes. Par A. de Humboldt. (Extrait)'. *Annales de chimie et de physique*, 5, 102–11.

Humboldt, Alexandre de, 1818. 'De l'influence de la déclinaison du Soleil sur le commencement des pluies équatoriales'. *Annales de chimie et de physique*, 8, 179–90.

Humboldt, Alexandre de, 1823. 'Sur les lois que l'on observe dans le distribution des formes végétales'. In Frédéric Cuvier ed., *Dictionnaire des sciences naturelles*, vol. XVIII. Strasbourg: Levrault et Schoell, 431.

Jankovic, Vladimir, 1998. 'Ideological Crests versus Empirical Troughs: John Herschel's and William Radcliffe Birt's Research on Atmospheric Waves, 1843–50'. *British Journal for the History of Science*, 31, 21–40.

Jungnickel, Christa, and McCormmach, Russell, 1990. *Intellectual Mastery of Nature: Theoretical Physics from Ohm to Einstein. The Torch of Mathematics, 1800–1870*, vol. 1. Chicago, IL: The University of Chicago Press.

Kämtz, Ludwig Friedrich, 1827. 'Vorläufige Notiz über Schwankungen des Barometers'. *Jahrbuch der Chemie und Physik*, 3, 168–71.

Kämtz, Ludwig Friedrich, 1832. *Lehrbuch der Meteorologie*, vol. 2. Halle: Gebauer.

Kämtz, Ludwig Friedrich, 1843. *Cours complet de météorologie*. Paris: Paulin.

Kämtz, Ludwig Friedrich, 1845. *A Complete Course of Meteorology*. London: H. Bailliere.

Koch, E., Demarée, G., Lipa, W., Zach, S., and Zimmermann, K., 2008. 'History of International Phenological Networks'. In Jiri Nekovář ed., *The History and Current Status of Plant Phenology in Europe. COST Action 725, COST Office*. Muhos: Finnish Forest Research Institute, 12–16.

Konvitz, Joseph W., 1987. *Cartography in France, 1660–1848. Science, Engineering and Statecraft*. Chicago, IL: Chicago University Press.

Lalanne, Léon, 1843. 'Sur la représentation graphique des tableaux météorologiques et des lois naturelles en général'. In Ludwig Friedrich Kämtz ed., *Cours complet de météorologie*. Paris: Paulin, appendix 1, 1–35.

Lalanne, Léon, 1845. 'Remarques à l'occasion du Mémoire de M. Morlet sur les centres de figures; et réflexions sur la représentation graphique de divers éléments relatifs à la population'. *Comptes rendus des séances de l'Académie des Sciences*, 20, 438–41.

Lamont, Johann von, 1839. 'Nachricht über die meteorologische Bestimmung des Königreiches Bayern'. *Jahrbuch der Königlichen Sternwarte bei München für 1839*, 256–64, 247–49.

Lamont, Johann von, 1867. 'Über die Bedeutung arithmetischer Mittelwerthe in der Meteorologie'. *Zeitschrift der Österreichischen Gesellschaft für Meteorologie*, 2 (11), 241–47.

Lamy, Jérôme, 2007. *L'observatoire de Toulouse aux XVIIIe et XIXe siècles: archéologie d'un espace savant*. Rennes: Presses universitaires de Rennes.

Landsberg, Helmut E., 1955. 'Weather "Normals" and Normal Weather'. *Weekly Weather and Crop Bulletin*, January 31, 1955, 7–8.

Landsberg, Helmut E., 1980. 'A Bicentenary of International Meteorological Observations'. *WMO Bulletin*, 29 (4), 235–38.

Laplace, Pierre Simon, 1814. *Essai philosophique sur les probabilités*. Paris: Mme. V^{e}. Courcier.

Lebrun, Marc, 1974. *Adolphe Quetelet: L'oeuvre sociologique et démographique. Choix des texts*. Brussels: Centre d'étude de la population et de la famille.

Leighly, John, 1949. 'Climatology since the Year 1800'. *Transactions, American Geophysical Union*, 30 (5), 658–72.

Lewis, Charles Lee, 1927. *Matthew Fontaine Maury. The Pathfinder of the Seas.* Annapolis, MD: The United States Naval Institute.

Locher, Fabien, 2007. 'The Observatory, the Land-Based Ship and the Crusades: Earth Sciences in the European Context, 1830–50'. *British Journal for the History of Science*, 40 (4), 491–504.

Locher, Fabien, 2008. *Le savant et la tempête. Étudier l'atmosphère et prévoir le temps au XIX siècle.* Rennes: Presses Universitaires de Rennes.

Loomis, Elias, 1845. 'On Two Storms Which Were Experienced throughout the United States in the Month of February, 1842'. *Transactions of the American Philosophical Society*, 9 (2), 161–84.

Lottin, Joseph, 1912. *Quetelet, statisticien et sociologue.* Louvain: Institut supérieur de philosophie.

Lüdecke, Cornelia, 2014. '… zur Erhaltung der nöthigen Gleichförmigkeit: Das weitsichtige meteorologische Netzwerk der Societas Meteorologica Palatina (1781–1892)'. In Ingrid Kästner, and Jürgen Kiefer eds., *Von Kometen, Windhosen, Hagelschlag und Wetterballons. Beiträge zur Geschichte der Meteorologie.* Aachen: Shaker, 8, 123–30.

Mahlmann, Wilhelm, 1841. 'Mittlere Verteilung der Warme auf der Edorberflache'. In Heinrich Wilhelm Dove ed., *Repertorium der Physik: enthaltend eine vollständige Zusammenstellung der neuern Fortschritte dieser Wissenschaft.* Berlin: Veit, 1–174.

Mailly, Nicolas Édouard, 1875. *Essai sur la vie et les ouvrages de L.-A.-J. Quetelet.* Bruxelles: F. Hayez.

Monmonier, Mark, 1999. *Air Apparent. How Meteorologists Learned to Map, Predict, and Dramatize Weather.* Chicago, IL: The University of Chicago Press.

Morrell, Jack, and Thackray, Arnold, 1981. *Gentlemen of Science: Early Years of the British Association for the Advancement of Science.* New York: Clarendon Press.

Muncke, Georg Wilhelm et al. 1827. *Johann Samuel Traugott Gehler's Physikalisches Wörterbuch.* Leipzig: bei E.B. Schwickert, 3.

Naylor, Simon, 2015. 'Log Books and the Law of Storms: Maritime Meteorology and the British Admiralty in the Nineteenth Century'. *Isis*, 106 (4), 771–97.

Palsky, Gilles, 1996. *Des chiffres et des cartes: Naissance et développement de la cartographie quantitative française au XIXe siècle.* Paris: Comité des Travaux Historiques et Scientifiques.

Palsky, Gilles, 1998. 'Origines et évolution de la Cartographie Thématique (XVIIe–XIXe siècles)'. *Revista da Faculdade de Letras–Geografia* (Porto), 14, 39–60.

Porter, Theodore M., 1986. *The Rise of Statistical Thinking 1820–1900.* Princeton, NJ: Princeton University Press.

Porter, Theodore M., 1995. *Trust in Numbers: The Pursuit of Objectivity in Science and Objective Life.* Princeton, NJ: Princeton University Press.

Prévost, Jean-Guy, and Beaud, Jean-Pierre, 2012. *Statistics, Public Debate and the State, 1800–1945: A Social, Political and Intellectual History of Numbers.* London: Pickering & Chatto.

Quetelet, Adolphe, 1835. *Sur l'Homme et le Développement de ses Facultés ou Essai de Physique Sociale*, vol. 1. Paris: Bachelier.

Quetelet, Adolphe, 1842. 'Instructions pour l'observation des phénomènes périodiques'. *Bulletins de l'Académie Royale des Sciences et Belles-Lettres de Bruxelles*, 9, 65–95.

Quetelet, Adolphe, 1843. 'Rapport sur les travaux de l'Académie royale des sciences et belles-lettres de Bruxelles, pendant l'année 1842–1843'. *Bulletins de l'Académie royale de Bruxelles*, 10 (2), 551–61.

Quetelet, Adolphe, 1849. *Letters Addressed to the Grand Duke of Saxe Coburg and Gotha on the Theory of Probability as Applied to the Moral and Political Sciences*. London: Charles and Edwin Layton. Translated by O. G. Downes (original 1846).

Quetelet, Adolphe, 1851. *Sur le climat de la Belgique. Quatrième partie. Pressions et ondes atmosphériques*. Bruxelles: M. Hayez.

Quetelet, Adolphe, 1853a. *Mémoire sur les variations périodiques et non périodiques de la température d'après les observations faites, pendant vingt ans, à l'Observatoire Royale de Bruxelles*. Bruxelles: Hayez.

Quetelet, Adolphe, 1853b. *Instructions pour l'observation des phénomènes périodiques*. Bruxelles: Académie royale des Sciences, des Lettres et des Beaux-Arts.

Quetelet, Adolphe, 1854. *Météorologie et Physique du globe*. [S.l.], [ca. 1854].

Quetelet, Adolphe, 1866. *Sciences mathématiques et physiques chez les belges, au commencement du XIXᵉ siècle*. Bruxelles: H. Thiryvan Buggenhoudt.

Quetelet, Adolphe, 1874. *Notice sur le Capitaine M.F. Maury, Associé de l'Académie Royale de Belgique*. Bruxelles: Académie royale des sciences, des lettres et des beaux-arts de Belgique.

Randeraad, Nico, 2011. 'The International Statistical Congress (1853–1876): Knowledge Transfers and their Limits'. *European History Quarterly*, 41 (1), 50–65.

Robinson, Arthur H., 1971. 'The Genealogy of the Isopleth'. *Cartographical Innovations*, 8, 49–53.

Robinson, Arthur H., 1982. *Early Thematic Mapping in the History of Cartography*. Chicago, IL: The University of Chicago Press.

Robinson, Arthur H., and Wallis, Helen, 1967. 'Humboldt's Map of Isothermal Lines: A Milestone in the History of Thematic Cartography'. *Cartographic Journal*, 2, 119–23.

Robinson, Arthur H., and Wallis, Helen, 1982. *Cartographical Innovations: An International Handbook of Mapping Terms to 1900*. Tring: Map Collector Publications.

Sachs, Aaron, 2007. *The Humboldt Current: A European Explorer and His American Disciples*. Oxford: Oxford University Press.

Schaffer, Simon, 1988. 'Astronomers Mark Time: Discipline and the Personal Equation'. *Science in Context*, 2, 115–46.

Schneider-Carius, Karl, 1975. *Weather Science, Weather Research: History of Their Problems and Findings from Documents during Three Thousand Years*. New Delhi: Indian National Scientific Documentation Centre; originally published as *Wetterkunde, Wetterforschung: Geschichte ihrer Probleme und Erkenntnisse in Dokumenten aus drei Jahrtausenden*. Freiburg: Verlag Karl Albert, 1955.

Scultetus, Hans Robert, 1943. 'Dove und Loomis als Wegbereiter der Synoptik'. *Meteorologische Zeitschrift*, 60, 419–22.

Seegel, Steven, 2012. *Mapping Europe's Borderlands: Russian Cartography in the Age of Empire*. Chicago, IL: The University of Chicago Press.

Sheynin, Oscar Boris, 1984a. 'On the History of the Statistical Method in Astronomy'. *Archive for History of Exact Sciences*, 29, 151–99.

Sheynin, Oscar Boris, 1984b. 'On the History of the Statistical Method in Meteorology'. *Archive for History of Exact Sciences*, 31, 53–95.

Sheynin, Oscar Boris, 1986. 'A. Quetelet as a Statistician'. *Archive for History of Exact Sciences*, 36, 281–325.

Sheynin, Oscar Boris, 1993. 'On the History of the Principle of Least Squares'. *Archive for History of Exact Sciences*, 46, 39–54.

Smith, Robert W., 1991. 'A National Observatory Transformed: Greenwich in the 19th Century'. *Journal for the History of Astronomy*, 45, 5–20.

Stigler, Stephen M., 1986. *The History of Statistics: The Measurement of Uncertainty before 1900*. Cambridge, MA: Harvard University Press.

Stigler, Stephen M., 1997. 'Adolphe Quetelet: Statistician, Scientist, Builder of Intellectual Institutions'. In Académie Royale des Sciences, des Lettres et des Beaux-Arts de Belgique eds., *Quetelet Actualité et universalité de la pensée scientifique d'Adolphe Quetelet*. Bruxelles: Académie royale de Belgique, 47–61.

Stigler, Stephen M., 1999. *Statistics on the Table: The History of Statistical Concepts and Methods*. Cambridge, MA: Harvard University Press.

Williams, Frances Leigh, 1963. *Matthew Fontaine Maury, Scientist of the Sea*. New Brunswick, NJ: Rutgers University Press.

Wise, M. Norton, 2010. 'What's in a Line?' In Moritz Epple, and Claus Zittel eds., *Science as Cultural Practice. Volume I: Cultures and Politics of Research from the Early Modern Period to the Age of Extremes*. Berlin: Akademie Verlag GmbH, 61–102.

Wolfschmidt, Gudrun, 2005. *Vom Magnetismus zur Elektrodynamik: herausgegeben anlässlich des 200. Geburtstags von Wilhelm Weber (1804–1891) und des 150. Todestages von Carl Friedrich Gauss (1777–1855)*. Hamburg: Schwerpunkt Geschichte der Naturwissenschaften, Mathematik und Technik.

Zogner, Lothar, 1983. 'Carl Ritter und Alexander von Humboldt: Ihr Beitrag zur Entwicklung der thematischen Kartographie'. In Wolfgang Eriksen, and Wilhelm Lauer eds., *Studia geographica: Festschrift Wilhelm Lauer zum 60. Geburtstag*. Bonn: Dümmler, 16, 394–406.

3   The adoption of an Eulerian approach by state interests

Introduction

In the mid-1850s, Quetelet's atmospheric waves occupied the forefront of meteorological cartography. In this limited field, the deep nature of the physical processes, or the precise relation between these waves and heat exchange, did not need to be known. Quetelet's French and Italian followers could confine themselves to the pressure field phenomenology favoured by the development of synoptic weather maps. They apparently did not expect the atmospheric structure to play a role in storm motion, in daily weather patterns, and in local and small-scale atmospheric phenomena. However, no one could ignore the theory on the opposite polar and tropical currents no matter how unpopular this air-mass theory had become in some scientific circles. Dove had championed this theory in Prussia. Yet, his French and Italian colleagues, who favoured a synoptic approach to meteorology, dispensed with structural considerations.

In the 1850s and 1860s, there were two conceptions of how air motion should be studied in meteorology. According to the so-called Eulerian approach, one should focus on air moving past fixed observation points. From a bird's eye, one might follow the patterns of air motion as recorded at all observation points. These patterns were then projected on a synoptic map. The only structure model used was the pressure field at 0-level and its supposed correlation with the weather. This approach was adopted by most French and Italian weather scientists.

Some British and German scientists took a different frame of reference: the Lagrangian approach. Here, one should follow an individual air parcel as it moved through space and time. By plotting its position through time, one obtained its pathway. Thus, the object of analysis was the trajectory—not the map. Unlike the Eulerian one-sided model, the Lagrangian structure models were based on the air-trajectory or the air-mass concept and usually depended more on non-routine measurements and mathematical analysis.

The Eulerian approach to the study of weather was easier to handle graphically and mathematically than the Lagrangian one. Moreover, its results were easily tractable in comparative studies involving international observations. These differences were confronted in the 1850s and 1860s. During these two decades, weather scientists managed to resolve the ensuing tensions both in Britain and on the continent. This took place mainly in two different contexts: the establishment of storm warning services in France and Italy, and the local system of weather forecasts in Britain. The present chapter deals with the former context, the next one the latter context.[1]

This chapter describes the adoption of the Eulerian approach in France and Italy, showing how the synoptic charts evolved into authoritative tools capable of revealing the storm motion and weather patterns, but also serving as a scientific means to achieve order and centralism in state-building. It begins with a detailed examination of the political and imperial context of France under Napoleon III, showing how the 1854 Black Sea storm was the catalyst that led the Paris Observatory's physicists to adopt this approach in the interest of security, order, and leadership. This adoption, I contend, resulted not from a conscientious deliberation derived from some mechanical or hydrodynamic theory, but rather from an accommodation of scientific thought to the state's interests. The next part of the chapter focuses on the establishment of a storm warning service in Pontifical Rome in the 1850s, which was also conceived on the basis of the Eulerian approach. An evolution in the forms of representation took place here: one that evolved from a graphic method aimed at studying atmospheric waves to a cartographic method aimed at storm warning. The last part describes how H.E. Marié-Davy developed a synoptic method of weather forecasting in 1863, how 'areas of low pressure' or depressions became the object of study, and how the French Navy Ministry meant the Lagrangian counterpoint in this confrontation of approaches.

Napoleon III: Empire, security, and order

In few countries were the revolutions of 1848 as momentous as in France. In most of Europe, the years following these revolts were of hunger and rebellion; urban discontent brought with it nationalist upsurges and the self-assertion of a new social class—the industrial proletariat. But in France, in addition to this, they gave rise to a short-lived republican government that soon resulted in an imperialist regime after the coup d'état by Louis-Napoleon Bonaparte in December 1851. The new emperor, Napoleon III,

---

1 The lack of American leadership in this matter is in good part a consequence of the American Civil War (1861–1865), which virtually paralyzed all efforts and initiatives. But not entirely. If in general they did not participate so fully in this confrontation of approaches as their European colleagues, that was also in part because they had other grounds to confront. The famous 'storm controversy' between Redfield and Espy polarized the scientific debate, in which the theory of atmospheric waves played a minor role. It was not until 1871 that the U.S. War Department officially justified the use of synoptic weather maps.

defended an imperial expansion that sharpened patriotic identity within France.[2] He quickly replaced the constitution of the Second Republic (of which he was president from 1848) with an authoritarian regime that sought to strengthen national security, social order, technical progress, and above all, the power of the Empire.[3]

Three aims were Napoleon III's strategic priorities for France: Empire, security, and order. In the imperial domain Napoleon III became the protector of Roman Catholic values. On one hand, he led an aggressive policy of missionary evangelization that embroiled France in wars with the kingdom of Vietnam. On the other hand, he played a similar protective role in the threat of the Russian Empire over the holy places of Palestine, then in the hands of Turks, which Tsar Nicholas I wanted to recover for the Orthodox faith. This fact was used as a pretext for tsarist troops to begin the expansion towards the Mediterranean, triggering the Crimean War of 1853 between the Russian Empire and the allied side, France and Britain.[4] By October 1855 Russia was vanquished. All parties at war signed the Treaty of Paris of 25 February 1856, whereby Russia lost its possessions in Bessarabia and the Danube while Britain safeguarded its interests in the Mediterranean and the Middle East. France re-emerged as the hegemonic power in continental Europe. It broke the old international order established in Vienna in 1815, which prevented France to reassert her authority in Europe.[5]

Socially, Napoleon III championed the return to order in France. During the Second Republic revolutionary forces fought against social privileges in the name of egalitarianism. Privileges and hierarchies were two of the battlegrounds on which republicans established the Constituent Assembly. Napoleon III believed France would restore her legitimate social order through an *élite de citoyens* able to monopolize political decision-making posts and social prerogatives. An example of this elite was the Senate, made up by the most loyal of his praetorian guard: cardinals, generals, and savants, among them astronomer Urbain Le Verrier. The return to order signified much more than a social aspiration: it was a return to political elite and hierarchical order—the pillars of Napoleonic counter-revolution.[6]

Notable changes affected security and safety as well. In 1853 the delegates in the Brussels Maritime Conference had urged governments to establish a uniform system of meteorological observations that would be useful for

---

2 Napoleon III was confirmed by a plebiscite in November 1852.
3 On the 1848 revolutions and reforms in Europe, and in France in particular, see Lévêque (2008, pp. 91–119), and the contributions to the work edited by Dowe, Haupt, Langewiesche, and Sperber (2008). On Louis-Napoleon's coup d'état and its political consequences, see Agulhon (1992, pp. 245–90).
4 On the dispute over the Holy Places as an excuse for the Crimean War, see Badem (2010, pp. 65–68).
5 Goldfrank (2014, pp. 102–04, 237–40, 293–96); Royle (2000).
6 Granger (2005, pp. 37–48).

marine safety, trade, and fishery. Within a year, this meeting had prompted the establishment of the Board of Trade's Meteorological Department in Britain and the *Koninklijk Nederlands Meteorologisch Instituut* in Holland. However, France's response was late, partly because here, as will be shown below, an astronomer under the Ministry of Public Instruction rather than a naval officer bore the responsibility for creating such a service.[7] It was not until 1859 that the *Dépôt de la Marine* (the French hydrographical centre) followed Brussels guidelines, ordering warships to collect data with the instruments supplied by the Navy and inviting all captains of merchant vessels to cooperate in the enterprise.[8] In this field, Britain was the mirror in which France looked at herself—in fact, from 1856 the Dépôt translated the Board of Trade wind charts.[9]

In this regard, telegraphy became a powerful security and safety tool with Britain as a benchmark. The first great World's Fair, held at the Crystal Palace in London in 1851, was a showcase for British inventiveness and telegraph industry. An example of this was the daily weather map published during the Fair, with weather reports received by telegraph. All this was a source of inspiration for France.[10]

In a context of colonial expansion, with the growing eagerness to communicate with Algeria (the richest of the French colonies, together with Indochina), and the demands from the French Army, in need of swift communications and preoccupied with the security of transmission, Napoleon III opted for a compromise: while the State held the monopoly of internal telegraphic communications and even extended it to Algeria, it should engage British telegraph engineers and companies because of the lack of a French cable industry. In light of the above, in 1853 the French government hired John Watkins Brett—the engineer who had laid the Channel cable in 1850—to connect France and Algeria via Corsica and Sardinia by a submarine cable.[11] At the same time it established a dense electric telegraph

---

7 Although in Holland this responsibility remained with a professor of mathematics (C.H.D. Buys Ballot), his close relationship with the Dutch naval officer Marin Henri Jansen made the *Instituut*'s priority to be geared towards maritime meteorology. See Lunteren (1998, pp. 232–39); Achbari (2015, pp. 262–66; 2017 pp. 21–68). By contrast, in France this responsibility fell upon an astronomer, under the Ministry of Public Instruction, unfamiliar with maritime matters.

8 On the meteorological activities at the *Dépôt des cartes et plans de la Marine*, see Rollet de l'Isle (1950), Locher (2008, pp. 148–51).

9 The Board of Trade wind charts and Maury's *Sailing directions* (translated into French by Édouard P. Vancechout in 1859 as *Instructions nautiques destinées à accompagner les cartes des vents et des courants*) became mandatory for French warships. On the Brussels Conference and the French Navy, see Locher (2008, pp. 151–54).

10 This telegraphic reporting scheme was devised by astronomer James Glaisher, who had organized a similar weather reporting service at the Greenwich Observatory in 1849. See Marriott (1904, pp. 12–14). See also Locher (2008, pp. 33–35; 2009, pp. 78–79).

11 The laying of communication cables between France and Algeria encountered serious technical difficulties, including cable breakage, and the operation was not performed until 1870.

network in Algeria that began with 249 kilometres of lines in 1854 and reached 3,179 kilometres by 1861.[12] Until 1860, in which a separate Post Office was expressly set up, the telegraph system was operated by the Army. Thus, the French government both ensured security in intra-imperial communications and trusted British technology for their colonial expansion.[13]

A meteorological service in Algeria

The conjunction of Napoleon III's three strategic priorities—Empire, order, and security—was evidenced in Algeria, particularly in the colonial meteorological service projected in 1853, a few months before the Crimean War.

Napoleon III's rise to power had a profound impact on Algeria's political and social evolution. Threatened by local rebellions and subject to an arduous colonization, the French colony soon became what the Emperor called 'a ball and chain attached to the feet of France'.[14] The elimination from the political arena of all the generals of Africa of the previous regime and the removal of Algeria's representatives from the Legislative Assembly helped strengthen the authority of the Emperor, and create an illusion of laboratory for experimentation and progress that could quell French settlers' discontent. Farmlands ostensibly breathed life into the colonies they purported to civilize, theoretically providing the possibility of undertaking a popular colonization. Or, as colonial authorities dreamed it, farmlands were expected to make Algeria the new Egypt.[15] In the 1850s the distribution of cultivable lands became the principle means for Algeria's popular colonization—especially in cotton and wheat plantations. In just ten years the French government distributed 250,000 hectares, with which over 15,000 new colonists could be settled and 85 new towns were created.[16]

Ironically, the Crimean War accelerated this process. The sending of the elite African forces to Crimea paralyzed military campaigns in Algeria due to concerns about the consequences the expansion of war in Europe might have on the control of native population—an insurrection might hold over 60,000 colonists settled there.[17] By 1854, it became clear that the practical advantages of land distribution for the cultivation of cotton and wheat ensured their widespread popularity among settlers and natives.

---

12 Headrick (1991, p. 15).

13 For the imperial telegraphic networks in the 1850s, in particular the French cable system and colonial telegraphs in North Africa and Asia, see Headrick (1988, pp. 110–11, 122–23).

14 Milza (2008, p. 359); Julien (1964, pp. 351–90). It was not until 2 June 1858 that a decree by Napoleon founded the Ministry of Algeria and Colonies, under the direction of the Emperor's cousin, Prince Jérôme Napoléon.

15 Martin (1963, p. 163).

16 Ibid.

17 Algerian shooters, known as the Turks, competed with the French Army's light infantry regiments (also called *Zouaves*).

By the start of 1853 the increasingly important role the science of meteorology began to play in the expansion of European imperial powers—highlighted by the recent creation of meteorological services in Prussia (1847) and Russia (1848)—led some authorities in Algeria to identify meteorology as another observational science worth being institutionalized.[18] But unlike previous initiatives in France, their proposal included the observational uniformity and standardization demanded by statistical science. The promoter of this initiative was Marshal Jean Baptiste Ph. Vaillant, an engineering officer decorated with the Legion of Honour. Napoleon III had good reason to trust him. Vaillant had conducted several military expeditions in Alger and Rome before being appointed Senator in January 1852, a few days after the *coup d'état*.[19] In 1854 he led the commission aimed at compiling, arranging, and publishing the correspondence of Napoleon I.[20] In recognition to his loyal services, Napoleon III named him Minister of War in March 1854, a post he held until 1859. As a minister, he published a report on the political and socio-economic situation in Algeria, showing that the fate of colonization was inexorably intertwined with trade and agricultural activity.[21] A true meteorology enthusiast, he became an elected member of the Paris Academy of Sciences in 1853.[22] In the 1860s, he would author a manuscript on storms and other meteorological phenomena, providing not only a storm theory based in part on the experiential method of his colleague Admiral Robert FitzRoy but also an appreciation of empirical methodology regarding other phenomena such as frosts, clouds, and atmospheric electricity.[23]

The organization of the Algerian meteorological service would be 'simple and inexpensive', resting on the collection of weather data from uniform observations and standardized instruments.[24] First, a dozen observation centres should be selected, placing special emphasis on those which were of greatest interest to colonization and had notable differences in position, climate, soil,

18 Barboza (2012, p. 34).

19 Vaillant's expeditions in Alger go back to 1830 and 1837.

20 As regards Vaillant, see Robert and Cougny (1889, pp. 468–69); Laurencin (1902); and the work on the exhibition entitled *Un président, un maréchal, un régiment…* (2001, pp. 81–149).

21 Vaillant (1854a, pp. 37–66).

22 Vaillant was a regular participant in the sessions of the Paris Academy of Science, with six contributions to the *Comptes Rendus* of said Academy from 1852 to 1856.

23 Jean Baptiste Philibert Vaillant, 'Des orages, et chemin faisant, de quelques autres phénomènes météorologiques', October 1865, 82 p. ADCO, Fonds du maréchal Vaillant, 11F16. For the abundant documentation regarding meteorology handled by Vaillant at that time, see ADCO, Fonds du maréchal Vaillant, 11F15, 11F17, 11F18, 11F25, and 11F26. It includes the draft 'Prévisions du temps. Système de Fitz-Roy', 26 June 1862, by Abbé François-Napoléon-Marie Moigno, written by Vaillant's hand (Ibid., 11F15).

24 Ducos (1855, p. 1129). The proposal of the Algerian meteorological service can be found in a letter of 21 April 1853, written by the Minister Secretary of State for the Marine and the Colonies, Théodore Ducos, and published in the *Comptes rendus de l'Academie des Sciences* in 1855.

and altitude. Next, these centres would receive instructions through which they would produce a field of data that could then be organized and tabulated. Finally, the results would be analyzed in a centre in Algiers which 'would become the branch office of Paris Observatory'.[25] In this way, the colonial administration would be able to uncover, as Vaillant admitted, causal relationships in weather and climate questions that were fundamental for cotton cultivation, the hygiene of French armed forces, and the success of military operations.[26] Previously unseen patterns or correlations could be identified while occasional phenomena like storms could be tracked through accurate observations on an hourly basis. In theory, repeated long-term observation of meteorological elements would show patterns of change over time.

If the meteorological service represented a valuable tool for colonization, it was equally helpful for the accomplishment of military operations. In a session at the Paris Academy of Science held in November 1855, a commission made up by a select group of savants discussed, at the request of the Ministry of War, the instructions needed to create this service.[27] Vaillant, who was already the minister of war and the author of a report addressed to Napoleon III on cotton cultivation in Algeria, participated in the discussion.[28]After listing the advantages for colonial agriculture and hygiene he underlined the prospects opened for the French Navy after the historic 1854 storm in Crimea—a fact that will be addressed below. As Vaillant succinctly put the matter:

> Through telegraphy and a set of barometric observations, we will be able to announce at a place, several hours or days ahead, the large atmospheric disturbances produced ten or fifteen hundred places away.[29]

However, not all were equally sanguine. In the next session, held in December 1855, physicist Jean-Baptiste Biot criticized the Russian service because, in spite of its magnificent resources and its Emperor's generosity, it had produced nothing more than 'an accumulation of unconnected facts, with no purpose of expected utility, either for theory or application'. 'In the absence of any success in the discovery of general laws, all the responsibility was dumped on the hope of practical applications', objected he.[30] To some extent, these disagreements were heavily influenced by their respective

25 Ducos (1855, p. 1129).
26 Vaillant (1855a, pp. 1143–44).
27 'Rapport sur les observatoires météorologiques proposés pour l'Algérie'. *CR*, 1855, 41, 1130–37. For the subsequent discussion by Henri Becquerel, César M. Despretz, Vaillant, Prince Bonaparte, Augustin-Louis Cauchy, and Jean-Baptiste Payer, see: *CR*, 1855, 41, 1130–49.
28 Vaillant (1855b).
29 Vaillant (1855a, p. 1145).
30 'Opinion de M. Biot sur les observatoires météorologiques permanents que l'on propose d'établir en divers points de l'Algérie'. *CR*, 1855, 41, 1177–90, on pp. 1180–81.

professional trainings as naval officers placed their own particular interests on storm warnings while physicists usually placed the focus on theory.

The meteorological service in Algeria, promoted by Vaillant and the Ministry of War, is a good example of the institutionalization of meteorology for both colonial and military purposes. Cotton and wheat production, military hygiene, public health, and marine safety were factors that did not go unnoticed by the French government. Nevertheless, it was the storm of Balaklava in Crimea on 14 November 1854 that epitomized the conjunction of Napoleon III's strategic priorities—Empire, order, and security.

The 1854 Black Sea storm as the trigger for order and security

The Balaklava storm and its catastrophic effects on the French and British fleets made French meteorologists to turn their attention to storm warning and forecasting. This conclusion is, of course, a commonplace in the historiography of meteorology.[31] However, storm progress was investigated for both scientific and imperial reasons, and because of this it is important to look at how the science policy of the Paris Observatory, until then closely linked to the Bureau of Longitudes, changed in 1854. The Balaklava storm was the catalyst for the Paris Observatory to turn its focus on storms and create an international meteorological network.

The Paris Observatory's astronomers and physicists substantially contributed to the understanding of storm movement and later to weather forecasting. Although they were in part inspired by an ingrained tradition at the Observatory (atmospheric observations as an aid for astronomical studies), their services were promoted by an observatory—not by a meteorological institute, like in Prussia, Austria, or Britain. Today, this fact may seem odd. It may be surprising that an essentially astronomical observatory becomes the world pioneer in storm warning service. Yet, it does make sense when taking into account the political and imperial context after Napoleon III's rise to power. The Paris Observatory was strongly affected by this change and to understand the implications of the Balaklava storm it is indispensable to understand the Paris Observatory's transformation.

In late 1853, the Paris Observatory was by no means the astronomical and geodetic centre that had grown up in the shadow of the Bureau of Longitudes since its creation in 1795. Worried by what they deemed as a dysfunctional centre of self-complacent astronomers and a lack of scientific productivity (except its director François Arago's personal genius), the new rulers decided the Observatory would gear its activities towards the theory and practice of astronomy instead of geodesy and the determination of time and longitudes which were, until then, two of astronomers' foremost duties.[32] The political

31 See, for example, Danjon (1946), Landsberg (1954), and Khrgian (1970, p. 154).
32 On Le Verrier's attempts to dismantle the Bureau and to show it as a declining institution, see Feurtet (2005, Chapter 3); Schiavon (2015, pp. 67–68); Lequeux (2013, pp. 84–87).

affinity towards the Republican side of the late Arago and his inner circle of astronomers undoubtedly contributed to this change of direction.[33] Political divergences about social order, power, and authoritarianism led Napoleon III's followers to drive both a deactivation of the Bureau's influence and the scientific and administrative reorganization of the Observatory.[34]

This reorganization was entrusted to one of Napoleon III's most devoted guardians, astronomer Urbain Le Verrier, whose career was marked by the seizure of power and authoritarian policy of the Emperor he always served.[35] Initially trained at the École Polytechnique de Paris, Le Verrier had joined the Bureau shortly after his memorable finding of the planet Neptune in 1846. Next he had held a chair expressly created for him at Sorbonne, celestial mechanics. Yet, he soon learned how to link his scientific career to a political career.[36] During the Second Republic, he supported the future emperor and joined the list of the *Amis de l'Ordre et de la Liberté*, before becoming a deputy in the Legislative Assembly. It was, however, after the coup-d'état that his name appeared among the first appointed senators. Napoleon III, he wrote with pride, 'has seized all the powers with a firm will to re-establish everywhere the principle of, and respect for, authority'.[37] His faithfulness was compensated: he was appointed Inspector general of higher education in 1853 and member of the Imperial council for public instruction in 1854.

Whatever the value of these political influences, it was an astronomical finding—the discovery of Neptune in 1846—that catapulted Le Verrier to fame.[38] By then, the stability of the solar system had become the foremost problem of celestial mechanics. Astronomers were unable to obtain satisfactory results for planetary eccentricities and inclinations. Yet in 1840, drawing on an analytic study of perturbations, Le Verrier deduced exact limits for the eccentricities and inclinations of the seven planets known until then. Despite that, no theory could explain satisfactorily the movement of Uranus. In 1845 Arago invited him to study the problem. Le Verrier proved a few basic facts: that Uranus's perturbations could not be explained as the result of the actions of Jupiter and Saturn; that there should be an unknown disturbing planet beyond them; and that this new planet should have the exact position and ap-

33 Among his circle of astronomers was his most likely successor, Claude-Louis Mathieu. But the death of Arago on 2 October 1853 precipitated the change. His brother-in-law Mathieu and many others of his close allies had to leave their posts and the Observatory.

34 Aubin (2003, pp. 87–88); Locher (2008, pp. 35–37).

35 As regards Le Verrier's life and personality, see Danjon (1946, pp. 363–82); Lamotte, Lantier, and Levert (1977); Barthalot (1978, pp. 67–71); Lévy (1973, pp. 276–79); Cox (2002, pp. 85–90); Locher (2007); Lequeux (2013).

36 Le Verrier's political career is described in Lamotte, Lantier, and Levert (1977, pp. 101–17); and Lequeux (2013, pp. 68–72).

37 In a letter addressed by Le Verrier to a friend—quoted by Lequeux (2013, p. 71).

38 On Le Verrier's contributions to the theory of astronomy, see Hind (1878, pp. 155–68); Lévy (1973, pp. 277–78). On the discovery of Neptune see Grant (1852, pp. 123–210); Lequeux (2013, pp. 21–54).

parent diameter he proposed.[39] From then until his death in 1877 he devoted himself to constructing planetary tables and theories and to determine their masses, all while taking all the mutual perturbations into consideration.[40]

Le Verrier's achievements in the analytic theory of perturbations made Napoleon III's government to project him as a man of national stature. On one hand, they provided a high degree of authority to those scientific projects that Napoleon III deemed to be a priority for the pursuit of order and security and that conformed to his previously formed ideological expectations. Thanks to this authority Le Verrier could benefit from the direct contact with Napoleon III, as his biographers often noted,[41] and cultivated good relations with influential ministers like Vaillant and Hippolyte Fortoul, as can be evinced from their correspondence.[42]

Yet at the same time, this authority provided him with a subtle weapon, strengthened by his political affinity with the Empire's regime. It made him seem like a sort of shadow minister, treating everyone as equals in dealing with other ministers. Indeed, this was the principal political goal behind the reorganization of the Paris Observatory—the appointment of a loyal minister that could impose order and authority. Otherwise, it is doubtful that he could have participated in the commission formed to that effect in October 1853, chaired by Vaillant and accompanied by other members whose loyalty to Napoleon was also absolute—among them, Biot and Jean-Baptiste Dumas.[43] The commission's report, whereby the Minister of Public Instruction signed the Decree of 30 January 1854 reorganizing the Observatory, marked a watershed in the history of this centre.[44] It broke all the traditional collegial rules that were applied theretofore, for the Director would be named by the Emperor upon recommendation of said Minister, and without the need for an introduction. This met not only Le Verrier's aspirations but also those of the rulers. For Napoleon III it was the return to order and absolute power not the minutiae of scientific method that mattered.[45]

On 14 November 1854, an unforeseen violent storm with winds of hurricane velocity broke through the strategic port of Balaklava in Crimea, at the

---

39 Le Verrier (1846, pp. 907–18).
40 Completed just one month before his death, this work occupied over 4,000 pages of the *Annales de l'Observatoire de Paris*. Lévy (1973, p. 278).
41 See, e.g., Lamotte, Lantier, and Levert (1977, p. 122). Le Verrier had an almost family link: he taught astronomy to Napoleon III's children; and helped him to calculate the moon dates and phases for Napoleon III's book *Histoire de Jules César* (1865–1866).
42 For the correspondence with Vaillant, see ADCO, Fonds du maréchal Vaillant, 11F 38/48; ABIF, Ms 3714; AOP, Ms 1047 / D7.
43 Le Verrier was the only active astronomer. For further details on the members of the commission, Lequeux (2013, pp. 78–79).
44 'Décret du 30 janvier 1854'. *Bulletin administratif de l'Instruction publique*, 1852, 3, 23–29. A draft of the commission's report of January 1854 can be found at the AOP, Ms 1047, D2.
45 Le Verrier was appointed to the directorship of the Observatory by decree on 31 January 1854.

same time as Le Verrier was assessing the scope and content of the scientific reorganization of the Observatory. Maritime and military safety was the ambitious objective set by Napoleon III's government after the catastrophe, and Le Verrier and the Observatory's staff spared no effort in this regard. A few days later, when the Ministry of War was assessing the casualties and damage caused by the storm, Vaillant urged Le Verrier to undertake a study on the conditions that produced the phenomenon in order to anticipate the arrival of future storms, while giving him assurances of the Ministry's assistance.[46] This climate of urgency and pressure forms the immediate context of the scientific reorganization of the Observatory.[47]

The impact of the 1854 storm on Balaklava was of the utmost importance for both military and strategic reasons. At that time, the main Russian naval base was in Sevastopol. In October of that year the allies sent a powerful French-British fleet to the Black Sea, and after arriving at destination, their land forces blockaded and besieged Sevastopol. In this scenario, the harbour of Balaklava was vital, for it soon became the supply port for the allies. The storm caused considerable damage: thirty-eight coalition ships were wracked, including the *Henry VI,* the finest French battleship, and much-needed stores that had just arrived from Britain, like food, clothes, and medical supplies were lost. These losses persuaded Vaillant that a service as coordinated and well connected as possible was the only way that could avoid repeating a similar tragedy.[48]

Thereafter, events followed one upon another, thick and fast. Le Verrier sent a circular to scientists in many countries requesting them to transmit the data they had gathered on the state of the atmosphere from 12 to 16 November, according to himself, 'in order to put ourselves into a position to respond to Vaillant's wishes'.[49] On 12 December, Vaillant addressed the French Foreign Minister, seeking his help for the circulars could arrive in those countries. In response, the Observatory received more than 250 documents coming from Europe and even India and the French colonies. Le Verrier requests are worthy of mention because they reveal not only the pressing

---

46 In a letter addressed to the Foreign Minister on 12 December 1854, Vaillant confirmed this request and noted the points that interested him: 'the magnitude of this atmospheric phenomenon [what he calls a "hurricane"], the expanse of the space on which the storm exerted its violence, its velocity, all these data can be very useful, in my view, for the next storm, if facts are well studied, well coordinated'. ADCO, Fonds du maréchal Vaillant, 11 F38/77. See also Vaillant (1855a, p. 1145).

47 Le Verrier's subsequent versions on this event agree with Vaillant's account, at least in this respect. See, e.g., Le Verrier (1865, pp. 1318–20; 1868). Le Verrier (1865, p. 1320): 'we wonder whether the presence of an electric telegraph between Wien and Crimea would not have served to warn our navy and fleets'.

48 On the storm of Balaklava and its effects on the allied forces, see Landsberg (1954, pp. 347–48); Arnold (2002, p. 73).

49 Le Verrier (1865, p. 1318).

need for atmospheric data but also the political urgency at the highest levels of government for the finding of a scientific response to the storm threat.

Having identified the political urgency, mainly within the Ministry of War, to seek scientific solutions to the storm problem, it will be easier to understand the great determination with which Le Verrier geared the Observatory's most efforts towards the establishment of a storm warning service, and to meteorology in general.

The decree for the Observatory's reorganization that the Minister of Public Instruction Hippolyte Fortoul signed on 30 January 1854 can be a good starting point.[50] In essence, its content is similar to the Commission's report that was mentioned earlier.[51] The aim of the decree was to define the objectives and scope of this reform. While the first part deals with the new scientific and administrative organization of the Observatory, the second part focuses on its separation from the Bureau of longitudes. In the former, it first deplores that the Observatory is in a disadvantageous situation compared with the observatories of Greenwich and St. Petersburg, which it aims to emulate. Then it concerns itself with astronomical activities and instruments. But the bulk of the decree addresses the powers of the director, whose authority becomes absolute and whose actions shall not be inhibited by the intervention of bodies such as the Bureau.[52] In this decree, as in the Commission's report, there is no mention of storms or meteorology at all. Unlike astronomy, meteorology was not the object of the reorganization.

Compare this with the project of scientific organization as appeared eleven months later, in a December 1854 report prepared by Le Verrier and submitted 'for the Ministry's approval'.[53] The original draft of thirteen pages, which is in his own handwriting, consists of two parts: one on the state of the Observatory; and the other on scientific organization, including astronomy and the various fields of terrestrial physics (here I will only deal with the section of meteorology). First, reiterating many of the conclusions of Vaillant's commission regarding the Observatory's backwardness when compared to Greenwich, Le Verrier complains of the lack of self-recording instruments and the low number of meteorological observations. However,

---

50 Hippolyte Fortoul, Jean-Baptiste Vaillant, Bonaparte Louis-Napoléon. 'Réorganisation de l'Observatoire de Paris et du bureau des longitudes. Rapport a l'empereur et décret'. *Bulletin administratif de l'instruction publique*, 5 (49), January 1854, 16–28.

51 A first draft of the commission's report can be found at the AOP, Ms 1047, D2. Several extracts of this draft are translated into English in Lequeux (2013, pp. 81–82). The final report of the commission was published by Vaillant (1854b), and reproduced by Beauchamp (1882, pp. 306–11).

52 With the new decree, the Director will regulate the staff's services, all the instruments will be at his disposal, and the scientific plan he devises for the Observatory will be obligatory for all the collaborators.

53 'Projet de rapport à l'Empereur rédigé par ordre du Ministre (décembre 1854). Rapport sur la situation de l'observatoire, projet de réorganisation scientifique'. AOP, Ms 1047, D4, 13 p.

he then emphasizes an issue that was not mentioned in the commission's report: that 'meteorology was an eminently practical science', aimed at 'the progress of navigation, agriculture, public works, and hygiene'.[54] Next, he stresses the importance of observing the variations of barometer to announce the approach of storms, and to warn of their arrival with the aid of electric telegraph. Le Verrier thereby implicitly establishes a relationship between barometric variations and storms' approach. Unlike in the commission's report or the decree of 30 January 1854, now storm warning was an essential component in the reorganization.

The meteorological component is still more present in the extended version of this draft, published by Le Verrier in December 1855, as part of the first issue of a new journal, *Annales de l'Observatoire Impérial de Paris*.[55] Here again meteorology deserves a chapter to itself. Although Le Verrier's name appears in the heading of the title, his assistant and head of the meteorological section, Emmanuel Liais, will claim authorship of the chapter on meteorology many years later.[56] In addition to the new journal, Le Verrier and Liais aimed to establish a regular system of observations, as well as relations with the principal French and foreign observatories. And most importantly, they anticipated precisely the relationship between barometric pressure and storm progress which was soon to be championed by the Paris Observatory and some foreign institutions.

> The barometer, with its variations, informs the navigator of approaching storms. [...] However, the indications deduced from isolated observations with the barometer have little value compared with those that can be obtained by examining simultaneous observations using all the meteorological instruments [...] By joining, with telegraphic lines, various stations at which meteorological observations are made, it should be possible to know from moment to moment the direction and velocity of propagating storms, and to announce several hours in advance on the coast high winds, especially the most dangerous ones [...] Already in the United States, high winds have been forecast on several occasions.[57]

---

54 AOP, Ms 1047, D4, p. 12.
55 Le Verrier (1855a, pp. 1–68). The heading of the title includes the date of December 1854. An extract of this article was previously published by Le Verrier (1855b).
56 Before accepting as fact that Le Verrier's proposals on storm warning were guided by his own knowledge about meteorological science, it is worth reading the paragraphs in which Liais spoke of the genesis of the project in retrospect: 'At a time in which, after having been qualified as a skilful physicist in a report, my meteorological works had received the highest accolades from Arago, Le Verrier was one of the worst enemies to this science, about which he had no idea'. Liais (1881, p. 36). However, while working for Le Verrier in those early years, Liais deemed his participation as 'mere corrections of facts'. On this point, see Barboza (2012, p. 55).
57 Le Verrier (1855a, p. 57), as translated into English by Bernard Sheehan in Lequeux (2013, p. 278).

Before discussing the implications of this relationship and meteoro-telegraphic networks, I will digress briefly to show how the Balaklava storm most likely acted as an incentive to Le Verrier to submit this ambitious project, which was in essence intended to obtain funds and resources from the government. To this end, his participations in the sessions of the Paris Academy of Science during 1854 are very significant, mainly for two reasons: first, because Le Verrier usually published his findings and projects in the *Comptes rendus* (remember that the *Annales de l'Observatoire* was launched in 1855); and second, because it was this journal that will report about the 1855 discussions on the meteorological service in Algeria, as well as Le Verrier and Liais's investigations on the Balaklava storm.

In examining the reasons for Le Verrier to decide to submit such an ambitious project, we must first search what meteorological concerns he might have had in the months prior to the November great storm that could lead to design a storm warning service. A perusal of his contributions to the *Comptes rendus* in 1854 seems to suggest that there was no concern at all. In fact, his main concerns in the summaries of meteorological observations from January to November 1854 were precision and discipline. He addressed these needs in several steps. First, the Observatory usually had published meteorological tables with gaps filled by additions and interpolations; he remedied this situation by introducing a strict discipline in observation times. Second, there were technical problems: the fixed thermometers used gave unreliable values of air temperature; he corrected these values by setting up gyratory thermometers. And finally, he increased the number of daily regular observations. During the eleven months before the storm, there was therefore in his publications no indication that he was concerned about the need for a storm warning service.[58]

So in seeking out the unexpressed reasons which could have led Le Verrier to justify the organization of an ambitious international system for storm warning, one cannot help but think of factors related to the Balaklava storm. Only by virtue of a generalized situation of political urgency, related to maritime and military safety, and its subsequent window of opportunity opened to Le Verrier, for the return to order and authority, can one understand the increasing weight of the predictive meteorology after the Observatory's reorganization in December 1854.

The Black Sea storm also represented a window of opportunity from a centralist viewpoint, for it allowed Le Verrier to take many of his core ideas about order, discipline, and authority, and project them as manifestations

---

58 In the two volumes of the *Comptes rendus* printed in 1854, Le Verrier published three summaries of the meteorological observations at the Paris Observatory from January to November 1854. Moreover, he presented the observations made by C. Dumas in Algeria. These articles leave no doubt: none of them mention the word 'storm'. Le Verrier (1854a, pp. 797–99; 1854b, pp. 771, 814–16, 1188). For a discussion on these works, see Barboza (2012, pp. 52–53).

of power into the framework of meteorology. Perhaps the most fundamental of these beliefs was the idea that centralization itself could be achieved through observational systems linked to telegraphy. More will be said about this in the following sections; here it will be sufficient to note that Le Verrier's project made centralization a priority and politically pertinent objective. While the Observatory had until then been tasked with collecting regular measurements of local temperature and pressure, after its reorganization it would become the centre of a state-controlled meteorotelegraphic service networked across the national geography. Similarly, Le Verrier's project was also centralist for the way it broke with the model of collaboration between amateur observers and professionals—a model that had traditionally formed the basis of the actual practice of meteorology in Europe. Here, its strained relations with the Meteorological Society of France were significant[59]; founded in 1852, this Society aimed to promote the exchange of data through the cooperation between amateur and professional observers.[60] It was no accident that Le Verrier's project ignored this proposal, under the pretext that observations must be made in a standardized way, opting instead to transform the civil servants working in telegraphic stations (i.e. telegraphists) into weather observers. Simply put, Le Verrier's project made any amateur-professional scheme vanish, replacing it with a centralized, government-controlled system, subject to political control.

Adaptation of knowledge to state interests

In the foregoing section, I have shown Le Verrier's readiness to adapt his scientific goals to the interests of the Ministry of War. Yet while this adaptation entailed a reorganization of the investigation and administration in the Observatory's meteorological section, it did not directly affect the foundations of storm theories, alter the basic concepts of meteorological practice, or reinforce the cognitive value of cartographical techniques. However, if the research focus is shifted from the late 1854 situation to the state of affairs in 1855, there clearly appears a new readiness to extend this adaptive ability to

---

59 As evidence of their strained relations, the founding member of the Society, Élie de Beaumont, claimed for the Society the originality of the idea about the use of telegraph in meteorology, before an audience composed of Le Verrier and other members of the Academy. See 'Meteorology', *CR*, 1855, 40, 439–40; and Barboza (2012, pp. 56–57). The Circular on 8 January 1852, published by the *Société* in February 1853, was very explicit: 'the whole of Europe will be soon criss-crossed with metal wires that will make distances to disappear and will enable us to recognize atmospheric phenomena as they occur and thus to foresee the most remote consequences'. *Bulletin des Séances de l'Annuaire de la Société météorologique de France*, 1853, 1, 5.

60 On the establishment of the *Société météorologique de France* and its goals and membership, the first number of its journal, *Annuaire de la Société Météorologique de France*, 1853, 1 (Paris: Imprimerie de Beau Jeune), and see Locher (2008, pp. 69–71).

scientific knowledge and adopt propositions which lacked solid theoretical grounding but were closely associated with Quetelet's atmospheric waves.

A most suggestive case of an accommodation of scientific thought to state interests can be found in Le Verrier's project submission to Napoleon III on 16 February 1855 and its subsequent defence at the Paris Academy of Science. Le Verrier's contention—or rather, temptation for political rulers— that 'our latest scientific findings point to certain crucial applications' created great expectancy in government circles. His account to the Minister of Public Instruction that 'these findings' had enabled them to take 'the first steps in the organisation of a storm-warning service' fulfilled governmental needs.[61] In this regard, Le Verrier would recall ten years later in what became known as the official version of the history of this service in France: on that very day, 'I had the honour of submitting a very complete project of a vast meteorological network aimed at warning of storm arrival to seamen'.[62] His strategy took effect, and the next day he received the Emperor's endorsement.[63] The Cabinet of the Emperor issued the following note: 'Propose with assurance what you deem appropriate. This matter is too important an issue to his Majesty for your efforts not to be entirely successful'.[64]

However, one need only glance at the minutes of the sessions at the Academy and the pages of its *Comptes Rendus* to realize that Le Verrier's project was far from complete. In two consecutive sessions, held on 19 and 26 February, he displayed several maps of the atmospheric state in France as evidence that his expected service could yield major results.[65] These maps, stated he, were part of a series of trials conducted with the assistance of the Administration of Telegraphs, whose stations connected Paris with departmental centres. Nevertheless, there is every indication that the initiative was not as welcome as might be expected. One of the attending members, Élie de Beaumont, hastened to add that the idea of using telegraph for those purposes was taken from the French Society of Meteorology.[66] Moreover, Le Verrier presented no atmospheric theory or cartographic method upon which his maps could have been based, and hence perhaps his announcement that Liais was entrusted with the task of continuing investigations. But everything suggests that, despite the mistrusts, Le Verrier was persuading the rationalist and antipositivist currents of his audience, trying to project an image of meteorology that would prove to be seductive for the academics

---

61 Le Verrier to the Ministry of Public Instruction, Hippolyte Fortoul, 19 January 1855. ANF, F17, 3728.
62 Le Verrier (1865, p. 1319). See also Fierro (1991, p. 110). For a critical discussion on Le Verrier's official account in the history of meteorology, see Barboza (2012, pp. 71–84).
63 'Nouvelles et faits diverses'. *Cosmos: revue encyclopédique hebdomadaire des progrès des sciences et de leurs applications aux arts et à l'industrie*, 1855, 6, p. 253.
64 Quoted by Le Verrier (1865, p. 1319). For the imperial decree approving Le Verrier's storm warning service, see *Le Moniteur Universel*, 23 February 1855.
65 *CR*, 1855, 40, 439–40, 454.
66 *CR*, 1855, 40, 439–40.

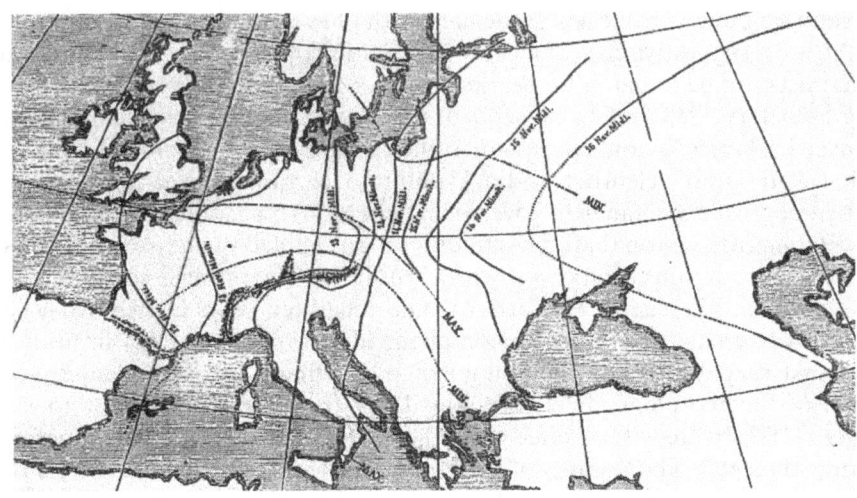

Figure 3.1 Chart illustrating the passage of an atmospheric disturbance across
Europe from 12 to 16 November 1854, drawn by Liais and presented by
Le Verrier to the Paris Academy of Sciences in his study on the Black Sea
storm on 31 December 1855. Source: Liais (1881, p. 392).

and convince them of the practical usefulness of this science. The most significant fact is that with his adaptive strategy he had achieved the Emperor's
endorsement, but not that of the academics.

Having observed the underwhelming response from academics to the viability of his project, it is easier to understand why the most striking example
of this same 'adaptive ability' is offered us by Le Verrier in his presentation
of the study on the Black Sea storm to the Academy on 31 December 1855.
There he presented the results of the study undertaken by Liais throughout
the year 1855, in which he followed the track of the storm by using data on
pressure and wind direction sent from different countries to the Observatory, in response to Le Verrier's circulars.[67] With these data, Liais drew a
map illustrating the passage of an atmospheric disturbance across Europe
(Figure 3.1).[68]

As Liais was not a member of the Academy, Le Verrier presented his
paper on his behalf. From his new post Le Verrier was the most authoritative official voice on meteorology in France. His and Liais's aim was to
gain the academic approval of their project. The audience were sceptical
academics, among them anti-empiricist Biot and A.L. Cauchy. In the paper, after describing the data sources, Liais contended that there was no
relationship between the storm progress and the direction and speed of its

67 Le Verrier (1855c).
68 Locher (2008, pp. 39–41); Barboza (2005, pp. 157–68; 2012, pp. 175–94).

accompanying winds. Next, he distinguished two kinds of pressure lines, the *atmospheric waves* and the *lines of transport of waves*. The atmospheric waves represented the displacement of air masses and were characterized by a barometric maximum; here he was inspired by Quetelet's waves, as he later acknowledged.[69] The lines of transport of waves represented the lines of the absolute values of the barometric maxima and minima in one single atmospheric wave; this idea was his and original. These lines, he believed, could indicate more accurately the trend of movement of the disturbance than atmospheric waves.[70] And last but not least, the bulk of the presentation concerned with a notion that would become basic in a short time and that Le Verrier treated with a realist slant: storms *travelled* through Europe in a general west-east direction.

Compare this presentation with the atmospheric maps as displayed by Le Verrier in the aforementioned sessions on 19 and 26 February. While on that occasion the maps had been traced from isolated data without following any methodology at all, the December map assumed Quetelet's premise that the lines of maximum values of pressure were continuous. Furthermore, the map showed that the maximum value lines propagated across the space in the days before and after the great storm. Thus, Le Verrier established a principle that he did not assume earlier: that the graphic representation of the isolines of atmospheric pressure was the most appropriate method to study the approach of storms for warning purposes. Le Verrier thereby established a correlation between local pressure variations and storm tracks. In the December presentation, unlike the previous one, there was an overemphasis on this correlation and a concealment of the risks assumed in theory—in fact, in the paper published by the *Comptes Rendus*, Le Verrier mentioned Liais's *lines of transport* in a footnote. Unlike before, he now showed to his audience the advantages of monitoring atmospheric disturbances in real time.

In this regard, Le Verrier invoked mathematical analysis as a means of academically legitimizing his project. Indeed, he sought the endorsement of

---

69 According to Liais, 'atmospheric waves do not move in isolation. The interval between two consecutive waves necessarily forms a hollow' [he also calls 'depression']. Le Verrier (1855c, p. 1199).

70 In a footnote, Le Verrier (1855c, p. 1201) explains how Liais examined three kinds of lines: (1) the line that delimits the points on which the barometer raises and those on which it falls—this line does not depend on the absolute value of pressure; (2) the isobarometric line that is not defined by barometric oscillations, but rather by its absolute value; and (3) the line of transport of waves, that depends both on the absolute value of pressure and barometric oscillations. To define these lines, added he, it is enough to observe the pressure variations on the crest of one single wave: by linking the points that successively represent the same maximum or same minimum on this crest, the lines of transport of waves will be derived. Davis (1984, p. 365) holds that Liais drew these three kinds of isobars on his charts. On the contrary, Barboza (2012, pp. 183–84) believes he never drew the isobarometric lines, and corrects the way Davis interpreted Liais's lines of transport of waves. Here I follow Barboza. I do not, however, attempt here to go into specifics.

a mathematician of great stature, like Cauchy, who in 1830 had formulated the propagation of transverse vibrations of light through an elastic solid ether. As Le Verrier noted, 'I certainly hope it will be possible to analyze the main circumstances of propagation, and I note with pleasure that M. Cauchy makes a sign of assent'.[71] Despite this tacit approval, there is no evidence that his proposal was subjected to any mathematical analysis.[72]

But before investigating how far the meteorological community in Europe was influenced by these ideas or how far state security interests affected their perception of the importance and scope of storm warning services, it is worthwhile to show how well disposed the Observatory's physicists were to alter or even abandon specific methods used in their discipline. For as the Black Sea storm became a fatality, it also became a golden opportunity to achieve immediate relevance, to bring their science closer to the priorities and needs of the French imperial regime.

The earliest to embrace Quetelet's notion of atmospheric waves was Emmanuel Liais. Liais was a young empiricist of quite a different sort from his employer Le Verrier, who recruited him in March 1854. A member of the Meteorological Society of France, Liais had been trained in the 'meteoric tradition' of the Cherbourg Observatory.[73] This phenomenological tradition, based upon empirical observation, simply aimed to describe the various types of meteors and the relationship of phenomena to each other. His earliest works on the behaviour of meteors, dating from 1849, never intended to ascertain any physico-mathematical law. However, this meteoric interest was soon to result in a concern for quantification in climatology. His 1852 study on Cherbourg's climate, comparing the temperature and pressure variations in his homeland and Paris, not only was lauded by Arago at the Paris Academy, but also persuaded him that scientific quantification was a precondition to join this select academe.[74] By 1851, he became the principal paladin of the mathematization of meteorology in France.[75] Liais saw an intimate connection between pressure variation and the Sun's calorific

---

71　Le Verrier (1855c, p. 1203).

72　Much more emphatic was the endorsement by Biot (1855, p. 1189): 'If, as Le Verrier suggests, the static state of the low atmosphere is simultaneously confirmed in several places, connected with a common centre where results will be comparatively discussed, then we do not think such a study would be sterile, for not being based on local barometric or thermometric observations carried out with the highest precision'. See also Le Verrier (1865, p. 1319).

73　The most comprehensive biography of Liais to date is Ancellin (1975–1978). See also Barboza (2008, pp. 293–94).

74　Liais (1852). A first version of this four-year long research project is his work, *Considérations sur le climat de Cherbourg*, published in 1849.

75　Liais's study on Cherbourg's climate cannot be separated from the context of the Paris Academy's scientific policy. In 1844, this academy had launched a prize for those who would be able to ascertain the laws of the general movement of the atmosphere, a prize in which it was assumed that this movement should be influenced by three factors: the Sun's calorific action, the Earth's rotation, and the attractive forces of the Sun and the Moon.

action, as can be viewed in his memoir, *Théorie mathématique des oscilla-tions du baromètre*, printed in 1851.[76] Is the variation of diurnal barometric oscillations with latitude subjected to mathematical laws?, he asked. Or go-ing further: is there any causal interconnection that governs the state of the atmosphere or, instead, chance should be recognized as the determining factor limiting the validity of laws? Liais opted for the mechanistic model, and formulated the equations governing the behaviour of barometric oscil-lations.[77] In March 1854, he took a further step and sought a mathematical formulation that would express the causal relationship between barometric pressure and wind direction. His equations were published by the *Annuaire de la Société Météorologique de France*.[78] Although he obtained a differen-tial equation that would not be resolved in his paper, he seemed to be firmly convinced that the general movement of the atmosphere could be mathe-matically determined by barometric pressures:

> The foregoing formula fully encapsulates the influence of latitude on the atmospheric pressure, and [shows] how the pressure differences on neighbouring points provoke movements of the atmosphere. The deriv-ative of this expression regarding latitude gives the intensity and direc-tion of the general resultant of wind [...] Intensity is given by the value of the derivative; direction (north or south) by the sign.[79]

This is the stance of the Observatory's young official Liais, in the middle of 1854. Let us compare it now with his 1855 study on the Black Sea storm that Le Verrier presented at the Paris Academy and that was published in its *Comptes rendus*. As shown earlier, here Liais changed his method, re-linquished every mathematical treatment, and opted for the graphic rep-resentation of atmospheric waves. Why this change? Admittedly, he had certain dissatisfactions with the pursuit of mathematical laws after being entrusted with that study, and, this may have been what led him to visit Que-telet and Buys-Ballot in the spring of 1855, taking advantage of a trip he had to make to London and Vienna to order instruments for the Observatory. But the complete account, he admitted in 1857, is that:

> The earlier works of Redfield, Reid and Espy on storm propagation and those of Quetelet on atmospheric waves had long attracted my atten-tion as regards the benefits the Navy, trade and agriculture could expect from the application of electric telegraphy to meteorology, the objective being the rapid collection of observations and, after their discussion,

---

76 Liais (1851). Liais mostly drew on Kaemtz's tables for this study.
77 Liais (1851, pp. 19–42).
78 Liais (1854, pp. 54–56). A thorough discussion on Liais's studies on Cherbourg climate and his mathematical works on barometric oscillations is Barboza (2012, pp. 115–75).
79 Liais (1854, p. 56).

the transmission of results in order to warn the localities threatened by atmospheric disturbances. In the summer of 1854, I communicated my ideas on this matter to Le Verrier [...], who approved them. Later, the study in which I was engaged, on the storm of November 1854, again called attention to this project and determined its immediate execution, at least in part, even before the preliminary studies, which, in my view, were essential to improve our understanding of the laws of propagation of atmospheric phenomena.[80]

Liais has been quoted at some length, both because this is the first and only time he gave his own version of the affair and because it differs from Le Verrier's official account. Yet, Liais's admission is extremely significant. Liais has resolved to abandon the ideal of the pursuit of physico-mathematical laws— for which he had devoted so much effort and made so many contributions— and has adopted a graphic method to follow storm tracks through the changes of certain barometric isolines. Or, in other words, he has adopted an Eulerian approach in which changes in the distribution of maximum values of pressure (at 0-level) allowed him to determine the approach of storms. Here, the Black Sea storm and the political urgencies created are not a mere backdrop, as it was only as a direct and precise consequence of these dramatic events that Liais immediately implemented the cartographic method for the Observatory's project. Thus, he managed to escape the mechanistic determinism which the mathematical laws governing the general movement of the atmosphere involved. Furthermore, Liais was a victim of the pressure that Le Verrier was exerting in the interest of his project, as the method was implemented even before having undertaken the preliminary studies necessary to assess the validity of its theoretical foundations.

In short, in seeking the reasons for this quite sudden adoption of an Eulerian approach to the study of storms throughout 1855, one cannot help but think of recent developments in the Observatory's reorganization. Only by reference to a situation of political urgency and opportunism, derived from the Black Sea storm, and geared towards maritime and military safety, can one explain the opportune adoption of Quetelet's cartographic method for the study of storm movement. For an approval by the Academy of the scientific grounds of the project was the necessary condition for Napoleon III's sanction to be effective to all intents and purposes. Not in vain was one of the functions of this institution the legitimation of those projects that satisfied both the criteria of theoretical soundness and their viability requirements.[81]

---

80 As quoted by Barboza (2012, pp. 62–63), and according from my own translation from Portuguese to English. This account was written by Liais as a part of a popular article on the Buys-Ballot's law in a journal edited by him and Théodose du Moncel, *La Science, Journal des Sciences pures et appliquées*. Paris, 1855–1857, vol. 3.

81 The importance of the legitimation of Le Verrier's project by the Academy is discussed by Barboza (2012, pp. 73, 196).

Moreover, in his study, Liais himself was indeed cautious about the causal correlation between pressure variations and storm tracks as upheld by Le Verrier, and warned the audience that the depression motion could undergo unforeseen changes. Le Verrier virtually ignored these warnings.[82]

The adoption of the Eulerian approach to the study of storms must be understood, therefore, as a consequence of the political opportunism and urgency for maritime and military safety in the months following the Black Sea storm. There is evidence that Le Verrier intentionally embraced Quetelet's graphic method, with its annexed atmospheric waves, to attract academics' attention for the endorsement of his ambitious project. Liais's readiness to abandon his lifelong physico-mathematical method, upon which his earlier studies on local barometric variations had rested, must be construed as a response to this opportunity opened. Their response was, to a large extent, strategic: by plotting the local pressure variations on a map the audience could easily infer the storm tracks. Perhaps the most remarkable aspect here is that the adoption of the graphic method not only reinforced the assumption of the correlation between the two variables, but also concealed the theoretical insufficiencies of atmospheric waves with which the project otherwise would not have passed the audience's critical scrutiny.

Pontifical Rome: centralization, order, and control

The Paris Observatory was not the only institution that promoted storm warnings on the basis of an Eulerian approach. The Pontifical Observatory of the Jesuit-run Collegio Romano, directed by Father Angelo Secchi from 1850 to 1878, adopted a similar approach. There was not, probably, any other institution in the 1850s.[83] Interestingly, the two cases had certain similarities. As I will show, like Napoleon III, Pope Pius IX's regime, with its support to this venture, sought centralization, order, and control within the Pontifical States.

As was shown, the Napoleonic regime and the nationalist political effects derived from the revolution of 1848 marked a highpoint in the collaboration between astronomers and the imperial government they nominally served under. In Italy, the revolution of 1848–49 brought the replacement of the Papal government by a short-lived Roman republic, which was defeated only

---

82 Although Liais seems to have refrained from publishing anything further on this matter over the next few years, he questioned the path taken in December 1855—what in this case is nothing but a further proof of the mentioned opportunism. In 1857, just before he left the Observatory, Liais stated that science could not forecast weather changes exclusively drawing on maps and equations. See Barboza (2012, p. 199).

83 Other two institutions involved in storm warnings were the Board of Trade's Meteorological Department in Britain and the *Koninklijk Nederlands Meteorologisch Instituut* in Holland. While the former adopted a Lagrangian approach, the latter developed a statistical method by 1860, as is shown in Chapters 4 and 5.

after French military intervention. Consequently, the Papal government was to be restored in 1850.[84] With restoration came a process of bureaucratic centralization and religious discipline, which when combined with the conservative policy of the Pontifical government towards territorial and population control would exploit the ability of men like Secchi to put their technoscientific innovations at the service of order and control. The general census of 1853 would produce statistical data on the conditions of the papal territories, and new infrastructure plans would criss-cross the entire state with telegraphic lines and railways. As a result, the use of technoscientific human capital in Rome shifted from the traditional mission of academic training to state centralization and communications.[85]

The primary institution for the organization and bureaucratic centralization of the technoscientific projects promoted by the papal government was the Observatory of the Collegio Romano. Here the traditional practice of observation within the loose framework of positional astronomy was replaced by a more hierarchical discipline geared towards physical astronomy and meteorology, as well as the state's infrastructure needs.[86] This observatory assumed responsibility for the implementation of all government-sponsored technoscientific projects dealing with communication infrastructures, such as the installation of telegraphic lines and lighthouses, including the determination of the trigonometric basis of the Appian Way (a calculation necessary for the triangulation of the Pontifical State).[87] Reflecting the centralizing nature of the renewed state, authority for the conduct of these projects was placed with the director Secchi, who became the principal advisor to Pope Pius IX in scientific matters.[88] Secchi was tasked to extend the frontiers of knowledge to the borders of the state, in turn, providing a model for how meteorology could be used to translate papal desires for centralization and control into a sensible program of political action.

As had been the case in countries like Holland and Britain, the surging interest in storm warning services experienced in the early 1850s was driven by developments on an international scale—in particular, the 1853 Brussels Maritime Conference. Here the needs for maritime safety and security, the legitimization of statistical procedures by states via uniform and standardizing measurements, and the calls for the search of the laws governing the sea wave and atmospheric motions by Maury (who had a direct influence on Secchi) all played a role in helping to convince the Pontifical government

---

84 Gooch (2001, pp. 10–30).
85 Mazzotti (2010, pp. 62–63); Coyne and Maffeo (2001, pp. 44–46).
86 Mangianti and Beltrano (1990, pp. 3–6).
87 Secchi (1858b). A similar role was played by the Paris Observatory in the establishment of railway and telegraphic networks in France during the Second Empire. See Aubin (2003, pp. 93–99).
88 This relationship was not extemporaneous as the pontiff himself had graduated with a dissertation on the construction of telescopes.

that their own state security was tied to the progress of meteorology. The laying of telegraphic lines and the increasing European competition for communication control also contributed to this conviction, adding a sense of urgency to ongoing security concerns.[89] For example, the professor of experimental physics at the University of Padova, Francesco Zantedeschi, had been urging Austrian authorities from 1849 to organize a storm warning service, taking advantage of the telegraphic network being laid across Central Europe (Vienna-Trieste, Vienna-Linz, etc.).[90] Although his plan was merely experimental, he informed Secchi of his intentions.[91] Thus, the perceived need to establish such a warning service increased notably. Yet rather than serving as a tool for meeting the imperialistic security needs triggered by a terrible storm (as had been the case for Le Verrier's project), the papal government's service would be a tool—created in response to the Brussels conference—for projecting a longed-for goal of centralization and control.

By the start of 1855 the increasingly important role telegraphic meteorology began to play in European scientific politics—highlighted by the Brussels conference and Le Verrier's recent project—led astronomer Caterina Fabbri Scarpellini from the Observatory of Campidoglio to suggest that a similar network should be established in Pontifical Rome.[92] Her husband Erasmo, the director of the journal she founded in 1848, *Corrispondenza scientifica in Roma,* requested the Minister for Trade and Public Works to allow stations for the daily transmission of their meteorological observations for the subsequent publication of the same.[93] The Minister of that branch, Monsignor Milesi, endorsed their initiative.[94] At the Minister's request and based on Scarpellini's idea, Secchi devised a plan of meteorological telegraphic transmission by June 1855, only four months later than the approval of Le Verrier's project.[95]

The idea that the Pontifical State and the Pontiff himself should promote a storm warning service was not obvious at that time. The main reason adduced by its authorities was that the state should strengthen the internal

89 For the establishment of an electro-magnetic telegraph connection Trieste-Vienna-Oderberg-Berlin-Hamburg, see Huurdeman (2003, p. 83).
90 See the correspondence between Zantedeschi and Austrian Minister De Bruck, 14 October 1849, 2 February 1850, in Zantedeschi (1864–65, pp. 1414–16).
91 A prolific author in studies on atmospheric electricity, telegraphy, and meteorology in Pavia, Verona, and other regions, Zantedeschi held intense discussions with Secchi on atmospheric electricity and storm forecasting in the 1860s. See Zantedeschi (1863, 1866), Cittadella (2014, pp. 322–23).
92 As regards Caterina Fabbri Scarpellini, see Mazzucato (2008, pp. 75–76).
93 The role played by Caterina Fabbri Scarpellini and her husband Erasmo in this initiative was acknowledged by Denza (1883, p. 416). For Secchi's personal testimony, see Secchi (1877b, pp. 44–45).
94 According to Iafrate (2011, p. 120), Pope Pius IX decided to fund personally the project.
95 Zantedeschi (1864–1865, pp. 1420–21). Secchi's plan was approved by Government Decree of 26 June 1856. See Besso (1871, pp. 447–49).

cohesion, both territorial and administrative, by implementing the latest techniques of telegraph transmission.[96] This was in part because electric telegraph opened new unifying prospects, and in part because of the geographical particularities of the state itself. Trapped between two seas and bifurcated by a high mountain range, its authorities soon realized that an immediate communication of the atmospheric conditions in one side could avoid or at least prevent damage on the opposite coast, provided the data exchanged between stations were collected and analyzed by a central observatory.[97] It was necessary to promote storm warnings because scientific instruments made possible infrastructures, and infrastructures made possible control and centralization.

Secchi's project was predicated on the daily telegraphic exchange of simultaneous observations. The Pontifical Observatory was the analysis centre coordinating its data with the stations in Ancona, Bologna, and Ferrara. Equipped with telegraphs, these stations were semiautonomous entities functioning under the centralizing umbrella of the Observatory. They exchanged data with Rome and among them, and alerted neighbouring ports to storm threats. From its foundation in June 1855, this network would conduct observations of temperature, pressure, humidity, wind, and the general state of the atmosphere.[98] By then its promoters could proudly claim that theirs was the first operational metereo-telegraphic system ever established in Europe.[99] But, here again, the differences with Paris were notable. For rather than resorting to a cartographic method as the basis for storm warnings as had been the case for the Paris Observatory, Secchi would initially draw on both local empirical relations and the extrapolation of pressure and temperature patterns in Europe to the papal territory for his *preavvisos*, or advance notices.[100]

Secchi's system shows both the prospects generated and difficulties encountered in early attempts to use telegraphy to face the challenge of how the motion of midlatitude storms could be foreseen. On one hand, it was a powerful tool for controlling atmospheric threats that allowed the Pontifical rulers to view all the meteorological elements affecting the national territory in a coordinated and synchronous way. It connected territory with discipline and order, creating an image of geographical and administrative

---

96 Angelo Secchi, 'Corrispondenza meteorologica telegrafica in Roma'. *Corrispondenza scientifica di Roma per l'avanzamento delle scienze*, 25 June 1855, 4 (7–8).
97 Eredia (1914, p. 1002).
98 Secchi (1855).
99 The claim of the 'priority' of the Roman telegraphic meteorological service is commonplace among both the promoters themselves' testimonies and Italian historians. See, e.g., Secchi to Francesco Denza, 9 March 1866, in Iafrate (2011, p. 128); Secchi (1861, p. 26); Eredia (1914, p. 1003); Besso (1871, p. 448); Zantedeschi (1864–65, pp. 1420–21); Iafrate (2008, p. 139).
100 Iafrate (2011, p. 120).

centralization that only very few projects could convey.[101] At the same time, however, his early storm warnings can also be viewed as failures, lacking the theoretical foundations and credibility to serve as the model for the redefinition of meteorological services then underway. Based on local observations and hypothetical empirical correlations between pressure and temperature, they were viewed as willing efforts to achieve a dream rather than as precise warnings of threats already in existence. Without accurate continuous recording instruments and standardized measurements, without the international telegraphic correspondence of data from many other European stations, such *preavvisos* could only be taken as a description of supposed empirical correlations from local observations. Yet everything began to change from the period 1857–59, as I will show below.

Atmospheric waves: Secchi and his synoptic method

By the early 1860s the dissonance between the prospects generated by Secchi's storm warnings and their believability as a service capable of evoking sufficient confidence in their recipients would disappear. This section charts the development of this process, showing how Secchi's storm studies evolved first into a graphic method capable of shaping the motion of atmospheric waves and then into a cartographic method capable of foreseeing the approach of storms.

The emphasis on the importance of cartography and the Eulerian approach for the study of storms can be seen in the work of Angelo Secchi,[102] whose career was marked by the recording instruments of the new observatory he installed.[103] A graduate in philosophy at the Roman College, he first became familiar with astronomy at the lectures of Francesco de Vico and Giovanni B. Pianciani. After the expulsion of the Jesuits from Rome in 1847, he travelled to Stonyhurst, England, where he not only completed his studies of theology but also was trained in observational astronomy. He would become an instrumentalist of the same school as his director at Stonyhurst, Alfred Weld. Next he taught physics at Georgetown University in Washington.

---

101 For the use of telegraph as a tool for governance, centralization, and social ordering at that time, see Eriksson (2011, pp. 1–13, 48–58).
102 Biographical studies on Angelo Secchi include Tricht (1878, pp. 353–402), Moigno (1879), Altamore and Maffeo (2012), and Udías (2015, pp. 193–98). For a complete inventory of Secchi's enormous and varied correspondence with personalities of his time see Chinnici and Gramatowski (2001). An exhaustive study on Secchi's contributions to meteorology and the study of atmospheric and magnetic phenomena and their relations with solar activity is Proverbio and Buffoni (2004, esp. pp. 125–45), though it is not based on archival sources. This may be supplemented by Brenni (1993), Maffeo (2001), and Iafrate (2008, 2011).
103 The first apparatus of the new Pontific Observatory (a new telescope and some meteorological and magnetic instruments) was installed on the roof of the Church of Saint Ignatius in 1854.

There he imbibed the latest works on experimental electricity and telegraphy, falling under the influence of Captain Maury and his dynamic view of nature. His own publications in the United States deal with researches on electrical rheometry based on the application of telegraph. This fact is important as his contact with Maury and his experience in telegraphy would later arouse his interest in dynamic meteorology.[104]

Upon his return, Secchi was appointed director of the new Pontifical Observatory, not without having visited the observatories of Greenwich and Paris. As he worked to implement his meteoro-telegraphic system in the Pontifical State, he also undertook the self-construction of apparatus. Here his efforts were directed at one of his major principles—attaining continuous recording, or making the data recording as precise, complete, and automatic a task as possible. Thus to record continuously and to show results visually, mechanical and electrical techniques were integrated into a single device and atmospheric data that had previously been displayed through numerical tables were now presented in the form of graphs—his 1857 *barometrografo a bilancia* is an example of this.[105] More importantly, Secchi also tried to transform the task of recording into one that constructed relations between phenomena. To effect this change he set aside the tradition of designing apparatus aimed at recording one single parameter, replacing it with the automatic registration of many meteorological parameters at the same time. The result was his *meteorograph*, developed from 1855 to 1865, a massive machine showing 'continuously in a graphic way on shreds of stripped paper the curve of barometer, thermometer, wind speed and direction, and rainfall time and length'.[106] This method of 'continuous recording', as Secchi puts it, not only increased the visualization of the results—replacing tables with graphs—but also organized and centralized the flow of data from distant stations, thereby facilitating the study of weather—recording could be done at long distances from instruments.[107]

104 In his 1877 book *L'astronomia in Roma nel Pontificato di Pio IX*, Secchi recalled that meteorology entered a new phase in the early 1850s—it thereafter included not only climatology but also the general atmospheric currents and storm tracks. In his view, Maury had much to do with it: 'the writer [Secchi] was in the USA at the time of the great discoveries of Maury. He saw his methods, and took his ideas from his own lips. Upon his return, he took care to inform his compatriots [about Maury's accomplishments] in a memoir printed in the *Annali* of Tortolini [A. Secchi, 'Guida dei naviganti a lungo corso Vincenzo Gallo'. *Annali di scienze matematiche e fisiche*, 1853, 4, 245–60]. Secchi (1877a, pp. 41–42) wrote an obituary as a posthumous tribute to Maury. For further discussion on the influence of Maury on Secchi, see Proverbio and Buffoni (2004, pp. 131–37).
105 Secchi (1857, pp. 139–45).
106 Secchi (1859)—as quoted by Proverbio and Buffoni (2004, p. 140).
107 This machine was presented to the 1867 Paris Universal Exhibition where it was awarded a gold medal. The Pontifical government used it to overcome the cultural and diplomatic isolation. For further details on this machine and its international impact, see Brenni (1993, 202–35); Mazzotti (2010, pp. 65–67).

Secchi's new graphic method was built around another main principle—
the availability of international meteorological information in near-real
time. Under Secchi's direction the Observatory joined the international
Bulletin quotidien that Le Verrier began to organize in Paris in 1857, which
symbolized an incipient European telegraphic meteorological network.[108]
Every morning telegrams with the observations made at 19 pm in the papal
state would be transmitted to Paris, where together with the information
collected from other stations they would be lithographed, and then dispatched back to observers through the *Bulletin*.[109] Observations from nearly
all of Europe would be almost immediately compared and visualized, with
observers taking decisions on an ad hoc basis.[110] This meant both an end
to the haphazard practice of using only local data displayed by numerical
tables, and the beginning of the study of storm laws by graphical methods
and large-scale data gathering. To facilitate the flow of information across
the country Secchi promoted the telegraphic centralization of the process of
data collection.

Armed with a technology that in theory reduced atmospheric phenomena
to an individual and objectively determinable characteristic visualizable by
graphs, and with devices capable of recording continuously and gathering
large-scale data in near-real time, by 1858 Secchi began to seek traces of
atmospheric waves that appeared to reflect the true motion of storms—an
issue that had attracted his interest from his letter exchanges with Quetelet.[111] By 1855, he had already constructed many barometric curves from observations made in several stations in Europe, and, after having compared
with those in Rome, he had realized that the atmospheric wave arrived in
the Holy See with a certain time lag.[112] However, a substantial shift in his assertions regarding the interpretation of atmospheric waves took place once
he installed his new self-registering barometrograph and joined Le Verrier's
network. Where in 1855 he outlined the possibilities afforded by barometric
curves, three years later he categorically asserted that the great atmospheric
disturbances moved in Europe from the North-West to the South-East in

108 By 1857, Le Verrier received dispatches from Rome (Secchi), Brussels (Quetelet), Geneva
(Émile Plantamour), Stockholm (Daniel Lindhagen), Vienna (Karl Kreil), and Lisbon
(Fradesso Da Silveira). The number of participants was eight in 1858, fifteen in 1859, and
nineteen in 1860. See Khrgian (1970, pp. 123–25, 130–31); Locher (2008, pp. 42–44).

109 Even though the *Bulletin* was at first intended only for the French public, it became international from 1 January 1858. From 1863 to 1872, it was known as *Bulletin international
de l'Observatoire [Impérial] de Paris*.

110 Secchi (1877b, p. 44).

111 APUG, Correspondence with Quetelet, 1853 /63.

112 Secchi (1856, pp. 143–44): 'This fact is very interesting to know the circulation of the barometric wave, and similar studies could explain the relationship between the barometer
and the state of the atmosphere'. He had come to this conclusion, using the old barometers of the Observatory.

less than a day, after examining several graphs showing pressure patterns.[113] Not only that, in 1860 he was also able to enunciate, under the same premises, a general relation between barometric variation and the direction and speed of wind.[114]

As someone acknowledgeable about the storm theories of Dove, Redfield, and Espy, and a Roman patriot concerned with the security, commerce, and agriculture in the Pontifical State, the issue of how best to ascertain storm motion was logically of great interest to Secchi. Troubled with the technical insufficiencies encountered in the use of the traditional instruments—which as noted earlier represented meteorological parameters individually and separately—Secchi proposed an ingenious yet crucial reorientation of the technique. Instead of recording parameters separately and in different sheets (which was the case with most contemporary instruments), his *meteorografo* would automatically register all the phenomena in one single sheet so that 'one could detect at a glance the course of the storm and discover all the circumstances that accompany it'.[115] Here the aim was to mechanize the recording of visual information. Secchi described it in almost cinematographic terms, saying that his machine 'showed at a glance the succession and concatenation of all the phenomena' by integrating graphic and mechanic-electrical techniques.[116] In the absence of such a machine, he said, if the observer had the bulletins with data from Europe's main stations, he could plot the respective barometric curves on sheets of graph paper. By comparing each curve with the rest of curves and arranging them in their geographical positions, he argued, one could ascertain the motion of atmospheric waves at a glance.[117]

Secchi's proposal, published in two articles in the influential proceedings of the *Accademia Pontificia dei Nuovi Lincei* in 1858 and 1860, was in some aspects revolutionary. On one hand, it in part broke with a current of statistical thinking in meteorology that extended from Humboldt to Quetelet and that was based on the representation of mean values rather than on the objectification of individual phenomena. On the other hand, and most importantly, Secchi placed the focus of storm research on the barometric curve as an instrument for analysis, thus prioritizing the pressure field over other fields. For as he admitted, 'the thermometer and wind direction are too subject to particular influences' and other local circumstances that can easily affect their course. Or to put this another way, Secchi opted to adopt an Eulerian approach to the study of storms for fundamentally technical

---

113 Secchi (1858a, p. 390) added these disturbances diminished in strength and increased in numbers as they moved toward the south.

114 Secchi (1860, pp. 237–44). He also described the distinctive elements of storms during their displacement. See Secchi (1860, pp. 245–49).

115 Secchi (1860, p. 234).

116 Secchi (1860, p. 235).

117 Secchi (1858a, p. 390).

reasons—rather than for motivations related to political or military urgencies, as was the case for Le Verrier.[118]

Part of the explanation of the attractiveness of Secchi's proposal was the visual simplicity of the concept and its seemingly original combination of graphic and mechanic-electrical techniques. But there were political and operational aspects of his plan that aroused its interest and attention among Pontifical authorities. To begin with, Secchi's system was a powerful tool for centralization: 'the only way to elucidate the law of the evolution of atmospheric disturbances', he stated in 1860, 'is to install some meteorographs located at a distance from a central station in order to study storms simultaneously in time and space'.[119] Second, and closely related, his system was extremely useful and practical in legitimating the construction of the Pontifical State, providing a basis for the territorial and administrative cohesion. His telegraphic network connecting stations distributed across the territory was especially beneficial in this regard. When Italy became a nation-state in 1861, Secchi was invited in recognition of his work to participate in the first official commission for the organization of the national meteorological service[120]; and in 1862, the papal authorities supported the launching of the *Bulletino dell' Osservatorio del Collegio Romano*.[121] Finally, it was highly useful in legitimating the construction of a supranational entity with the centre being in Rome. The meteorograph was installed in places as distant as Havana, Manila, Washington, and Shanghai, constituting the centrepiece of an observational network of Jesuit stations distributed worldwide and guided by Secchi's plan.[122] Given its role in legitimating Rome's aspirations it is hardly surprising that the name of Secchi has been historically linked to the meteorograph.

Yet at the same time, Secchi's proposal did not yet represent the most sophisticated application of cartographic method to the problem of foreseeing the motion of storms—as expressed by Liais with his atmospheric waves and lines of transport in 1855. Although the foundation of Secchi's proposal was the analysis of barometric curves as indicators of the motion of atmospheric waves, at the heart of his methodology was the aim of showing at a glance the succession and concatenation of these events, whether by graphic or cartographic methods. The latter step was taken in 1862. During that year, Secchi and his assistant Giuseppe Serra-Carpi, a twenty-one-year-old

---

118 In fact, Secchi admitted his preference for the barometer because, in addition of being an essential instrument and having a close relationship with the state of the atmosphere, it was less exposed to local circumstances related to its location. Secchi (1858a, p. 359).

119 Secchi (1860, p. 236).

120 Chinnici and Gramatowski (2001, p. 574); Eredia (1914, p. 1005).

121 On the Pope's financial support to the *Bulletino*, see Iafrate (2008, p. 140).

122 For the international expansion of Secchi's meteorograph, see Brenni (1993, 236–40); Anduaga (2017, pp. 75, 205–12). For the transnational, imperial, and local dimensions of the Jesuit scientific research in Manila and Havana, see Anduaga (2017, 26–77, 194–293). On nineteenth-century Jesuit observatories, see Udías (2003, 2015).

engineer familiar with electric telegraphy, would work to turn this challenge into reality.[123] Laying out a large map of Europe on a table and driving a 20 cm stylet—on the points where stations were located—to represent the barometric and thermometric heights, Serra-Carpi devised a method to show the distribution of pressure and temperature in Europe. Pencils of different colours (red for maxima, blue for minima, and green or yellow for average values) were used to trace the isobaric lines, as well as the crests and depressions (also called *pozzi* or holes) of waves. This created an aerial overview of the general distribution of the atmospheric pressure in Europe at a given interval. At the same time, a series of banderoles and lines of various lengths with arrowheads indicated the direction and strength of wind, reminding observers of the important role winds played within the overall sequence of the weather. Isothermal lines were thickened to make them stand out against other lines on the map.[124] In his own way, Serra-Carpi transformed the graphic image of Secchi's barometric curves into a cartographical image, a map of isolines showing the succession and concatenation of phenomena.

Secchi and Serra-Carpi's synoptic method brought the illusion of visualization meteorologists had long desired. Although they admitted to having followed the footsteps of Quetelet (but not those of Liais), and although they published no map at that time, they left their legacy, as I will show below, in the way in which the isolines were constructed and interpreted. By tracing isobaric and isothermal lines, Serra-Carpi was able to discern the direction cyclonic storms usually followed in Europe, running from North-West to South-East.[125] He noticed that when a great depression was formed in Scotland, it usually took two or three days to arrive in Italy, while if the depression appeared below or above that place, only the indirect effects of its environs were felt in Rome.[126] It was these results, Secchi would later say, that would have probably lead Le Verrier and his assistants to emulate their

---

123 Serra-Carpi published several works on apparatus and electrostatic induction in the following decades, including a study entitled 'Sulle linee isotermiche dell'Italia de' suoi mari ed isole adiacenti' in 1865.

124 This method was presented in a study on the distribution of the atmospheric pressure in Europe during January and February 1862, and was published in the third issue of the *Bulletino* of the Pontifical Observatory on 31 March 1862. See Secchi (1862a, pp. 20–22). The rest of results appeared in the *Bulletino* on 15 August 1862. See Secchi (1862b, pp. 93–95).

125 Secchi (1862b, pp. 93–95).

126 The predisposition to formulate an empirical law of the motion of storms is surely the single most common element of the meteorological studies by Secchi at that time. Thus, giving shape to an idea already thought in 1857, Secchi stated in 1867 that 'each great depression produced in Ireland or Scotland arrives [in Rome] two days later; if the depression is accompanied by a great storm, it will unfailingly arrive after the time specified'. Indeed, 'for nine years after these studies, I never found false this rule'—Secchi (1867, pp. 1–4), as quoted by Proverbio and Buffoni (2004, p. 145).

method and subsequently publish maps, using 'the resources available in a rich nation like France'.[127]

The die was cast. Anchored in the language of cartography, legitimated by the Pontifical State through the Pope's logistical and financial support, and recognized by colleagues abroad through the diffusion of the meteorograph, Secchi's synoptic method became an authoritative technique for the location and tracking of storms in the European skies. Although he had insufficient resources to organize an international bulletin or a worldwide meteorological network, his method did leave its mark. In essence, it would make it possible for meteorologists and concerned observers and authorities to see the atmosphere in motion, transforming the ethereal and visionary nature of atmospheric waves into an apparently tangible and visualizable object: a depression encircled by isobaric lines. They could now determine its temporal origins, identify its spatial contours, track its motion, analyze the causes for its formation, and most importantly, forecast the places and dates of its arrival. This is what the Paris Observatory's astronomers will do.

Marié-Davy's synoptic view: maps and depressions

By the early 1860s, the official meteorology in France had experienced considerable expansion. A continental weather observation system had been organized under Le Verrier's direction, with Paris as the international operational centre. Between 1858 and 1864 the number of European weather stations involved in this network multiplied sixfold, rising from eight to sixty-four.[128] Paris's position was further enhanced by the publication of the *Bulletin international de l'Observatoire de Paris* in November 1857, a project entailing agreements between stations and the public and private telegraph companies in Europe.[129]

Despite these ambitious accomplishments, however, Le Verrier and the Napoleonic regime failed to establish a national storm warning service which would go beyond the mere observation of the weather. Their criteria oscillated between various positions based on the role and views of the two leading institutions—the Paris Observatory and the French Navy. Thus the former followed the baton of Le Verrier, who aspired to organize a service run by the Navy but whose data would be collected by the Observatory.[130] The plan, however, received little support from the Navy Ministry and the

---

127 Secchi (1877b, p. 44).
128 The network also included twenty-four French meteorological stations. See Fierro (1991, p. 111); Locher (2008, pp. 42–44); Lequeux (2013, pp. 282–88).
129 Shaw (1926, pp. 303–04). By 1867, the *Bulletin* was being received by the most important Paris dailies. See Locher (2008, p. 43; 2009, pp. 83–84).
130 See Le Verrier (1866, p. 1046). See also the report presented by Le Verrier to the Minister of Public Instruction, 'Service météorologique des ports', 22 May 1861. ANF, box F17/3728.

War office despite their great concern over maritime security and meteorology. In 1860 the Navy Minister opted to discard Le Verrier's plan, though he had previously entered into negotiations.[131] In addition to the consequences of this setback, Liais had left the Observatory in 1857 after having serious frictions with Le Verrier.[132] With Liais's resignation, it was unclear whether the future storm warning service would continue to be rested upon the conceptual framework of atmospheric waves.

It is worth noting, however, that these institutional differences were driven not only by domestic power struggles but also by the international rivalry and methodological debates that followed from the establishment of storm warning services in other countries. Thus the decision by the Minister of Public Instruction to make such a service viable in France was caused in large part by Le Verrier's concerns over the effect a similar foreign service might have on the security of the nation and the Imperial Navy. In fact, the British had just instituted a storm warning service—under the direction of FitzRoy and the Board of Trade's Meteorological Department—based not only on barometric and thermometric observations but also on the application of certain empirical rules.[133] Seven months earlier, on 1 June 1860, the Dutch issued the first telegraphic warning based on the simultaneous difference of the barometer 'readings at five well-arranged meteorological stations'.[134] Within the French Navy the idea of receiving storm warnings via FitzRoy's service was strongly welcomed. To his annoyance and exasperation Le Verrier saw how the British warnings of gales were dispatched to his own Ministry of Marine.[135]

The man charged with the responsibility of remedying this situation and bringing Le Verrier's pretensions into fruition was physicist Hippolyte Marié-Davy. Admittedly, Le Verrier's new recruit was not in the best

---

131 The disagreements between Le Verrier and the Navy officers were much more profound and complex than what has been outlined here. A principal consideration underlying the Navy's hesitant and variable support for Le Verrier's storm warning service was, once again, power—in particular, his ambition for power on all the fields linked to observatory sciences. Between 1854 and 1862, Le Verrier sought to marginalize the *Bureau des longitudes*, and did everything he could to align the *Dépôts de la Guerre* with his interests and with the Navy. This behaviour brought him many enemies. For further details, see Locher (2008, pp. 58–61).

132 For further details, as well as for the complaints of other astronomers against Le Verrier, see Barboza (2012, pp. 8, 63–64); Lequeux (2013, p. 291).

133 Its first warning was issued on 6 February 1861. FitzRoy (1862, pp. 444–56). See also Halford (2004, p. 205); Walker (2012, pp. 42–45).

134 Buys Ballot (1864, p. 20).

135 By 1864, FitzRoy's warnings seemed to have gained acceptance in the French Navy: 'Reports from various ports on our north shore testify frequently to the perfect agreement between the forecasts and the observed weather'. 'I constantly receive new requests from ports asking for the communication of your telegrams'. Captain E. Mouchez to FitzRoy, 19 and 26 January 1864, in *Report of the Meteorological Department of the Board of Trade 1864*, appendix, p. 9—quoted by Davis (1984, p. 370).

position to assume such a challenge. He had studied at the *École Normale Supérieure* in Paris, and earned a doctorate in science in 1844 with a thesis on the theory of batteries.[136] Next he taught physics in Montpellier, and began to publish his early works dealing with electricity, cells, and the physiology of vision. In fact, his first and almost only contact with meteorology was his 1851 study on the climate of Montpellier, whereby he earned a doctorate in medicine.[137] Upon his return to Paris, he taught physics at a *lycée*, where he prepared students for the entrance examinations of the *Grandes Ecoles*. Indeed, he was preparing students and performing electrical experiments when Le Verrier offered him charge of terrestrial magnetism work in early 1862.[138] This fact shows that his first assignment was not meteorology, and neither was it his speciality.

The synoptic method that Marié-Davy will apply to the study of storms owed much to the meteorological accomplishments of Secchi and Serra-Carpi in Italy. In February 1863, the president of the *Société Météorologique de France* Eugène Grellois published a detailed description of Secchi's method in the *Annuaire* of this society.[139] Grellois was impressed. Secchi, he stated, has envisaged a highly ingenious method: 'a large map of Europe is laid out on a table', thus beginning his long account, and he then depicted how a 20 cm stylet was used to represent the heights of barometer and thermometer. And he concluded: 'by using pencils of different colours, he was able to trace the isobarometric lines, maxima, minima and averages, in brief, the crests and depressions of atmospheric waves'.[140] It is most likely that Marié-Davy read Grellois's article, as he himself admitted that, after joining the Observatory, he read 'what European meteorologists had achieved' and it was only then when he 'began to be captivated by a service that had such marvellous potential'.[141] Moreover, the *Annuaire* was an indispensable reference source for anyone interested in the topic, and there atmospheric waves received more attention than might be expected.[142]

---

136 The source containing personal information about his life and work is Marié-Davy (1868). See also Davis (1984, pp. 368–69); Locher (2008, pp. 109–10).

137 Marié-Davy (1851).

138 Although he would become responsible for the section of geomagnetism, Marié-Davy was free to conduct research on any other field of his interest.

139 Eugène Grellois was a military physician, hydrologist, and amateur meteorologist, who published a book on *Météorologie religieuse et mystique* in 1870.

140 Grellois (1863, p. 22).

141 Marié-Davy to Quetelet, 12 June 1864, ARSB, box M1720: 'I joined the Observatory because I felt it would be more conductive to carrying out the electrical experiments that I had been conducting for the past eighteen years but when I saw what European meteorologists has achieved, I began to be captivated by a service that had such marvelous potential'—as translated by Locher (2009, p. 85).

142 Typical to this regard is the circumstance that most historians accepted the 'death' of atmospheric waves when Birt acknowledged their 'non-agreement with hydrodynamical principles' around 1850. Thus Jankovic (1998, p. 39): 'The meteorology of atmospheric waves received relatively little attention among nineteenth-century scientists.

Marié-Davy had several motivations. Like Le Verrier, he was aware of the practical importance of a storm warning service for agriculture and navigation. He also emphasized the scientific viability and feasibility of the service, which not only had a 'marvellous potential' but also could bring insights into the dynamics of atmospheric currents.[143] Most importantly, neither Marié-Davy nor Le Verrier wanted the Observatory and the French Navy to depend on British storm warnings and the mistrust was mutual. As Marié-Davy asserted later, the French Navy was not in favour of depending on the Observatory's warnings because they felt the service would not be operational in the medium term.[144]

Following Secchi and Serra-Carpi, Marié-Davy traced on a map of Europe the curves of equal pressure from which storms are studied, and named them *lignes isobarométriques* in 1863.[145] By means of isobaric lines that were not allowed to intersect, he devised a method based on judgement and the draughtsman's leeway. According to his method, the examination of the deformations of these lines over the surface of Europe is in principle sufficient to determine the isobaric patterns and, at best, to predict the approach of storms. In practice, Marié-Davy first sought to identify the atmospheric waves through which pressure fronts propagate. But he soon dismissed this way, for, as he stated in April 1863, 'it is difficult to trace any isolated wave with distinct patterns of movement'.[146] So he changed tactics, and, drawing

---

Birt's definitions were generally ignored'. However, the persistence of a 'subterranean' current of support to this type of wave is suggested not only by some graphic demonstrations of the concept in the 1850s (see, e.g., Quetelet, Liais, and Secchi) but also by some statements published in the *Annuaire de la Société Météorologique de France* (also in journals like *Cosmos* and *Les Mondes*) in the 1850s and 1860s. Thus, in June 1862, describing the instructions for meteorological observations, the vice-secretary of the *Société* and author of the first isobaric chart (France, 1864), Émilien Renou (1862, p. 80), acknowledged that data tables were indispensable 'to know the progress of atmospheric waves during storms'. In the same journal, Adolphe Bérigny asked Liandier if he had not highlighted 'the correlations between the progress of atmospheric waves and that of shooting stars, as observed by Coulvier-Gravier'. Liandier (1862, p. 120).

Last but not least important, in his project on the establishment of a meteorological observatory in Havana under the auspices of the Spanish government, polymath Andrés Poey (1855, p. 203) emphasized that phenomena such as 'hurricanes, storms, atmospheric waves, and earthquakes' could be studied only by a timely and efficient correspondence between the stations of the Antilles and those of the United States. For Poey, atmospheric waves were as quantifiable, measurable, and real as hurricanes or earthquakes. On Andrés Poey and his positivist view, see Anduaga (2017, pp. 175–83).

143 Marié-Davy to Quetelet, 12 June 1864. ARSB, box M1720.

144 Marié-Davy (1866b).

145 One of his first meteorological works was read by Le Verrier at the Paris Academy of Sciences on 17 August 1863, and deals with the state of the atmosphere during the first fortnight of August. Marié-Davy (1863b, p. 385): 'The form of the isobarometric lines that cross the north of France and Holland allow us to presume that the centre of the phenomenon is on Scotland'.

146 Marié-Davy (1863a, p. 233)—as translated by Locher (2009, p. 86).

on Redfield's storm theory and especially on Henry Piddington's studies on cyclones, he searched for the areas of low pressure around which the wind direction is 'almost tangential to isobarometric lines'.[147] He named these areas *cyclonoïdes*. The phenomenon known as *bourrasque* (or gale) thus emerged as the last structure of air masses following an order of magnitude, covering a much wider area than cyclones and *cyclonoïdes*. Finally, not satisfied with the elucidation of atmospheric dynamics, Marié-Davy penetrated into the field of prediction. As he asserted in a presentation given to the Academy of Sciences on 12 October 1863, 'by analysing our weather charts it is generally possible to predict the arrival of relatively strong gales along our coasts twenty-four or forty-eight hours beforehand'.[148] Or to put it simply, atmospheric disturbances are preceded by some particularly visible signs: the deformations in isobarometric lines. With this he transformed the 'area of low pressure' from a purely graphic description of enormous air masses into a more analytical tool aimed at forecasting the weather.

Marié-Davy had most of these results by August 1863, although he first issued a warning of strong winds on 17 September. By 10 September, the *Bulletin international* included, with the traditional columns of figures, a chart of Europe with the isobaric patterns of that day, and devoted a separate section to the weather forecast for the next day and the meteorological state (Figure 3.2). By December 1863, the Paris Observatory daily telegraphed their weather and wind forecasts to the French press and ports, as well as to Belgium, Denmark, Italy, and Spain; and named them *probabilités*—rather than *avertissements* (Le Verrier) or *forecasts* (FitzRoy). Here Marié-Davy was putting the synoptic technique to the service of weather forecasting.[149]

Marié-Davy and Secchi are among the founders of synoptic meteorology. In many respects, their cartographical methods are not only complementary, but also concatenating: Secchi and Serra-Carpi drew isobars and isotherms and showed the direction of cyclonic storms in Europe by identifying atmospheric waves with pressure fronts, while Marié-Davy first searched for these waves, and after failing that, took the isobaric technique one step further, and identified areas of low pressure or depressions by quasi-circular isobaric curves. However, their theoretical foundations were different insofar as the constraints of the former impelled and fed the expectations of the latter.

147 English merchant captain Henry Piddington first used the term 'cyclone' in his 1848 book *The Sailor's Horn-Book for the Law of Storms*, to refer to all winds turning around a common centre. Drawing on observations collected by seamen during their voyages through the Indian Ocean and the Caribbean waters, he examined the structure of tropical storms. Piddington's ideas were widely disseminated in France, especially after the translation of the book into French in 1859. For the influence of Piddington's ideas on Marié-Davy, see Locher (2008, pp. 112–14).

148 Marié-Davy (1863c, p. 643)—as translated by Locher (2009, p. 90).

149 For further details about the newspaper companies, French coastal towns, and foreign administrations receiving the Observatory's daily dispatches, see Locher (2008, pp. 119–21).

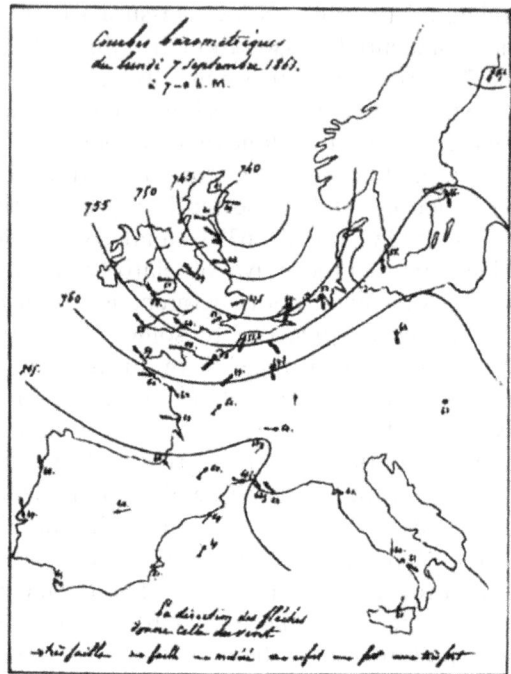

Figure 3.2 Chart of barometric curves and winds, 7 September 1863, drawn by Marié-Davy and published in the *Bulletin international de l'Observatoire impérial de Paris*, 10 September 1863.

Secchi's synoptic meteorology was not dominated by theory. Secchi embraced what he called a 'healthy experimentalism' and strove for minimizing theoretical prejudice. He regarded Birt's and Quetelet's atmospheric waves as detectable and measurable, and identified the crests and depressions of these waves by evolving self-registering devices and an isolinear technique. He did not aim at a storm theory but instead sought the empirical correlations between atmospheric waves and pressure fronts. His deep conviction of the intimate nature of electricity, magnetism, and mechanics focused him on the correlations of the various atmospheric phenomena rather than on their physical causes or processes. His mechanistic conception of physical reality led him to conceive the atmosphere as essentially unified and comprehensible in terms of wave motion and matter. The most essential notion was the idea of a unified universe. Phenomena (atmospheric waves, storms, aurorae) were correlated by the unitary laws that govern them.[150]

150 A philosophical essay by Secchi just published in 1864, *L'unità delle forze fisiche saggio di filosofia naturale*, reinforced his view on the unified nature underlying his scientific and meteorological practice. On the roots of this view in Catholic culture, see Redondi

In contrast, Marié-Davy's synoptic meteorology was dominated by both theory and intuitive judgement. Although he first embraced atmospheric waves, he later constructed his own storm theory from a combination of theoretical and practical resources, including Piddington's cyclonic concept, personal judgement, and the qualitative interpretation based on intuition and experience. Thus, in contrast to Quetelet's waves (traced by parallel and equidistant lines) and Secchi's correlations between wave forms and barometric depressions, Marié-Davy defined the 'areas of low pressure' as compact physical entities in the form of quasi-circular isobaric curves. His predictive interpretations of isobaric curves followed no empirical rule, nor were they subject to algorithm or statistical law. They were highly intuitive and individual in their construction and reflected the weakness of his theoretical physical foundations. As he himself acknowledged in 1864:

> It is not easy to explain how we predict the weather as we change tack day to day. The best way of explaining how the method works is to analyse successive series of barometric charts and the resulting forecasts.[151]

Only after his first weather forecasts based on synoptic charts did Marié-Davy offer a synthetic view of storm formation and motion. In fact, his interest in theory had a practical origin, as it was expounded in a book on 'the movements of the atmosphere and seas considered from the viewpoint of weather forecasting'.[152] Marié-Davy's view can be called a theory because of its ability at explaining the complex. Yet it was different from Espy's convective theory or even Dove's and Redfield's mechanical theories. It was not mathematical, and it was geared towards prediction. It was a mechanistic theory in which a storm was defined as a gigantic rotating air disc at a certain altitude. The central concept, that of storm formation and its forms of motion, was a balance of forces: the inflow of air at the two ends of the cyclone axis compensated the outflow of air at the centre caused by centrifugal forces. Later, he would differentiate formation from motion: while the primary cause of its formation was the condensation of water vapour, its motion depended on the large currents in the atmosphere.[153]

Thus unlike other synoptic precursors such as Quetelet's atmospheric waves, Liais's map of pressure lines, or Secchi's isolinear method—whose

---

(1980, pp. 797–811). On Secchi's overall picture of a unified universe, see Mazzotti (2010, pp. 67–71).

151 Marié-Davy in his intervention in the session held on 10 May 1864, *Annuaire de la Société Météorologique de France*, 1864, 12, p. 208—as translated by Locher (2009, p. 94).

152 Marié-Davy, *Météorologie. Les mouvements de l'atmosphère et des mers considérés au point de vue de la prévision du temps*. Paris: Victor Masson et Fils, 1866a. Marié-Davy's notion was later developed by astronomer Hevre Faye, for whom cyclones formed in the upper strata of the atmosphere.

153 See the second edition of his book, Marié-Davy (1877, pp. 226–27).

influences were somewhat limited by their unsatisfactory realization or speculative basis—Marié-Davy's method was able to use institutional channels like the Paris Observatory and the *Bulletin* to establish its authority as a technique for weather forecasting. Even his storm theory served as an inspiration to other Observatory's physicists such as Émile Fron and Léon Sonrel, who wrote two doctoral theses on the general movement of the atmosphere, both dedicated to Marié-Davy.[154] In time, his synoptic method became a powerful tool, not only for identifying and tracking the new physical entities known as barometric depressions but also as a means of scientifically legitimating both the use of telegraph and mapping for weather forecasting and the need for centralized meteorological services. In the near future, however, it suffered major setbacks: Le Verrier suppressed twenty-four hour forecasts, relieved Marié-Davy of his functions, and abolished the storm warning service—which would be reinstituted in May 1866 only to warn of the bad weather and its duration.[155] Behind these setbacks lay tensions between Le Verrier and Vaillant who contended for different methodologies and interpretative practices, as shown below.

Le Verrier versus Vaillant

For Vaillant, a storm warning service must satisfy two requirements: centralization and the non-dependence on isobaric maps. According to him, Le Verrier's isobaric maps provided a 'somewhat posthumous forecast'[156]— or to express this in his own words: isobarometric maps can never provide 'a snapshot of the atmosphere', since observations are made in the early morning and meanwhile pressures can evolve significantly.[157] Vaillant strongly advocated a non-predictive warning service based on an empirical, inductive, and non-geometrical method. This illustrates an important conclusion—that synoptic weather forecasting was the result not so much of an imperial thrust by state-backed authorities imposing an Eulerian ap-

---

154 Léon Sonrel, *Étude sur les mouvements généraux de l'atmosphère*. Versailles: Beau jeune, 1867; Émile Fron, *Étude sur les mouvements généraux de l'atmosphère dans leurs rapports avec les orages*. Versailles: Imprimeur Libraire Beau, 1868.

155 The second half of the 1860s was a period of institutional instability at the Paris Observatory in which internal conflicts, ministerial tensions, and international rivalries strongly converged. Part of this crisis, however, was due to Le Verrier's authoritarian nature, which led its astronomers to sign a letter of collective resignation in 1870 as a sign of protest against him and his dictatorial style. The litany of his onslaughts is endless: according to Camille Flammarion, 104 astronomers and calculators quit the Paris Observatory between 1854 and 1870. For more on the history of this turbulent period, see Locher (2008, pp. 130–39); Lequeux (2013, pp. 135–42; 291–93).

156 Marié-Davy made these statements in a meeting of the Academy of Sciences and referred to the Paris Observatory's maps on 28, 29, and 30 November published in the days preceding the famous storm on 2 and 3 December 1863. See Vaillant (1863, p. 1003).

157 *Bulletin international de l'Observatoire impérial de Paris*, 8 February 1864.

proach 'from above', but rather of tensions between the various agents involved (the Navy, the Observatory, etc.) that were resolved in favour of the adoption of this approach.

The isobaric method provoked feelings of frustration and doubt in Vaillant. These were expressed in the form of public debates and private correspondence around 1864. For example, the note submitted to the Paris Academy for its publication in the *Comptes rendus* in December 1863: 'the speed with which the drafting work has to be carried out' at the Observatory 'leads us to regard the [isobarometric] curves as a mere summary indication, to which one should not bestow too much credibility'.[158] This note irritated Le Verrier, who wrote a letter to Vaillant asking him to withdraw it. The note meant a frontal attack on the scientific authority of Le Verrier and his staff. But Vaillant did not withdraw nor did he want to withdraw it: 'I am open to being beaten, I will not reply; but I will send some other articles of meteorology. This part of physics is still obscure; I would not want to die before putting my two cents'.[159]

Vaillant's anti-isobaric feelings could become radical. For example, when Secchi defended *à outrance* the existence of storms from the barometric pressure drops observed in Rome and Greenwich about twenty hours apart, Vaillant asked a tremendously mordant question: 'Who can honestly believe that, because a rough weather was observed at Greenwich and because a strong gale appeared in Rome twenty hours later, there is necessarily a concomitance between the two phenomena?'[160] Recalling other entities long ago fallen into oblivion, such as Descartes's *tourbillons* and the epicycloids of the Ptolemaic system, Vaillant expressed scepticism about the existence of those isobaric entities known as storms—those storms 'that promenade on the surface of our globe like wandering Jews doing from 10 to 12 leagues per hour'.[161]

Vaillant's critical view linked the existence of storms to the isobaric forms of synoptic maps. Forecasts, he thought, have a potentially negative component, for the evolution of bad weather was inexorably linked to the motion of geometric bodies on isobaric charts. Everything would change if, instead of forecasts, one could continuously monitor atmospheric conditions and warn of bad weather as soon as it happened by sending some indications on how the situation might evolve. Like his own 1854 proposal, transmitted to Le Verrier on the occasion of the Black Sea storm, Vaillant again advocated

158 Vaillant (1863, p. 1003).
159 Vaillant to Le Verrier, 22 December 1863. ABIF, Ms 3714.
160 In the *Bulletin international* on 15 January 1864, Vaillant anticipated the approach of a storm from barometric observations made in Rome and Greenwich on 2, 3, and 4 December 1863.
161 *Bulletin international de l'Observatoire impérial de Paris*, 8 February 1864. For Vaillant, storms originated from '*pôles de refroidissement*' (cooling poles) around which the condensate air would have begun to move and blow.

for a non-predictive and centralized warning service capable of detecting a phenomenon and following its course.[162]

Both in the following meetings of the Academy and his correspondence with Vaillant, Le Verrier was cautious and gentle.[163] He kindly rebutted Vaillant's critics by showing letters of appreciation from some port authorities in which the value of his storm warnings was recognized.[164] Le Verrier's continence in this issue is important in its historical context, for it was at the end of 1863 that the struggle for the control of the *Bureau des longitudes* reached its peak; a struggle between, on one side, Le Verrier, current *Bureau*'s chairman Vaillant, and the former Minister of Public Instruction Gustave Rouland, and, on the other, the allied side of the *Bureau*, led by astronomer Charles-Eugène Delaunay and the new minister Victor Duruy (Rouland's successor) (Figure 3.3). Hence the debate on the advisability of promoting forecasts or warnings was subordinated to a broader power struggle for control and centralization.[165]

Historians have emphasized the importance of both technology and institutional dynamics in the emergence of synoptic meteorology.[166] Thus, the process of learning to forecast the weather was enabled, shaped, and even constrained by the various techniques and sociocultural concerns through which it was achieved. We cannot doubt the influential role that factors such as the establishment of storm warning services, the expansion of telegraphic networks, and improvements in graphic technology (especially in mapping and printing), had. However, this type of analysis often forgets that the methods for weather prediction cannot be evaluated in the abstract and be dissociated from the local context. These methods were usually conditioned by what interests served to, who promoted, and by the dominant values within the political and scientific community; this is why the methods chosen could be relevant to the ruling regime or the established social order. The true promoters of prediction methods were both scientific and political authorities, and these would not have embraced them if they did not satisfy the vested interests. Placing the full weight on technological advancements leaves out the responsibility of social and political factors. Likewise, these advancements are as important as their cultural 'environment', namely, the context in which they are interpreted and implemented. Hence, the tensions

---

162 *Bulletin international de l'Observatoire impérial de Paris*, 8 February 1864.

163 For a full and detailed account of the controversy between Vaillant on the one side, and Le Verrier and Marié-Davy on the other side (which also included in part FitzRoy and Matteucci, aligned with Vaillant's views in many respects), see Locher (2008, pp. 123–29).

164 Le Verrier (1864, pp. 19–21); Vaillant to Le Verrier, 22 December 1863, 4 January 1864. ABIF, Ms 3714.

165 Davis (1984, p. 372). The persistence of these tensions between the two sides in part resulted in Le Verrier's dismissal and the brief reign of Delaunay as director of the Paris Observatory in 1870. See Lequeux (2013, pp. 193–99).

166 See, e.g., Fierro (1991, pp. 93–113); Khrgian (1970, pp. 63–183); Monmonier (1999, pp. 39–56); Nebeker (1995, pp. 16–21, 36–40); Walker (2012, pp. 23–92).

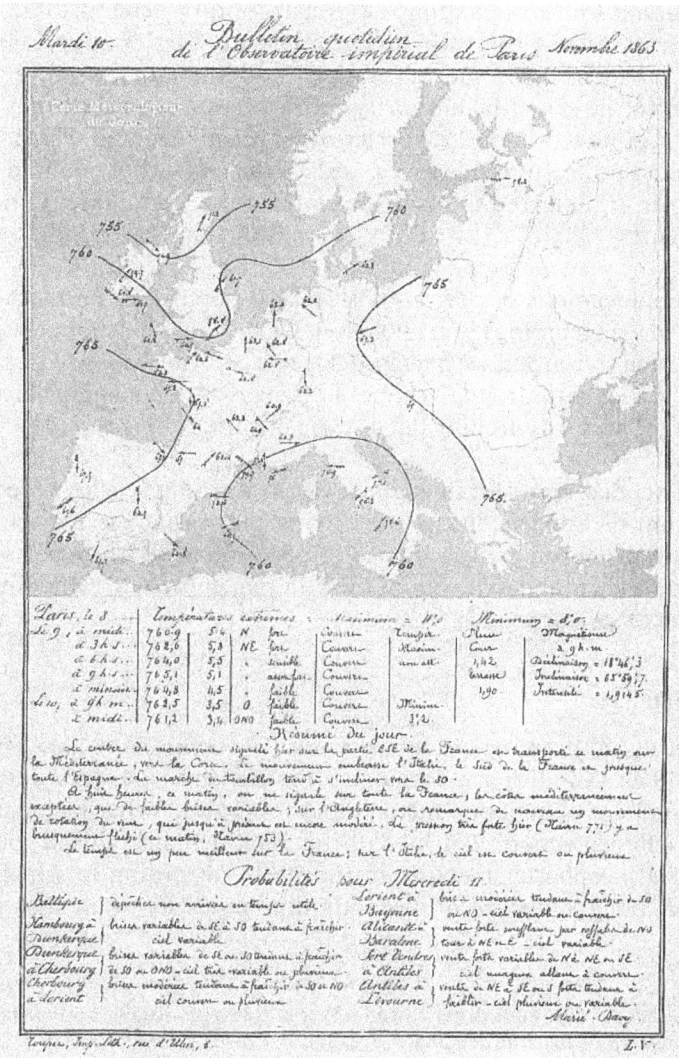

Figure 3.3 Illustration from the *Bulletin quotidien de l'Observatoire impérial de Paris,* 10 November 1863. Below forecasts (or *probabilités*) for European coasts on 11 November 1863.

between the expectations about the national security and safety these methods were supposed to produce and the results obtained could prevent a systematic or strict implementation of the same, thereby opening space for the search for common ground—like centralization and nation-state building.

This is what Le Verrier had discovered after having some disputes with Vaillant in the pages of the *Bulletin* and the *Comptes rendus*, which had dealt with not only the existence of storms but also the forecast service.[167] Vaillant urged him to abandon the preparation of general forecasts and focus instead on warnings for bad weather. Le Verrier suppressed these forecasts, not without some reluctance.[168] Vaillant's views surely meant an affront to the scientific authority of Le Verrier. However, they formed part of a regime to which Le Verrier owed allegiance and where Vaillant was the most powerful of his allies, and all that was what counted. That is why Le Verrier preferred to place the emphasis on common ground rather than dissent:

> The appreciations of port authorities and population on the service of maritime meteorology [were positive]. We believe their support would not have been so warm if we had announced a calm and reliable weather; the ships that departed with the hope of this promise would have been thrown ashore. There was nothing but a string of curses along the coast.[169]

Le Verrier ended up re-establishing a storm warning service based on the examination of isobaric charts—the living proof that the materialization of synoptic meteorology was the result of tensions which were essentially resolved on the ground in favour of the Eulerian approach, rather that of the imposition of said approach from above.

Conclusion

Contributing with an article to the work *The Atmosphere and the Sea in Motion* (1959), Swedish meteorologist Tor Bergeron outlined the historical development of synoptic meteorology.[170] He and other members of the so-called Bergen school, led by Vilhelm Bjerknes, introduced the frontal notion of atmospheric structure in the 1910s, according to which storms developed along a surface of discontinuity between two air masses—that

---

167 For further details of this dispute, see Le Verrier (1864, pp. 16–19); *Bulletin international de l'Observatoire impérial de Paris*, 3, 5, 8, and 11 February 1864.

168 In contrast to Le Verrier's moderate attitude, Marié-Davy took a radical rationalist-deductionist stance, and in a letter to Quetelet, he called Vaillant's theories as 'false and puerile'. Marié-Davy to Quetelet, 18 June 1864. ARSB, M1720. As regards the cause-effect relationship between snow and storm, as held by Vaillant to explain the storm on 2 December 1863, Marié-Davy (1864, p. 67) demurred: 'It is not the snow that produces the storm, but rather the storm that determines the snow and rain falling'.

169 Le Verrier (1864, p. 20). Under the cover of this utilitarian language, however, Le Verrier stand firm upon the identification of depressions with isobaric curves and the predictive potentiality of the Eulerian approach.

170 Tor Bergeron, 'Methods in Scientific Weather Analysis and Forecasting. An Outline in the History of Ideas and Hints at a Program'. It was printed in the Rossby Memorial Volume, edited by Bert Bolin, *The Atmosphere and the Sea in Motion* (New York, 1959); and its historical contents were reprinted in the journal PAGEOPH, 1980–81, 119, 443–73.

is, there was a front separating polar and equatorial air.[171] In his article, Bergeron asserted that a new era dawned with the polar-front theory, an era resting upon Bjerknes's following principles: (1) the weather-forecast problem should be treated by a 'rational physical approach'; (2) the physical laws should be applied to 'the atmospheric states at any moment, and not only to average values'; and (3) the synoptic tool should be improved by including orography, stream-lines, and other wind isolines on the maps.[172] Bjerknes was not alone in recognizing the need for atmospheric structure models subject to the laws of mechanics and thermodynamics; Bergeron himself stated that 'the very names of Dove's ingenious structure models, *Polar current* and *Equatorial current*, became the symbols of a repugnant school of thought', these being 'strongly rejected by the new weathermen' in the 1860s. The common characteristic of what was called the 'new era' is that many meteorologists, especially those of the Bergen school, expressed their misgivings about, or even their rejection of, baric-synoptic approaches in treating the weather-forecast problem.

Besides being highly transgressive, Bjerknes's and Bergeron's assertions were typical of the theoretical meteorologists, who placed great emphasis on the need for three-dimensional and dynamical physical models. Yet, the allusions to the insufficiencies of synoptic maps were equally explicit and reiterated. In his 1959 article Bergeron asserted that after 1865, the synoptic map 'was employed to treat only a minor part of the observations available':

> Just one rather poor and one-sided structure model was used: the pressure field at 0–level and its alleged correlation with the weather. Together observations, synoptic tool and structure model provided a most convenient way of handling weather forecasting. So comfortable was this Eulerian method, and so well founded physically it seemed, that its use became world-wide within a short time. Anyone could learn to draw circular or oval shaped isobars around the centers of low pressure, thus showing the Lows, which were supposed to be the real *carriers* of the bad weather, and therefore also hailed as the main object of study for the weather scientist and the practical forecaster.[173]

Bergeron's last considerations are most significant, for he openly asserts that the Eulerian method lived in a very comfortable setting in which it easily attained a position of hegemony, despite the weakness of its physical

---

171 On the polar-front theory and the shaping of a new meteorology, see Friedman (1978, 1989, pp. 81–96, 179–95); Fleming (2016, pp. 13–74).

172 As quoted by Bergeron (1959, p. 454). Bjerknes included other two principles: (4) the necessity of a dense network of observations that could be applied to the calculation of derivatives; and (5) the idea of specialized atmospheric structure models. These principles can be found in Bjerknes (1913).

173 Bergeron (1959, p. 445).

foundations, and he points to meteorologists' readiness to make practical forecasting a routine, graphic, and even geometric task.

This circumstance is hardly surprising if one bears in mind that the synoptic method was not, according to a contemporary meteorologist, 'a set of rules or a dogma, but rather a set of instructions': 'the weather only can be forecasted if one studies it and understands it—not by the automatic application of finished recipes'.[174] Indeed, many meteorologists viewed synoptic charts as the end product of the application of the Eulerian method, as the outcome of a laudable way of understanding weather patterns. As I showed, this view dates back to the 1840s. The theory of atmospheric waves did not solve the challenges posed and proved to be only the half-way house to the modern Eulerian view, which is fairly clear in Quetelet and even more explicit in Liais, Le Verrier, and Secchi, though it received its fullest formulation only in Marié-Davy: the view that the pressure distribution was the primary cause of all weather changes, including those in wind direction.

The very fact that the transition to this Eulerian view experienced an abrupt leap circa 1855 and was driven by no new physical atmospheric theory suggests that other factors—closely linked to the political and social environments in which meteorologists interacted—played a decisive role in this shift. Or, to express this in more precise and concrete terms: in the mid-1850s, the Paris Observatory's weather scientists—led by Le Verrier, and guided by Liais—induced by the political opportunism and needs for maritime and military security in the months following the Black Sea storm altered their methods and adopted an Eulerian approach to the predictive study of storms. Practical needs and aims, rather than theoretical motivations, became the reason for adopting such an approach. For those scientists, the Eulerian approach became a functional tool aimed at achieving the goals of order, centralization, and control, rather than an epistemic end in itself. With this shift materialized, many meteorologists, for similar aims, subsequently embraced the cross of the Eulerian method.

Le Verrier adopted the Eulerian approach as part of his efforts to reorganize the Paris Observatory and create an ambitious international system for storm warning; that is, he adopted it as a means to attain order and scientific centralism, not the other way round. The 1854 Black Sea storm was the trigger for this adoption, as it generated a situation of political urgency regarding maritime and military safety in the Napoleonic regime. A series of technical and national considerations, such as advantages of continuous recording, search for correlations, and interest in both administrative centralization and territorial cohesion, all contributed to Secchi's adoption of the Eulerian method for the Pontifical State's storm warning service. In this process, Secchi moved from his early graphic method aimed at revealing the motion of atmospheric waves to a cartographic method aimed at foreseeing

---

174 Chromow (1940, p. 7).

the approach of storms. Finally, Marié-Davy gave it its fullest formulation by transforming the undetectable atmospheric waves into a discernible and foreseeable object: a depression encircled by isobaric curves.

There is every indication that the diffusion of the Eulerian approach among weather scientists was due not to an imposition by state-backed authorities from above, but rather to the result of tensions between forces associated with the nation-building process. In fact, the most ardent and firm opponent to the isobaric method in France was Vaillant himself, whose ideas were basically in line with the Lagrangian approach defended first by Dove and later by FitzRoy, as will be shown in the next chapter.

Therefore, the failure by historians to identify the confrontation between two currents of thought in nineteenth-century weather forecasting is by no means negligible. While national cultural factors may well have played a role in the conception of weather forecast methods, they do not seem to be determining factors in the evolution of weather prediction in that century. Rather, it is only by reference to the confrontation of two approaches— that is, the tension between the expectations about the national and maritime security the Eulerian and Lagrangian method-based forecasts were supposed to produce and the results obtained—that one can explain this evolution.

References

Achbari, Azadeh, 2015. 'Building Networks for Science: Conflict and Cooperation in Nineteenth-Century Global Marine Studies'. *Isis*, 106 (2), 257–82.

Achbari, Azadeh, 2017. *Rulers of the Winds. How Academics Came to Dominate the Science of the Weather, 1830–1870*. Amsterdam: Vrije Universiteit.

Agulhon, Maurice, 1992. *1848 ou l'apprentissage de la république. 1848–1852*. Paris: Seuil.

Altamore, Aldo, and Maffeo, Sabino eds., 2012. *Angelo Secchi. L'avventura scientifica del Collegio Romano*. Vaticano: Specola Vaticana.

Ancellin, Jacques, 1975–1978. 'Un homme de science du XIXème siècle: l'astronome Emmanuel Liais'. In M. Maurice Durchon (dir.). *Mémoires de la Société Nationale des Sciences Naturelles et Mathématiques de Cherbourg*, vol. 57. Coutances: Imprimerie OCEP.

Anduaga, Aitor, 2017. *Cyclones and Earthquakes: The Jesuits, Prediction, Trade and Spanish Dominion in Cuba and the Philippines, 1850–1898*. Quezon City: Ateneo de Manila University Press.

Arnold, Guy, 2002. *Historical Dictionary of the Crimean War*. London: The Scarecrow Press.

Aubin, David, 2003. 'The Fading Star of the Paris Observatory in the Nineteenth Century: Astronomer's Urban Culture of Circulation and Observation'. *Osiris*, 18, 79–101.

Badem, Candan, 2010. *The Ottoman Crimean War (1853–1856)*. Leiden: Brill.

Barboza, Christina Helena da Motta, 2005. 'Nice Weather, Meteors at the End of the Day'. In Stefan Emeis, and Cornelia Lüdecke eds., *From Beaufort to Bjerknes* and Beyond: Critical Perspectives on Observing, Analyzing, and Predicting Weather and Climate. Augsburg: Rauner, 157–68.

Barboza, Christina Helena da Motta, 2008. 'Liais, Emmanuel-Bernardin'. In Noretta Koertge ed., *New Dictionary of Scientific Biography*. Detroit, MI: Scribner, 4, 293–94.

Barboza, Christina Helena da Motta, 2012. *As viagens do tempo. Uma história da meteorologia em meados do século XIX*. Rio de Janeiro: E-papers Serviços Editoriais Ltda.

Barthalot, Raymonde, 1978. *Pour une histoire de l'Observatoire de Paris au XIXème siècle*. Paris: Pantheon-Sorbonne.

Beauchamp, Alphonse de ed., 1882. *Recueil des lois et règlements sur l'enseignement supérieur: comprenant les décisions de la jurisprudence et les avis des conseils de l'instruction publique et du conseil d'Etat*. Paris: Typographie de Delalain Frères, 2, 306–11.

Bergeron, Tor, 1959. 'Methods in Scientific Weather Analysis and Forecasting. An Outline in the History of Ideas and Hints at a Program'. In Bert Bolin ed., *The Atmosphere and the Sea in Motion*. New York: Rockefeller Institute Press, Oxford University Press. Reprinted as 'Synoptic Meteorology: An Historical Review'. *Pageoph*, 1980–81, 119, 443–73.

Besso, Beniamino, 1871. *L'elettricità e le sue applicazioni*. Milano: E. Treves.

Biot, Jean-Baptiste, 1855. 'Opinion de M. Biot sur les observatoires météorologiques permanents que l'on propose d'établir en divers points de l'Algérie'. *CR*, 41, 1177–90.

Bjerknes, Vilhelm, 1913. *Die Meteorologie als exakte Wissenschaft. Antrittsvorlesung gehalten am 8. Januar 1913 in der Aula der Universität Leipzig*. Braunschweig: F. Vieweg & Sohn.

Brenni, Paolo, 1993. 'Il meteorografo di Padre Angelo Secchi'. *Nuncius*, 8 (1), 196–247.

Buys Ballot, C.H.D., 1864. 'On the System of Forecasting the Weather as Pursued in Holland'. *Report of the British Association for the Advancement of Science*, 33, 20–21.

Chinnici, Ileana, and Gramatowski, Wiktor, 2001. 'Le carte di Angelo Secchi S.J. (1818–1878) conservate presso la Pontificia Università Gregoriana. Un inventario inedito rivisitato'. *Nuncius*, 16 (2), 571–627.

Chromow, Swoboda P., 1940. *Einführung in die synoptische Wetteranalyse*. Berlin: Springer-Verlag.

Cittadella, Alex, 2014. 'Dalla meteorologia agraria alla climatologia scientifica. Spunti di ricerca da Giuseppe Toaldo a Renato Biasutti'. In Alessio Fornasin, and Claudio Povolo eds., *Per Furio. Studi in onore di Furio Bianco*. Udine: Forum, 317–28.

Cox, John D., 2002. *Storm Watchers: The Turbulent History of Weather Prediction from Franklin's Kite to El Niño*. Hoboken, NJ: John Wiley & Sons, Inc.

Coyne, George S. J., and Maffeo, Sabino S. J., 2001. 'Astronomia al Collegio Romano'. In Domenico Vento ed., *Presenze scientifiche illustri al Collegio Romano: Celebrazioni del 125° anno di istituzione dell'Ufficio Centrale di Ecologia Agraria*. Roma: Istituzione dell'Ufficio Centrale di Ecologia Agraria, 40–46.

Danjon, André, 1946. 'Le Verrier, créateur de la météorologie'. *La Météorologie*, 1 (1046), 863–82.

Davis, John L., 1984. 'Weather Forecasting and the Development of Meteorological Theory at the Paris Observatory, 1853–1878'. *Annals of Science*, 41 (4), 359–82.

'Décret du 30 janvier 1854'. *Bulletin administratif de l'Instruction publique*, 1852, 3, 23–29.

Denza, Francesco, 1883. 'La meteorologia in Italia'. In *Gli Studi in Italia*. Roma: Befani, 6, 405–49.

Dowe, Dieter, Haupt, Heinz-Gerhard, Langewiesche, Dieter, and Sperber, Jonathan eds., 2008. *Europe in 1848. Revolution and Reform*. New York: Berghahn Books.

Ducos, Théodore, 1855. 'Rapport sur les observatoires météorologiques proposés pour l'Algérie'. *CR*, 1855, 41, 1130–37.

Eredia, Filipo, 1914. 'L'organizzazione del servizio dei presagi del tempo in Italia'. *Rivista Meteorico-Agraria*, 35, 1001–48.

Eriksson, Kai, 2011. *Communication in Modern Social Ordering: History and Philosophy*. London: Continuum International Pub. Group.

Feurtet, Jean-Marie, 2005. *Le Bureau des longitudes (1795–1854). De Lalande à Le Verrier. Thèse pour le diplôme d'archiviste-paléographe*. Paris. https://halshs.archives-ouvertes.fr/halshs-00087963/en/

Fierro, Alfred, 1991. *Histoire de la météorologie*. Paris: Éditions Denoël.

FitzRoy, Robert, 1862. 'An Explanation of the Meteorological Telegraph and its Basis now under Trial at the B.o.T.'. *Proceedings of the Royal Institution of Great Britain*, 444–56.

Fleming, James Rodger, 2016. *Inventing Atmospheric Science: Bjerknes, Rossby, Wexler, and the Foundations of Modern Meteorology*. Cambridge, MA: The MIT Press.

Fortoul, Hippolyte, Vaillant, Jean-Baptiste P., and Bonaparte Louis-Napoléon, 1854. 'Réorganisation de l'Observatoire de Paris et du bureau des longitudes. Rapport a l'empereur et décret'. *Bulletin administratif de l'instruction publique*, 5 (49), 16–28.

Friedman, Robert Marc, 1978. *Vilhelm Bjerknes and the Bergen School of Meteorology, 1918–1923: A Study of the Economic and Military Foundations for the Transformation of Atmospheric Science*. Baltimore, MD: Johns Hopkins University Press.

Friedman, Robert Marc, 1989. *Appropriating the Weather. Vilhelm Bjerknes and the Construction of Modern Meteorology*. Ithaca: Cornell University Press.

Fron, Émile, 1868. *Étude sur les mouvements généraux de l'atmosphère dans leurs rapports avec les orages*. Versailles: Imprimeur Libraire Beau.

Goldfrank, David M., 2014. *The Origins of the Crimean War*. London: Routledge.

Gooch, John, 2001. *The Reunification of Italy*. London: Routledge.

Granger, Catherine, 2005. *L'empereur et les arts: la liste civile de Napoléon III*. Paris: École des Chartes.

Grant, Robert, 1852. *History of Physical Astronomy*. London: Henry G. Bohn.

Grellois, Eugène, 1863. 'Analyse des bulletins météorologiques de l'Observatoire du Collége Romain'. *Annuaire de la Société Météorologique de France*, 11, 20–26.

Grellois, Eugène, 1870. *Météorologie religieuse et mystique*. Metz: F. Blanc.

Halford, Pauline, 2004. *Storm Warning. The Origins of the Weather Forecast*. Stroud: Sutton Publishing.

Headrick, Daniel R., 1988. *The Tentacles of Progress. Technology Transfer in the Age of Imperialism, 1850–1940*. Oxford: Oxford University Press.

Headrick, Daniel R., 1991. *The Invisible Weapon: Telecommunications and International Politics, 1851–1945*. Oxford: Oxford University Press.

Hind, J. R., 1878. 'Le Verrier'. *Monthly Notices of the Royal Astronomical Society*, 38, 155–68.

Huurdeman, Anton A., 2003. *The Worldwide History of Telecommunications*. Hoboken, NJ: John Wiley & Sons.

Iafrate, Luigi, 2008. *Fede e scienza: un incontro proficuo. Origini e sviluppo della meteorologia fino agli inizi del '900*. Roma: Ateneo Pontificio Regina Apostolorum.

Iafrate, Luigi, 2011. 'Padre Angelo Secchi: l'astronomo italiano che inventò le previsioni del tempo'. *Nova Historica: rivista internazionale di storia*, 36, 110–30.

Jankovic, Vladimir, 1998. 'Ideological Crests versus Empirical Troughs: John Herschel's and William Radcliffe Birt's Research on Atmospheric Waves, 1843–50'. *British Journal for the History of Science*, 31, 21–40.

Julien, Charles André, 1964. *Histoire de l'Algérie contemporaine. La conquête et les débuts de la colonisation (1827–1871)*. Paris: Presses universitaires de France.

Khrgian, Aleksandr Khristoforovich, 1970. *Meteorology: A Historical Survey*, 2nd ed. Jerusalem: Israel Program for Scientific Translations.

Lamotte, Françoise, Lantier, Maurice, and Levert, Paul, 1977. *Urbain Le Verrier, savant universel, gloire nationale, personnalité contentine*. Coutances: OCEP.

Landsberg, Helmut, 1954. 'Storm of Balaklava and the Daily Weather Forecast'. *The Scientific Monthly*, 79 (6), 347–52.

Laurencin, Paul, 1902. *Le Maréchal Vaillant (1790–1872)*. Paris: Les Contemporaines.

Le Verrier, Urbain Jean Joseph, 1846. 'Recherches sur les mouvements d'Uranus'. *CR*, 22, 907–18.

Le Verrier, Urbain Jean Joseph, 1854a. 'M. Le Verrier présente à l'Académie un résumé des observations de la pression barométrique et de la température, faites à l'Observatoire impérial de Paris pendant les mois de Janvier, Février, Mars et Avril 1854, et accompagne cette communication des remarques suivantes'. *CR*, 38, 797–99.

Le Verrier, Urbain Jean Joseph, 1854b. 'M. Le Verrier présente les observations météorologiques faites à l'Observatoire impérial pendant les mois de juillet, août et septembre 1854'. *CR*, 39, 771, 814–16.

Le Verrier, Urbain Jean Joseph, 1855a. 'Rapport sur l'Observatoire impérial de Paris et projet d'organisation'. *Annales de l'Observatoire Impérial de Paris*, 1, 1–68.

Le Verrier, Urbain Jean Joseph, 1855b. 'Note sur le développement des études météorologiques en France'. *CR*, 40, 620–26.

Le Verrier, Urbain Jean Joseph, 1855c. 'M. Le Verrier présente un travail fait à l'Observatoire impérial, par M. Liais, sur la tempête de la mer Noire, en novembre 1854'. *CR*, 41, 1197–1204.

Le Verrier, Urbain Jean Joseph, 1864. 'Remarques as sujet d'une note de M. le Maréchal Vaillant, sur la tempête des 2 et 3 décembre 1863'. *CR*, 58, 16–21.

Le Verrier, Urbain Jean Joseph, 1865. 'Réponse de M. Le Verrier à la nouvelle Note de M. Matteucci'. *CR*, 60, 1317–26.

Le Verrier, Urbain Jean Joseph, 1866. 'Avertissements donnés aux côtes sur l'approche des tempêtes. État present de la question'. *CR*, 62, 1045–52.

Le Verrier, Urbain Jean Joseph, 1868. *Historique des Entreprises Météorologiques de l'Observatoire Impériale de Paris, 1854–1867*. Paris: Gauthier-Villars.

Lequeux, James, 2013. *Le Verrier – Magnificent and Detestable Astronomer.* New York: Springer.

Lévêque, Pierre, 2008. 'The Revolutionary Crisis of 1848–1851 in France: Origins and Course of Events'. In Dieter Dowe, Heinz-Gerhard Haupt, Dieter Langewiesche, and Jonathan Sperber eds., *Europe in 1848. Revolution and Reform.* New York: Berghahn Books, 91–119.

Lévy, Jacques R., 1973. 'Le Verrier, Urbain Jean Joseph'. In Charles Coulston Gillispie ed., *Dictionary of Scientific Biography.* New York: Charles Scribner's Sons, 8, 276–79.

Liais, Emmanuel-Bernardin, 1849. *Considérations sur le Climat de Cherbourg.* Cherbourg: Imprimerie de Thomine.

Liais, Emmanuel-Bernardin, 1851. *Théorie mathématique des oscillations du baromètre et recherche de la loi de la variation moyenne de la température avec la latitude.* Paris: Bachelier.

Liais, Emmanuel-Bernardin, 1852. 'Résultats des observations météorologiques faites à Cherbourg pendant les années 1848, 1849, 1850 et 1851'. *CR,* 35, 349–53.

Liais, Emmanuel-Bernardin, 1854. 'De l'influence de la latitude sur la pression moyenne du baromètre et sur les directions générales du vent'. *Annuaire de la Société Météorologique de France,* 2, 51–57.

Liais, Emmanuel, 1881. *L'Espace Céleste, ou Description de l'Univers, suivi de récits de voyages entrepris pour en compléter l'étude.* 2. ed. Paris: Garnier Frères, Libraires-Éditeurs.

Liandier, M. 1862. 'Notice sur les courants des hautes régions. Observation du 26 mai 1862, à 10 heures du soir'. *Annuaire de la Société Météorologique de France,* 10, 119–20.

Locher, Fabien, 2007. 'L'empire de l'astronome : Urbain Le Verrier, l'Ordre et le Pouvoir'. *Cahiers d'histoire. Revue d'histoire critique,* 102, 33–48.

Locher, Fabien, 2008. *Le savant et la tempête. Étudier l'atmosphère et prévoir le temps au XIX siècle.* Rennes: Presses Universitaires de Rennes.

Locher, Fabien, 2009. 'Les météores de la modernité : la dépression, le télégraphe et la prévision savante du temps (1850–1914)'. *Revue d'histoire moderne et contemporaine,* 56 (4), 77–103.

Locher, Fabien, 2009. 'Atmosphere of Globalisation. Depressions, the Astronomer and the Telegraph (1850–1914)'. *Revue d'histoire moderne et contemporaine,* 56(4), 77–103.

Lunteren, Frans van, 1998. 'De oprichting van het Koninklijk Nederlands Meteorologisch Instituut: Humboldtiaanse wetenschap, internationale samenwerking en praktisch nut'. *Gewina,* 21 (4), 216–43.

Maffeo, Sabino S. J., 2001. 'Padre Angelo Secchi e la meteorologia'. In Domenico Vento ed., *Presenze scientifiche illustri al Collegio Romano: Celebrazioni del 125° anno di istituzione dell'Ufficio Centrale di Ecologia Agraria.* Roma: Istituzione dell'Ufficio Centrale di Ecologia Agraria, 21–31.

Mangianti, Franca, and Beltrano, Maria Carmen, 1990. *Il Collegio Romano: 100 anni di osservazioni meteorologiche.* Roma: Ufficio Centrale di Ecologia Agraria.

Marié-Davy, Edme Hippolyte, 1851. *Considération sur le climat de Montpellier.* Montpellier: Ricard.

Marié-Davy, Edme Hippolyte, 1863a. 'Sur la météorologie'. *Revue des sociétés savantes,* 3, 232–35.

Marié-Davy, Edme Hippolyte, 1863b. 'Sur l'état de la atmosphère pendant la première quinzaine d'août, d'après les renseignements recueillis à l'Observatoire impérial de Paris'. *CR*, 57, 384–86.

Marié-Davy, Edme Hippolyte, 1863c. 'Sur les tempêtes de l'équinoxe'. *CR*, 57, 640–44.

Marié-Davy, Edme Hippolyte, 1864. 'Tempêtes des 2 et 3 décembre 1863'. *CR*, 58, 65–67.

Marié-Davy, Edme Hippolyte, 1866a. *Météorologie. Les mouvements de l'atmosphère et des mers considérés au point de vue de la prévision du temps*. Paris: Victor Masson et Fils.

Marié-Davy, Edme Hippolyte, 1866b. *Mémoire adressé à Sa Majesté L'Empereur*. Paris.

Marié-Davy, Edme Hippolyte, 1868. *Notice sur les travaux scientifiques de M. Marié-Davy*. Paris: Gauthier-Villars.

Marié-Davy, Edme Hippolyte, 1877. *Météorologie Générale. Les Mouvements de l'atmosphère et les variations du temps: Avec 24 cartes tirées en couleur et de nombreux figures dans le texte*. Paris: G. Masson.

Marriott, William, 1904. 'Some Account of the Meteorological Work of the Late James Glaisher, F.R.S.'. *Quarterly Journal of the Royal Meteorological Society*, 30 (129), 1–28.

Martin, Claude, 1963. *Histoire de l'Algérie française, 1830–1962*. Paris: Ed. des 4 Fils Aymon.

Mazzotti, Massimo, 2010. 'The Jesuit on the Roof: Observatory Sciences, Metaphysics, and Nation Building'. In David Aubin, Charlotte Bigg, and H. Otto Sibum eds., *Observatories and Astronomy in Nineteenth-Century Science and Culture*. Durham, NC: Duke University Press, 58–85.

Mazzucato, Michele T., 2008. *Italiani nel sistema solare*. Milano: Maggioli editore.

'Meteorology', *CR*, 1855, 40, 439–40.

Milza, Pierre ed., 2008. *Napoléon III, l'homme, le politique: actes du colloque organisé par la Fondation Napoléon, Collège de France, amphithéâtre Marguerite de Navarre, 19–20 mai 2008*. Saint-Cloud: Napoléon III Éditions.

Moigno, François, 1879. *Le révérend père Secchi sa vie, son observatoire, ses travaux, ses écrits, ses titres à la gloire*. Paris: Gauthier-Villars.

Moncel, Théodose du (ed.), 1855–1857. *La Science, Journal des Sciences pures et appliquées et des découvertes et inventions*, vol. 3. Paris: [s.n.].

Monmonier, Mark, 1999. *Air Apparent. How Meteorologists Learned to Map, Predict, and Dramatize Weather*. Chicago, IL: The University of Chicago Press.

Nebeker, Frederik, 1995. *Calculating the Weather. Meteorology in the 20th Century*. New York: Academic Press.

'Nouvelles et faits diverses'. *Cosmos: revue encyclopédique hebdomadaire des progrès des sciences et de leurs applications aux arts et à l'industrie*, 1855, 6, 253.

Piddington, Henry, 1848. *The Sailor's Horn-Book for the Law of Storms, being a Practical Exposition of the Theory of the Law of Storms, and its Uses to Mariners of All Classes in All Parts of the World, Shewn by Transparent Storm Cards and Useful Lessons*. London: Smith, Elder.

Poey, André, 1855. 'Projet d'installation d'un Observatoire météorologique à la Havane, sous les auspices du gouvernement espagnol et de S. E. le capitaine général de l'île dé Cuba'. *Annuaire de la Société Météorologique de France*, 3, 202–08.

Proverbio, Edoardo, and Buffoni, Letizia, 2004. 'La meteorologia dei fenomeni estremi alla metà del XIX secolo e il contributo di Angelo Secchi allo studio dei rapporti tra fenomeni meteorologici, magnetici e solari'. *Physis; Rivista Internazionale di Storia della Scienza*, 41 (1), 125–65.

Redondi, Pietro, 1980. 'Cultura e scienza dall'illuminismo al positivismo'. In Gianni Micheli ed., *Storia d'Italia, annali 3: scienza e tecnica nella cultura e nella società dal Rinascimento a oggi*. Turin: Einaudi, 797–811.

Renou, Émilien, 1862. 'Instructions sommaires pour les observations météorologiques'. *Annuaire de la Société Météorologique de France*, 10, 75–81.

Robert, Adolphe, and Cougny, Gaston, 1889–91. *Dictionnaire des parlementaires français, comprenant tous les membres des Assemblées françaises et tous les ministres français depuis... 1789 jusqu'au 1er mai, 1889*. Paris.

Rollet de l'Isle, Maurice, 1950. 'Étude historique sur les ingénieurs-hydrographes et le Service hydrographique de la Marine (1814–1914)'. *Annales hydrographyques*, special issue (also published in Paris: Imprimerie Nationale, 1951).

Royle, Trevor, 2000. *Crimea: The Great Crimean War, 1854–1856*. New York: Palgrave Macmillan.

Schiavon, Martina, 2015. 'The Bureau des Longitudes: An Institutional Study'. In Richard Dunn and Rebekah Higgitt eds., *Navigational Enterprises in Europe and its Empires, 1730–1850*. Basingstoke and Hampshire: Palgrave Macmillan, 65–88.

Secchi, Angelo, 1853. 'Guida dei naviganti a lungo corso Vincenzo Gallo'. *Annali di scienze matematiche e fisiche*, 4, 245–60.

Secchi, Angelo, 1855. 'Corrispondenza meteorologica telegrafica in Roma'. *Corrispondenza scientifica di Roma per l'avanzamento delle scienze*, 25 June 1855, 4 (7–8).

Secchi, Angelo, 1856. *Memorie dell'Osservatorio del Collegio Romano, 1852–55*. Roma: Tipografia delle Belle Arti.

Secchi, Angelo, 1857. 'Intorno a un nuovo barometrografo'. *Atti Accademia Pontificia dei Nuovi Lincei*, 1 February 1857, 139–45.

Secchi, Angelo, 1858a. 'Di alcuni risultati ottenuti dalla corrispondenza meteorologica telegrafica, e dal barometrografo a bilancia'. *Atti Accademia Pontificia dei Nuovi Lincei*, 11, 389–95.

Secchi, Angelo, 1858b. *Misura della base trigonometrica eseguita sulla via Appia per ordine del governo pontificio nel 1854–55*. Rome: Rev. Camera Apostolica.

Secchi, Angelo, 1859. 'Descrizione di un meteorografo, ossia registratore meteorologico universale all'Osservatorio'. In *Memorie dell'Osservatorio del Collegio Romano, 1857–59*. Roma: Tip. delle Belle Arti, 1–8.

Secchi, Angelo, 1860. 'Alcune ricerche meteorologiche sulle tempeste occorse nel 1859–60'. *Atti Accademia Pontificia dei Nuovi Lincei*, 13, 231–49.

Secchi, Angelo, 1861. *Sui recenti progressi della meteorologia*. Roma: Tipografia delle Belle Arti.

Secchi, Angelo, 1862a. 'Distribuzione della pressione atmosferica sull' Europa durante il gennajo e il febrajo del corrente anno 1862'. *Bulletino Meteorologico dell'Osservatorio del Collegio Romano*, 1(3), 19–24.

Secchi, Angelo, 1862b. 'Sulle onde atmosferiche'. *Bulletino Meteorologico dell'Osservatorio del Collegio Romano*, 1(12), 93–95.

Secchi, Angelo, 1864. *L'unità delle forze fisiche saggio di filosofia naturale*. Roma: Tipografia Forense.

Secchi, Angelo, 1867. 'Sulla burrasca dal 14 al 15 gennaio 1867'. *Bullettino Meteorologico dell'Osservatorio del Collegio Romano*, 4, 1–4.

Secchi, Angelo, 1877a. 'Un omaggio alla memoria del Commodoro Maury'. *Bulletino Meteorologico del Collegio Romano*, 16, 41–42.

Secchi, Angelo, 1877b. *L'astronomia in Roma nel Pontificato di Pio IX. Memoria.* Roma: Tipografia della Pace.

Serra-Carpi, Giuseppe, 1865. *Sulle linee isotermiche dell'Italia de' suoi mari ed isole adiacenti.* Roma: Tipografia delle belle arti.

Shaw, William Napier, 1926. *Manual of Meteorology. Vol. I: Meteorology in History.* Cambridge: Cambridge University Press.

Sonrel, Léon, 1867. *Étude sur les mouvements généraux de l'atmosphère.* Versailles: Beau jeune.

Tricht, V. van, 1878. 'Le Père Secchi'. *Revue des questions scientifiques*, 4, 353–402.

Udías, Agustín, 2003. *Searching the Heavens and the Earth: The History of Jesuit Observatories.* Dordrecht: Kluwer Academic Publishers.

Udías, Agustín, 2015. *Jesuit Contribution to Science: A History.* New York: Springer.

Un président, un maréchal, un régiment, le 27ème Régiment d'infanterie ou 70 ans de phaléristique. Dijon: Musée de la vie bourguignonne, 2001.

Vaillant, Jean-Baptiste Philibert, 1854a. *Rapport présenté à l'Empereur sur la situation de l'Algérie en 1853.* Paris: Imprimerie impériale.

Vaillant, Jean-Baptiste Philibert, 1854b. 'Rapport fait au nom de la Commission, chargée par le ministre de l'Instruction publique d'examiner les améliorations à apporter dans l'organisation scientifique et administrative de l'Observatoire de Paris et du Bureau des Longitudes'. *Moniteur universel*, 3 February 1854.

Vaillant, Jean-Baptiste Philibert, 1855a. 'Opinion de M. le Maréchal Vaillant'. *CR*, 41, 1142–47.

Vaillant, Jean-Baptiste Philibert, 1855b. *Rapport adressé à l'Empereur sur la culture du coton de l'Algérie, 1854.* Paris: Typ. Panckoucke.

Vaillant, Jean-Baptiste Philibert, 1863. 'Sur la tempête des 2 et 3 décembre 1863'. *CR*, 57, 1001–07.

Walker, J. Malcolm, 2012. *History of the Meteorological Office.* Cambridge: Cambridge University Press.

Zantedeschi, Francesco, 1863. *Riflessioni del prof. Francesco Zantedeschi all'articolo del R.P.A. Secchi intitolato: Altri studi di elettricità atmosferica.* Padova: Tip. A. Bianchi.

Zantedeschi, Francesco, 1864–1865. 'Dei documenti comprovanti la proposta e l'applicazione da lui fatte negli anni 1849, 1850, 1853 e 1854, del telegrafo elettro-magnetico alla meteorologia e agli avvisi delle burrasche'. *Atti dell'I.R. Istituto veneto di scienze, lettere ed arti*, 23, 1414–23.

Zantedeschi, Francesco, 1866. *Risposta documentata del professore Francesco Zantedeschi all'articolo del P. A. Secchi… intorno ai presagi delle meteore e delle burrasche con documenti storici.* Padova: Tip. A. Bianchi.

4 The Lagrangian approach as a counterweight

Introduction

On 1 August 1861, the UK's Met Department issued what is regarded as the first general 'weather forecast'—a prediction of the weather and wind direction for the next day in the United Kingdom. It was so called by its author, Robert FitzRoy. That day, the empirically conceived body of the weather forecast (not to be confused with storm warning) came to life, gestated from a series of rules of thumb and instrumental readings. One year later the Department organized a complete forecasting service to the ports, and dispatched daily forecasts for up to forty-eight hours to several newspapers and institutions. By then FitzRoy was convinced that by combining barometric and thermometric observations with wind data, and by drawing on a series of empirical rules, he could foretell weather changes. His purpose was not unreasonable, according to historian James Burton, who counted the coastal recipients and classified them by sea regions. He concluded that the accuracy of FitzRoy's storm warnings fluctuated between 76% and 90%.[1]

FitzRoy's forecasts were well received, both inside and outside the British seafaring community they were purportedly aimed at. Britain's poorest fishing communities overwhelmingly endorsed his forecasts and storm warnings, as evinced by a questionnaire circulated to fifty-six coastal authorities in the early 1860s. According to the results, only an exiguous minority of three were extremely critical.[2] The public and the most widely read London newspapers enthusiastically welcomed this practice.[3] Yet they also found a

---

1 Burton (1988, pp. 56–59). For the differences between warning and prediction as perceived by mid-nineteenth-century weather scientists, as well as the polemics on this question, see Williamson (2017, pp. 777–79). See also the introduction by Barbot and Favino (2017) to the monographic issue of the journal *Quaderni Storici* on 'Forecast'. The issue gathers the contributions to the workshop on 'Prévision et incertitude de la physique à l'histoire' held in 2016 at the École Normale Supérieure de Cachan, France.

2 Halford (2004, p. 172).

3 The initial euphoria unleashed after FitzRoy's warnings about an 1861 gale was corroborated by the newspaper *The Times*: 'The event was predicted with as much certainty as an

warm reception in foreign navy and meteorological departments as well.[4] French ship captain Vincent-Alfred Moulac, commandant-in-chief of the fishery districts of the English Channel and the North Sea, was so impressed by his warnings that he not only requested his weather observations but also promoted the establishment of an autonomous weather service in the French Navy that would rely on FitzRoy's warnings.[5]

Unfortunately for FitzRoy, however, historical events would converge to reverse this early euphoria and high consideration for his forecast. Viewed through the lens of the official and elite science, both the author and his methods were often deemed as examples of an extreme empiricism that helped pave the way for speculation and non-scientific prognostication. FitzRoy was generally portrayed as an anarchic and radical empiricist whose forecast method was not capable of being conveyed in the form of instructions or logical calculations. This characterization was particularly dominant among the Royal Society Fellows, whose members wrote a devastating report against him in 1866. Many astro-meteorologists and lunarists viewed him as an unwelcome challenger, and bureaucrats like Thomas Farrer and the vociferous group led by James Glaisher from the Greenwich Observatory viewed him negatively as well.[6] Indeed, so strong was this opposition that most of historians agree that it was the cause that led FitzRoy to commit suicide in 1865. In this regard, historians have often tended to focus on this empiricism and its confrontation with the scientific values of Victorian culture, whether through the forecasting controversy or through public debates about scientific authority and responsibility.[7]

The point in broaching the historical legacy of FitzRoy is neither to defend his method as scientifically well-founded and legitimate, nor to stress its obvious linkages with empiricism or its discordances with Victorian cultural values; rather, I want to show here an aspect that has virtually gone

---

eclipse [...] Meteorology now rests upon evidence as palpable as that which confirms our theories of astronomy'. *The Times*, 13 February 1861, 8–9. Quoted by Dry (2009, p. 48).

4 Warnings of gales were telegraphed to weather departments in Sweden, Denmark, and some German states.

5 We may read this as a veiled criticism of the warning service that Le Verrier sought to establish in France and that was perceived with reluctance by the French Navy Ministry. The French Navy's consideration for English warnings and the irritation that it caused Le Verrier are corroborated by Captain Ernest Mouchez, from the French Navy's hydrographic office: 'I constantly receive new requests from ports asking for the communication of your telegrams'. Mouchez to FitzRoy, 26 January 1864. *Report of the Meteorological Department of the Board of Trade*, 1864, appendix, p. 9—quoted by Davis (1984, p. 370). On the French Navy and the reception of FitzRoy's storm warnings, Locher (2008, pp. 49–51).

6 Halford (2004, p. 193) holds that, according to FitzRoy, his critics could be grouped into four categories: certain individuals who objected to his method on scientific bases (like Francis Galton); broad sections of the uninformed public; the small group commanded by Glaisher; and 'those pecuniarily interested individuals, heedless of the precarious occupation of the Coasters and fishermen' (like big shipowners).

7 See, e.g., Halford (2004) and Anderson (2005).

unnoticed in historiography: FitzRoy's Lagrangian approach to weather forecasting and its associated observational system and world view. For it was his local system of coastal barometric observations and his aims of autonomy and freedom of individual judgement that largely drove him to adopt this approach, thus enabling him to establish rapport between his forecasts and the seafaring community. In 1861, his method underscored the importance of the wind flow and air masses, basing his forecasts on the authority of coastal observers. The result was a set of empirical rules for the weather prediction, which were not based on any representation of isobars, curves, or wave lines. And yet it was also the product of a deeply social view that prioritized the local over the central, autonomy over state power. In essence, it pursued the self-rule and the freedom of individual judgement of local coastal observers, which not coincidentally would go against the centralism and authority that later prevailed in national weather services. Simply put, then, the Lagrangian form weather forecast as approached by FitzRoy went hand in hand with his quest for autonomy and individual judgement.

This chapter describes the continuities and differences of what may be regarded as the first 'generation of weather scientists' who chose the Lagrangian frame of reference for its analysis, as represented by FitzRoy, George Jinman, and Francis Galton (though the latter indeed adopted a semi-Lagrangian approach). In particular, it examines Jinman's cyclone flow-model, FitzRoy's wind flow-model, and Galton's glyph and iconic maps and schematic micromaps of weather. I argue that the basic tenets of the Lagrangian weather analysis largely fell into oblivion after the 1860s, with the forecasters of national meteorological offices employing largely the isobaric-synoptic techniques developed earlier in France and Italy. Yet, their work was not in vain. A part of the language of the front and frontal surface theories that emerged so successfully for weather analysis in the 1900s and 1910s had its origins in the works of this generation of Lagrangian precursors.

Sea trade, routes, and wind statistics in the 1850s

The expansion of world maritime trade and colonial communication was stimulated by the growth of steam propulsion. In the 1850s and 1860s, and under the pressure of the growing colonial trade, shipbuilders and naval officers began to devise steam-powered technologies in order to reduce costs and optimize resources. These new technologies basically consisted of the screw-propeller, the surface condenser, the iron hull, and the compound engine, and enabled steamers to overcome their rivals, sailing ships, in many respects. By the 1850s, steam-powered freighters were used across the Atlantic and in the Mediterranean, though they at first transported only expensive cargos.[8]

---

8 On the characteristics of the world shipping and the technological and commercial development of steamships before 1869, see Headrick (1988, pp. 18–25).

New sea routes, together with improvements in steam shipping lines—speed, safety, and reliability—brought falling costs into the world trade network. Transit times between Britain and India, for example, fell to less than a month in the 1850s, to the satisfaction of British merchants.[9] The trade of the prized and perishable Chinese tea similarly benefited: while clippers (the swiftest sailing ships) needed about one hundred days in the years of the Opium War (1839–42), steamers made the trip in sixty-eight days in 1868. However, not only the growth of steam propulsion but also the very rivalry between steam and sail favoured the pursuit of new sea routes.[10] Evidence of this demand is the gold rush in California in the late 1840s, which stimulated the production of new navigation charts capable of shortening passage through the sailing route round Cape Horn. These charts not only benefited the flourishing clipper industry but also challenged those gold-seekers who used to take the steamer-railroad route across the Isthmus of Panama to sail, through the waters of the Pacific, to their golden destination.[11]

Be that as it may, it is important to keep in mind that the growth of navigation charts resulted from the growth of steam propulsion and its increasing competition with sail ships. Figures bear witness to this: the volume of the world's merchant fleet almost quadrupled from 1850 to 1910, going from nine to thirty-five million net registered tons. Throughout this period, Britain maintained a hegemonic position between 40% and 50% of the world volume; and the Dutch fleet became the third largest in the world.[12]

Under these circumstances, it is easy to understand why naval officers involved in nautical innovations and the pursuit of sea routes played an important role in governmental navy and trade departments. The pioneering figures in sea routes and charts tended to be qualified naval officers combining navigational experience with meteorological and hydrographic insight, many of whom cooperated with professors to gain public visibility and scientific credentials.[13] A good example is Matthew F. Maury, the superintendent of the U.S. Depot of Charts and Instruments, who was in charge of preserving charts and conducting routine meteorological studies.[14] From 1842, he devoted himself to preparing ocean charts, and by analyzing the

---

9 The same applies to other routes. The trip between England and Sydney moved went from a duration of 125 days in the early nineteenth century to only ninety-two days in the 1850s.

10 For the ascendancy of the sailing ship from 1850 to 1885, see Graham (1956, pp. 74–88).

11 Bruce (1987, p. 181).

12 Kirkaldy (1914, appendix 17), for Britain's share and the history of British shipping; Charlier (2004, p. 350), for the Dutch commercial fleet.

13 For the mutually beneficial relations between naval officers and professors in building maritime networks for data collection for commercial, national and military purposes, see Achbari (2015, pp. 257–61).

14 As regards Maury, see Williams (1963); Hearn (2002, pp. esp. 37–176); Burstyn (1981, pp. 195–97); Cotter (1979, pp. 75–83); Cox (2002, pp. 57–64); Rozwadowski (2016, pp. 288–93).

data from the large amount of logbooks kept at the Depot, he was able to trace charts and sailing directions for frequent shipping routes.[15] From 1847 until well into the 1850s, he published his results in a series of works entitled *Wind and Current Charts*,[16] freely issuing them to mariners, provided that they would supply him with the observations of winds and currents of their voyages.[17] These works complemented the *Pilot Charts* for the different oceans that Maury issued from 1848 to 1873 in which he synthesized wind frequencies in the form of circular numerical tables known as 'wind roses'. Maury's wind roses showed the average frequency and strength of the wind in each five-degree square during a given month, and were used to determine the optimal—i.e. the statistically fastest—route between two ports, for a sailing ship (Figure 4.1).[18] By 1855, he published a compilation of his early works, *Physical Geography of the Sea*, which, in spite of its lack of rigour, enjoyed great popularity, going through eight editions in its first five years.

Likewise, and perhaps most importantly, he posited a theory of the general circulation of the atmosphere from the data on winds and currents extracted from these logbooks. According to this theory, there were two different vertical circulation cells in each hemisphere—the trade wind between the equator and the belt of calms; and other cell between the poles and the subtropical highs.

This leads directly to the wider question of naval officers' studies on the general circulation of the atmosphere, which were instrumental for the adoption of a Lagrangian approach to the issue of weather forecasting. Most of these naval and military officers, at least in the early stages of their careers, received scientific training in meteorology and hydrography, and through sea voyages or autodidactism, were very familiar with the distinct theories on atmospheric circulation. Vaillant, for instance, in his study on the storm of 2 December 1863, alluded to the struggle between two opposing wind currents as the origin of the storm, before expressing his doubts about isobaric charts.[19] Other officials, more associated with marine affairs or working for

15 Maury's interest in the preparation of his own oceanic charts resulted, in part, from the fact that American ships depended on foreign charts for both open sea routes and inland waterways. Bruce (1987, pp. 170–74).

16 Matthew Fontaine Maury, *Explanations and Sailing Directions to Accompany the Winds and Currents Charts*. Washington, DC: C. Alexander, 1851.

17 He included emblematic routes like that from New York to Rio de Janeiro, and subsequently prepared the best routes to California, South America, and the West Indies. Schlee (1973, p. 38).

18 Each wind rose was divided into sixteen angle sections representing the sixteen possible wind directions. In each section, the reader could see the number of times that the wind was blowing in that direction for each month in the year. Information was geometrically distributed: while a central disc provided statistics during calm periods, concentric rings reported the total number of monthly wind observations. For further details, see Locher (2008, pp. 150–51).

19 Vaillant (1863, p. 1003). He did not mention Dove.

Figure 4.1 Maury's wind rose.
Source: Maury (1851, plate V).

naval or merchant fleets, also played a significant role in nurturing the Lagrangian frame of reference with their studies on cyclones and storms. For
example, E.E. Miller from the Pavlovsk Magnetic-Meteorological Observatory attempted to identify the approach and departure of Dove's tropical
and polar currents in his synoptic charts in 1864.[20] There were some others.[21] Many seamen held storm theories based on the conflict between these
two currents. In all these cases, at the centre of these officials' and seamen's
concerns was the idea that the observer who studied the air motion in the atmosphere had to follow the individual currents or air parcels as they moved

20 Miller published charts of lines of equal pressure variation (or *isallobars*) in his work 'On
    the Prediction of Storms, Especially Concerning the Storms which Ranged from 1 to 4
    December 1863'. For further details on Miller's work, see Khrgian (1970, p. 173, 197, 205).
21 Among others, we find the names of Captain Langford, Henry Piddington, and Mikhail
    A. Rykachev.

through space and time—which is to say that the Lagrangian approach was the most appropriate. This distinct approach shows how misleading it can be to view the history of weather forecasting simply in terms of the Eulerian approach and the longing for centralization and order that the states that promoted it had.

Jinman's cyclone flow-model

In the foregoing section I have given some idea of the relations between maritime trade and the search for new sea routes and wind statistics. This search was stimulated by naval officers and mariners who gathered weather observations, in part to produce charts with which to design the most economical routes, and in part to ascertain the laws of storms and the rules to detect their approach.[22] The present section is the first in a series that describes the Lagrangian approach adopted by those naval officers and mariners who were able to propose storm and cyclone models, and even in one specific case, to foretell their approach and forecast the weather. The earliest of them was George Jinman and his flow-model of middle-latitude cyclones.[23]

Marine expertise, intuitive knowledge, and induction from observation were all aspects that were part of the daily practice of the seamen interested in storm theories, well before the advent of the weather telegraphy in the 1850s. Redfield's whirlwind theory had been one of the models proposed for explaining tropical storms. According to his theory, storms were like great whirlwinds with calm centres around which the air moved in circles. They originated between the equator and the tropic, and from there they moved westward to enter the temperate zones and become ordinary Atlantic storms.[24] However, the storm observations collected by ships in the Bay of Bengal and the Arabian Sea in the 1830s had led Piddington to introduce a 'cyclonic' effect in Redfield's circular theory.[25] Thus, rather than tangential, Piddington found the winds to be 'incurved', or slightly inclined inwards. Indeed, his main concern was not so much the dynamics of storms, but rather the production of practical rules to allow seamen to avoid tropical storms. In fact, he showed them how the track of a cyclone could be foreseen by drawing on both wind and barometric observations: while the former provided the bearing of the centre, the latter gave its probable distance by applying a radial scheme of average pressure.[26]

---

22 From Maury's first pilot charts in 1848, the sailing route between England and Australia witnessed this change, and thus a trip that ordinarily lasted three or four months could be reduced by as much as a month. This fact was, according to Brunt (1951, p. 115), 'perhaps the most dramatic achievement of meteorology in all its history'.

23 On the early history of flow-models of cyclones, see Ludlam (1967, pp. 23–26).

24 Redfield (1831, p. 333).

25 Cushman (2013, pp. 137–38); Locher (2008, pp. 115–16); Longshore (2010, pp. 347–48).

26 Piddington's 1848 work, *The Sailor's Horn-Book for the Law of Storms*, became a fundamental reference book for institutions such as the British Board of Trade and not only to those concerned with tropical storms.

Whatever theories were formulated to explain the formation and motion of tropical storms, it is clear that the storms' structure was not only similar to that of middle-latitude storms but also was characterized —albeit in varying ways and to varying degrees— by both pressure pattern and wind pattern.

The importance of differentiating their structures and stressing the model of fluid (or air stream) was emphasized by George Jinman in his book *Winds and Their Courses*, first published in 1859.[27] This English master mariner had just conducted an Atlantic expedition from Newfoundland to Brazil in 1858. His model rested on three legs: his long sea experience among tropical and extratropical cyclones; his detailed knowledge of winds and clouds; and his reading about the storm theories of Redfield, Reid, and Piddington. Observations from ship's logbooks indicated that the principles laid down by them were 'far from being correct', and they could be even highly misleading. Drawing on his own experience and the various logs and data hitherto published, Jinman asserted that 'there never has been such a thing as a really *circular storm*'.[28] The apparently 'established facts' by these theories were simply untrue: the centre, he said, was not the most dangerous part of the storm, nor are the wind direction or the barometer a good guide to the overall movement of the centre. In contrast to the circular storm theories, Jinman propounded that the storms in the temperate zone of the North Atlantic were composed of two distinct air currents crossing each other at two 'confluences'.

> Every gale or hurricane has two distinct sides, formed by two distinct currents of air flowing in opposite directions, and crossing each other at two points, one on each side of the centre [...] The two sides are not always equally developed at the earth's surface —that is, the one often blows harder than the other.[29]

In Jinman's view, the western frontal line was the sharper of the two, and here the wind blew from southwest to northwest; Dove's opposing current theory also had suggested that midlatitude storms arose from the conflict of two opposite currents which alternately displaced each other, weather change being the result of the predominance of one of them.[30] Although the influence of these ideas on Jinman's storm theory is clear, nowhere in his book is Dove mentioned.

---

27 George Jinman, *Winds and Their Courses; or a Practical Exposition of the Laws Which Govern the Movements of Hurricanes and Gales. With an Examination of the Circular Theory of Storms, as Propounded by Redfield, Sir William Reid, Piddington, and Others.* London: George Philip and Son, 1861.

28 Jinman (1861, p. 62).

29 Ibid., p. 1, 3.

30 Dove (1841, p. 7). See also Chapter 1 of this book.

In contrast to circular theories, a consequence of Jinman's theory was that midlatitude storms had an *asymmetric* structure in which the western confluence would be always 'more marked'—or, in his own words, 'the *meeting* of the two winds' would 'be seen more distinctly'.[31] Jinman's theory also suggested that the most dangerous part of a storm was not its centre, and this, too, seemed to be confirmed by the experience of many steamships, which passed from one side of the storm to the other, across the centre, encountering slight winds. Because of the existence of two distinct currents of air flowing with different forces, the theory implied that the form or shape of extratropical storms would be elongated or almost elliptical; this did not conflict with the fact that, in his view, tropical storms had a circular form (Figure 4.2). On the other hand, unlike earlier circular theories, Jinman's theory had only vague and speculative explanations for the physical processes originating the formation of such currents of air; he stated, for example, that the atmospheric circulation depended on 'the electric or magnetic currents produced by the influence of celestial bodies', like the sun.[32]

Between 1858 and 1863, two British naval officers—master mariner Jinman and Admiral Robert FitzRoy—independently developed flow-models of cyclones covering temperate-latitude storms and gales. Furthermore, FitzRoy was able to trace weather systems on British synoptic maps and produce weather forecasts within a Lagrangian frame of reference. Dove's theory of the conflict of two opposite air currents lay behind these advances. Yet, before showing how and to what extent FitzRoy was able to achieve these—at that time controversial—accomplishments, I will first examine the social and institutional frameworks in which they were conceived and evolved. These frameworks had a direct influence on the way in which he approached the problem of weather prediction.

The British counterpoint: autonomy, localism, and safety

In France and Italy, Napoleon III and Pope Pius IX, with their support to the storm warning services, sought centralization, order, and control within their territories. This is in stark contrast to the situation in Britain, where there was a symbiotic relationship between an elite form of science (incarnated by the Board of Trade's Meteorological Department) and the individual authority of local mariners (through the so-called 'fishery barometer program'). All this resulted in the pursuit of autonomy, localism, and freedom of individual judgement—as opposed to the state control, centralism, and authority sought in France and Italy. The Met Department, having

31 Jinman (1861, p. 4).
32 Ibid., pp. 92–93. According to the Bergen School's meteorologist and defender of the polar front theory of cyclones, Bergeron (1981, p. 459): 'Jinman's excellent observations [...] were mingled with absurd notions of the physical cause of atmospheric motion—another regrettable case of lacking contact empirie-theory in our science'.

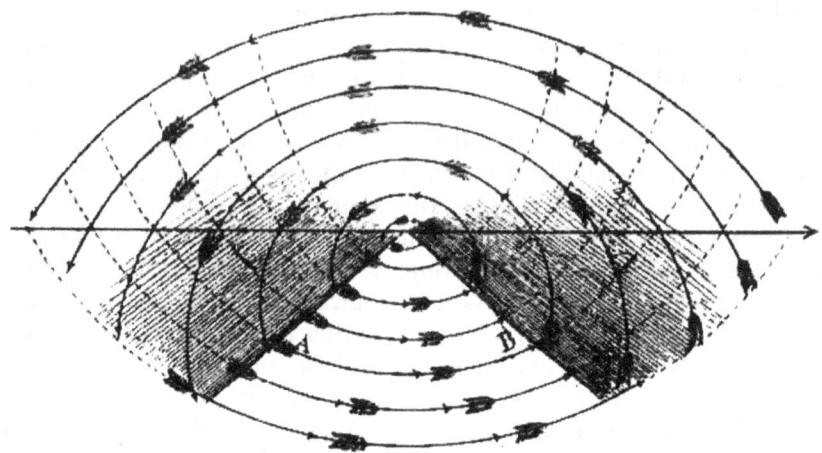

Figure 4.2 George Jinman's flow-model of cyclone, according to which two distinct
air currents crossed each other at two confluences (shown above by the
rear and forward confluence lines).
Source: Jinman (1861, p. 97).

strong links with the state and the fisheries sector, provides an interesting
counterpoint to this history.

The British Navy of the 1850s had, unlike the retrenched and disarmed
navy of the 1810s and 1820s, the global exploration geared towards science
and trade as the driving force of its recovery. Persuaded by latest advances
in instrumentation, the secretary of the Admiralty had no qualms about
stating that 'exploration would increase scientific knowledge, that it would
be a boon to national commerce', and 'a terrible blow to national pride if
other countries should open up a globe over which Britain ruled supreme'.[33]
Favoured by the political pragmatism of Lord Palmerston, the merchant
marine and the Royal Navy soon became interested in global exploration,
which is to say the study of physical and oceanic features on a global scale.[34]
It was the scientific and commercial interests of this period that made it
plausible for Britain to participate in the first International Maritime Conference held in Brussels in 1853.

Although it was not committed to any expenditure, the British Government made arrangements for important officials of the merchant marine

33 The Second Secretary of the Admiralty, John Barrow—in the words of Fleming (1999, p. 11).
34 The Royal Navy's involvement in science was by no means insignificant; according to
   Friendly (1977, p. 289), 'the Navy was the principal governmental subsidizer of science'
   in the first half of the nineteenth century. For the role played by Admiralty Hydrographer
   Francis Beaufort in the promotion of meteorological observations on board Royal Naval
   ships, see Naylor (2015, pp. 88–92).

and the Royal Navy to assist this event.[35] The measures adopted there made the importance of the safety of sailors and navigation clear. The conference aimed at creating a meteorological register for use on naval and, if at all possible, merchant ships. After the event and its subsequent discussions, in which politicians and members of the Royal Society participated,[36] Parliament resolved to sanction funds for both the Board of Trade and the Admiralty to establish a 'uniform system of meteorological observations at sea', with the aim of finding the 'very best tracks for ships to follow' and making 'the quickest as well as safest passages'.[37]

The Met Department had to wait for more than a year for the government to implement a scientific program, a full staff, and the allocation for budgeted expenditures. However, this program dealt not so much with the safety of sailors or the optimization of sea routes as with the compilation of meteorological statistics (wind, temperature, and pressure measurements), which the government authorities and members of the Royal Society regarded as an essential data source for an understanding of the laws of the weather. This understanding would not be an immediate process, but rather a slow and gradual one.[38]

Before the Met Department could implement the double task of compiling meteorological statistics and supplying instruments and instructions to ships, however, the Royal Society recommended Robert FitzRoy for its leadership. Born into an aristocratic family known for its association with seafaring, FitzRoy's naval and scientific concerns soon converged.[39] Trained at the Royal Naval College at Portsmouth, he obtained the rank of lieutenant

---

35 The British delegates were the head of the Board of Trade's Meteorological Department, Captain Frederick Beechey, and Captain Henry James of the Ordnance Survey. For the repercussions of Brussels Conference in London, see Naylor (2015, pp. 92–93); Walker (2012, pp. 21–22); Burton (1986, pp. 150–57).

36 A detailed account of the discussions in the parliamentary sessions and other forums can be found in Burton (1986, pp. 150–52).

37 To be precise, a vote of £3,200 to the Board of Trade and £1,000 for the Admiralty. 'Report of the Meteorological Department for 1857', pp. 1857 XX, 283–372—quoted by Dry (2009, p. 36).

38 At the request of the government, in 1855 the Royal Society advised the aims to which the Met Department should be directed. The president of the Royal Society, Edward Sabine, held that this new office should pursue, 'in addition to the information required for the purpose of navigation', data 'for the investigation and establishment of great atmospheric and oceanic laws', obtainable 'by observations either on land or on sea'. The office should also publish and circulate these statistical results. See Royal Society letter from Earl of Rosse, 19 June 1854, NAUK, Public Record Office, BJ 7/4 iv. See also 'Report of a Committee appointed to consider questions relating to the Meteorological Department of the Board of Trade', PP, 1866, XLV, vii.

39 Mellersh (1968) and Basalla (1972, pp. 16–18) are the main sources for FitzRoy's biography. Apart from the *Beagle* expedition, for which there is a vast literature, some particularly relevant works dealing with FitzRoy's contributions to meteorology include Agnew (2004), Anderson (2005, pp. 48–49, 84–85, 105–08, 114–28, 191–95), Barlow (1994, pp. 123–28), Brunt (1956, pp. 193–95), Burton (1986), Cox (2002, pp. 75–84), Dry (2009),

in 1824, before undertaking his first voyages in Mediterranean and South American waters. From 1828 to 1836, he participated in the Beagle expeditions, in the second as the commander, accompanied by the young Charles Darwin on perhaps the most famous scientific expedition in history. This by no means prevented him for joining, because of his conservative views on religion and slavery, the opposition to Darwin's *Origin of Species*.[40] After his return, he was elected a member of the Parliament for Durham, and then appointed Governor of New Zealand. By the early 1850s he was a rising star in naval circles who was known for his cosmopolitanism and his dedication to hydrography by his determinations of longitude around the globe.[41] In 1851, just after retiring from active naval service, he was elected a fellow of the Royal Society of London. In 1854, he was invited to head the Met Office, serving as statist or collector of weather information—a role he would far exceed in a few years, but not without polemics and opposition.

The compilation of meaningful statistics described earlier was but one side of the larger quest to scientifically register marine meteorology; for any statistical enterprise was only as valuable as its capability to gather data. Without the physical instruments and their pertinent instructions to carry out the completed register of observations, no ship of the merchant marine and the Royal Navy would have been able to deliver the information necessary from which the laws of the weather could be subsequently derived. Indeed, the Met Department would claim its authority through the argument that data had been generated through a double and long process of 'collecting' and 'grouping' —a process that demanded the most precise and versatile recording instruments possible.[42] It was precisely in this context of statistical compilation that FitzRoy began to design and distribute standard instruments for merchant marine ships and the Royal Navy. By 1855, he had arranged for one hundred of the reliable marine barometers, previously tested at Kew Observatory, to be delivered to coastal agents for their distribution among Navy ships.[43] Over the course of the following years coastal

---

Halford (2004, pp. 74–124), Hughes (1988), Shaw (1926, pp. 149–53, 302–03), and Walker (2012, pp. 23–70).

40 FitzRoy published a three-volume *Narrative of the Surveying Voyage of His Majesty's Ships Adventure and Beagle*.

41 The British Hydrographic Department's director, Francis Beaufort, entrusted FitzRoy with the task of continuing the South American charting program. Naylor (2015, pp. 88–89).

42 Data 'collecting' consisted of recording and tabulating separately each element of observation (e.g. pressure, temperature, wind, ocean currents, storms, ice, meteors, aurora, etc.) in specific data books, each page of which covering a five degree square of ocean. Data 'grouping' consisted of re-copying the observations gathered into sheets, each devoted to a specific month. For further details, see Burton (1988, pp. 35–36).

43 For the standard Kew marine barometer and the advantages and limitations of its distribution to merchant marine ships, see Dry (2009, pp. 40–41). For standardization of meteorological instruments at Kew, see MacDonald (2018, pp. 8–9, 246).

agents would receive fifty shillings for each vessel serviced, while ships' masters would be supplied with the instruments (loaned free of charge) and with copies of Maury's charts, provided they committed at returning their meteorological observations.[44]

As an important member of the naval service with close ties with many ship captains and a British patriot concerned with the safety of sailors and fishermen at sea, the issue of how best to foretell bad weather was obviously of great interest to FitzRoy. Troubled by seafarers' safety demands and the lack of immediate utility of meteorological statistics—which, as noted earlier, would begin to produce something positive only some years after observations had been conducted—FitzRoy introduced a subtle yet critical change in the observational system. In addition to distributing standardized instruments to ships' masters and naval officials, a vertical top-down form of dictating rules and requesting data, ordinary and easy-to-read barometers would be sent to illiterate fishermen through the so-called 'fishery barometer program', a means of horizontally networking observers.[45] Here the illiterate fisherman was regarded as an autonomous observer, attached to the program voluntarily. FitzRoy defined it as a supplementary enterprise, covering the other side of the coin, that of 'poor men, whose lives are more or less dependent on weather'.[46] No matter whether this program was excluded from the statistical project, no matter whether the standard barometers were not officially intended for any predictive use, the fundamental aspect of fishery barometers, FitzRoy argued, would be their prognostic function.[47]

FitzRoy's plan, advanced first to the instrument makers Negretti & Zambra in December 1857 and approved by the Board of Trade in January 1858, was in many respects quite groundbreaking.[48] In essence, it broke with a deep-rooted modus operandi in European meteorology that ran from Paris to Rome and that was predicated on the central control of production rather than the autonomous judgement of local observers. As FitzRoy himself wrote to Negretti & Zambra in 1857, his plan intended to 'place ordinary land barometers as weather glasses solely, at some of the more exposed Fishing stations and coasting harbours in Great Britain and Ireland' so that they will be able to forewarn of impending storms.[49]

---

44 By May 1855, fifty merchantmen and thirty men-of-war had been equipped with scientific instruments for collecting data. Burton (1986, p. 155).

45 FitzRoy's fishery barometer program has been most thoroughly and forcefully examined by Dry (2009).

46 FitzRoy to Negretti & Zambra, 9 December 1857. NAUK, Public Record Office, BJ 7/615.

47 On the shift of the Met Department's work towards a much larger audience, and the balance between FitzRoy's authority and that of his Department, see Anderson (2005, pp. 108–09).

48 For a justification of the fishery barometer program to FitzRoy's superior at the Board of Trade, see FitzRoy to Thomas Henry Farrer, undated. NAUK, Public Record Office, BJ 7/19.

49 FitzRoy to Negretti & Zambra, 9 December 1857. NAUK, Public Record Office, BJ 7/615.

FitzRoy's fishery barometer program moved away from the Met Department's focus on the statistics as the single subject of study, broadening it instead by inviting local observers to forecast and disseminate the weather. Or to put this another way, FitzRoy was more interested in promoting the individual authority of local observers than in imposing centralism and control at a national level.[50] In fact, by 1859 he had begun to send cheap barometers with accompanying instructions to many coastal fishermen in Scotland, which evinced the existence of high awareness of the risks affecting fishing industry.[51]

At this point, we must ask what there was in the British seafaring community, both military and civilian, that made their cooperation with the Met Department's program so advisable to FitzRoy. First was the old tradition of prognostic weather-watching: FitzRoy was first and foremost a sailor, forged within the traditions of the Royal Navy, based on intuition and empirical insight. In fact, in his early letters on the program he often referred to fishery barometers as weather glasses.[52] Second, there was an administrative factor of considerable importance: the Northcote-Trevelyan reforms of 1854.[53] These reforms catalyzed the development of civil service in Britain, by creating a scientific civil bureaucracy that would replace the traditional practice of amateur civil servants. This fact created some unease and qualms among the seafaring community, including FitzRoy, who saw how the government placed their trust in scientific bureaucrats rather than in experimented amateurs and officials. And third, there was a geographic factor of tactical importance: the island situation of Britain. In Eastern United States, for example, where the weather came across the land, a system of many land-based observers could telegraphically track the progress of the storms approaching from the west and foresee which regions of the coast they would hit. In Britain, however, such tracking was not feasible.[54] Nearly all storms approached from the Atlantic, where there was no effective telegraph service or network of land-based observers. All these reasons made the seafaring community the most suited, for FitzRoy, to judge what the coming weather might be.

---

50 FitzRoy published a manual of instructions—*Barometer and Weather Guide* (London, 1858)—with the aim of explaining the fundamentals of the barometer to fishermen.

51 By 1866, ninety-five of the 'smaller and less affluent ports of the British Isles' had been equipped with barometers and instructions. See Walker (2012, p. 36). For further details about the distribution of fishery barometers and the importance of local control for the success of the program, see Dry (2009, pp. 40–46). See also Halford (2004, pp. 118–22).

52 See, e.g., FitzRoy to Negretti & Zambra, 9 December 1857. NAUK, Public Record Office, BJ 7/615; Dry (2009, pp. 42–48).

53 These reforms were reflected in the report prepared by Stafford H. Northcore and Charles E. Trevelyan in 1854. For further details, see O'Toole (2006, pp. 51–58).

54 The existence of a disadvantageous geographical situation of Britain vis-à-vis the United States and other Western countries as the main (if not deciding) reason that impelled FitzRoy to go a step further in storm warnings and focus on weather forecasting is suggested by Halford (2004, p. 146).

FitzRoy's fishery barometer program found widespread (though not unanimous) appeal within the British seafaring community, to the extent that he published a small book of instructions, *Barometer and Weather Guide*, which reached four editions in just three years. In this book, FitzRoy explained not just how to record the figures and how to interpret them, but also how to properly use the fishery barometer so that seamen could infer wind information, and, thereby, keep away from storms. As he noted, neither the exclusive use of barometer nor just one pressure reading was sufficient; rather, an overview was required:

> It is not from the point at which the mercury may stand that we are alone to form a judgement of the state of the weather, but from its *rising* or *falling*, and from the movements of immediately preceding days as well as hours, keeping in mind the effects of change of *direction*, and dryness, or moisture, as well as alteration of force or strength of wind.[55]

Now, if FitzRoy presupposed that judging the state of the weather was a task for seamen, how and why was he fully involved in weather forecasting? How could the Met Department accommodate his forecasts and seamen's weather judgements? A first answer can be found in the following episode that took place in September 1859, on the occasion of the annual meeting of the British Association for the Advancement of Science (BAAS) held in Aberdeen. Here a motion was presented by Colonel Sykes on behalf of John Locke, an amateur weather writer and a regular contributor to the Royal Dublin Society. Locke, who had engaged in extensive correspondence with Maury and knew first-hand how a system of telegraphed weather reports had been successfully operating in the United States from at least 1856, suggested that a similar system should be established in Britain. Colonel Sykes had met FitzRoy in August 1859 and it is very likely that he agreed with Locke's plan, albeit adapted to the British barometer program.[56] Be that as it may, and not without opposition, the motion was accepted. The Council of the Association passed a resolution, 'praying the Board of Trade to consider the possibility of watching the rise, force and direction of storms and the means for sending, in case of sudden danger, a series of storm warnings along the coast'.[57] With all due caution, this resolution resembled the appeal made by Vaillant to Le Verrier after the Black Sea storm of 1854.

As with the effect of the Black Sea storm in France examined in the previous chapter, a 'tragic' element was also present in the Board of Trade's final decision. The sinking of the legendary iron vessel *Royal Charter* on the night of 26 October 1859, with the ensuing loss of her 500 passengers and crew, as a result of a strong storm that, paradoxically, was not able

55 FitzRoy (1858, p. 8).
56 Halford (2004, pp. 144–45).
57 Quoted in Burton (1988, p. 40).

to damage a small American sail ship, aroused ire and controversy in England.[58] Apparently, the American ship had followed Maury's instructions for ships facing storms. Deeply affected by the tragedy, FitzRoy conducted a thorough investigation and concluded that had there been a proper storm warning system in place, the tragedy could have been avoided.[59] This time his appeal did have an effect. In June 1860, the Board of Trade, through its president, authorized the Met Department to establish a storm warning service around the British coasts.[60] Ironically, a sudden, calamitous event had eliminated a whole host of inertia and resistance that had inhibited the decision favourable to any warning or forecasting service. The close relationship between centralism, order, and safety that had so fascinated Le Verrier and the French imperial government in 1855 was now absent in Britain, as FitzRoy either would opt for his own local and empirical forecasting system or would cooperate with the seamen and fishermen engaged with his barometer program, thus authorizing them to interpret his forecasts with autonomy and self-responsibility.

FitzRoy's forecast method

Isolines (including isobars) are a simple, large-scale, graphic device whose deformations can be drawn in a map and subjected to synoptic analysis. They are the result of routine measurements taken in a framework with spatially fixed coordinates. The emergence of synoptic forecast depended on the focus on a framework of this kind—Eulerian; so did the later progress of weather forecasting. Yet the study of small-scale, local, weather systems took a different form in Britain, especially by FitzRoy. It constituted what might be called a 'Lagrangian' study method, combining experience with ocular observations and empirical rules. In this framework, the application of physical-mathematical laws was rarely attempted and the approach was qualitative, synoptic, and non-isolinear.

In the late 1850s the methodological gap between atmospheric wave maps and isolinear maps was still small, as evinced by the efforts of continental meteorologists. One important novelty had been the distinction, introduced by Liais, between atmospheric waves and lines of transport of waves, the pressure lines representing the trend of movement of the storm. Another was the supposed *linear* relationship between barometric height and the wind force and direction; in fact, as Kämtz held in a note published in the *Comptes Rendus* in 1858, it was only by tracing isobarometric lines that this relationship could be deduced.[61] Weather forecasting would indeed retain

---

58 For further details on this event, see Burton (1988, pp. 40–41), Halford (2004, pp. 145–51).

59 The preliminary results of this investigation were published by FitzRoy (1860, pp. 222–24); a full account of the same was given by FitzRoy (1861, pp. 39–44) at the annual meeting of the BAAS at Oxford.

60 FitzRoy (1861, p. 42).

61 Kämtz (1858, pp. 944–47).

its empirical, cartographic character. Attempts to subject it to physical-mathematical laws were rare. As late as 1903, the director of the Austrian weather service, Jesuit Joseph Maria Pernter, still defended its absolute dependence on mapped information: 'the knowledge of the weather conditions for every place and for every type of pressure distribution offers the only entirely satisfactory empirical basis for weather predictions'.[62]

But before inquiring how and why FitzRoy developed his own forecast method, it is worthwhile to stress how ready FitzRoy was to distance himself from this Eulerian approach, adopting instead a method that would allow engaging with his local audience. For as the notion of the 'synoptic chart'—first defined by him, as the graphic expression of 'consecutive simultaneous states of the atmosphere'[63]—became a 'creed', it also became a pretext, a reason to separate himself from an approach seeking for centralism and order, and to establish rapport between the scientist and his seafaring and fishing community. By distancing himself from a (centralizing) Eulerian approach, he not only made contact with his audience, but *ipso facto* showed that he and they themselves shared the spirit of the barometer program, while respecting the authority of local observers and individual judgement. His forecast method will thus project all those characteristics which the fishing community finds most motivating, but which the scientific elite finds most objectionable.

Let us see how FitzRoy's methodological ideas about weather forecast evolved throughout the 1850s. That FitzRoy at first took for granted, or at least showed enthusiasm for, the notion of atmospheric waves emerges from the letter that a young sea captain, Henry Clifton Sorby of Sheffield, sent in June of 1856. In this letter, Sorby made reference to the theory that the air moved over the atmosphere in a series of pressure waves, whose 'lows' and 'highs' would be comparable to the troughs and crests of sea waves. During the second part of 1856, FitzRoy was devoted to detecting such waves in the columns of barometric readings, attempting to show their connection to storms.

This is FitzRoy's stance just a year before the launch of the fishery barometer program. Let us compare it now with the views expressed in his *Notes on Meteorology* that he published three years later, when he began to distribute barometers in fishing stations and coasting harbours. The several pages dealing with 'atmospheric waves' denote change of mindset. Drawing on barometric and thermometric local observations, FitzRoy gave his first warning:

> The effect of the wind shifting round when traced upon paper by a curve seems certainly wave-like to the eye; but I believe it to be simply consequent on the wind shifting round the compass, and indicating alteration in the barometric column.

---

62 Pernter (1903, p. 160)—as quoted by Nebeker (1995, p. 38).
63 The notion appears in his *Weather book*. FitzRoy (1863, p. 103).

An important warning followed: 'If the wind remained north-east', 'there would be no wave at all—there would be almost a straight line along a diagram'. Thus, 'such *atmospheric waves* may be an optical delusion'.[64] By way of recapitulation, he summed up his new stance as follows:

> Birt drew particular attention to these supposed undulations of atmosphere [...] Sir John Herschel, Le Verrier, and other great authorities then countenanced Mr. Birt's theory, and apparently sanctioned his opinion. Yet there is so much argument against those views [...] That there must be undulations in the atmosphere constituted as it is cannot be doubted, but that the curve traced on paper, representing the oscillations of a barometer, as the wind veers round the compass, corresponds to a mechanical, wavelike, undulation of the body of atmosphere is not perhaps sufficiently proved.[65]

And the further one follows his arguments, the clearer it becomes that FitzRoy's forecast method was based not on the detection of atmospheric waves but rather on the observation of barometer and thermometer changes and wind shifts. These diagrams show 'how the barometer and thermometer may be used in connection with each other in foretelling wind and consequently weather that is coming on, because as the one rises the other generally falls'.[66]

In this regard, FitzRoy seems to have abandoned the search for atmospheric waves, an idea he had pursued so tenaciously in 1856, and instead opted for the combination of barometric and thermometric readings, together with wind observations, as the most proper approach. Here, two factors seem to be involved in this change. On one hand, the fishing and seafaring community and their barometer program, which are no mere backcloth for it is only by precise reference to fishermen's and coasters' barometer and wind observations that FitzRoy came to question that the curves traced on paper really represented mechanical undulations of the atmosphere. Moreover, he was fully aware of their interest in the search for local empirical solutions to the threat of storms, and, hence, their overwhelming endorsement of FitzRoy's forecast method.[67] On the other hand, there was the influence of Dove's ideas, whose book *The Law of Storms* had just been translated and published by the Board of Trade in 1858.[68] Perhaps the most fundamental of this influence was the idea that atmospheric changes stemmed from the confrontation of

64 FitzRoy (1859, p. 13).
65 Ibid., p. 34.
66 Ibid., p. 13.
67 In the early 1860s, a questionnaire was distributed to fifty-six coastal authorities to sound out their opinion about FitzRoy's storm warnings: forty-six were highly favourable, nine endorsed with some doubts, and only three were clearly disapproving. See Halford (2004, p. 172); Dry (2009, pp. 52–53).
68 Dove's work was the third of a series of *Meteorological Papers*, compiled by FitzRoy, which largely consisted of the translation—made by Robert Henry Scott, at least in its second edition— of Dove's *Ueber das Gesetz der Stürme*.

polar and equatorial currents. More will be said about this in the following section; here it will be sufficient to note that FitzRoy's predictive approach was in part indebted to Dove's storm theory.

In response to the challenge posed, FitzRoy and his assistant Thomas Henry Babington produced a series of synoptic charts that he believed would accurately show the 'consecutive simultaneous states of the atmosphere'. These charts, he said, would reflect states 'as if an eye in space looked down on the whole North Atlantic' at regular periods.[69] Variables related to these states were featured, for example, on the map of the *Royal Charter* Storm, including wind vectors showing the direction and force of the wind, symbols for different forms of precipitation, and shadings indicating cloud, rain, snow, hail, or fog.[70] However, no effort was made to trace clear isobaric shapes.[71] As a result, the reader encountered confusing maps—such as the *Royal Charter* Storm of 1859, which revealed such visual complexity that one could hardly foretell the weather from these charts. FitzRoy's synoptic charts also tended to show the contest of two types of air currents (warm and cold), in which the position of wind arrows seemed to suggest the whirling counterclockwise motion of air (Figure 4.3).[72]

By 1864, FitzRoy had become even more sceptical about the use of isobars and the chart preparation in general, in the forecast process. In his 1864 *Report* of the Met Department, he had no qualms about admitting that 'neither isobars, curves or any kind of wave or crest lines seem to show the direction of the wind'.[73] Similar criticism can be found in his private correspondence. In a letter written to Vaillant on 4 February 1864, he stated that the maps of the Paris *Bulletin* did not show how depressions propagated across Europe (as Marié-Davy claimed), but rather how winds evolved locally and then advanced beyond the local realm.[74] In a further letter, this time to professor James David Forbes, he was perplexed by the strong predisposition of meteorologists to extrapolate values from very few data:

> Isobarometric curves—to be really useful, ought to depend on many more stations of observation than [Francis Galton], or even M. Marié Davy have used: I think you will allow when duly considering effects of high land—depressions—snowy ranges of mountains, and local

---

69 FitzRoy (1863, p. 103).

70 Ibid., appendix. Although the Met Department had been preparing synoptic charts since 1857, everything suggests that these charts were retrospective rather than operational for an immediate use.

71 FitzRoy's pressure lines showed a separation from the corresponding parallel of latitude, this departure being equivalent to the pressure at each point along the parallel. Burton (1988, p. 47) justifies FitzRoy's omission of isobarometric lines in his weather charts by invoking the argument that the isobaric technique was still in its infancy.

72 Anderson (2006, pp. 72–74); Kutzbach (1979, pp. 82–83); Monmonier (1999, pp. 44–45).

73 FitzRoy (1864, p. 23).

74 FitzRoy to Vaillant, 4 February 1864. ABIF, Ms 3714.

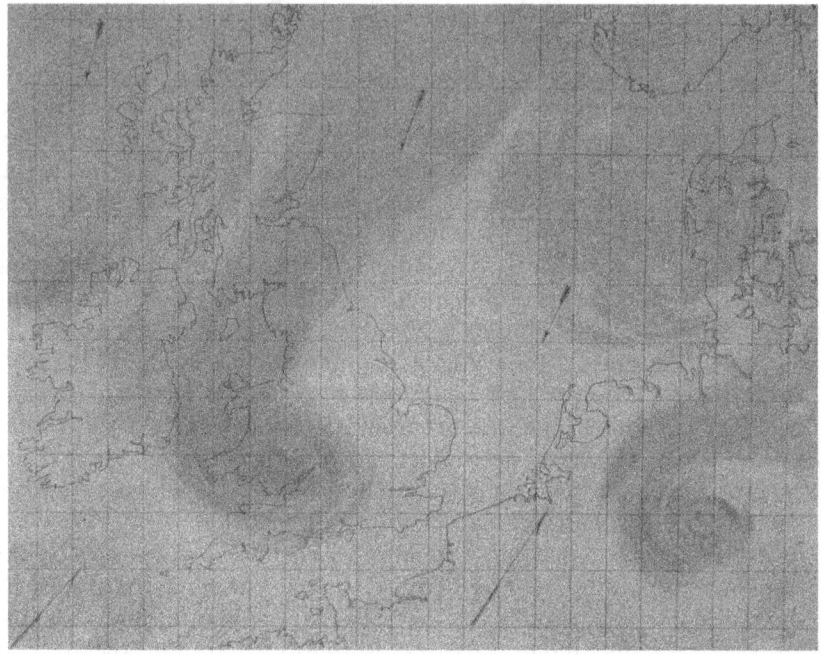

Figure 4.3 FitzRoy's wind flow-model, showing the alternation of polar and tropical currents on a map of the British Isles and the North Sea. Shaded
yellow refers to tropical air; shaded blue to polar air; the direction of the
current is indicated by straight arrows.

Source: FitzRoy (1863, plates XII and XIII).

precipitations: besides the direction of winds. Pray consider how little is
known of the elevations (above a sea or other normal level –) of almost
all continental *stations inland.*[75]

I have quoted FitzRoy so extensively because it would be hard to find, in discussions before 1865, criticism as well-founded and detail-oriented as these
to the practice of his predecessors—that is, the practice of plotting a line
between two sets of observations, very distant among them, and presuppos-
ing that the pressure is the same at any point along that line. Yet, these criticisms are highly significant. FitzRoy seems to have discarded first the idea
of atmospheric waves, and then the technique of isobars and has decided

75 Quoted by Halford (2004, pp. 185–86). In his letter, FitzRoy's main aim was to criticize
Francis Galton's views about the notion of anticyclones. I will discuss this issue in the
next section.

instead to prioritize the search for an overview of wind flow patterns. One can view this evolution as part of a broader shift in the study of atmospheric dynamics from an Eulerian frame of reference to a Lagrangian one. Here, once again, the seafaring community and the barometer program are no mere backdrop. For it is by clear reference to their wind and barometer observations—as well as to Dove's ideas (but not to theorists like Loomis and William Ferrel, whose storm theories seem to have had no influence on his thought, even if he knew them)—that FitzRoy managed to make this methodological shift.[76]

By late 1860 FitzRoy was receiving telegraphic reports from twenty-two coastal observers, which enabled him to gather real-time observations. On 6 February 1861, he issued his first storm warning.[77] In just under a year, he established a complete forecasting service, the number of ports attached to the service rising from 50 to 130. Through this network of observers and his fishery barometer program, he gained an empiricistic sensitivity that had not been expressed so deeply in his earlier meteorological statistics. 'An outline chart, with wind-makers', he stated in 1863, 'is useful' for forecasting; 'likewise a transparent horn, or a glass, with circles'. Yet, 'a certain amount of practice' enables one 'to work out the questions mentally (like a chess-player who need not look at the board)'.[78] His empirical method based on simple rules of thumb and instrumental readings was practically similar to fishermen's method resulting from the interpretation of barometric changes and wind shifts. The primacy of individual judgement was a common denominator of the two methods.[79]

---

76 One can find earlier precedents for this 'questioning' of the areas or lines of barometric pressure in FitzRoy's academic addresses. So, for example, FitzRoy's contribution to the 1860 meeting of the BAAS, in which he read the results of his study 'On British Storms, Illustrated with Diagrams and Charts':

> The principal object of [using these isolinear techniques], as it were soundings, of the atmosphere, shown in the diagrams, was to prove whether lines of pressure, or whether areas of pressure prevailed; and I think, when they are all closely looked into, they go to prove that while the atmosphere in the British islands varied in its pressure from time to time, such variation was not on a particular line, but extended over a large area. I may say, [however], that at some places there was little or no wind; the barometer fell much, but there was no storm, for the wind circulating around these districts did not affect them, while at other places the storm was tremendous.
>
> FitzRoy (1861, p. 41)

77 For his storm warnings, FitzRoy devised a dual system of visual signals: cautionary signals (by conical-shaped patterns) and night signals (by lanterns). See Burton (1988, pp. 44–45, 327, 336).

78 FitzRoy (1863, p. 218).

79 FitzRoy's forecast method is described in detail in his work *The Weather-Book: A Manual of Practical Meteorology* (London, 1863), and his 1862 *Report of the Met Department of the Board of Trade*. There is abundant secondary literature on this issue; see, e.g., Anderson (2005, pp. 110–15); Barlow (1994, pp. 126–28); Burton (1986, pp. 161–64); Halford (2004, pp. 167–76, 189–229); Hughes (1988, pp. 201–02); Walker (2012, pp. 38–45).

FitzRoy's main reason for discarding atmospheric waves and isolines was not their technical difficulty nor any mathematical theory of atmospheric dynamics, but the practical impossibility of deriving wind flow patterns from those graphic methods. His own approach was based on the specification of the wind flow field. This directly concerned the motion of individual air masses. Accordingly, FitzRoy began with an empirical interpretation of barometric changes in combination with wind and temperature indications. Marié-Davy's weather charts were too ambiguous, since they at times showed winds flowing normal to isobars and at times parallel to them. FitzRoy's method was based on the Lagrangian scheme of fluid motion, as was Dove's opposing current theory. He posited that a weather system experienced a steady west to east movement, in the same way as a storm would affect regions placed at its eastwards in due course.[80]

Despite these innovations, however, FitzRoy's forecast method failed to provide solid foundations for the scientific legitimization of his work, oscillating between various positions ranging from a radical empiricism to a too individual proceeding. This is, at least, the view expressed by the formal investigation promoted by the Board of Trade into the Met Department in 1866, which was chaired by Francis Galton—hence, it was better known as the *Galton report*. To begin with, FitzRoy's forecasts had disfigured and altered the primary objectives of the Met Department. 'There is no indication', asserted the report, that publishing 'undiscussed observations', speculating on 'the theory of meteorology', and still less prognosticating the weather were 'a part of the functions of the Department'.[81] His method was, according to the report, too empiric, based on varying rules and devoid of any internal logical process—'no notes or calculations are made', it stated—while his rules were not 'reduced to any definite and intelligible form of expression', nor were they, as they were then used, 'capable of being communicated in the form of instructions'.[82] To end with, FitzRoy's use of 'pseudo-scientific rules' as sources of legitimacy inevitably met with opposi-

---

80 The construction of a too empiricistic, rules of thumb-based image of FitzRoy's forecast method is however suggested by some studies on weather forecasting from that time. For example, in his 1863 opuscule on *The Foretelling of the Weather*, Dutch Captain F.H. Klein compared the forecast methods of FitzRoy and Christoph H.D. Buys Ballot, the director of the Royal Netherlands Meteorological Institute, Utrecht. Klein acknowledged that FitzRoy's rules 'are useful to isolated observers', though Buys Ballot's system of deviations—that will be discussed in the next chapter—was 'more serviceable to promote the interests of meteorology'. In his enthusiasm to reduce FitzRoy's method to a set of simple rules, largely based on barometric readings, he distinguished eleven rules of thumb, e.g. the third said: 'when during winter time the barometer falls and the thermometer reads relatively low, a fall of snow be expected' Klein (1863, p. 22). Just how widespread this empiricistic image was, and just what its scope and repercussions were, I am not able to say, but it is my impression that it was more pervasive among his detractors (Galton, Le Verrier) than his followers.

81 *Report of a Committee Appointed to Consider Certain Questions Relating to the Meteorological*
    Department of the Board of Trade [Galton Report]. London: Eyre & Spottiswoode, 6.

82 Ibid., p. 20.

tion from the authors of the report embodying the elite science of the Royal Society. 'The practice of issuing daily official notices of the weather', they concluded, 'is inconsistent with the position and functions of a Government department, and must be prejudicial to the advancement of true science'.[83]

In a more technical sense, FitzRoy's forecast method also inherited some of the historical problems of meteorology in establishing certain equivalence between weather forecasts and storm warnings.[84] The Galton report agreed that there was a clear-cut separation between the two concepts, as was proved by the testimony of clear examples in France—thus, Le Verrier had opted to communicate the actual state of the weather, and, at best, the approach of storms known to exist elsewhere.[85] But this separation was more problematic for FitzRoy: he always held that the foundations underlying the two cases were identical, a position that the Royal Society had maintained in 1863. In any event, the report recommended stopping his daily weather forecasts and restricting the duration of storm warning. This had immediate consequences: forecasts and the storm warning service were suspended, the former being reinstituted in 1879 and the latter being resumed in a modified form in 1868. By then, the shadow of the late forecaster (who committed suicide in 1865) was too long.[86]

In short, FitzRoy's forecast method emerges as a counterweight to the methods developed in France and Italy. FitzRoy's approach was Lagrangian; Le Verrier and Marié-Davy's approach Eulerian. FitzRoy's aims were autonomy, localism, and freedom of individual judgement —just the opposite of Le Verrier's aims, which were state control, centralism, and authority. FitzRoy used the seafaring community as both reference and audience. For this community, the weather was the summation of barometer and thermometer changes and wind shifts. Hence, there was no place for the primacy of the pressure field (whether through barometric waves or isobars) over other fields.

Fitzroy's wind flow-model

The confluence of polar and tropical air currents was propounded by Dove in the theoretical framework of the midlatitude circulation, but meteorologists soon considered its forecasting implications. Among these was FitzRoy, perhaps the staunchest supporter of Dove, in part because by then it was widely acknowledged that midlatitude storms formed vortices.[87] As was shown, FitzRoy attempted to identify atmospheric waves in his weather

---

83　Ibid., p. 34. For a discussion on the Galton report, see Walker (2012, pp. 60–65); Anderson (2005, pp. 124–25); Burton (1986, pp. 55–59).

84　Williamson (2017, pp. 777–79).

85　Gillham (2001, p. 148); Halford (2004, pp. 202–03).

86　For the consequences of the Galton report, see Burton (1986, pp. 59–97); Walker (2012, pp. 65–89).

87　FitzRoy and Dove indeed professed mutual admiration, for Dove had dedicated his seminal book *Ueber das Gesetz der Stürme* to FitzRoy.

charts and diagrams and thereby demonstrate their utility for the forecast of storms and the weather in general. Instead, he proved the reverse: graphic analysis showed that the existence of mechanical, wavelike, atmospheric undulations could not be proved. Empirical knowledge showed, therefore, the utter inadequacy of the hypothesis of atmospheric waves.

The assumption of atmospheric waves as a forecasting tool was premised on the arguments of Quetelet that barometric waves would show the approach of storms. Upon failure of this hypothesis, FitzRoy worked with isobars, crest lines, and pressure curves, but none of these graphic devices appeared to be able to show the direction of winds. In 1858, his office published the English translation of Dove's seminal work, *Ueber das Gesetz der Stürme* (Berlin, 1857)—the book that had the greatest influence on FitzRoy.[88] Dove's law of storms was based on the argument that the confluence of currents of warm and cold air would cause atmospheric changes. Therefore the British weather had to be evaluated in the context of opposing currents of warm and cold air.[89]

FitzRoy was sensitive to the constraints of Dove's theory, and he tried to differentiate between those principles that were fundamental, such as the concept of polar and tropical currents of air, and those that were purely guesswork—e.g. that storms were normally associated with linear air currents. An example of common element was the notion of opposing currents. Dove had forcefully argued that the atmosphere was dominated by two basic air currents. Their movement was depicted by a vivid image: while one was cold, dry, and northeasterly, the other was warm, humid, and southwesterly; and they replaced one another consecutively as the atmospheric system as a whole moved slowly eastwards. FitzRoy retained this image and followed his ideas almost *au pied de la lettre*, though he preferred to speak of an alternation (instead of conflict) of currents. FitzRoy pointed out that in moving towards the tropics, the polar air

> divides, diverges, or splits into streams, interspersing with those advancing from more or less opposite directions as parallel currents. Hence in middle latitudes the continuous alternation of polar and tropical currents, which in their innumerable modes and degrees of opposition or combination, occasion every variety of mixed wind.[90]

FitzRoy applied this model to his analysis of the Royal Charter Storm, and drew straight lines to show average directions of polar and tropical currents (Figure 4.4).

88   Dove, *The Law of Storms*. [London], 1858. A revised and considerably enlarged edition was published in 1862 (translated by R.H. Scott).
89   For Dove's influence on FitzRoy's storm model, see Chromow (1940, pp. 304–05); Kutzbach (1979, p. 82).
90   FitzRoy (1863, p. 237).

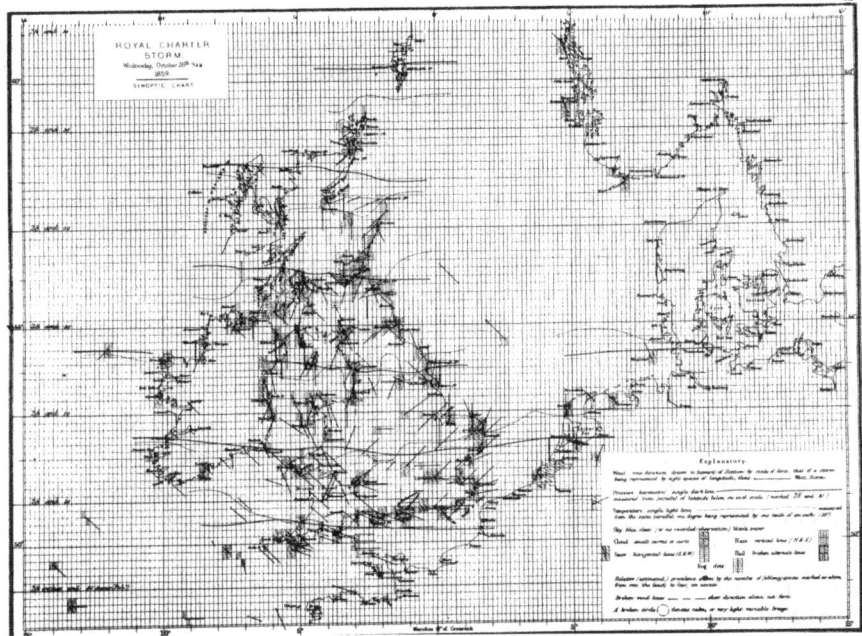

Figure 4.4 Synoptic chart of the storm of 26 October 1859 (also known as the Royal
Charter Storm) by FitzRoy. Blue (thick) and red (thin) lines represent
barometric pressure and temperature, respectively; wind strength is
shown by the length of the line; and hatched boxes show weather conditions such as cloud and rain.

Source: FitzRoy (1863).

At this point in his study, FitzRoy turned from theoretical and physical
considerations to empirical evidence. Did seamen's observations support
the notion of parallel currents? Yes, without a doubt. Common experience says, he argued, that wind direction within air streams is practically
the same before there is a sudden shift caused by the arrival of a different
stream. In fact,

> in most, if not all the storms to which I can bear testimony, currents of
> air arriving from different directions appeared to succeed each other,
> or combine together. One usually brought the "dirt" (to use a sailor's
> term), and the other cleared it away.

FitzRoy wrote emphatically[91]

---

91 'One of these currents was warm and moist, another cold and dry, comparatively speak-
ing. While one lasted the barometer fell or was stationary; with the other it rose'. FitzRoy
(1863, p. 101).

Understanding the global circulation system was crucial, for storms were a consequence of the meeting of alternating currents.[92] However, how could storms be formed within these currents? The answer lay with Fitz-Roy's 'boundaries'. The most conspicuous feature of the circulation system, he argued, was the widespread evidence of eddies between the two main currents: these eddies were the origin of what was known as cyclones or gyratory movements. Yet, here too, the pupil distanced himself from his *maestro*: while for Dove cyclones originated from the tongue on an incursion of warm air, for FitzRoy they arose at the boundary between the two air currents. 'Ordinarily, there are two, three, or more atmospheric nodes having irregular curvilinear shapes; these nodes are situated between the main currents, as the latter approach one another or collide forcibly'.[93] Here, what FitzRoy meant by the term 'atmospheric nodes' was the successive cyclones of a cyclonic series or family. Once the notion of a boundary between two air masses was accepted, FitzRoy concluded, the occurrence of 'circuitous eddies, storms, or cyclones' is more easily explainable. This very idea of the formation of cyclones as boundary phenomena between a polar and a tropical current has often been viewed as a precursor of the polar front theory of cyclones.[94]

Despite this precursory role, however, constraints remained that undermined the scientific credibility of FitzRoy's ideas before academic audiences. To begin with, the theoretical basis of his wind flow-model was not nearly as clear and specific as its empirical depiction, which he had colourfully showed in his map of the 1859 Royal Charter Storm. Not a single word was said in his book about the source of energy in storms. Nor did he include Espy's theory of precipitation by adiabatic cooling in ascending the currents of air.[95] The issue of the effect of gravitational forces also presented

92 FitzRoy (1863, p. 101):

   A quarter of a century's attention to the subject *since*, has convinced the writer that consideration ought to be given *first* to the great general order of circulation, with alternating, and more or less "parallel" currents, and afterwards to *their* consequences, when disturbed —namely storms, and other occasional phenomena.

93 FitzRoy—quoted by Khrgian (1970, p. 171).

94 Poulter (1934, p. 342) has emphasized the small margin by which FitzRoy 'failed to establish the polar-front conception'. In fact, he found close similarities between FitzRoy's ideas and the cyclone notion defined by Jacob Bjerknes and Halvor Solberg in their seminal paper 'Life Cycle of Cyclones and the Polar Front Theory of Atmospheric Circulation' in 1922:

   The polar front is generally a wavy line, in continual motion through all latitudes of the temperate zone, bordering large tongues of polar and tropical air. The tongues of tropical air form the warm sectors of young travelling cyclones and the intermediate tongues of polar air constitute the moving wedges of high pressure between successive cyclones.

95 Kutzbach (1979, p. 82) holds that Fitzroy's book is 'one of the last major meteorological treatises that did not incorporate the new concepts of energy conservation and conversion'.

challenges, particularly with regard to the main air currents. Lunar and solar tides, he thought, could intensify a current, to the point of making it dominant. Yet, he had only a vague idea of the effect of the deflecting force of the earth's rotation—an issue that had just been formulated by Ferrel.[96] Finally, he had a bifocal, rather than a three-dimensional, view of the boundaries between air masses. All he could say is that these masses, whether warm or cold, evolved in the air first and at the surface last. FitzRoy said nothing about the structures that would result from this model.[97] For all these reasons, FitzRoy's wind flow-model was regarded to be very limited, and, after his death in 1865, was nearly forgotten. Indeed, it was not until the beginning of the twentieth century that forecasters were able to detect the polar and tropical air currents in their synoptic charts.[98]

Dove's conception of opposing air currents provided a theoretical basis to FitzRoy's weather forecast method. Here the comparison with Dove is instructive. For both, storms were the result of a mechanical interference of currents, rather than of a dynamic process. They achieved a clear overview of the meeting of two air masses. Yet, they lacked the notion of front as a boundary surface between air masses.[99]

However, their approach to weather forecast was very different. Dove emphasized his law of rotation (*Drehungsgesetz*) and the dependence of weather on wind direction, while FitzRoy favoured the synoptic, but Lagrangian, approach. This difference may in part reflect their early forays into meteorology: Dove began with the periodical variations of climate and venerated Humboldt, whereas FitzRoy started out in oceanic statistics, of which Maury had given a chart model that he soon modified by devising his *wind stars*.[100] While Dove rejected the use of synoptic charts and prioritized instead local observations and their statistical treatment, FitzRoy drew on Dove's air current theory to defend a synoptic (but not isobaric) view of weather forecast. Dove's approach can be seen as an extension of Humboldtian climatology, and FitzRoy's as an adaptation of Dove's ideas to the local weather problem.[101]

96 The name of Ferrel is mentioned twice in FitzRoy's *Weather Book*, although none of them refers to his theory of geophysical hydrodynamics. FitzRoy (1863, p. 48, 139).

97 Ludlam (1967, pp. 24–25).

98 For the genesis of the polar front theory of cyclones and the detection of air currents in synoptic charts, see Fleming (2016, pp. 13–50); Friedman (1989); Jewell (1981, pp. 824–30); Kutzbach (1979, pp. 159–218).

99 Dieckmann (1931, pp. 750–51); Chromow (1940, p. 304).

100 On Dove as a precursor of dynamic climatology, see Chromow (1931, pp. 312–14).

101 It would be more accurate, perhaps, to say that although Dove sporadically used it and even pioneered the synoptic method in some regards, as emphasized by Scultetus (1943), he ended up opposing its use for weather forecast. In 1868, Dove stated:

    it must be emphasized that a cartographic representation of a storm using isobaric lines leads us into error and it gives the impression that almost all storms are cyclonic

This interpretation is somewhat fragile, however, for it overlooks the relations between their theoretical and observational practices, as well as the influence of their immediate environment. Dove's predilection for phenomenological schemes should rather be seen as a reflection of his method of local observations. He deduced the meeting of conflicting air currents from the shape of wind roses, and his law of rotation from local wind statistics. For him, theories should be based on already available observational data. For a long time, his work was a one-man business covering from the inspection of observation places to data collection, analysis, and publication. In contrast, through a group of coastal observers and his fishery barometer program, FitzRoy conceived a new horizontal form of networking autonomous observers. This drive towards autonomy and localism prompted him to discard any solution within the Eulerian framework and adapt Dove's scheme to weather forecast.

Galton's semi-Lagrangian approach

Between FitzRoy's empirical rules and busy charts and Marié-Davy's isobaric maps of the weather, there was a middle ground which seems to have attracted its main promoter, Francis Galton. Basically, it was about the construction of glyph and iconic maps and schematic micromaps representing streamlines (flow field), while also showing temperature and barometric pressure. His scheme combined the virtues of the two methods, the Eulerian and the Lagrangian: while he drew on data from measurements taken at fixed points near the ground, he sought to obtain an instantaneous picture of the overall flow field and the rest of meteorological elements. Galton, who is credited with coining the term *anticyclone*, reached the culmination of this semi-Lagrangian scheme in his work *Meteorographica*.

Known in retrospect as one of the greatest contributors to statistics, eugenics, and inheritance, and as the creator of the correlation method (aimed at determining the degree of interdependence of two or more variables), Galton was known at that time as a botanist and an outstanding geographer.[102]

---

in form. Actually, an equatorial flow, propagated at storm velocity from SW to NE, lowers the barometer over its entire extent.

—as quoted by Khrgian (1970, pp. 168–70)

Before we accept as a proven fact that he became the staunchest enemy of the synoptic method (as has been historically suggested by his numerous detractors), we should hear what Karl-Heinz Bernhardt (2004, p. 3; 2013, p. 79) has written on this score: Dove 'graphically presented qualitative wind observations ("currents") of the same synoptic situation (Dec. 24, 1821) as described by Brandes's in 1828. Moreover, he drew 'an idealized circular vortex (streamlines) over western Europe' as soon as 1841. Finally, in 1873 he published 'a synoptic map (isobars, wind, arrows, thunderstorms) for Jan. 20, 1863'.

102 Indispensable biographical works on Francis Galton are Forrest (1974), Pearson (1914–30), and Waller (2001).

After graduating from Cambridge and receiving a large inheritance from his late father, he conducted several explorations in South West Africa from 1850 to 1852. In recognition of his talents he was awarded the gold medal from the Royal Geography Society of London. He also joined the managing committee of the Kew Observatory in 1858, where he could deepen his commitment to instrumental precision and statistical knowledge. A cousin of Charles Darwin, his interest in meteorological mapping derived from his African explorations, in which he cultivated geographical cartography.[103] His first foray into meteorology led him to privately print a retrospective chart of the English weather in 1861.[104]

In Kew, Galton was deeply impressed by the visual and geometrical language developed by Halley, Lalanne, and others, much of which overcame the traditional techniques used in meteorological cartography.[105] His concerns were focused on questions related to astronomy, geodesy, and meteorology, which required first the use of reliable and systematic data, and then the pursuit of coherent patterns that could be deduced from a graphic treatment of data. He realized that, when lists of observations were 'printed in line and column', they were 'in too crude a state for employment in weather investigations'.[106] Galton was particularly impressed by the use of isopleth lines for three-dimensional surfaces, and by how grids of schematic contour maps could show the changes of relations between two or more variables over space and time.[107]

In July 1861, Galton sent a circular letter to meteorologists throughout Europe, inviting them to daily telegraph him reports of their observations in order to compile and print weather maps throughout December 1861.[108] He received responses from over 300 observers, from which he undertook a process of graphic abstraction, culminated two years later in his work *Meteorographica or methods of mapping the weather*. In a first stage, he produced a series of ninety-three weather maps for Europe for December 1861 (three per day). Yet, his maps were by no means conventional. Through his maps, he not only highlighted the meagre information in ordinary weather maps but also urged map-makers to find better ways to express the power of visual sense.[109]

103 For Galton's early geographical mapping work and his relationship with the Royal Geographical Society, see Pearson (1914–30, vol. 2, pp. 21–32).

104 Galton's first map was printed by his friend the mathematician William Spottiswoode on 12 June 1861. Gillham (2001, p. 142).

105 For Galton's contributions to the science of meteorology, see Friendly (2008, pp. 514–16), Gillham (2001, pp. 140–51), Tickell (1993, pp. 54–61), Anderson (2005, pp. 200–04).

106 Galton (1863a, p. 3).

107 As was shown in Chapter 2, isopleths were used for distributions belonging to complex geographical concepts including functions of two or more variables (such as density, spacing, correlation, or ratios). See Robinson (1971, pp. 49–50).

108 Gillham (2001, pp. 141–42), Pearson (1914–30, vol. 2, pp. 37–38), Shaw (1926, p. 306).

109 In 1865, Galton suggested that a stereoscopic photograph of a model would better represent mountainous landscapes. In fact, he was able to publish some stereoscopic maps by photographing some physical models with a double-view camera. See Galton (1865, pp. 99–106); Pearson (1914–30, vol. 2, pp. 33–34).

Or, as he sagely stressed it, 'a few judicious sweeps and shadings of a draughtsman's pen may embody the simultaneous observations of hundreds of meteorologists'.[110]

By then, however, Galton had already realized the inherent limitations of isobars used in ordinary maps, a fact that had impelled him to distance himself from the conceptions prioritizing the baric field in weather forecasting. In a brief article on the recent weather published in December 1863, just a few months after the printout of Marié-Davy's first maps, Galton depicted the procedural constraints of the Paris Observatory's weather maps:

> The lines of equal barometric pressure which had formed a series of concentric curvatures of small radius round the area of lowest pressure, almost always deployed themselves into nearly straight lines that stretched across Europe, sloping slightly towards the north or east. It seems evident to ordinary observers that any prevision of the weather founded on the westerly translation of an aerial eddy must have been wholly in error in these cases. There is an undoubted tendency to a westerly movement in the storms and in the weather generally that is pictured by these charts, but its character is so modified during the process of translation that the barometric lines of any one day have grown out of resemblance to those of the day previous.[111]

The inadequacy of isobaric lines to timely represent the westerly movement of storms—and therefore to serve as a basis for weather prevention—led him to extend information contents of maps by devising new symbols and figures. His aim was not so much to debunk the value of isobars as to expand the limits of information by means of a schematic representation. As he stated in the same article, 'our first impression on examining [the Paris] charts is surprise at the narrow limits of what is really known compared to the audacious dogmatism too common among meteorologists'.[112] To overcome these limitations, he introduced symbols and figures in such a way that each set of observations occupied a small rectangular label whose position indicated the station where the observations were made. While the thermometric and barometric readings were given in figures, the amount of cloud, the wind direction, and its force were expressed by symbols—shaded types, arrows, and marks, respectively. Symbols were circular and hexagonal (Figure 4.5).[113]

110 Galton (1863a, p. 3).
111 Galton (1863b, p. 730).
112 Ibid.
113 Galton (1863a, p. 4) emphasized the visual power of symbols or icons: they 'have the advantage of telling their tale directly to the eye'. For an explanation of these symbols and images of circular stamps, see Pearson (1914–30, vol. 2, pp. 36–37).

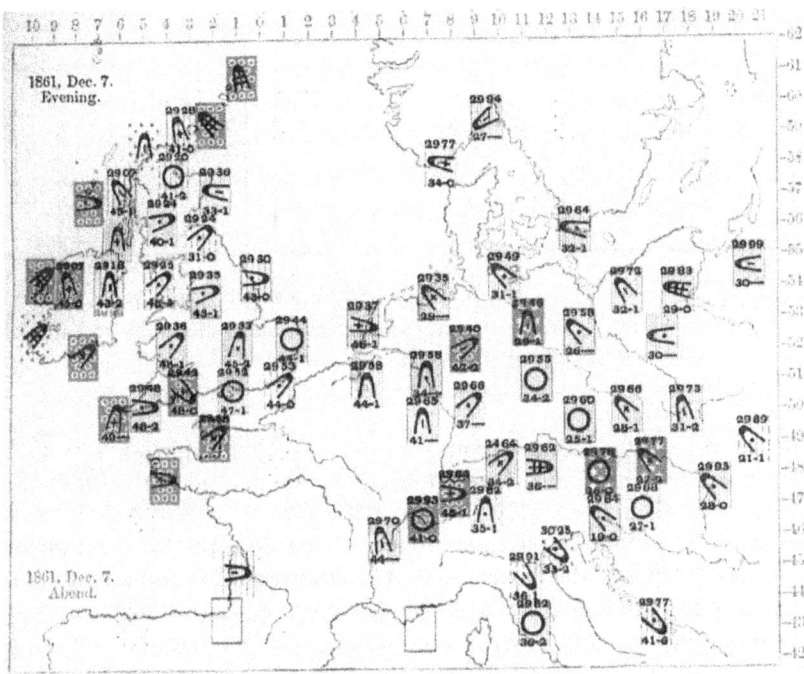

Figure 4.5 Galton glyph map of wind, cloud cover, and rain, December 1861.
Source: Galton (1863a).

In *Meteorographica*, he chose iconic maps of barometric pressure instead of traditional contour lines or isobars. The reasons for this choice were that, while wind direction could be easily represented (as its movements 'are on the plane of the earth' and can be 'described by directive marks'), the rest of elements required 'a third dimension'.[114] Or, as he lucidly expressed, they required 'a similar pictorial treatment to that employed for geographical elevations; in other words, areas of elevation and depression of the barometer, etc., must be pictured by contour lines and shadings, on the same principles as mountains and valleys'.[115] Although he mentioned contour lines, he eventually opted for barometric charts or grids containing symbols with shaded types, in which the intensity of shape represented the deviation from an average (Figure 4.6).[116] The end result of all this is clear: whereas in the Eulerian specification of the baric field the barometric pressure was usually depicted by isolines in a two-dimensional plane, Galton's barometric maps were three-dimensional and iconic (not linear).

114 Galton (1863a, p. 5).
115 Ibid.
116 Galton also used two colors: one for excess; and the other for deficiency.

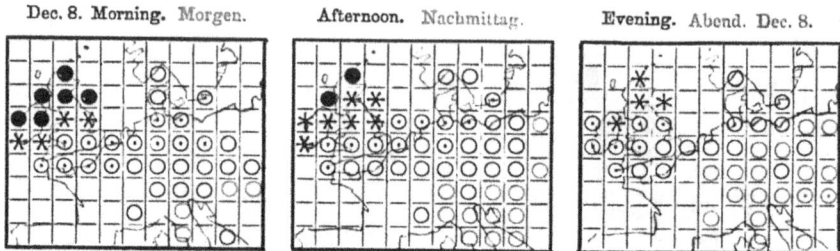

Figure 4.6 Galton's iconic 3D barometric maps for 6 December 1861. Lower and higher atmospheric pressures than average are shown by red and black symbols, respectively.

Source: Galton (1863a).

The adoption of a schematic and semi-Lagrangian approach by Galton is even more perceptible when compared the sequential maps of *Meteorographica* with his first chart for the British weather in 1861. As can be seen in Figure 4.7, his 1861 chart gave an aerial view of separate local observations on an outline of the British Isles. Circular and hexagonal stamps were scattered within this outline, and, by arrows and shapes engraved within the stamps, they indicated wind direction and barometric changes. In contrast, the maps of *Meteorographica* were much more schematic (Figure 4.6). In addition to a sequence of six synoptic charts of Europe with three daily observations, it included other six iconic 3D maps showing the deviation from a barometric average of 29.95 inches. Each map was a chessboard-like grid of small cells, with no reference to the land outline. The outcome was a set of charts and grids providing an abstract representation of weather conditions, unconnected to the observer and point of observation.[117] This lack of connection with systems fixed at a particular point in space is one of the characteristics of the non-Eulerian frames of reference.[118]

In his set of charts and grids represented in Figure 4.6, there seems to be a clear disengagement from the observer and point of observation. Yet, a specification of streamlines was still lacking. In the same *Meteorographica*, Galton undertook a second phase of abstraction. He reduced the data for each day to a 3 × 3 grid of multivariate schematic micromaps, in which barometric pressure, temperature, wind direction, and rain were shown by a range of colours, textures, shapes, and arrows (Figure 4.8). In particular, he drew streamlines (or lines tangent to the velocity vector of wind) in his attempt to show how one air current could be replaced by another (Figure 4.9). From these micromaps, he observed a phenomenon totally unforeseen: while in

---

117 This feature is discussed in more detail by Anderson (2006, pp. 79–80).
118 Typical of this difference, between the Eulerian and Lagrangian frames of reference, is the circumstance (that is noticeable here) that while the isobaric maps such as those of Marié-Davy showed discontinuities in the fields traced, iconic maps such as those of Galton represented continuous changes.

low-pressure areas winds rotated counterclockwise in inward spirals (as expected in cyclones), in high-pressure areas winds rotated clockwise in outward spirals. He termed these calm areas *anticyclones*, as opposed to the cyclone pattern. In an article on the development of the theory of cyclones, he defined this term in one sentence: an area of high pressure

> is usually a locus of dense descending currents, and therefore of a dispersion of a cold dry atmosphere, plunging from the higher regions upon the surface of the earth, which, flowing away radially on all sides, becomes at length imbued with lateral motion.[119]

Indeed, if there is a calm centre around which currents flow in radial lines, then they would rapidly curve to the right and form a sort of anticyclone. This phenomenon is explained with reference to Dove's law of rotation (*Drehungsgesetz*). Galton seems to have made this finding intuitively, through the analysis of high-dimensional graphs and streamline micromaps, on the basis of very limited surface data.

Galton's work exemplifies a symbiotic evolution of visual thinking and physical interpretation in meteorology. The need to elucidate the weather patterns such as air currents and barometric changes inspired new graphic devices such as iconic 3D maps and multivariate schematic micromaps. In turn, the application of these graphic forms to a wide spectrum of atmospheric variables provided him with physical interpretations that subsequently opened up new avenues and prospects. Thus, for example, in 1870 Galton sought to develop what some readers regarded as a way towards correlation—i.e. a formula to determine wind velocity from three variables: barometric reading, temperature, and humidity.[120] While in their more primitive guise Galton's weather charts betrayed an Eulerian approach, in their mature form they revealed the adoption of a semi-Lagrangian approach.

This evolution largely explains Galton's frustration in dealing with the isobaric synoptic maps of weather in the 1870s. That Galton could rely upon the isobar technique of weather maps depended on another important attribute, namely, its ability to represent the movement of air masses, whether through the paths followed by air currents (trajectories) or through the instantaneous wind pattern (streamlines). Yet, this was not the case. In 1876, after reading the work *Weather Charts and Storm Warnings*, published by the Meteorological Office, Galton bemoaned that atmospheric change seemed to be 'mainly regulated by the distribution of barometrical pressure over the globe'. And he added the following sarcastic remark: the author of the work 'seems to regard the pressure as a sort of wild beast having volition of its own'.[121]

---

119 Galton (1863c, p. 385).
120 Galton (1870, pp. 31–33). Pearson (1914–30, vol. 2, p. 55): 'the interest of his paper lies in the evidence that he was feeling his way towards "correlation"'.
121 Galton Papers, copy of Scott's *Weather Charts and Storm Warnings*, 121, in a note in the margin, handwritten by Galton—quoted by Anderson (2005, p. 286).

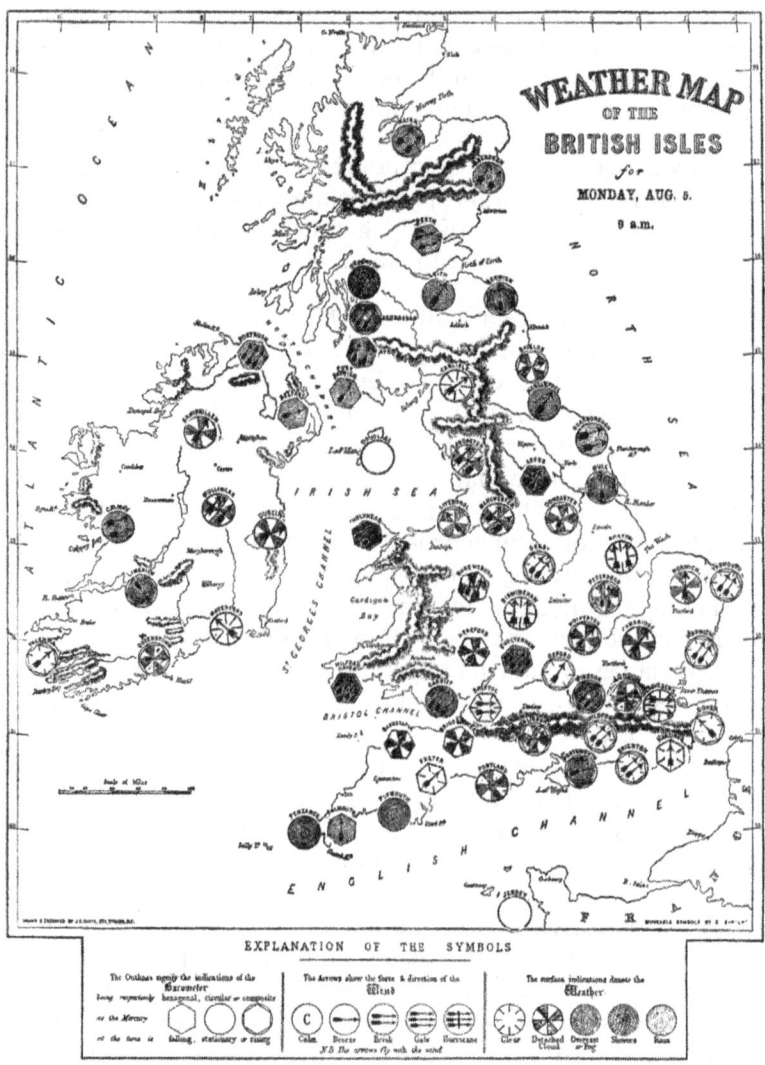

Figure 4.7 One of Galton's earliest synchronous weather maps, 5 August 1861. This
       map accompanied the prospectus of the Daily Weather Map Company.
Source: Shaw (1926, vol. 1, p. 309).

These suspicions did not prevent him from resolutely opposing FitzRoy's
forecasting method in the famous 1866 Galton report, on the grounds that it
had 'as yet no scientific basis' and its results had not been deduced 'by means
of accurate induction from observed facts'.[122]

122 Galton Report, 1866 (London: Eyre & Spottiswoode)—quoted by Gillham (2001, p. 148).

Conclusion

Following Dove's narrative on the conflict of two opposed air currents, we began with Jinman's cyclone flow-model, spent the most time on FitzRoy's forecast method, and ended with Galton's schematic micromaps and his streamlines. They all addressed the question of storms and weather forecast through a broad series of empirical rules and graphic innovations. Most importantly, they identified the wind field as a fundamental atmospheric variable that could be described in terms of trajectories and streamlines, and inaugurated a Lagrangian approach to weather forecast in which atmospheric changes were no longer exclusively determined by barometric distribution. With this new perspective, they anticipated basic processes of storm formation, frontogenesis, and dynamic meteorology.

An equally important feature: they all provided an interesting counterpoint to the Eulerian approach adopted in France and Italy. However, unlike Le Verrier and Marié-Davy's isobarometric maps, FitzRoy's charts and his forecast method had little follow-up. Their historical legacy was almost limited to the concept of 'boundary' between two air currents as the origin of cyclones and his notion of cyclonic family, which only became a reality five decades later. Forecasters paid little attention to his forecast method, as can be judged from the fact that forecasts, suspended after the Galton report, were not resumed until 1879, and then on the basis of the isobaric-synoptic method. Subsequent treatises on weather forecasting simply ignored it.[123] Theorists only had a passing interest in his work. Napier Shaw seems to have been one of the few important physicists to take up some of his ideas, though not before the 1900s.[124]

The reasons for this neglect have often been described by historians. One was that FitzRoy produced confused weather maps based on varying empirical rules, and followed no calculation or logical process. Others point to factors related to the articulation of scientific values in Victorian culture and public expectations. However, the rivalry between the two opposing approaches to the problem of weather forecasting and their disparate 'world views' has gone unnoticed up to now. Beyond the Victorian cultural specificity and from a

---

123 For instance, FitzRoy's name is mentioned only four times, and none in a positive sense, in R.H. Scott's popular book *Weather Charts and Storm Warnings* (London: Henry S. King & Co., 1876), pp. 15, 118–19, 124; and Ralph Abercromby's work, *Principles of Forecasting by Means of Weather Charts* (London, 1885, 2nd ed.), contains but one passing mention of FitzRoy (on p. 53).

124 The rejection of the storm analysis focused on trajectories of air rather than isobaric patterns was surely one of the most common stances in the operational practice of weather forecasting. Thus, Shaw and Rudolf Lempfert's 1906 *magnum opus*, *The Life History of Surface Air Currents. A Study of the Surface Trajectories of Moving Air*, which has been regarded as 'a worthy successor to those of Dove, FitzRoy and Jinman' [Walker (2012, p. 144)], and anticipated by over a decade the frontal cyclone model of Bergen meteorologists, went virtually unnoticed at the time. See Kutzbach (1979, pp. 180–81). For the popularization of FitzRoy's achievements by Shaw, see Shaw (1911, pp. 5, 7, 53, 262; 1926, vol. 1, pp. 9, 149–53, 162, 296, 303, 311).

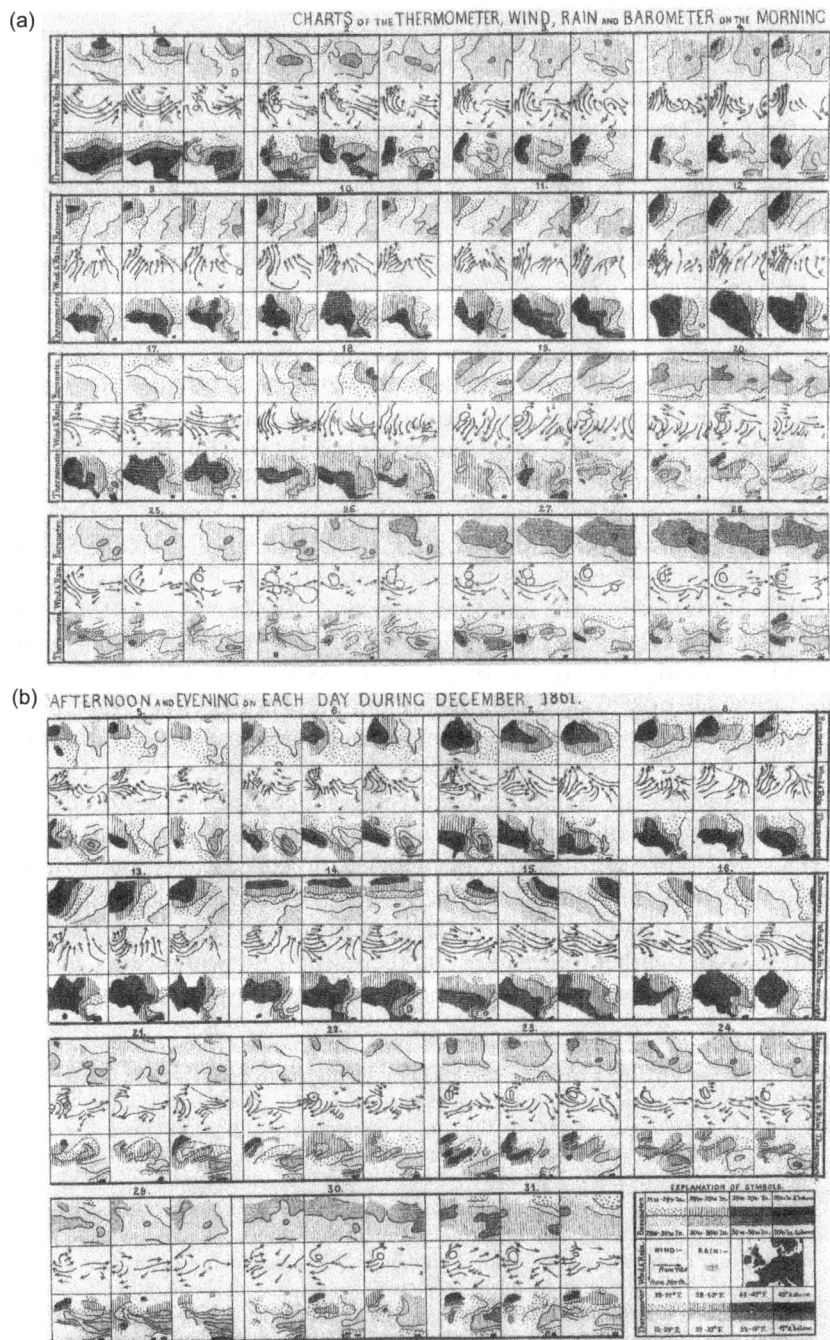

Figure 4.8 Galton's schematic micromaps. 'Charts of the thermometer, wind, rain and barometer on the morning, afternoon and evening on each day during December 1861'. Galton (1863a).

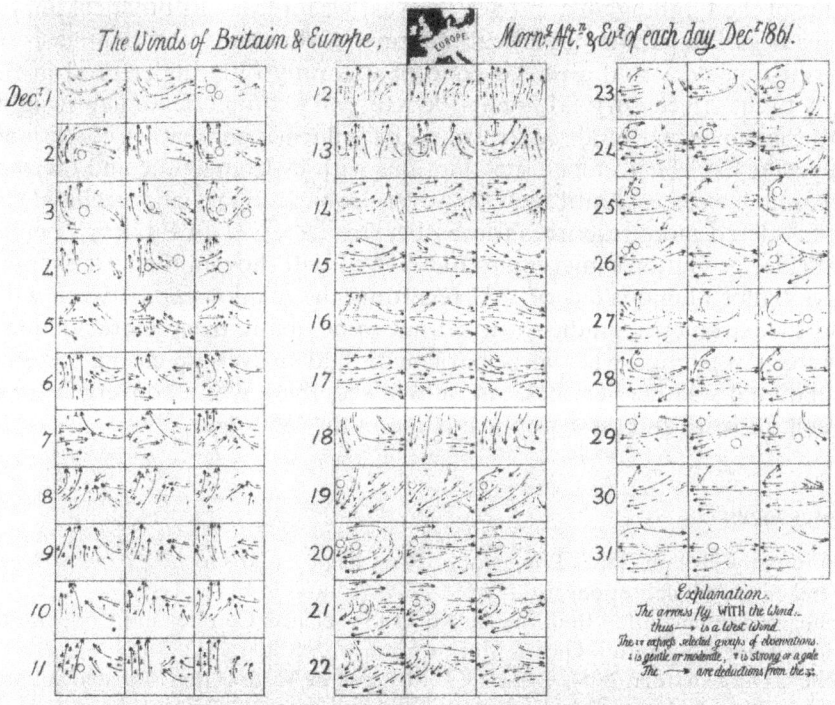

Figure 4.9 Galton's schematic micromaps. 'The Winds of Britain & Europe on the morning, afternoon and evening on each day during December 1861'. Galton (1863a).

wider perspective, a fundamental reason to ignore FitzRoy's forecast method was that it embodied, or at least carried with it, a set of *problematic* social aims. He pursued autonomy, local power, and the freedom of individual judgement—instead of the state control, centralism, and authority that would end up prevailing in national weather services. He distrusted isobars and isolinear maps, in part as a consequence of his own observational system. His audience was mainly the seafaring community, whereas national weather services tended to cover broader sectors and interests. Finally, he took the storm structures to be two-dimensional and said no word about the source of energy in storms. Three-dimensional cyclone models including the thermal theory and energetic aspects later became known, which made his model become obsolete in important respects. All of this largely explains why FitzRoy's forecast method and his wind flow-model soon fell into oblivion.[125]

125 Three-dimensional cyclone models were propounded in the 1870s by Clement Ley from synoptic-statistical studies and cloud observations, and by Hugo H. Hildebrandsson from cirrus observations; and by Nils Ekholm, Frank H. Bigelow, and Napier Shaw in the 1900s, based on synoptic studies of horizontal temperature contrasts. Kutzbach (1979, pp. 120–23, 128–34, 171–80).

Although late nineteenth-century meteorologists barely recognized the value of the Lagrangian approach for weather forecast, its practical importance began to gain strength with the turn of the century. Its emphasis on the air-trajectory and air-mass concept was only fully understood in the early twentieth century. Although FitzRoy and Galton offered new insights into the movement of air currents and its influence on weather conditions, they were not able to relate such motions with hydrodynamic and thermodynamic processes. Until then, however, isobaric charts and probabilistic rules offered more resources to weather forecasters than FitzRoy's empirical rules or Galton's micromaps did it. As I will show in the next chapter, Buys Ballot's famous law of 1857 regarding the relationship between wind and the barometric gradient turned out to be qualitatively more adequate for storm warnings, whereas isobaric maps had an ever-increasing presence in national weather services. As we will see, there was a convergence between statistics and the synoptic method in the 1850s and 1860s.

References

Abercromby, Ralph, 1885. *Principles of Forecasting by Means of Weather Charts*, 2nd ed. London: Stationery Office.

Achbari, Azadeh, 2015. 'Building Networks for Science: Conflict and Cooperation in Nineteenth-Century Global Marine Studies'. *Isis*, 106 (2), 257–82.

Agnew, Duncan Carr, 2004. 'Robert FitzRoy and the Myth of the 'Marsden Square': Transatlantic Rivalries in Early Marine Meteorology'. *Notes and Records of the Royal Society of London*, 58 (1), 21–46.

Anderson, Katharine, 2005. *Predicting the Weather: Victorians and the Science of Meteorology*. Chicago: The University of Chicago Press.

Anderson, Katharine, 2006. 'Mapping Meteorology'. In James Rodger Fleming, Vladimir Jankovic, and Deborah R. Coen eds., *Intimate Universality. Local and Global Themes in the History of Weather and Climate*. Sagamore Beach, MA: Watson Publishing International, 69–91.

Barbot, Michela, and Favino, Federica, 2017. 'Premessa'. *Quaderni Storici*, 52 (3), 643–53.

Barlow, Derek, 1994. 'From Wind Stars to Weather Forecasts: The Last Voyage of Admiral Robert FitzRoy'. *Weather*, 49 (4), 123–32.

Basalla, George, 1972. 'Robert FitzRoy'. In Charles Coulston Gillispie ed., *Dictionary of Scientific Biography*. New York: Charles Scribner's Sons, 5, 16–18.

Bergeron, Tor, 1981. 'Synoptic Meteorology: An Historical Review'. *Pure and Applied Geophysics*, 119, 443–73.

Bernhardt, Karl-Heinz, 2004. 'Heinrich Wilhelm Dove's Position in the History of Meteorology of the 19th Century'. In *International Commission on History of Meteorology*. Polling, Germany, 28 May 2004.

Bjerknes, Jacob, and Solberg, Halvor, 1922. 'Life Cycle of Cyclones and the Polar Front Theory of Atmospheric Circulation'. *Kristiania, Geophysisks Publikationer*, 3 (1), 3–18.

Bruce, Robert V., 1987. *The Launching of Modern American Science, 1846–1876*. New York: Knopf.

Brunt, David, 1951. 'A Hundred Years of Meteorology (1851–1951)'. *The Advancement of Science*, 30, 114–24.

Brunt, David, 1956. 'The Centenary of the Meteorological Office: Retrospect and Prospect'. *Science Progress*, 44 (174), 193–207.

Burstyn, Harold L., 1981. 'Maury, Matthew Fontaine'. In Charles Coulston Gillispie ed., *Dictionary of Scientific Biography*. New York: Charles Scribner's Sons, 9, 195–97.

Burton, James, 1986. 'Robert FitzRoy and the Early History of the Meteorological Office'. *British Journal for the History of Science*, 19, 147–76.

Burton, James, 1988. 'The History of the British Meteorological Office to 1905'. The Open University, PhD diss.

Charlier, Roger H., 2004. 'Fratres in Maribus 150 Years Ago: The First International Ocean–Science Conference'. *Journal of Coastal Research*, 20, 347–50.

Chromow, Swoboda P., 1931. '"Dynamische klimatologie" und Dove'. *Zeitschrift für angewandte Meteorologie*, 48, 312–14.

Chromow, Swoboda P., 1940. *Einführung in die synoptische Wetteranalyse*. Berlin: Springer-Verlag.

Cotter, Charles H., 1979. 'Matthew Fontaine Maury (1806–1873): "Pathfinder of the Sea"'. *Journal of Navigation*, 32, 75–83.

Cox, John D., 2002. *Storm Watchers: The Turbulent History of Weather Prediction from Franklin's Kite to El Niño*. Hoboken, NJ: John Wiley & Sons, Inc.

Cushman, Gregory, 2013. 'The Imperial Politics of Hurricane Prediction: From Calcutta and Havana to Manila and Galveston, 1839–1900'. In Erika Marie Bsumek, David Kinkela, and Mark Atwood Lawrence eds., *New Approaches to International Environmental History*. Nueva York: Oxford University Press, 137–62.

Davis, John L., 1984. 'Weather Forecasting and the Development of Meteorological Theory at the Paris Observatory, 1853–1878'. *Annals of Science*, 41 (4), 359–82.

Dieckmann, Alexander, 1931. 'Fitz Roy. Ein Beitrag zur Geschichte der Polarfronttheorie'. *Die Naturwissenschaften*, 36, 748–52.

Dove, Heinrich Wilhelm, 1841. 'Über das Gesetz der Stürme'. *Annalen der Physik*, 52, 1–41.

Dove, Heinrich Wilhelm, 1858. *The Law of Storms*. [London], [G.E. Eyre & W. Spottiswoode]. Second edition, entirely revised and considerably enlarged, by R.H. Scott in 1862, *The Law of Storms Considered in Connection with the Ordinary Movements of the Atmosphere*. London.

Dry, Sarah, 2009. 'Safety Networks: Fishery Barometers and the Outsourcing of Judgement at the Early Meteorological Department'. *The British Journal for the History of Science*, 42 (1), 35–56.

FitzRoy, Robert, 1858. *Barometer and Weather Guide*. London: Board of Trade.

FitzRoy, Robert, 1859. *Notes on Meteorology*. London: Board of Trade.

FitzRoy, Robert, 1860. 'Remarks on the Late Storms of 25–26 Oct and 1 Nov'. *Proceedings of the Royal Society of London*, 10, 222–24.

FitzRoy, Robert, 1861. 'On British Storms, Illustrated with Diagrams and Charts'. In *Report of the Thirtieth Meeting of the British Association for the Advancement of Science Held at Oxford in June and July 1860*. London: John Murray, 39–44.

FitzRoy, Robert, 1863. *The Weather-Book: A Manual of Practical Meteorology*. London: Longman, Green, Longman, Roberts, & Green.

FitzRoy, Robert, 1864. *Report of the Meteorological Department of the Board of Trade*. London: H.M.S.O.

Fleming, Fergus, 1999. *Barrow's Boys: A Stirring Story of Daring, Fortitude and Outright Lunacy*. London: Granta.

Fleming, James Rodger, 2016. *Inventing Atmospheric Science. Bjerknes, Rossby, Wexler, and the Foundations of Modern Meteorology*. Cambridge, MA: The MIT Press.

Forrest, Derek, 1974. *Francis Galton: The Life of a Victorian Genius*. London: Paul Elek.

Friedman, Robert Marc, 1989. *Appropriating the Weather: Vilhelm Bjerknes and the Construction of a Modern Meteorology*. Ithaca, NY: Cornell University Press.

Friendly, Alfred, 1977. *Beaufort of the Admiralty: The Life of Sir Francis Beaufort 1774–1857*. New York: Random House.

Friendly, Michael, 2008. 'The Golden Age of Statistical Graphics'. *Statistical Science*, 29 (4), 502–35.

Galton, Francis, 1863a. *Meteorographica, or, Methods of Mapping the Weather: Illustrated by Upwards of 600 Printed and Lithographed Diagrams Referring to the Weather of a Large Part of Europe, During the Month of December 1861*. Cambridge: Macmillan.

Galton, Francis, 1863b. 'Recent Weather'. *The Reader*, 2 (19 December), 730.

Galton, Francis, 1863c. 'A Development of the Theory of Cyclones'. *Proceedings of the Royal Society of London*, 12, 385–86.

Galton, Francis, 1865. 'On Stereoscopic Maps Taken from Models of Mountainous Countries'. *Journal of the Royal Geographical Society of London*, 35, 99–106.

Galton, Francis, 1870. 'Barometric Predictions of Weather'. *British Association Reported*, 31–33. Also in *Nature*, 2 (1870), 501–03.

Gillham, Nicholas Wright, 2001. *A Life of Sir Francis Galton: From African Exploration to the Birth of Eugenics*. Oxford: Oxford University Press.

Graham, Gerald S., 1956. 'The Ascendancy of the Sailing Ship, 1850–85'. *Economic History Review*, 9, 74–88.

Halford, Pauline, 2004. *Storm Warning. The Origins of the Weather Forecast*. Stroud: Sutton Publishing.

Headrick, Daniel R., 1988. *The Tentacles of Progress. Technology Transfer in the Age of Imperialism, 1850–1940*. Oxford: Oxford University Press.

Hearn, Chester G., 2002. *Tracks in the Sea: Matthew Fontaine Maury and the Mapping of the Oceans*. Camden, ME: International Marine / McGraw-Hill.

Hughes, Patrick, 1988. 'FitzRoy the Forecaster: Prophet without Honor'. *Weatherwise*, 41 (4), 200–04.

Jewell, Ralph, 1981. 'The Bergen School of Meteorology. The Cradle of Modern Weather-Forecasting'. *Bulletin of the American Meteorological Society*, 62 (6), 824–30.

Jinman, George, 1861. *Winds and Their Courses; or a Practical Exposition of the Laws Which Govern the Movements of Hurricanes and Gales. With an Examination of the Circular Theory of Storms, as Propounded by Redfield, Sir William Reid, Piddington, and Others*. London: George Philip and Son.

Kämtz, Ludwig Friedrich, 1858. 'Extrait d'une lettre de M. Kaemtz, concernant les relations qui existent entre les indications du baromètre, la direction et la force du vent'. *CR*, 46, 944–47.

Khrgian, Aleksandr Khristoforovich, 1970. *Meteorology: A Historical Survey*, 2nd ed. Jerusalem: Israel Program for Scientific Translations.

Kirkaldy, Adam W., 1914. *British Shipping: Its History, Organisation and Importance*. London: K. Paul, Trench, Trübner and Co.

Klein, F.H., 1863. *The Foretelling of the Weather in Connection with Meteorological Observations by F.H. Klein together with a Description of the Telegraphic Warning System Introduced in The Netherlands, June, 1860, as proposed by the Director of the Royal Netherlands Meteorological Institute Professor Dr. Buys-Ballot*. London: Benjamin Pardon. Translated from the original Dutch by A. Adriani.

Kutzbach, Gisela, 1979. *The Thermal Theory of Cyclones: A History of Meteorological Thought in the Nineteenth Century*. Boston, MA: American Meteorological Society.

Locher, Fabien, 2008. *Le savant et la tempête. Étudier l'atmosphère et prévoir le temps au XIX siècle*. Rennes: Presses Universitaires de Rennes.

Longshore, David, 2010. *Encyclopedia of Hurricanes, Typhoons, and Cyclones*, 3rd ed. New York: Facts on File.

Ludlam, Frank Henry, 1967. *The Cyclone Problem: A History of Models of the Cyclonic Storm*. London: Imperial College of Science and Technology.

MacDonald, Lee T., 2018. *Kew Observatory and the Evolution of Victorian Science, 1840–1910*. Pittsburgh, PA: The University of Pittsburgh Press.

Maury, Matthew Fontaine, 1851. *Explanations and Sailing Directions to Accompany the Winds and Currents Charts*. Washington, DC: C. Alexander.

Mellersh, Harold Edward L., 1968. *FitzRoy of the Beagle*. London: Rupert Hart-Davies.

Monmonier, Mark, 1999. *Air Apparent. How Meteorologists Learned to Map, Predict, and Dramatize Weather*. Chicago: The University of Chicago Press.

Naylor, Simon, 2015. 'Weather Instruments all at Sea: Meteorology and the Royal Navy in the Nineteenth Century'. In Fraser MacDonald, and Charles W.J. Withers eds., *Geography, Technology and Instruments of Exploration*. London: Routledge, 77–96.

Nebeker, Frederik, 1995. *Calculating the Weather. Meteorology in the 20th Century*. New York: Academic Press.

O'Toole, Barry J., 2006. The *Ideal of Public Service. Reflections on the Higher Civil Service in Britain*. London: Routledge.

Pearson, Karl, 1914–30. *Life, Letters and Labours of Francis Galton*. Cambridge: Cambridge University Press, 3 vols.

Pernter, Joseph Maria, 1903. 'Methods of Forecasting the Weather'. *Annual Report of the Board of Regents of the Smithsonian Institution*, 151–65. Reprinted in *Monthly Weather Review*, 1903, 31, 576–82.

Piddington, Henry, 1848. *The Sailor's Horn-Book for the Law of Storms, being a Practical Exposition of the Theory of the Law of Storms, and its Uses to Mariners of All Classes in All Parts of the World, Shewn by Transparent Storm Cards and Useful Lessons*. London: Smith, Elder.

Poulter, R.M., 1934. 'Cycles in Meteorological Knowledge'. *The Quarterly Journal of the Royal Meteorological Society*, 60, 341–44.

Redfield, William Charles, 1831. 'Remarks on the Prevailing Storms of the Atlantic Coast of the North American States'. *American Journal of Science*, 20, 17–51.

Report of a Committee Appointed to Consider Certain Questions Relating to the Meteorological Department of the Board of Trade [Galton Report], vol. 65. London: Eyre & Spottiswoode, 1866. Also in *Parliamentary Papers*, 1866.

Robinson, Arthur H., 1971. 'The Genealogy of the Isopleth'. *Cartographical Innovations*, 8, 49–53.

Rozwadowski, Helen M., 2016. 'Introduction: Reconsidering Matthew Fontaine Maury'. *The International Journal of Maritime History*, 28 (2), 388–93.

Schlee, Susan, 1973. *The Edge of an Unfamiliar World: A History of Oceanography.* New York: Dutton.

Scott, Robert Henry, 1876. *Weather Charts and Storm Warnings.* London: Henry S. King & Co.

Scultetus, H.R., 1943. 'Dove und Loomis als Wegbereiter der Synopsis'. *Meteorologische Zeitschrift*, 60, 419–22.

Shaw, William Napier, 1911. *Forecasting Weather.* London: Constable.

Shaw, William Napier, 1926. *Manual of Meteorology. Vol. I: Meteorology in History.* Cambridge: Cambridge University Press.

Shaw, William Napier, and Lempfert, Rudolf Gustav Karl, 1906. *The Life History of Surface Air Currents. A Study of the Surface Trajectories of Moving Air.* London: Darling.

Tickell, Sir Crispin, 1993. 'Meteorographica and Weather'. In Milo Keynes ed., *Sir Francis Galton, FRS. The Legacy of His Ideas.* New York: Palgrave Macmillan, 140–51.

Vaillant, Jean-Baptiste Philibert, 1863. 'Sur la tempête des 2 et 3 décembre 1863'. *CR*, 57, 1001–07.

Walker, J. Malcolm, 2012. *History of the Meteorological Office.* Cambridge: Cambridge University Press.

Waller, John C., 2001. 'Gentlemanly Men of Science: Sir Francis Galton and the Professionalization of the British Life-Sciences'. *Journal of the History of Biology*, 34, 83–114.

Williams, Frances Leigh, 1963. *Matthew Fontaine Maury: Scientist of the Sea.* New Brunswick, NJ: Rutgers University Press.

Williamson, Fiona, 2017. 'Uncertain Skies: 'Forecasting' Typhoons in Hong Kong c. 1874–1906'. *Quaderni Storici*, 52 (3), 777–802.

5 The convergence between statistics and synoptic method

Introduction

In September 1867, Jean Cantoni and Pietro Maestri, the rector of the University of Pavia and the director of the Italian Bureau of Statistics, respectively, successfully persuaded the organizers of the sixth International Statistical Congress in Florence that meteorology was admitted to the section of topography. Well aware of the rising interest in local meteorological data experienced as part of the surge of statistical censuses and atlases, they noted the lack of presence of meteorological science in international forums. The reasons behind this, Cantoni argued, lay not in any lack of observations or a shortage of 'very respectable savants who made meteorology the object of their studies'.[1] Rather, it was largely attributable to the stance of most astronomers, who were not willing to grant the title of science to meteorology. This problem, he announced in the Congress, could be soon solved by his plan for the organization of meteorological stations in Europe. By the early 1870s, this plan and other similar initiatives had received enough attention to move forward, culminating with the first international meteorological meetings in Leipzig and Vienna—the first and last one had taken place in Brussels in 1853.[2]

In his report read in the section of topography, Cantoni distinguished three ways to do meteorology. The first was the 'prévision à court délai' (short-term forecasting), which had been long studied and organized in France, Britain, the United States, and Italy by Le Verrier, FitzRoy, Maury,

---

1 'Séance du 29 Septembre' in Maestri (1868, p. 371). See also Cantoni's address in this Congress, 'Organisation des stations météorologiques et formation d'une carte diurne de l'Europe', Cantoni (1867).

2 After a period of two decades from the first International Maritime Conference in Brussels, the Leipzig international Meteorological Conference (1872) opened the way to a series of congresses that were held in European capitals every few years: Vienna (1873), Paris (1878), Rome (1879), Berne (1880), Copenhagen (1882), and Paris (1885). For a review of the resolutions adopted, see Hildebrandsson and Hellmann (1909). Cannegieter (1963) provides an overview of the early years. For a detailed account on the preparation of the Leipzig Conference, see Fedorov and Böhme (1972).

and Carlo Matteucci, proving to be very useful for navigation and commerce. This practical, daily meteorology of urgency aimed at providing storm warnings shortly before their arrival. The second was the extensive study of meteorological phenomena, by 'patient observations, repeated several times and continued for a long number of years'. This variant, known as climatology, aimed at determining the laws of variations of each element, and especially the influence of local conditions on these elements. Finally, there was the meteorological statistics presented at the Congress, a 'meteorology of simultaneous facts' based on space, which was in stark contrast to climatology or 'meteorology of successive facts', based on time. This third way required a large number of observations, made simultaneously every day on a vast extension of the earth's surface, and aimed to determine the laws governing meteorological phenomena.[3]

The landscape of forms painted by Cantoni including the daily meteorology, climatology, and meteorological statistics reflects well the many and varied paths taken by meteorologists at that time. However, this portrait did not distinguish between the two types of approaches, Eulerian and Lagrangian, used in short-term forecasting, or discerned differences in method in the last group.

There were essentially two conceptions of how statistics should be introduced in meteorology. According to the mathematical perspective adopted by Buys Ballot, physical causes had always to be sought in the *deviations* from mean values. Deviations were calculated from the difference between mean values and actual measurements. Deviations, rather than direct measurements, informed us of the state of meteorological elements such as pressure and temperature. Direct measurements were not trustworthy values.

A different view was held by cultivators of vital statistics in Scotland. Alexander Buchan drew on statistical rigour to collate and refine maps of average pressure, temperature, and rainfall for Britain, and even envisaged methods to correct data and extend incomplete records.[4] Yet as to daily meteorology, while demanding simultaneous observations to prepare synchronous charts of large areas of the globe, he relied on the pressure reduced to sea level, rather than daily barometric means. Deviations from means did not adequately reflect the pressure distribution.

By the early 1870s, the distinction of paths found by Cantoni in 1867 between Le Verrier's daily synoptic meteorology and the third form known as meteorological statistics would disappear. This chapter depicts the

---

3 See Pietro Maestri, 1868. *Compte-rendu des travaux de la vie session du Congrès international de statistique réuni à Florence les 29, 30 septembre, 1, 2, 3, 4 et 5 octobre 1867.* Florence: Impr. de G. Barbèra, pp. 371–72.

4 Buchan's name became well-known in climatology for his postulation on certain singularities (*Buchan spells*) or departures from the normally expected annual transition of temperatures, which he attributed to changing pressure patterns at certain seasons of year.

development of this convergence, showing how meteorological science was acquiring from the 1850s certain 'statistical' forms that affected not only its cognitive realm (e.g. procedures and graphic methods) but also organizational aspects related to the standardization and coordination of observing modes. It begins with a detailed examination of Buys Ballot's calculation method based on deviations from means, showing how this method allowed him to derive a relationship—later known as Buys Ballot's baric wind law—between wind direction and the areas of high and low pressure. Next it describes how vital statistics and meteorological science intersected in Britain. Shortage of simultaneous observations and the lack of precision of the calculation of exact means, for example, pushed Scottish meteorologists to demand strict observing regulations and statistical rigour in their investigations. These needs and self-discipline also led Buchan to devise an analytical method of synoptic weather charts. Buchan's method, I contend, shared essential features of Marié-Davy's Eulerian approach: the pressure field determined the temperature, wind, and precipitation patterns associated with a storm.

The second part of the chapter focuses on the organizational influence of statistics on meteorology, which took shape in the international meteorological meetings held in Leipzig and Vienna in 1872 and 1873, respectively. A new type of synoptic meteorology would be projected in these meetings: an official, state-sponsored one aspiring to the standardization of methods and definitions, and coordination of observations, in which the *barometric gradient* emerged as indicator of the degree of approach of storms.[5] When combined with the language of statistical quantification, the language of synoptic maps contributed to the legitimization and consolidation of the national weather forecast services.

Deviations from normals

In 1850, the professor of mathematics Christophorus Buys Ballot reviewed the state of meteorology and its evolution in a study on temperature changes.[6] Although he praised Humboldt's pioneering isolinear depictions and welcomed the rise of climatologic charts, he claimed the need to predict weather events, advocating for the separation of climatology and meteorology. He distinguished three stages in this history. Humboldt took the first step, paving the way for the knowledge of average states of the atmosphere and their laws. Next Dove promoted the search for deviations from these

---

5 The *barometric gradient* is a physical quantity that denotes which direction and at what rate the barometric pressure changes the most rapidly around a specific place. It shows the slope of the isobaric surface on a map. The concept was first proposed by Thomas Stevenson (1867). See also Khrgian (1970, p. 178).

6 His first attempt to review the modern history of meteorology can be found in Buys Ballot (1848, pp. 106–08).

mean values, with the aim of finding the laws of such deviations. Yet the last stage was still to come. In particular, Buys Ballot referred to the prediction of 'meteorological events', and emphasized the role the self-registering instruments and electrical telegraphs could play in this goal. To achieve it, Buys Ballot applied the statistical method to meteorology while distancing himself from Quetelet's dynamic notion of atmospheric waves.[7]

The application of statistical methods was actually quite widespread in observatory sciences, especially with the use of the method of least squares in astronomy and geodesy. This method, based on the theory of the properties of large numbers, was used to assess the measurements made by trustworthy means—i.e. by precise instruments. With this method, the astronomer and the geodist were able to choose the best value that could be taken on for the measure of a physical observation that was presumed to remain unchanged and when one had a great amount of independent factors. However, the method of least squares found little sympathy in meteorology. Indeed, the meteorologist of the 1850s was faced with a paradoxical situation, for while he borrowed from the science of statistics the methods that had been devised to treat errors, he almost never took account of errors in the measurement of the periodical changes of elements to which such methods were applied.

Among those who saw the need to develop a specific statistical method for meteorology—unrelated to that of least squares—was Buys Ballot. Professor of Mathematics at the University of Utrecht and founder and director of the Royal Dutch Meteorological Institute from 1854 to his death in 1873, Buys Ballot was one of the most influential persons in the history of synoptic meteorology in that century.[8] Buys Ballot was a follower of Herschel and of the idea of a worldwide network of observations—in fact, he had published a call 'to all friends of meteorology' to participate in a common project in the Netherlands in 1848—but he nevertheless disagreed with the instructions given by the British astronomer in 1836. Herschel had suggested that observations should be as precise as they were in astronomy, as well as being comparable with each other and with those called standards. In his 1848 call, Buys Ballot detached himself from the preciseness and correctness required in astronomical measurements, and quoted an adversary as basis for his instructions:

> For me it is impossible to suppress a feeling of distrust, whenever I consider these immense series of observations, from which one hopes to achieve knowledge about laws and causes of the phenomena in our atmosphere. Nobody is going to assert that *those* will fully meet the

---

7 Buys Ballot (1850a, p. 629)—quoted by Sheynin (1984b, p. 55). Buys Ballot did not mention the statistical method.
8 As regards Buys Ballot's life and work, see Everdingen (1953), Stok (1899), and Burstyn (1970, p. 628). For a clear explanation of Buys Ballot's statistical method in meteorology, see Boumans (2015, pp. 68–75). See also Sheynin (1984b, pp. 55–56, 77).

requirements of precision, and also that no greater correctness is now already achievable and necessary.[9]

Like many meteorologists of his time, Buys Ballot provided more specific details in his scientific articles than in his public claims. In a paper published in the *Philosophical Magazine* in 1850, 'On the great importance of *deviations* from the mean state of the atmosphere', Buys Ballot declared unequivocally, 'the most important causes are always to be sought in the variations (deviations)'.[10] According to him, deviations from the mean values calculated over long times were more significant than direct measurements; the latter were not trustworthy. It would be, therefore, from deviations that one must derive the exhibition of the state of elements such as temperature and pressure, not from absolute measurements. Indeed, Buys Ballot was not sure how reliable direct measurements were. On one hand, they came from different instruments operating at different places and under various conditions all over the world. On the other hand, the method of means (so successful in astronomy) was only applicable and efficient when one had many direct and various observations conducted under similar conditions. This was evidently not the case in meteorology. Hence in his 1850 article Buys Ballot devised a calculus of observations in order to turn unreliable measurements into reliable estimates; his calculus was predicated not upon the method of means, but that of residues.

The mechanics of his calculus of observations was quite simple, resting mainly on the method of residues developed by William Whewell in 1847.[11] First, one chose a meteorological element, such as temperature, and inferred its daily variations from readings of unreliable instruments. Next one used the method of arithmetical mean to treat instrumental errors, comparing the mean with the actual measurement. The two values included instrumental errors. Finally, the meteorologist would then calculate the difference between these two values (that is, their *deviation*), thereby eliminating the measurement error. In this way, the meteorologist was able to use the arithmetical mean to capture instrumental errors rather than to erase or neutralize them. Last but not least, the meteorologist could uncover causal relationships where individual elements were highly variable and subject to a host of influences operating simultaneously. Previously unseen variables, such as the normal, could function and be regarded as a standard. Indeed, Buys Ballot's calculus was not intended to attain precision. Calculated normals such as the mean theoretical pressure were assumed to prevail at any

9 In the words of Volkert Van der Willigen, a pupil of the Leiden astronomer Frederik Kaiser, internationally recognized for his precise measurements. Quoted by Buys Ballot (1848, p. 380), as translated from the original language (Dutch) by Boumans (2015, p. 73).

10 Buys Ballot (1850b, p. 45). For an early study on periodical temperature changes aimed at weather forecasting ('pronostic du temps'), Buys Ballot (1847).

11 For a reconstruction of Buys Ballot's method of residues, see Boumans (2015, pp. 75–76), who also points to Whewell's influence on Buys Ballot.

specific place and time, but this fact was the result of a calibration procedure rather than a higher level of precision.

As Buys Ballot articulated his notion of normal by the method of residues, he began publishing yearbooks including tables with the daily deviations from the normals. These were more trustworthy than direct measurements, because they reduced the effect of systematic errors, and therefore, remedied instrumental deficiencies. In the first yearbook of 1851, which contained the daily deviations for the years 1849 and 1850, he reminded his colleagues that normals could be useful to unmask influences. 'The averages', he allowed, 'cover all influences that do not manifest themselves each year on the same date; deviations make them conspicuous'. Moreover, he urged them to distinguish climatologic observations from meteorological ones. In this regard, observers should learn from their own experience: 'it is good that at some observatories', Buys Ballot argued pointedly, 'very extensive observations are made that fill bulky volumes: that is important for *Climatology*'. However, he added, 'it is preposterous to think that with them one meets the needs of *Meteorology*'. The reason was simple: given that 'the weather situation at one place depends on those at the surrounding places', observations at one place cannot learn us very much. Accordingly, he concluded, 'it is infinite better to make simple observations at hundred places not too far from each other, than at ten places very complete ones'.[12] Here, then was the demarcation between climatology and meteorology reflected in the deviations: one could accept their immediate usefulness in the study of weather changes while reserving judgement on their use in the field of climatology.

Buys Ballot's principal inference—or law as it was known in the 1870s, as will be explained below in more detail—was that wind direction was correlated with the location of areas of high and low pressure.[13] This relationship was derived from his method based on the deviations from normal in 1857, when he already had at his disposal the measurements of an extensive network of stations coordinated by the Dutch Meteorological Institute, among them some from foreign observatories. By 1863, his accomplishments had been made known in Britain when Buys Ballot presented his predictive method at the annual meeting of the British Association for the Advancement of Science. In 1865, with the method of deviations from normals receiving widespread attention in British meteorological circles, Buys Ballot summarized the foundations of what he called the *Dutch system*:

> It is obvious, but not generally acknowledged, that no absolute reading of the barometer has any significance, but only the difference (called

---

12 Buys Ballot (1851, p. 2).

13 Within a little over one year, Buys Ballot (1864) read a paper 'On the System of Forecasting the Weather pursued in Holland' in the 1863 BAAS meeting and A. Adriani translated into English Captain F.H. Klein's book, *The Foretelling of the Weather in Connection with Meteorological Observations* (London, 1863, original in Dutch), in which he compared the forecast methods of Buys Ballot and FitzRoy.

departure) of an actual reading with the average reading of *that* instrument at the *same* place, at that latitude, longitude and height above the sea on the same day. The departure is the true and accurate measure of the perturbance, and intimately connected, but [...] not identical, with the force that tries to restore the equilibrium. The single reading of the barometer, on the contrary, is an arbitrary number of no signification at all, unless you substitute an accurate approximation of the average height of the readings. I therefore call the determination of the average or normal height *the* characteristic of the Dutch system, and this base is adopted in the Dutch meteorological Annuaires, which have appeared regularly since 1852, and wherein for every day of the year a tabular view is given, representing the departure of temperature and atmospheric pressure for the whole of Europe.[14]

By linking departures to the true measure of perturbations a door was opening to the problem of weather forecasting. The idea of using deviations from normals was perhaps not as great a conceptual leap as might otherwise appear. Dove and others had written explicitly of 'normals' and had even plotted local observations as deviations against latitudinal averages.[15] Here the crux of the issue was the field of this application; namely, whether normals represented a stable value towards which the long-term averages of climatic elements converged, as climatologists seemed to suggest, or whether they could be plotted and used for weather forecasting purposes, as Buys Ballot hoped to achieve.

Buys Ballot's baric wind law

Buys Ballot successfully devised a specific statistical method that could be very functional to a variety of users. Yet he wanted to provide practical rules for storm warnings. These could offer new approaches and valuable techniques for solving the issue of wind circulation even if its physical meaning was unclear for him. Since his work on deviations from normals, Buys Ballot had admired the originality of Dove's statistical thought and his law of wind rotation. The discovery of the baric wind law increased his sympathy for the statistical-empirical approach.[16]

---

14 Buys Ballot (1865, p, 246).
15 As was shown in Chapter 2, Dove used for the first time the word 'normal' in 1840. When analyzing Dove's loose use of the word 'normal', Landsberg (1955, p. 7) distinguished three meanings: '(1) "normal" as a reference value obtained for a given geographical latitude by averaging all observations in that latitude; (2) "normal" station as a reference locality with a long observational record for comparisons in time and space; (3) "normal" as equivalent to the average or mean value of a long series of observations'. Only the last meaning would triumph in the science of climatology.
16 This law was at first known as Buys Ballot's rule and then law, until in 1878 Wladimir Köppen from the Central Physical Observatory in St. Petersburg gave it the name of 'baric wind law'.

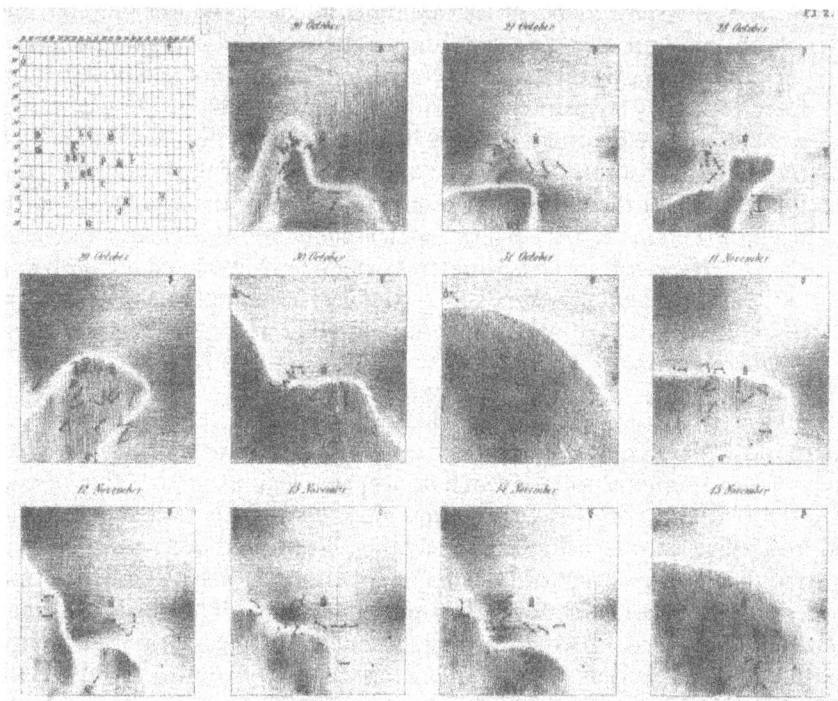

Figure 5.1 Series of weather charts from 1852 covering most of Europe; reproduced by Buys Ballot (1854). Charts show wind directions and temperature deviations from normal: on one hand, arrows and curved lines
indicate direction of wind and its changes during the day, respectively;
on the other hand, hatching shows temperature deviations (horizontal for those below normal, and vertical for those above normal). The
distribution of warm and cold air masses is indicated by the different
shading of areas.
Source: Buys Ballot (1854).

Buys Ballot's method was soon to bear fruit. Since the publication of his
first tables with the daily deviations from normals, he felt inclined to plot
such deviations on weather maps. In 1852, he plotted temperature deviations from normals and wind directions on a series of rather rudimentary
weather maps (Figure 5.1).[17] His graphic technique entailed a certain degree
of novelty: while arrows indicated wind direction (as was the practice thus

17 Since 1852, the *Yearbook* of the Royal Dutch Meteorological Institute published Buys
Ballot's daily weather maps. A series of weather maps was also published by Buys Ballot
(1854) in the journal *Poggendorf's Annalen der Physik und Chemie. See also Kutzbach
(1979, p. 68).*

far), hatching showed temperature deviation—horizontal for deviations below normal, and vertical for those above normal. The denser and shorter the lines of the shaded areas, the greater the temperature deviation from normal. These shaded areas, whose boundaries were clearly defined on his maps, showed the distribution of cold and warm air masses. Buys Ballot became inspired by this technique when he saw the daily weather charts in the Great Exhibition of 1851 in London.[18]

However, these maps did not fulfil his expectations. His representations could not be easily applied to the problem of storm warning, because they provided no practical rule on the relationship between disturbances and surface winds. Since 1855, Buys Ballot sought some signs that could foresee strong winds. These signs could be either statistic (deviations) or dynamic (barometric maxima and minima). He was aware of Quetelet's studies on atmospheric waves, and even flirted with this possibility. Yet by 1857 he was convinced that the statistical option was the only one workable. Comparing the pressure difference in three Dutch stations, he explained how the wind strength was related to the magnitude of this difference and asserted that a statistical relationship could be enunciated, i.e. while in the Northern Hemisphere wind travels anticlockwise around low-pressure zones, in the Southern Hemisphere the reverse is true.[19] Fortunately for laypersons, he gave a clear example of this crucial relationship. A careful reading of the data from those three stations leads to the following conclusion. If the pressure is at least four mm higher in Maastricht in the south than in Groningen in the north, a storm would arrive from the west; if lower in the south, the storm would arrive from the east.[20]

Buys Ballot expressed his local rule of thumb in a language aimed at general public. As his observations shown, the best indication of the prospect of strong winds is not the barometric height at a given time and place or its sudden rise or fall, but rather the differences between absolute barometer readings. In plain language, the rule for the direction of the wind is as follows: *If one places oneself in the direction of the wind with one's back towards the place from where it arrives, then one will have the place of lowest pressure on the left side, just as in the case of hurricanes.*[21] Buys Ballot soon applied

---

18 Dettwiller (1982, p. 66).
19 Buys Ballot revealed this relationship in a note presented at the Academy of Sciences in Paris in 1857, entitled 'Note sur le rapport de l'intensité et de la direction du vent avec les écarts simultanés du baromètre'.
20 Previously, Buys Ballot had presented a note at the Amsterdam Academy of Sciences, in which he provided this relationship as an empirical basis for a warning system in Holland. For further details, see Achbari and Lunteren (2016, pp. 12–15).
21 Buys Ballot (1860, p. 50), as quoted and translated by Achbari and Lunteren (2016, p. 15). This known phrasing of what would become Buys Ballot's law was first published in a brochure for the general public, entitled *A Few Rules for Forthcoming Weather Change in the Netherlands* (1860).

the rule to a particular case, storm warnings in the Netherlands deduced from local and therefore partial indications.[22]

Indeed, his baric wind law helped him reformulate the relation between storm warning and weather forecast. In his vision, warnings should be based not only on already existing announced storms but on storm predictions as well. This was in stark contrast to the practice by Le Verrier in France and Henry in the United States, whose warnings provided information about storms already known to exist and their estimated trajectories. Although he restricted his public announcements to storm warnings (as opposed to weather forecasts), he needed to forecast the weather everyday to prepare his warnings.[23] In order to implement a warning service, Buys Ballot resorted to the Inspector of the Dutch Telegraphic Service in 1859, and reasoned as follows.[24]

If clerks were instructed on his rule of the direction of the wind and the differences of barometer readings, they would be able to decide for themselves whether storm warnings should be prepared and communicated by telegraph. The corresponding safety measures would be taken, if that were the case. With an empirical eye for statistical relations, Buys Ballot insisted that there was a correlation between the strength of the wind and the difference between barometric readings.[25] As a result, his storm warning service was approved by the Dutch Minister of the Interior, coming in full operation in 1860.[26] Three years after the finding of the rule for the direction of winds, he was the first to issue a storm warning based on weather prediction.

Buys Ballot's and FitzRoy's accomplishments had certain similarities. The two introduced the empirical prediction of the weather, and thus started two important traditions of North European meteorology. On the observational side, they focused on empirical rules, whereas Le Verrier and Secchi rarely used rules, focusing instead on precision measurement and synoptic maps. On the predictive side, the former aimed at storm warnings based on predictions, while the latter were content to report on actual storms and their tracking. Inspired—although to varying degrees—by an outstanding experimental physicist, Dove, they brought empirical prediction to the new field of telegraphic meteorology.

---

22 At that time, seamen conceived of Buys Ballot's rule in much more mundane terms, although not entirely divorced from the early idea of experiencing something with their own eyes or senses. That is how the rule, turned into a simple logo, appeared in a German maritime journal in 1871, in an anonymous article on the use of the barometer on board ships: 'If a person stands with his back to the wind, the low pressure area will be on his left'. In 'Der Barometer an Bord'. *Hansa. Zeitschrift für Seewesen*, 1871, 8, 197–98, on p. 98—quoted by Achbari and Lunteren (2016, p. 42).

23 Halford (2004, p. 175).

24 On this point, see Achbari and Lunteren (2016, pp. 13–14).

25 Klein (1863, p. 11).

26 Buys Ballot (1860, pp. 75–76).

However, their initiatives were largely independent, and their methods differed widely. Buys Ballot's method was empirical-statistical, whereas FitzRoy used to combine rules of thumb with individual judgement and local weather observation. Buys Ballot's method was based on the simultaneous difference of barometric readings in at least five stations, as well as their comparison with the average reading of each. FitzRoy took notice of solely the reading and the oscillations of the barometer in one place. The statistical treatment of deviations from normals was largely responsible for Buys Ballot's originality: it implied the elimination of measurement error, regardless of the higher or lower level of precision. Although FitzRoy was a strong advocate of the construction and manipulation of apparatus, his method relied on instrumental precision. Thus one might say that Buys Ballot's challenge was storm warning and FitzRoy's challenge was weather forecast.[27]

Interestingly, the tragic fate of FitzRoy and his method played a key role in the transformation of Buys Ballot's wind rule into a scientific law.[28] After FitzRoy's death in 1865, several factors began to make Buys Ballot's rule more appealing in Britain. On one hand, there was a surge in demand for storm warnings after the suspension of FitzRoy's service, which required meteorologists to announce atmospheric disturbances (gales, storms, fogs, etc.) in predictive rather than descriptive terms. Here rules of thumb and synoptic maps seemed to prove the most effective methods. The academic acceptance of the rule was favoured by the explicit endorsement of Herschel, who in his 1867 article on barometric waves deemed the wind rule as a 'general feature' of nature.[29] In his *Discourse of the Study of Natural Philosophy* Herschel had already linked the progress of science to 'the discovery of general laws', laws that should be regarded as objects of enquiry rather than isolated facts.[30] Finally, the 1866 Galton report calling for the search for laws, which place 'the practice of foretelling weather on a sound basis' while rejecting FitzRoy's method for not having 'scientific basis', also increased the academic receptiveness to Buys Ballot's rule and his method.[31]

27 Klein (1863, pp. 17–19) describes in detail the differences between the two methods. Klein is generally critical of the lack of connectedness that FitzRoy's method entails, since 'a large number of valuable and good observations remain necessarily isolated and unconnected'. In his view, it is 'useful to isolated observers'. But this view is inconsistent with the facts; otherwise, how may one explain FitzRoy's fishery barometer program? Ironically, Klein's view is remarkably like the supporters of the pro-Eulerian approach—i.e. FitzRoy's system entails localism and unconnectedness; the best system is centralism, order, and state control—but exactly with the opposite intention, namely to strengthen rather than undermine autonomous empiricism.

28 On the dissemination of Buys Ballot's wind rule in Britain and its transformation into a scientific law, see Achbari and Lunteren (2016, pp. 16–40). See also Achbari (2017, pp. 69–118).

29 Herschel (1867, p. 91).

30 Herschel (1831, p. 360).

31 *Report of a Committee Appointed to Consider Certain Questions Relating to the Meteorological Department of the Board of Trade*. London: Eyre & Spottiswoode, 1866 (also PP, 1866), p. 37.

The man responsible for transforming the wind rule into Buys Ballot's law was Robert Henry Scott, director of the Meteorological Office from 1867 to 1900.[32] A graduate in physics and a former student of Dove in Berlin (whose book *Das Gesetz der Stürme* he translated into English in 1862), he worked as keeper of minerals to the Royal Dublin Society when the president of the Royal Society of London, Edward Sabine, proposed his name for the Office's directorship, mainly for reasons of intimate friendship. Imagining himself as following the precepts of his protector Sabine—he actually had no experience in meteorology, excepting his translation—he conducted an inquiry into the connection between strong winds and barometric differences as soon as he took up the post. At the request of the Royal Society's Meteorological Committee he played an active role in this investigation, first travelling to the Netherlands to learn in situ how the Dutch warning service worked, and later examining old records of daily weather reports to test the reliability of the rule. Its results seemed to be satisfactory: in 60% of the cases, storms were preceded by the calculated barometric differences, whereas in over 90% the wind direction was correctly predicted.[33]

As the director of the Meteorological Office involved in the reform of the weather service in Britain and an outstanding state representative as secretary of the International Meteorological Committee from 1873 to 1900, the issue of how best to place storm warnings on a scientific basis was of great interest to Scott. Troubled both by the persistence of public protests for the restoration of storm warnings and by the pressure of the Board of Trade, Scott introduced a subtle yet significant redefinition of the wind rule. Instead of referring to it as a mere rule of thumb, essentially the designation of an empirical correlation, he would speak about a law or a general principle, whose 'prima facie confirmation' had been provided by his 1867 inquiry.[34] Here, the wind rule was envisioned as a natural law, regarded as an object of enquiry and discovery. This was the conception of the rule adopted by the Meteorological Committee of the Royal Society in an official report presented to Parliament in 1868. The Committee described it in founding terms, asserting that their investigations corroborated 'the value of the law as the foundation of a practical principle for the issue of cautionary telegrams'.[35]

In brief, Buys Ballot was first to enunciate the wind rule, which was improperly called the Buys Ballot's (baric wind) law. It indeed subsumed both Dove's 'law of gyration of the wind', also known as *Drehungsgesetz*, and

32 As regards Robert Scott, see Burton (2008); 'Robert Henry Scott, F. R. S. 1833–1916' [Obituary], *Quarterly Journal of the Royal Meteorological Society*, 42 (1916), 301–04.
33 Scott (1869, pp. 7–11).
34 Ibid., p. 17.
35 *Report of the Meteorological Committee of the Royal Society, For the Year ending 31st December 1868*. London: Eyre & Spottiswoode, 1869, p. 18.

the so-called law of storms governing the direction of winds of a cyclone.[36] However, it did not appear as a central, explicit statement in Buys Ballot's papers. Having been trained in the Netherlands, a seafaring nation and at once a colonial empire, he gave central importance to its practical implementation. In contrast, Scott, spurred by a situation of tension and institutional change in British meteorology, led the transformation of the wind rule into a scientific law. He regarded Buys Ballot's law as a founding principle, conceived as the basis for the legitimatization of the practice of storm warnings in Britain.

Vital statistics and the weather

Buys Ballot was not alone in his belief that statistics could be used to produce storm warnings and weather forecasts. Quetelet and others believed that as well. This belief originated from their experience in the fields of mathematics, astronomy, and oceanography. However, these were not the only fields in which statistics and meteorology intersected. The years before and after the 1848 revolutions witnessed a proliferation of publications on medical meteorology as statisticians from national institutions began to attach weather and climate data to their censuses and registers of births and deaths. In Belgium, under the impulse of Quetelet, the *Commission Centrale de Statistique* promoted the publication of general statistics, which contained numerous references to meteorology in its volume devoted to territory and climate.[37] The US statistics periodically compiled under the direction of the Secretary of the Interior included climate data and its effects on mortality, diseases, and consumption.[38] Before the outbreak of revolution across Europe and the subsequent efforts to bureaucratize the modern states, the desire for vital statistics had received special impetus in Britain, resulting in the establishment of the General Register Office (GRO) in 1837.[39] Few bodies reflected the synergy between vital statistics and weather as well as the GRO did it.

Vital statistics was an integral and essential part of state bureaucratization. 'Since bureaucracy has a "rational" character, with rules, means-ends calculus, and matter-of-factness predominating', Max Weber wrote regarding the formation of the rational state, 'its rise and expansion has everywhere had "revolutionary" results, in a special sense still to be discussed, as had the advance of *rationalism* in general'.[40] One of its effects is that bureaucratic

---

36 The law of storms stated that winds of a cyclone in the Northern hemisphere blow in an anticlockwise direction around a centre or an area of low pressure.

37 See 'Résumé des procès-verbaux des séances de la Commission *Centrale de Statistique*. No 189. Du 7 février 1851'. *Bulletin de la Commission centrale de statistique*, 1853, 5: 1–105, p. 7.

38 See, e.g., *Statistics of the United States: (Including Mortality, Property, etc.) in 1860*. Washington, DC: Government Printing Office, 1866, vol. 4.

39 On the beginnings of the GRO, see Higgs (2004, pp. 28–32).

40 Weber (1978, p. 1002).

offices became the tools of domination and surveillance for governors. English sociologist Anthony Giddens has defined the modern bureaucratic nation-state as a 'bounded power container' in which surveillance functioned as the driving force of integration.[41] For him, the bureaucratic surveillance is not only one of the defining features of modernity but also 'marks a historical rupture'.[42] Following the religious registration of births, deaths, and marriages, which had been collected by the Church for centuries, the GRO had become by the middle of the nineteenth century into a powerful tool for the civil, secular, state-controlled registration of vital statistics. Historian Gabriel K. Wolfenstein from Stanford University has written that 'the formation of the GRO must be seen as one of the seminal moments in modern bureaucracies': the GRO was 'not only emblematic of the modern state but constitutive of it'.[43]

The strategy of statistical publications developed by the GRO was two-tier, resting on the dissemination of numerical data relating to four main fields: property, life tables, public health, and population growth. On one hand, it promoted weekly and quarterly bulletins of comparative death-rates, which would serve to determine the extent of preventable death and the measures that should be taken to that end. On the other hand, it published a parallel series of annual and decennial reports, which would serve to look for correlations between independent variables and analyze the incidence of diseases.[44] In this way, statisticians were able to describe 'all the principal movements of the population of which knowledge is required for immediate administrative purposes'.[45] Moreover, given that the causes of many diseases were still unknown, they might find answers by analyzing their geographical incidence. Little known meteorological factor could be identified and formulated by numerical expressions.

Because of their association with the vital statistics, meteorological observations soon acquired a degree of unprecedented attention. The emphasis on the influence of climate on diseases and the professionalism of medical officers helped to create an illusion of objectivity between observer and observation, increasing the veracity of the results derived from them. The numbers associated with meteorological elements were accepted as true indicators of the diseases physicians purported to diagnose, presumably revealing deep physiological changes. By the 1850s, the statistical method had become the

41 Giddens (1985, p. 13).
42 Wolfenstein (2007, p. 263) draws on Giddens's concept to show that the role of the GRO as a 'power container' is clear. For an analysis on surveillance as a means of administrative power, see Dandeker (1990).
43 Wolfenstein (2007, p. 263, 280). For the rise of the rational state in Britain: see Silberman (1993).
44 Szreter (1991, pp. 437–38).
45 *Supplement to 25th Annual Report of the Registrar General of Births, Marriages and Deaths for England and Wales* (PP, 1865, XII), p. iii—quoted by Szreter (1991, p. 437).

most important means of investigating phenomena and aspects related to public health and population growth. In the United Kingdom, this turn towards meteorological statistics took time to evolve because of the heated debates on the causes of diseases between contagionist and anti-contagionist parties—that is, whether the occurrence of epidemics such as cholera should be attributed to the susceptibility to the weather or was caused by a miasma (i.e. a noxious form of 'bad air').[46] Yet by 1857, a Meteorological Committee of the Medical Officers of Health of London had been formed, so that daily meteorological observations were made at Fulham, Hackney, Piddington, and St. Thomas hospitals, the results being published in the weekly return of sickness made to the GRO.[47] The Committee's chairman, John William Tripe, the Medical Officer for Hackney, studied the variation of mortality with temperature, by drawing on the observations recorded at Greenwich, in combination with the weekly mortality rates.[48]

If the vital statistics represented a useful tool for analyzing the relationship between climate and disease, it was equally valuable for their wide variety of information. As a 'power container', the GRO generated information about a wide range of meteorological elements, including air temperature, pressure, wind strength and direction, humidity, and the amount of clouds and water vapour. In 1847, at the request of the Registrar-General, James Glaisher from the Royal Greenwich Observatory began to collect observations and produce reports for inclusion in the GRO's *Quarterly Returns of Marriages, Births, and Deaths*.[49] Although he was reputedly an intolerant and petulant man, the later secretary of the British Meteorological Society was able to establish a nationwide network of voluntary and unpaid observers which he instructed, with more or less success, over decades.[50] Almost all observers could be trained by instructions, their equipment inspected quarterly, tested, and in certain cases duly repaired. All that was asked was to use good instruments and ascertain their errors by comparison with standards.[51] Then, the results could be reduced to mean values by using

---

46 For the debate on the support or not of quarantine measures against epidemics in the 1840s, see Worboys (2000, pp. 39–40).

47 Burton (1990, p. 303).

48 Tripe (1862, p. 186), on the temperatures in London during the years 1859, 1860, and 1861. Another medical officer, Arthur Mitchell (1866), argued that, among all elements of climate, it was temperature that had greatest direct influence on mortality. On the basis of pathogenic criteria, he identified two distinct climatic types: climate of a 'thoracic constitution' and an 'abdominal constitution'. See also Burton (1990, pp. 303–04).

49 For biographical notes of James Glaisher, see Hunt (1978, 1996), Marriott (1903), and Nash (1903).

50 Because of his role in this network of voluntary observers, Glaisher became the heartbeat of the British Meteorological Society.

51 At the end of the first report (for 1847) of the Registrar-General, Glaisher described what he expected from the observers who would participate in the network of stations. See Marriott (1904, pp. 3–4).

Glaisher's diurnal range tables. Within the field of meteorology, Glaisher highlighted several elements that needed to be tracked and were worthy of being included in vital statistics. These included temperature, pressure, wind, humidity, cloudiness, precipitation, and the occurrence of phenomena such as lightning and thunder—all of which were regularly reported by Glaisher in the *Quarterly Returns* for the rest of his life.[52]

By the early 1850s, the increasingly important role vital statistics began to play in the evolution of British demography—amplified by the cholera epidemic of 1854 in England—led some meteorologists to consider mean value as a statistical measure worth calculating.[53] But unlike other observatory sciences, mean values in meteorology lacked the necessary methodological uniformity required by statistical science. In his 1847 report of the Registrar-General, Glaisher had noted he was persuaded by the method of reducing observations and determining mean values would 'same day lead to empirical laws'.[54] However, it was unclear what particular procedure should be adopted for the determination of mean values, given the many and varied methods used to estimate mean values in climate studies. Thus, for example, in his meteorological tables for England published in the *Quarterly Reports*, Glaisher estimated the mean temperature as follows: first, he compiled and examined the observations made with the self-registering and dry-bulb thermometers; then, he prepared certain tables to correct the exact means of these instruments for diurnal and monthly range.[55]

Hence methods on how to estimate mean values of temperature or pressure in the 1850s were varied and often uncoordinated, determined by local observers rather than by any standardize rule. In Scotland, when the Scottish Registration Act passed in 1854, the first Superintendent of Statistics, James Stark, devised a method for producing statistical tables that was not based on the English pattern.[56] A fellow of the Royal College of Physicians of Edinburgh and curator of its museum, Stark had long been active in the fields of public health and vital statistics. Within the specific field of meteorology Stark published in 1862 a study on 'the fallacy of the present mode

52 Glaisher prepared reports for the *Quarterly Returns* during fifty-four years. For a review of Glaisher's work for the Registrar-General, see Burton (1990, pp. 305–07).
53 In 1854, at the request of the General Board of Health, Glaisher conducted an investigation on the meteorology of London in relation to the epidemic of cholera. He concluded that the London climate had accelerated the epidemic, even though its outbreak could not be attributed only to atmospheric factors. See Glaisher (1855).
54 Marriott (1904, pp. 3–4).
55 In 1859, Glaisher (1859–1860) determined the most probable mean pressure for every day in the year from a curve previously obtained. The curve was traced from the tables showing the mean daily readings of the barometer for a period of eighteen years.
56 The English pattern was introduced by William Farr, the Registrar-General's statistician, who decided until then how these tables should be presented. Farr's purpose was to construct a predictive social science based on numerical relationships in the fields of vital statistics, epidemiology, and demography. See Eyler (1979).

of estimating the mean temperature in England'.[57] He strongly denounced Glaisher's practice of 'withholding the strict means of the different series of dry and wet bulb readings', and publishing only deductions rather than calculating exact means, which, according to him, proved to be erroneous. Instead, Stark contained that the 'only safe rule for the estimation of mean temperature was to take the strict mean of the self-registering thermometers', provided these were reliable and of proper construction.[58]

Stark's method had practical advantages. First, the exact mean seemed to be a measure that was more reliable and really obtainable. Accordingly, the voluntary observers of the Scottish stations—a network of meteorological stations established in 1856—had to follow strict regulations on the methods of observation and the submission of monthly returns.[59] Equally important was this methodology's predictive potentiality, which could draw on Atlantic island observations and ships logs as a source of knowledge of the coming weather. Not in vain were the voluntary observers located not only in Scotland, but also in Iceland, the Faroes, and the Near East. If one assumed that exact means could lead to ascertain the general laws regulating atmospheric changes—as Stark and his successors did—then it therefore followed that weather forecasting was an achievable goal with the resources of the Scottish network.

Buchan's analysis of synoptic charts

In the period 1863–69 the Scottish schoolmaster Alexander Buchan devised an analytical method of synoptic weather charts which in its final form shared essential features of Marié-Davy's: the baric field associated with a storm determined its temperature, wind, and precipitation patterns. Like Marié-Davy, but unlike Buys Ballot and other European meteorologists, Buchan asserted that the best method was to calculate the sea-level pressure rather than plotting deviations from mean. However, a look at his laborious studies suffices to reveal important differences. In his studies on synoptic charts Buchan meant to develop an analytical method, not a forecasting system. He provided general rules describing the relationships between pressure and winds. His rules meant an extension of Buys Ballot's law. Moreover, he enunciated a hypothesis of the thermal nature of cyclones. As discussed in this section, his analytical method was a first and important step towards combining synoptic charts and Buys Ballot's law.

Buchan's interests and methods—not his forecasting contribution—owed much to Stark and the Scottish observational practice. A graduate in liberal

---

57 Stark was analyzing the temperature of the sea around the coasts of Scotland during the years 1857 and 1858.
58 Stark (1862, p. 265).
59 For the establishment of this network at the initiative and under the impetus of the Scottish Meteorological Society, see Walker (2012, pp. 32–34).

arts at the University of Edinburgh, Buchan had to leave the teaching profession in 1860 because of a weakness of his larynx, and became the secretary of the Scottish Meteorological Society—a position inaugurated by Stark in 1856.[60] Buchan revered Stark's statistical rigour and insight as the appropriate way for the discovery of not only the general laws of the atmosphere (one of the Society's objects) but also the interdependence of phenomena which were separately affected by independent causes.[61] This insight would be well illustrated in his late works with Arthur Mitchell on the relations of weather and public health in London.[62] He also praised Stark's efforts to 'collect, arrange and reduce on one uniform plan the observations' made in Scotland.[63]

After being appointed as secretary of the Society, Buchan became interested in the relationship of the pressure field to weather changes. The problem of the storm's progress was in part a methodological problem: in constructing synoptic charts, should one plot the departure of pressure from mean as was the practice in Holland and Sweden? Buchan thus extended the inquiry into graphic methodological questions. In an analysis on storms of wind occurred in Europe in 1863, published in the *Transactions of the Royal Society of Edinburgh* in 1865, he discussed the implications of synoptic charts and deduced physical relationships from the pressure distribution. Drawing on data collected—but not without difficulty—from '135 places scattered over Europe', Buchan furthered the argument, earlier made in part by Stark, that deviations from the mean provided a distorted picture of the pressure distribution.[64] Instead, he opted to reduce barometric observations to sea level, before being entered in their place on the map. Only then should isobarometric lines be drawn through all the points where the pressure was equal. The pressure distribution, he reasoned, determined the progress of storms:

> The observations of the barometer are the most important of all the observations, since it is within the area where the barometer falls to some extent below the average that storms occur [...] Hence, while we

---

60 Buchan held this post until his death from pneumonia in 1907. For Buchan's life and publications, see Mitchell et al. (1908, 101–18), the works edited by Roy (2007), Kutzbach (1979, p. 227), and Marjory G. Roy, 'Buchan, Alexander'. *Oxford Dictionary of National Biography*, online.

61 Shaw (1907, p. 83) contrasted this statistical insight with the physicists' method of investigation, who, in having the conditions under their own control, and unlike meteorology, were 'accustomed to trace the direct connection between cause and effect in each separate experiment'.

62 Buchan and Mitchell (1875–1878).

63 As stated by Stark in a note published in the *Quarterly Report of the Registrar-General for Scotland*, 31 March 1856—as quoted by Walker (2012, p. 33).

64 The difficulties arose in part from the fact that original data were largely in the form of station-level pressure, and the station height was needed in order to convert these data to sea level.

trace the progress of these low pressures over Europe from day to day, we trace at the same time the progress of the storms.[65]

Buchan performed what in modern parlance might be called a synoptic analysis and found that means and direct observations were misleading in relation to how the barometric pressure was distributed. He plotted eighteen charts including observations of surface pressure and wind speed and direction, and showed that the barometric pressure around a stormy area was not randomly distributed (Figure 5.2); it was in fact contoured. The forms of the areas circumscribed by the isobarometric lines were either circular or slightly elliptical. 'It follows from this', argued Buchan, 'that the storms most commonly assumed' such forms: they would be circular or oval as prefigured by the isobarometric lines, and only this kind of line could portray storms. On the contrary, isotherms were *not* appropriate to depict storms; in fact, the evidence of synoptic charts showed their forms were entirely irregular. As Buchan puts it[66]:

> The observations of the thermometer do not equal in importance those of the barometer [because] while the barometer measures the weight of the whole atmosphere pressing on it, the thermometer gives only the temperature of that portion of the air which is in immediate contact with the earth.

Buchan prioritized the baric field. For him, the barometer, not the thermometer, was the indicator *par excellence* of the atmospheric state and its impending changes. For predictive purposes, the remaining elements were not indicative enough, and, moreover, they were subjected to air pressure changes. 'The temperature rose as the barometer fell, and fell as the barometer rose', he wrote in 1867.[67] On the contrary, the isobarometric lines were critical, because they revealed the areas of low pressure (depressions) and the extent to which the storms were represented and limited by such areas. The areas where the form of these lines was circular or oval were the areas where storm or cyclone phenomena were taking place. What mattered most for meteorologists was what happened with the horizontal pressure distribution; it determined the weather conditions during the passage of a cyclone:

> When the barometer fell, the sky began to be obscured, and rain to fall at intervals; and as the central depression advanced, the rain became more general, heavy and continuous. After the center of the storm had passed,

65 Buchan (1865, p. 194).
66 Ibid., p. 200.
67 Buchan (1867, p. 200).

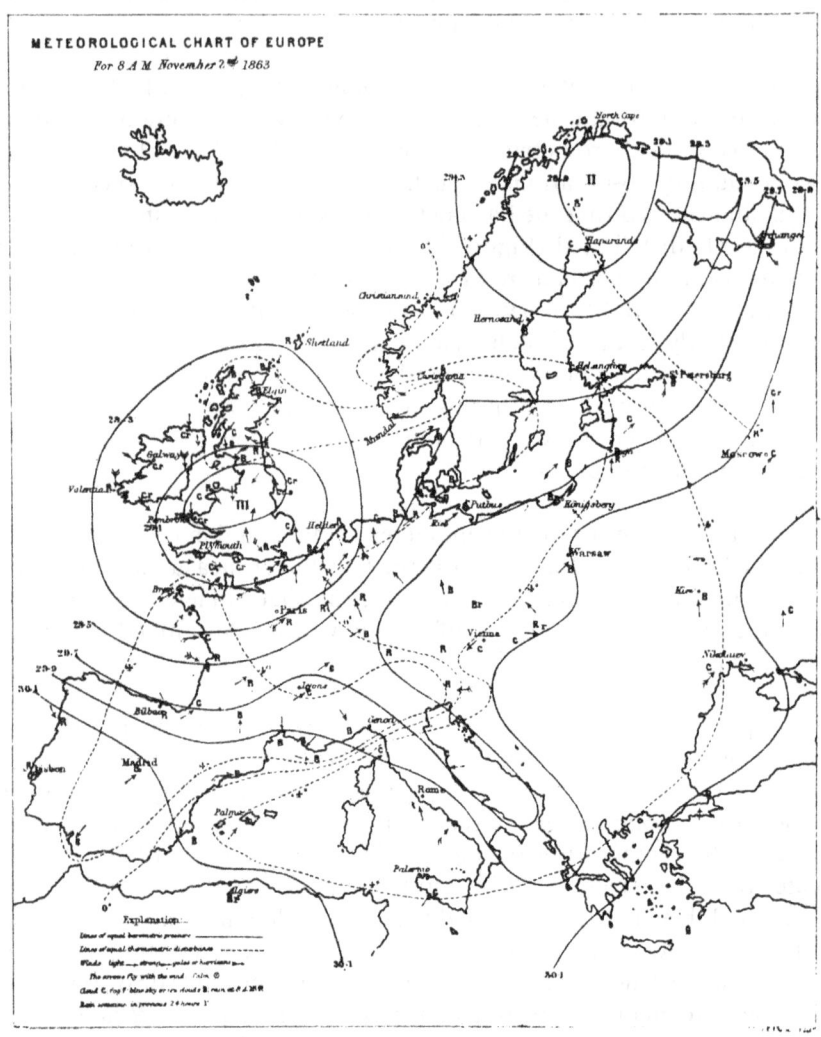

Figure 5.2 Synoptic weather chart of Europe; reproduced from Buchan (1865). Solid
lines indicate isobars (mm Hg, reduced at sea level); dashed lines indicate isotherms (deviations from mean); and wind strength and direction
are shown by symbols (calm O, light →). Moreover, C symbolizes clouds;
F, fog; R, rain; and B, clear or fair skies.

Source: Buchan (1865).

or when the barometer had begun to rise, the rain generally became less
heavy, falling more in showers than continuously; the clouds began to break
up, and fine weather, ushered in with cold breezes, ultimately prevailed.[68]

68 Ibid., p. 201.

If the baric field was determining, how did it affect wind? Buchan closely examined this question in a second study published in 1869 on the storms which passed over the United States and the Atlantic in March 1859. Reaffirming his earlier results of wind speed and direction, Buchan pointed out that winds always blow anticlockwise around a centre of barometric depression in the Northern Hemisphere, as they ought to be according to Buys Ballot's law.[69] However, his results showed that the air movement had a strong component towards the centre of the depression because the closer each other the isobarometric lines, the higher the wind speed. This suggested that, against Buys Ballot's claims, the air movement in storms could not be explained exclusively, or even primarily, by the sense of rotation of winds. In fact, it could not be explained without taking into account its spiral movement towards the centre. The cause of this spiral movement laid in the influence of the earth's rotation: an air mass driven towards the centre, because of the vacuum created, is deviated towards the left in the Northern Hemisphere, hence its rotation in an anticlockwise direction around the centre.[70] For Buchan, Buys Ballot's rule was but a local expression of two general rules: the rotation of winds and the progress of storms.[71]

The practical result of all these synoptic analyses was the production of a numerical base that could serve as the reference point for forecasting projections. For Buchan, wind speed was proportional to the closeness of the isobars—or to use his own words, to the concept of *barometric gradient* borrowed from his colleague Thomas Stevenson.[72] This concept defined the relationship between the difference of barometer reading in two stations placed on a line at right angles to isobars and the distance between these two stations.[73] The suggestive graphic nature of early synoptic charts like Marié-Davy's was replaced by numerical specificity, so that one could numerically express the relative intensity of storms by calculating barometric gradients—namely, by dividing the distance between any two barometers by their difference in reading (both being reduced to 32° and sea level). It also had predictive potentiality. A barometer indicating a pressure of two inches of mercury more in Edinburgh than in London would likely be accompanied by a wind moving between both cities with a velocity of 63 miles an hour.

While Buchan promoted a vision of weather forecast from an Eulerian frame of reference, he also believed that the formation and maintenance of

69 Buchan (1869a, pp. 207–11).

70 Hildebrandsson and Teisserenc de Bort (1898, pp. 94–104).

71 This point—that is, the argument that Buchan's formulation was an extension of Buys Ballot's rule—was emphasized by American meteorologist Cleveland Abbe. See J.P., 'Scrap Album of Alexander Buchan'. *Weather*, March 1950, 119–22, p. 122.

72 Buchan mentioned Stevenson's barometric gradient in the second edition of his work, *Handy Book of Meteorology*. Edinburgh and London: William Blackwood and Sons, 1868, pp. 531–32.

73 Thomas Stevenson introduced this concept in a paper entitled 'On ascertaining the intensity of storms by the calculation of barometric gradients', read at the general meeting of the Scottish Meteorological Society in June 1867.

storm systems could be explained by a thermal theory. As with some of the cyclone models examined in previous chapters, an explanation of the thermal nature of cyclones was also present in his work. In his *Handy Book of Meteorology*, first published in 1867, he aligned himself with Loomis's and Espy's ideas and concluded that strong ascending currents arose within and about the centre of storms and maintained the low pressure centre despite new air currents converging continuously.[74]

For the explanation of the process of cyclone formation and maintenance Buchan offered an empiricist version of thermal convection, which was predicated on surface data and cloud observations. 'All observation', he said, 'shows that in front of storms the air is warm and moist'. It is 'there that most of the rain accompanying storms falls'.[75] That is where observations ended. Yet next he assumed, as a working hypothesis, that the heat of condensation released in these currents of moist air caused pressure fall and formed 'the motive power' of the storm. As available observations were too scarce and too incomplete, Buchan never deemed his explanation as a theory, but a working hypothesis.[76]

As the Baconian overtones of Buchan's explanations seem to suggest, his theoretical framework was derived from the British empiricist tradition. The custom of putting facts first, building up theories subsequently, and finally proposing inductively reasoned conclusions were compatible with the empiricist inheritance that Scottish meteorologists shared with the rest of their British colleagues. That Buchan paid special attention to the physical processes accompanying adiabatic changes of dry and moist air had an explanation; he had been a student of mathematical physicist Peter G. Tait at Edinburgh and had heard first-hand the first law of thermodynamics and its applications. He was also acquainted with Espy's theory, whom he quoted in favour of his convective hypothesis. However, Buchan clearly followed an inductive logic of justification. In fact, he took the question of induction seriously as a methodological issue and laid out a course from which he would hardly veer in the years ahead.[77] Paraphrasing what late nineteenth-century physicist Ferdinand Rosenberger said about his colleagues' stance on this issue, we might say that hypothesis constituted for Buchan 'only aids of construction for a better synopsis of results'.[78] They were valuable only to the extent that they could facilitate the induction of laws and the derivation of rules from observations.

74 Buchan (1867, pp. 104, 143, 150, 158).

75 Ibid., p. 158.

76 As Kutzbach (1979, p. 74) notes, Buchan stated in the second edition of his *Handy Book of Meteorology* (Edinburgh: Blackwood, 1868, p. 291) that a theory could be only 'vague and unsatisfactory' in absence of reliable facts.

77 On Buchan's empiricist approach and his familiarity with the laws of thermodynamics, see Kutzbach (1979, pp. 74–75).

78 Rosenberger (1884, vol. 2, p. 327)—as quoted by Kutzbach (1979, p. 75).

By embracing the hypothesis of the thermal nature of cyclones Buchan distanced himself from the advocates of mechanical theories. He did not doubt that the polar and equatorial currents played a role in storm formation. In fact, he thought that the dry polar current, by flowing under the moist equatorial current, thrust it into upper regions, thereby producing the disturbance that constitutes a storm. However, Buchan rejected Dove's view of conflicting currents. In his opinion, 'it is difficult to imagine how the polar and equatorial currents could be brought to affect each other as they flow in opposite directions'.[79] He would also have disapproved FitzRoy's storm model, even if he did not do it explicitly.[80] His convective approach required that the wind field should be depended upon thermodynamic processes only.

Buchan's *Handy Book of Meteorology* was greatly appreciated, both inside and outside of the English-speaking audience it was nominally aimed at. Meteorologists Hugo H. Hildebrandsson and Léon P. Teisserenc de Bort, in their classic book on the basis of dynamical meteorology (published in 1898), were so impressed by its contents that they did not hesitate to regard it as the work that most accurately reflected the new ideas of modern meteorology. In fact, they added, 'the first investigations conducted in different countries were at first nothing more than a verification of those ideas'.[81] Yet it also found a warm reception in British academic and scientific establishments as well, with its use as a textbook in universities, through the edition of an extended version in 1871, *Introductory Textbook in Meteorology*.[82] As a supposedly up-to-date synthesis of knowledge about modern meteorology, Buchan felt impelled to improve his *Handy Book* in a second edition in 1868, considerably extending the chapter on storms. These books, along with what was described as an 'epoch making work' in climatology[83]—the first global charts of mean sea-level barometric pressure and global wind charts (1869)—enhanced Buchan's reputation as a meteorologist of the first order.[84]

In sum, Buchan's influential work of 1863–69 was not just the reunion of synoptic charts, the Buys Ballot's law, and the thermal theory of storms. It also brought with it an array of synoptic analysis tools which successors

---

79 Buchan (1867, p. 163).

80 Although Buchan commended FitzRoy's efforts to establish a storm warning service, he did not mention his popular treatise of 1863, *The Weather Book*.

81 Hildebrandsson and Teisserenc de Bort (1898, p. 105).

82 Buchan's book 'immediately became the standard text in English', and 'was widely read', according to J.P., 'Scrap Album of Alexander Buchan'. *Weather*, March 1950, 119–22, p. 120.

83 As defined by meteorologist Julius Hann. See J.P., 'Scrap Album of Alexander Buchan'. *Weather*, March 1950, 119–22, p. 121.

84 These climatic charts, published by Buchan (1869b) in the *Transactions of the Royal Society of Edinburgh* in 1869, represented the mean pressure and prevailing winds over the Earth for months and the year.

could use and imitate. Moreover, it prefigured two ways of deepening the forecasting of weather changes: by examining the circular or oval forms of isobars and by calculating the barometric gradients indicating the relative intensity of storms.

The majestic building of his work, compiled in *A Handy Book of Meteorology*, rested on a three-storey structure. In the first floor, and most importantly for forecast purposes, the pressure distribution determined weather conditions during the passage of a storm. In the second, some general rules on the relationship between pressure and winds were inferred from synoptic analysis; these rules meant an extension of Buys Ballot's law. In the third, the formation and maintenance of storm systems could be explained by a thermal theory, provided that facts corroborated initial hypotheses. This building provided a strong structure for advocates of the Eulerian approach, and helped convince forecasters of the advantages of combining synoptic charts and Buys Ballot's law.

Constructing synoptic charts, ca. 1870

In the late 1860s, two methods quickly stood out as the most effective solution to the problem of constructing synoptic charts. The first of these was the conversion of barometric pressure to sea level, which assumed a direct correlation between the pressure distribution and winds. This method was a particular favourite of French meteorologists, especially the influential Le Verrier and Marié-Davy, but was also adopted by Buchan and then by Henrik Mohn in Norway. The second, calculation of means, followed a different line of reasoning, consisting of plotting the deviations of pressure from mean for each observing station. This method assumed the trustworthiness of deviations from normals instead of direct measurements, and was promoted by Buys Ballot in Holland and then adopted in Sweden. A variant of this method, however, was proposed by Carl Jelinek in Austria; mountainous conditions in places like Austria and Switzerland led to the view that the calculation of pressure at sea level and the plotting of isobars were not most appropriate for them.

Jelinek was one of the great personalities of Austrian meteorology. A graduate of law but also a passionate mathematician and physicist, he had been educated in Vienna and had worked as assistant at observatories in Vienna and Prague from 1843 to 1852, conducting magnetic and meteorological observations under Karl Kreil's supervision.[85] After a decade as a professor of mathematics at the Polytechnic Institute in Prague, he became director of the Central Meteorological Institute at Vienna in 1863, where he succeeded Kreil and tried to institute a national system of daily weather

---

85 On Jelinek and his early scientific career, see Steinhauser (1974, p. 389); Hammerl, Lenhardt, Steinacker, and Steinhauser (2001, pp. 42–61).

telegrams. He started receiving fourteen daily dispatches in 1865 and by 1869 was producing daily weather forecasts.[86] Yet he soon realized that, in order to arouse the interest of voluntary observers, their periodical collaboration with the Institute was not enough. So, emulating its counterpart in France, he founded the Austrian Society of Meteorology in 1865. An active entrepreneur, he also founded with Julius von Hann a monthly journal supported by the society, the prestigious *Zeitschrift* der *Österreichische Gesellschaft für Meteorologie*.[87]

Early in his career Jelinek had achieved some renown for a book on instrumental construction illustrating the efficacy of self-registering meteorological instruments, devised to produce massive data in a continuous and systematic manner.[88] This mass production had helped meteorologists in France and Scotland to regularly convert the pressure to sea level. But with time Buchan came to realize that this kind of conversion was not an easy task in countries like Austria. In a memoir on storms, presented at the Academy of Vienna in 1867, Jelinek proposed instead the calculation of deviations from normals, as has been the common practice in Holland with Buys Ballot, as well as the plotting of the lines of constant deviation (rather than isobars).[89] Using the data gathered in fifteen stations from Central Europe and Italy he drew two kinds of maps: one showed curves of equal barometric deviation from normals and the direction and strength of winds; the other presented lines of equal temperature deviation and weather conditions (Figure 5.3).[90] The analysis of these maps led him to doubt that European storms were indeed turbulent air movements like inter-tropical cyclones; rather, it confirmed that the wind successively changed direction from south to southwest, west, or northwest, every time the storm swept north a station. Or, in other words, unlike other places, the wind regime in Austria was subject to large local disturbances.[91]

Jelinek's method was especially useful from a local predictive standpoint, for it took core beliefs about the relationship between pressure and wind and questioned them through statistical techniques employed in a synoptic map. Perhaps the most fundamental of these beliefs was the idea that there

---

86 Jelinek's daily weather reports including synoptic charts began to be published in 1877. See C.A., 1904. 'Meteorology in Austria', *Monthly Weather Review*, 32 (6), 277–78, p. 277.
87 Marié-Davy himself, in an article reporting the scientific mission carried out with Émilien Renou across six European countries in 1868 in order to study the national meteorological establishments in those places, remarked on Jelinek's notable enterprising ability to attract free and voluntary observers through the foundation of a society and a meteorological review. See Marié-Davy (1868, p. 459); Renou (1868). Moreover, they highly commended Italy for the 'advanced level of internal organisation'.
88 Jelinek (1850).
89 Jelinek (1867, pp. 369–400).
90 Ibid., pp. 388–89.
91 For a valuable analysis of Jelinek's synoptic charts, see Hildebrandsson and Teisserenc de Bort (1898, pp. 106–07).

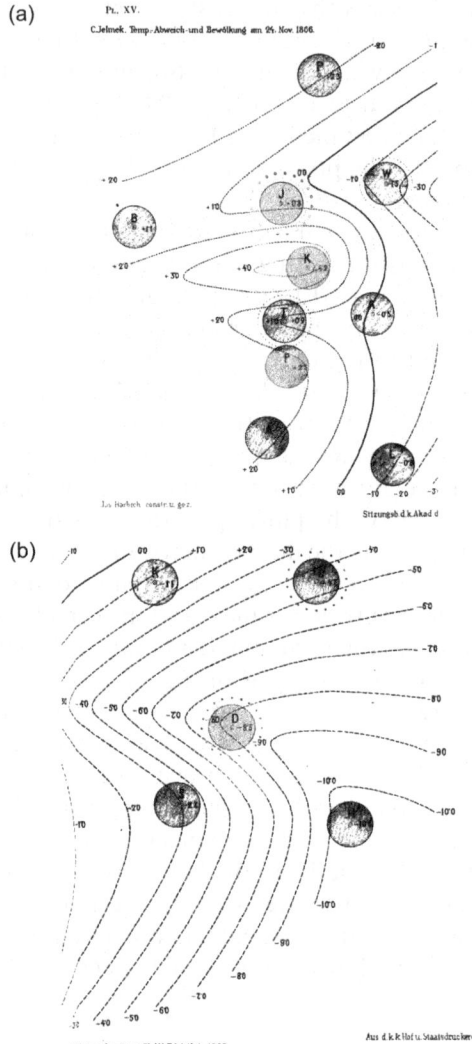

Figure 5.3 Weather maps, from Carl Jelinek's memoir on the storms of November and December 1866, presented at the Academy of Vienna in 1867. One map includes curves of equal barometric deviation from normal. Arrows and their length and fletching indicate the direction and strength of wind, respectively. The other map includes curves of equal temperature deviation in °R (Reaumur scale). Circles of different blue tint show sky conditions.

Source: Jelinek (1867, pp. 388–89).

was always a clear relationship between wind direction and the centre of the barometric depression. Buchan had just published his first monthly isobaric charts in 1867, from which one could infer that the barometric pressure gradually decreased from south to north in Europe. As in Buys Ballot's case, for Jelinek normal values were fundamental and highly significant, objectively identifiable as the functions of the atmospheric movement. One should then search for absolute causes in barometric deviations from normals. Yet, by cartographically representing such deviations, Jelinek went one step further than Buys Ballot. It was no accident, after all, that it was his synoptic maps of 1867—rather than numerical tables—which showed that there was no relationship between wind direction and centre of depression in Vienna. They also served to show that geographical accidents were the cause of local changes of wind in mountainous regions. Simply put, Jelinek's method went beyond Buys Ballot's statistical-empirical approach, broadening it through an updated, Eulerian, statistical-synoptic approach.

The method of calculation of sea-level pressure was different from the method of deviations from mean. The former had exploited synoptic cartography from the very beginning, first by Marié-Davy and Le Verrier in Paris, also by Buchan in Scotland, and then by Norwegian meteorologist Henrik Mohn. Mohn was unusual among meteorologists in being a consummate mathematical physicist: he would later become famous for his memoir *Études sur les mouvements de l'atmosphère* (written in 1876 along with mathematician Cato M. Guldberg) in which they formulated the equations of motion for a cyclone driven by a central convective current.[92] But in the early 1860s he was a novice assistant at the astronomical observatory in Christiania, who had originally studied theology and then a master in mineralogy.[93] Having been appointed director of the *Norske Meteorologiske Institut*, to the foundation of which he contributed in 1866, at first as a part of the University of Christiania, Mohn became highly interested in the study of storms, the biggest threat from the Atlantic to the long coast of Norway.[94] His first two works in 1868 made it clear what his intentions and approach were for the new venture: while the first was an article intended to draw public attention to the importance of weather forecasting, the second— *Stormenes Love* (law of storms)—was a detailed study on several storms of 1867 illustrated by ten synoptic charts, through which he demonstrated the

---

92 Guldberg and Mohn (1876–1880). Kutzbach (1979, p. 100): 'It was primarily this memoir which determined the development of dynamic meteorology in Europe during the following decades'.

93 On Mohn and the Norske Meteorologiske Institut, see Pedersen (1981, 442–43); Arntzen (2005, p. 204); Kutzbach (1979, pp. 240–41); and N.N., 'Professor Mohn og det norske meteorologiske institute'. *Naturen*, April–May 1985.

94 Mohn was not only director of the Institute but also held the first chair of meteorology at the University of Christiania from 1866 to his retirement in 1913.

validity of Buchan's results—i.e. barometric depressions are represented by circular or oval isobaric forms.[95]

By studying the weather charts of Le Verrier and their subsequent comparison with FitzRoy's ones, Mohn accepted 'the reduction of pressure to sea-level', that is, the conversion of the barometric data in the form of station-level pressure into sea-level pressure. In other words, he applied Marié-Davy's and Buchan's method of barometric reduction to the Scandinavian storms. Mohn justified his application of the conversion method by basing on his firm belief that deviations from normals did not allow us to know the most immediate absolute causes of phenomena. Normal values are themselves functions of the movements of the atmosphere, he held, and we search for their causes in the changes of barometric pressure. On the contrary, by reducing pressures at sea level, he concluded, an absolute view is obtained.[96]

Like Buchan, his desire to highlight isobaric forms led Mohn to pay special attention to the quality of weather charts and their physical interpretation. Using the method for the construction of maps introduced by Marié-Davy in Paris, Mohn would patiently plot a large number of synoptic charts of the storms in the winter of 1867–68, from data gathered in 210 stations. By 1870 he had published a seminal book, entitled *Det Norske Meteorologiske Instituts Storm-Atlas*,[97] which included both synoptic charts and explanations of some of the basic characteristics of cyclones. In this second part, he proposed a graphic, qualitative cyclone model.[98] The use of synoptic charts—four per day (two pressure charts, one depicting pressure variations and other of temperature variations)—was largely descriptive, but in the case of temperature variations it played an important role in his theory of cyclone formation. Water vapour content was represented in a separate chart. Wind direction was indicated as usual by arrows, while sky conditions, rain, snow, etc. were shown in the same way that French charts did it. The contrast in these elements between the front and rear part of cyclones could be seen in some of his charts. A tongue of warm and moist air predominated in the front part of cyclones. From these analyses Mohn deduced that different types of air currents converged in cyclones.[99]

95 The works were originally published in Norwegian, *Stormen den 7de til 8de februar 1868* (Christiania, H. Tonsbergs bogtrykkeri) and *Stormenes love* (Christiania, P.T. Mallings bogtrykkeri). See Mohn (1868a, b). See also Hildebrandsson and Teisserenc de Bort (1898, p. 109).

96 Mohn (1870)—quoted by Hildebrandsson and Teisserenc de Bort (1898, p. 110).

97 Originally published in Norwegian and French, *Det Norske Meteorologiske Instituts Storm-Atlas—Atlas des Tempêtes de l'Institut météorologique de Norvège*. Christiania: Bentzen, 1870.

98 Mohn's *Storm-Atlas* provided the basis for his textbook published in 1872, *Om Vin og Vejr*, which was translated into several languages (German, Russian, Polish, Finnish, French, Italian, and Spanish).

99 Hildebrandsson and Teisserenc de Bort (1898, pp. 109–22); Kutzbach (1979, p. 76).

Mohn emphasized that the maintenance and propagation of barometric depressions were not the result of mechanical forces but of essentially thermal processes. The movement of translation of the depression centre, he stated, 'is caused by the contrast in properties of the currents between the front part and the back part'. He also noted that this translation consisted of 'an incessant formation through barometric fall in the front part of the cyclone', and assumed that the same causes would be present 'at the formation of the origin of a depression centre'.[100] Mohn thus came to the same conclusions as Buchan. In fact, a few years earlier Buchan had explained the maintenance of cyclones on the basis of the temperature and moisture contrast between the front and rear parts of a cyclone.[101] Yet Mohn went a step further, with his emphasis on the convergence of two air currents of different thermal properties in a cyclone. Both Mohn and Buchan corroborated most results of Loomis's synoptic studies of the 1840s.

Mohn's ideas on air circulation were by no means original. As shown in Chapter 1, Dove had already identified the warm and cold air currents participating in cyclone formation—the so-called polar and equatorial currents. We also saw that FitzRoy retained Dove's air current concept when he developed his storm model in 1863. For Dove, the origin and propagation of midlatitude cyclones were a result of the meeting of these two conflicting currents. For Mohn, however, the incessant air renewal of the cyclone as its centre progresses was due to the inflow of air currents of different thermal properties. The cause responsible for the cyclone formation and propagation was no mechanical conflict; it only entailed a thermal, physical process. Despite these differences, Dove's views remained alive in meteorological memory. Even Mohn and Buchan, two of the staunchest supporters of the Eulerian view in weather forecasting, identified the warm and cold air currents of their storm charts with Dove's polar and equatorial currents.[102]

If Dove's views on weather forecasts were rejected but his cyclone model was accepted, one might think that FitzRoy's ideas faced the same divisive fate. From a purely mechanical viewpoint such an account would be indisputable: in some respects, FitzRoy's storm model of 1863 illustrated the confluence of polar and tropical air currents in vertical storms in a way that Dove's model itself was not able. But from a physical viewpoint, there was a problem: FitzRoy's model and Mohn's system were mutually exclusive. In fact, FitzRoy's storm model is not even mentioned in Mohn's *Storm-Atlas* or his subsequent textbook on meteorology. Mohn ignored FitzRoy's wind flow-model for at least three reasons. FitzRoy's model involved only the mechanical interference of two currents and thus excluded any thermal causation. His concept of interference omitted any mention of adiabatic cooling,

---

100 Mohn (1870, pp. 21–22), as translated by Kutzbach (1979, p. 78).
101 Buchan (1867, pp. 200–01).
102 See, e.g., Mohn (1870, p. 18); and Buchan (1867, p. 163).

which was proposed by Espy to explicate precipitation in ascending currents of air. And third, at no point did FitzRoy deal with the source of energy in storms or the new concepts of energy conversion and conservation. After all, Mohn regarded the dynamical explanation of cyclones as the highest aim of the new modern meteorology.[103]

In brief, by 1870 meteorologists had begun to not only lay the foundations of the dynamical theory of cyclones but also take serious steps on the path of weather forecast from the Eulerian frame of reference. They began to study the influence of pressure distribution on weather conditions during the passage of a storm by different methods of construction of synoptic charts, whether converting pressure to sea level (this was the case in Scotland and France, but particularly in France, where Le Verrier exploited the potential of the Eulerian approach to achieve his goals of order, centralization, and control), whether plotting pressure deviations from mean (as was the case in Holland, Sweden, and Austria). The pressure distribution began to be regarded as the element determining the progress of storms. Geometric forms of isobars were taken as true representations of storms, convincing the public up to even the present day of the equivalence between the progress of low pressure centres and the progress of storms. However, setting common goals and standards and coordinating observations among different countries continued to be a major challenge.

Proximity of system of values in statistics and meteorology

The foregoing sections have shown the way in which over the last two decades the meteorological science acquired more-or-less 'statistical' forms. Quetelet's and Buys Ballot's works, for example, show the importance of the *cognitive* influence of statistics—namely, the influence concerning procedures and graphic methods. However, this influence generally operated in two dimensions, *cognitive* and *organizational*. Studying how weather forecast services were implemented and consolidated in different countries poses the question of the link between the two dimensions of this influence. Historians agree that throughout the nineteenth century statistical bureaus demanded a system of values based on both the standardization of definitions and observing modes and the exhaustiveness of records within a delimited space (national and/or regional). Yet these values were similar to those required in the meteorological institutes that were being founded at that time on uniform rules and the spirit of coordination.[104] In this section, I will show that the proximity of such systems of values was reflected in the international congresses of statistics and meteorology—a fact that

---

103 For a discussion on the mutual exclusion between FitzRoy storm model and Mohn's empirical cyclone model, see Kutzbach (1979, pp. 82–83).

104 Desrosières (1994, pp. 48–49) has laid the emphasis on the similarity between the values required in the state administration in general and those of statistical bureaus.

contributed to legitimize the weather forecast services by combining statistical quantification and the language of synoptic charts.

This organizational and institutional influence of statistics on meteorology would have surprised an observer from the 1840s. At that time the influences between them had a single dimension—cognitive—and indeed could be best described as one of mutual relationship and reinforcement. Humboldt was well aware of the need to introduce statistical tools into meteorology when he published worldwide depictions of average thermometric conditions with the help of isotherms, but his innovations were limited to the fields of methodology and graphic procedures. It is also true that the Prussian Meteorological Institute was established in 1847, originally as a division of the Prussian Statistical Bureau, on the initiative and under the impetus of Humboldt himself.[105] But the fact remains that it seemed to vie as an exception within the European observatory-based data recording tradition. Meteorology, in fact, had to rely on other kinds of organizations to achieve its institutional emancipation. Here the astronomical observatory, with its well kept records of atmospheric observations (e.g. of air temperature to determine astronomical refraction) played a key role, offering the weather-scientists of the era the niches where they could develop meteorological divisions and programs and even central institutions. Such was the case, for example, with Quetelet at the Royal Observatory in Brussels (1831), Buys Ballot at the Observatory in Sonnenborgh (1854), and Le Verrier at the Paris Observatory (1855).[106] National academies of sciences also promoted their own networks of meteorological stations, often coordinated and equipped by members of these academies, though only in a few cases, such as those of Vienna and Sweden, these initiatives resulted in the establishment of central meteorological institutes.[107]

The change in the nature of the statistical influence on meteorology was driven largely by Quetelet and his determination to lead an international scientific collectivism that could elevate Belgium to the rank of elite nations.

105 The following letter by one of Prussia's most distinguished scientists (Humboldt), addressed to the government, seems to have drawn considerable attention: 'Heat and moisture are the most important elements for plantlife, and without numerical data on the variability of these two factors during different years [...] any discussions of the reasons for crop failure will be futile'—quoted by Khrgian (1970, p. 125). The Prussian government took it into full consideration, without citing it, in establishing an institute primarily aimed at addressing immediate problems in agricultural meteorology. For further details, see Boeckh (1863, pp. 65–66); Dove (1873, pp. 275–77); Körber (1993, pp. 9–18).

106 For meteorological institutionalization in Belgium, see Dufour (1947), Dufour and Defrise (1981, pp. 5–18); in the Netherlands, see Van Kerbel (1999, pp. 116–17); *Koninklijk Nederlands Meteorologisch Instituut, 1854–1954*. Den Haag: Staatsdrukkerij- en Uitgeverijbedrijf, 1954.

107 On the establishment of the Austrian *Central-anstalt für Meteorologie und Erdmagnetismus*, see Hammerl, Lenhardt, Steinacker, and Steinhauser (2001, pp. 30–41); on the role of the Royal Swedish Academy of Sciences in the foundation of the *Meteorologiska Centralanstalt*, see Frängsmyr (1989, pp. 76–80).

This idea would be materialized in the double meeting held in Brussels in 1853, the first General Statistical Congress and the first International Maritime Conference.[108] According to Quetelet, the former aimed at starting up 'the unity between the official documents intended for the administration and science' while the latter aimed at coordinating the works of different countries 'to determine the laws governing the movements of the seas and the atmosphere'.[109] This parallelism, he thought, would lead to encourage the standardization of measurement procedures and implement observing networks through the direct participation and support of the states.[110] In Britain and Holland, important efforts were made to create centres for (ocean and atmospheric) statistical data collection for the sake of navigators, resulting in the foundation of the Board of Trade's Meteorological Department (1854) and the Royal Dutch Meteorological Institute (1854), respectively. These and other meteorological institutes tried to develop a more collaborative relationship among them and with observatories worldwide, whose work they thought could be beneficial for boosting confidence in the organization of new international meteorological meetings.

However, these initial lofty intentions were for the most part failures. Mired in internecine strife on a national scale—even the most ambitious project, Le Verrier's *Bulletin international*, was born in that context—and historical rivalries, none was able to organize an international meteorological meeting and boost universal standardization until the early 1870s. Maury's initial plan for a universal system of meteorological observations on land and at sea had to be limited to the project approved at Brussels meeting, a uniform system of observations aboard warships. But here, rather than using universal standards, the warships were to collect data somewhat arbitrarily, often relying on local naval authorities or at times on the heads of the national meteorological institutes, instead of being coordinated by an international body directly.[111] Measurements were often made according to local or national rules rather than unified and global standards—a messy process that too frequently led to recording errors. Problems existed with the analysis of synoptic charts as well, as each country used its preferred method and its observing systems and scales. Thus, the British and Americans used the Fahrenheit and the English scales for their thermometric and barometric observations, while their European counterparts preferred the Centigrade scale or the Réaumur scale (or octogesimal division) for temperature and

---

108 Desrosières (1992, p. 80); and Prévost and Beaud (2012, p. 50).
109 Quetelet (1860, p. 2)—quoted by Brian (1989, p. 122). Translation mine from French original.
110 On this parallelism, see Chapter 2. See also Desrosières (2002, pp. 6–7).
111 According to one of the representatives, 'the participants in the Brussels meeting returned to their countries and each one proceeded to do as he pleased'. Lamont [Bavière] to Maury, n.d. ca. 1853, Naval Observatory Records, US National Archives—quoted by Fleming (1997, p. 277).

the Metrical scale for pressure.[112] This confusion pushed Dove to suggest that an international meteorological meeting be held on the occasion of the Swiss Naturalist's Society Assembly in Geneva in 1863 with the aim of creating a common organization.[113] But his proposal did not succeed. As a result of all these issues, the meteorological observations of that time were simply not capable of fulfilling the requirements of procedural standardization and notational uniformity meteorologists needed in order to draw general laws about the changes of the weather and atmospheric phenomena.

Interestingly, debates over which observing systems and notations were the most appropriate for national meteorological institutes played out within the framework of the International Statistical Congresses between 1853 and 1872. Originally promoted by Quetelet and with some objectives being similar to meteorological meetings, the Congresses were conceived to foster the organization of official statistical bureaus in Europe and adopt common procedural practices, thereby facilitating international or transnational statistical comparisons.[114] Apart from leading statisticians, the meetings sometimes gathered authorities from statistical offices with meteorological responsibilities, who not only debated methodologies for censuses but also presented information about how to gather the elements of a meteorological statistics. To a great extent, these debates were strongly influenced by their own experiences as the attendees included their meteorological activities at the heart of their agendas. Thus the fourth Congress in London in 1860 included an extensive communication by FitzRoy on meteorological statistics in Britain, in which he focused on questions of practicality and organization—key concerns for the British seafaring community.[115] The London Congress also brought the case of the Academy of Sciences in Stockholm and their efforts to organize a network of meteorological stations in Sweden.[116]

Despite these communications, the participants failed to promote a serious discussion on a uniform criterion for meteorological measurements until the sixth International Statistical Congress in Florence in 1867. The

---

112 For a contemporary view on the lack of uniformity in notation in meteorology in the 1860s, see Dove (1873, pp. 45–46).

113 Dove invited meteorological representatives from Austria, France, Italy, and Spain to come to Geneva to deliberate on the establishment of a 'common organization for terrestrial meteorological Observations'. See Dove (1872, pp. 297–98); Fedorov and Böhme (1972, p. 137).

114 See Adolphe Quételet, 1873. *Congrès international de statistique. Sessions de Bruxelles (1853), Paris (1855), Vienne (1857), Londres (1860), Berlin (1863), Florence (1867), La Haye (1869), et St-Pétersbourg (1872).* Bruxelles: F. Hayez; Engel (1863). See also Westergaard (1932, Chapter 14).

115 A. Quetelet, 'Sur le Congrès International de Statistique, tenu à Londres le 16 Juillet 1860'. *Bulletin de la Commission Centrale de Statistique de Belgique,* 1860, 9, 1–31, p. 23; Engel (1863, p. 221).

116 Engel (1863, pp. 222–23).

issue on the organization of meteorological stations and the construction of a daily chart of Europe received great attention at Florence, after one of the Italian participants, Giovanni Cantoni, had read a report on this matter in the section of topography.[117] This professor and rector of the University of Pavia bemoaned that Le Verrier's *Bulletin international* gathered so few data from the northern and eastern regions of Europe and argued that the isobars traced from these points 'were largely reduced to simple conjectures'. He asked the section of topography to establish the regulations whereby each central meteorological institute should prepare their partial charts so that a large daily chart could be formed. Cantoni also asked this section to solve the two following issues:

1   In order to give a more secure basis for meteorological forecasts, might not the major states of Europe do a convention on the choice of main stations, the form of transmission to a single centre for all the meteorological network in Europe, and the way of representing facts (pressure, temperature, sky conditions, wind direction, etc.)?
2   If France wished to assume the daily publication of Europe's weather chart, might not the other main states publish special charts for each region?[118]

It is worth noting, however, that these matters were raised not only by methodological concerns but also by the administrative implications that the choice of one type of meteorological organization or another could entail. Thus Cantoni's proposal to promote the coordination between the chiefs of the central meteorological institutes tallied very well with Italian plans to leave the control of the *Ufficio Centrale Meteorologico* under the Italian *Direzione di Statistica*.[119] The Italians held that statistics should deal with meteorology in all its dimensions, cognitive, organizational, and institutional. They saw this dependence on statistics as something that was appropriate for their own circumstances (for example, to avoid duplication of efforts and institutional overstretch).[120] By contrast, German delegates advocated for this relationship be confined exclusively to the cognitive field. Endorsing

---

117 See Cantoni (1867, pp. 31–3); Maestri (1868, pp. 370–75).
118 Cantoni (1867, p. 33).
119 Through Circular of 14 January 1865, the Italian Ministry of Agriculture, Industry and Trade had established the first official meteorological organization (or *Ufficio Centrale*) under the control and supervision of the *Direzione di Statistica* of this Ministry. See Iafrate (2001, pp. 53–56).
120 A proof of the interest that the *Direzione di Statistica* (led by Pietro Maestri, one of the commissioners at the Florence Congress) had in the management and organization of meteorology is the periodical publication of meteorological observations gathered in the journal, *Meteorologia italiana, pubblicata per cura del Ministero di Agricultura, Industria e Commercio, Direzione di Statistica, Anni 1865 (Marzo a Dicembre), 1866, 1867, 1868.* [It continued until 1882].

Dove's ideas, demographer Ernst Engel argued that statistics 'should leave to natural sciences their domain and not to seek to invade them'.[121] According to them, coordination should be promoted between the directors of meteorological observatories—rather than the chiefs of central institutes, as suggested by Cantoni. Finally, Cantoni's proposal was passed. As a result, the Italian *Direzione di Statistica* was requested to invite the directors of the central meteorological institutes to arrange among themselves the bases upon which they could interchange observations.[122]

These debates coincided with an increasing concern for standardization of data collection and their statistical treatment by states. This was in part a consequence of agricultural development, which made farmers and producers request information about climate and the weather changes in order to assess their impact on production—a relationship that could easier be established through numerical and graphic representation. But it was also an effect of industrialization, which increased the need to manage the dispensation, trade, and transport of consumer goods and raw materials across regions and beyond national boundaries. For governments the systematic exchange of meteorological data among countries also presented the opportunity to increase national productivity, as both those data and meteorological statistics could be used to produce agricultural and industrial goods in more favourable physical conditions. This perception was quite widespread, as Marié-Davy and Renou could verify during a visit to some European meteorological institutes in 1868[123]; in his survey addressed to the French government, Marié-Davy concluded that there was general agreement on 'the desirability of setting up a permanent Congress for Terrestrial Physics along the lines of the Statistical Congress'. National interests would make all countries 'send representatives to this congress', which would meet once a year for 'joint discussion of the use of instruments, methods of observation, publications, and the results' and problems of scientific research.[124]

The value systems debated in statistical congresses surfaced in the organization of the international Meteorological Conference in Leipzig in 1872, which not only debated coordination and standardization, but also formed the basis for the agenda of the first international Meteorological Congress (Vienna, 1873).[125] There three prominent meteorologists—Carl Jelinek

121 Maestri (1868, p. 374).
122 Brown (1868, p. 19); *Solutions arretées dans la session du Congrès International de Statistique rénui à Florence en 1867 et recommandées par la junte organisatrice aux différents états*. [Florence]: [G. Barbèra], [1868], p. 7.
123 Marié-Davy (1868, pp. 457–68); Renou (1868, pp. 449–55).
124 Marié-Davy (1868, p. 467)—as translated from original French by Fedorov and Böhme (1972, p. 137).
125 On the Leipzig Conference, see Scott (1873); 'Meteorological Conference at Leipzig during August 1872'. *Nature*, 28 August 1873, 341–43; Bruhns, Jelinek, and Wild (1872, pp. 193–96); Fedorov and Böhme (1972, p. 136–41); Schröder and Wiederkehr (1992, pp. 49–51).

(Austria), Carl Bruhns (Saxony), and Heinrich Wild (Russia)—organized the event and prepared the analysis and standardization of methods agenda for further discussion. Thus, the organizers largely followed the lead of Buys Ballot as regards standardization, who during that same year published in Utrecht his booklet *Suggestions on a Uniform System of Meteorological Observations*.[126] Professional experiences were shared and a list of twenty-six items on technical questions was prepared for discussion and reflection.[127] The first item, for example, dealt with the introduction of the metric system into meteorology, a matter that remained unsettled. Questions on very specific observing methods, such as the use of aneroid barometers at second-order stations and the introduction of thermometer screens, were widely discussed in items 2 to 21 on the agenda. The responsibility of the central institutes for the collection, exchange, and publication of data was also a discussion point. However, some delegates criticized the lack of agreement on the coordination of investigations.

The new emphasis on the importance of isobaric charts and weather prediction for the states can be seen at the Vienna Congress, whose agenda was marked by its predecessor (Leipzig) and by the monopolizing control of the national meteorological institutes.[128] Unlike Leipzig, where meteorologists from outside the institutes participated, in Vienna only official representatives could attend the Congress. These delegates essentially endorsed the control policies of their states, including the formation of a permanent committee responsible for supervising international cooperation as well as the rejection of any plan for an independent international institute.[129] At this Congress, delegates also opted to debate thorny questions related to standardization, coordination, and instrumental uniformity. Yet just a few months earlier these issues had surfaced again as part of the commission entrusted to a committee appointed at Leipzig, which had prepared a

126 In this work, Buys Ballot (1872) reviewed some of the questions that could be discussed at the Vienna Congress in 1873. In a second booklet published just before this Congress, Buys Ballot (1873, p. 3) posed the following question regarding the unification of observations: 'Ought there be introduced in all countries the same unity of measure, or is it thought sufficient to accept some rules for the reduction of the measures [into a single system]? The general opinion was favourable to the desirability of a universal system of unities'.

127 For the list of items discussed at Leipzig Conference, see Bruhns, Jelinek, and Wild (1872, pp. 194–95).

128 For the Vienna Congress, see *Report of the Proceedings of the Meteorological Congress at Vienna. Protocols and Appendices*. London: G.E. Eyre & W. Spottiswoode for H.M. Stationery Office, 1874; 'The Meteorological Congress at Vienna'. *Nature*, 7 May 1874, 17–18; Cannegieter (1963, pp. 10–11); Anderson (2005, pp. 245–46); Walker (2012, pp. 109–12).

129 The Vienna's agenda included twenty-eight items, twenty-six of which had been discussed in Leipzig. A last additional question addressed the foundation of an international central institute for collecting all meteorological observations with the aim of publishing them for climatologic investigations. On the 'permanent committee', see Cannegieter (1963, pp. 10–11).

questionnaire on the practicality and utility of weather and storm warnings in Europe. The members of this committee—Buys Ballot (Holland), Scott (Britain), and Georg von Neumayer (Germany)—had propounded six questions to the principal meteorologists of the world, and submitted the received answers at Vienna for discussion; in particular, they received written answers to the following two questions:

I   Is it your opinion that in addition to communications of the direction and force of the wind which is actually blowing, the Barometric 'Gradients' should be given, in order to give warning of approaching wind? To what should these gradients be referred? viz.: (a) to the differences between the actual readings at the different stations; (b) to 30 inches, at sea level; (c) to the mean normal heights of the barometer at the stations taken into consideration, etc.?

II   In what way should we take into consideration temperature, vapour, etc.?[130]

The appointed 'committee on weather-telegraphy and storm-signalling' reported on the answers received, and deduced that this service was considered to be a necessity for all the countries.[131] Moreover, all the respondents agreed that the barometric gradient should be given; yet as to the second part of the question, some preferred the mode of reduction at sea level, and the remainder to the mean normal heights (none of them was in favour of *a*). As a result, it was accepted that for stations not above 300 metres in height the reduction of the barometer reading to the mean sea level would be admissible.[132] According to the committee, this agreement 'was attained with special reference to the drawing of isobars for the purposes of storm warnings'.[133] The answers on the role of the rest of elements (temperature, humidity, clouds, etc.) were equally significant. Indeed, most of them recommended these for consideration, but no one said in what way they should be considered. The committee declared that 'this question should be considered as not answered', and recommended them 'for the investigation of the causes of storms'.[134] With the success to develop consensus about how best to produce storm warnings, the drawing of isobars received an important boost while the rest of elements was relegated to the field of research.

130 *Report Weather Telegraphy and Storm Warnings, Presented to the Meteorological Congress at Vienna, by a Committee Appointed at the Leipzig Conference.* London: Her Majesty's Stationery Office, 1874, p. 8.

131 The report, made through Georg von Neumayer, was edited and published by Georg von Boguslawski (1874) in Berlin.

132 For greater heights, the barometric gradients would be referred to the mean normal heights.

133 *Report Weather Telegraphy and Storm Warnings* (1874), p. 13.

134 Ibid.

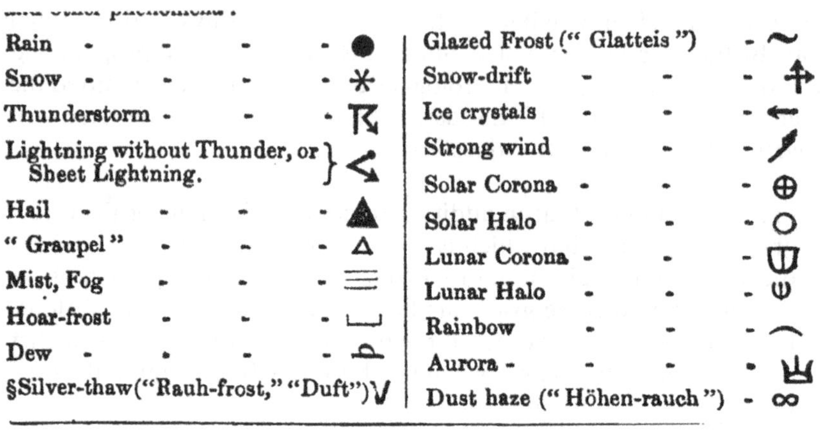

Rain · · · · ●	Glazed Frost (" Glatteis ") · ∼
Snow · · · · ✳	Snow-drift · · · ⤲
Thunderstorm · · · ℞	Ice crystals · · · ←
Lightning without Thunder, or } Sheet Lightning. ⪦	Strong wind · · · ⌀
	Solar Corona · · · ⊕
Hail · · · · ▲	Solar Halo · · · ○
" Graupel " · · · △	Lunar Corona · · · ⋓
Mist, Fog · · · ≡	Lunar Halo · · · ⋓
Hoar-frost · · · ⊔	Rainbow · · · ⌒
Dew · · · · ⌓	Aurora · · · · ⚎
§Silver-thaw("Rauh-frost," "Duft") V	Dust haze (" Höhen-rauch ") · ∞

Figure 5.4 Symbols proposed at Leipzig Conference (1872) and approved at Vienna Congress (1873) for the designation of hydrometeors and other phenomena.
Source: Schoder (1874, pp. 48–49).

While the impetus to produce weather forecasts and storm warnings met with the scientific goals of the meteorologists who regarded the pressure field as the prime indicator, practical concerns mandated a linguistic refinement before this Eulerian approach could be used to achieve credible results. In fact, in their report the committee recommended that each director should give his opinion on the probable course of atmospheric disturbances, not as predictions or prophecies, but as *probabilities*. Unlike the categories of prediction or prophecy—which denoted at that time a certain idea of infallibility and which could be easily produced by unscientific methods—probability was a statistical, quantifiable measure of the likelihood that an event such as the approach of a storm could occur. Thus, only opinions should be given on the weather; and warnings only when the most severe storms were indicated by the barometric gradients. This greatly facilitated efforts to standardize the gradient as indicator of the extent of the approach of storms.[135]

Despite this 'barometric' standardization, however, the Vienna Congress failed to establish a criterion for the uniformity of instruments, oscillating between several stances based on national scientific traditions and domestic interests. The British representative Robert Scott defended the use of Kew Observatory standards, but his Belgian and Russian counterparts, Quetelet and Wild, refused, and the issue was postponed.[136] Likewise, the metric

135 Ibid., pp. 15–16.
136 Scott and Buchan recommended the Kew barometer as 'an excellent and cheap barometer'. See *Report of the Proceedings of the Meteorological Congress at Vienna* (1874, pp. 20 and 36).

system was deemed as the most desirable, and with the best prospect of universal adoption, but this desire met British opposition, for fear of the invasion of the 'French system' in their country. The uniformity of symbols, however, received almost a unanimous approval, despite the disparity of codes and rules in participating countries. Thus, delegates agreed upon a list of symbols for hydrometeors to be used on weather charts and in climatological tables (Figure 5.4).[137] Similarly, they agreed to classify meteorological stations into three classes (first, second, or third order), according to their equipment.[138] These agreements remained practically unchanged for many decades.

Conclusion

Buys Ballot and Buchan contributed importantly to the convergence of statistics and meteorology and helped merge two important traditions in weather forecast. On one hand, Buys Ballot devised a method of residues for meteorology, and derived a relationship between wind direction and barometric pressure (later known as the *baric wind law*). On the other hand, Marié-Davy and Le Verrier had promoted a predictive synoptic method. Inspired in part by the two French physicists, but also by Buys Ballot, Buchan succeeded in combining the synoptic method and Buys Ballot's law in its generalized version.

Buys Ballot's and Buchan's careers were independent and their methods differed substantially. Buys Ballot's measurements consisted of the difference of barometric readings at a few stations, they were shown by statistical graphic methods, and their aim was storm warnings. Buchan had no access to real-time observations, instead using data from the Paris *Bulletin international*; and he opted for synoptic charts. Buys Ballot calculated barometric deviations from mean values, whereas Buchan reduced barometric pressure to sea level. For Buys Ballot, deviations from normals were more significant than direct measurements; for Buchan, reduction at sea level provided a more reliable picture of the pressure distribution than deviations from normals. Buys Ballot derived the correlation between winds and pressure from his statistical tables, whereas Buchan inferred an extension of Buys Ballot's relationship from the analysis of synoptic charts. Simply put, Buys Ballot's method was empirical-statistical and Buchan's method was synoptic-statistical.

Buys Ballot's and Buchan's contributions exemplified the cognitive influence of statistics on meteorology. However, the organizational dimension of this influence played a special role in the consolidation of weather forecast services. The international meteorological meetings in Leipzig and Vienna in 1872 and 1873 marked a watershed in the history of predictive meteorology, proving to be the transitional period when the rivalry between the

137 Schoder (1874, pp. 48–49).
138 *Report of the Proceedings of the Meteorological Congress at Vienna* (1874, p. 61).

Eulerian and Lagrangian approaches to weather analysis was surpassed by one hegemonic Eulerian thinking founded on cartographic and statistical sciences. As I showed earlier, the value system based on procedural standardization, coordination, and the exhaustiveness of records was endorsed in international statistical congresses from 1853 to 1872. In fact, it was its proximity to the values long claimed in meteorology that drove the states to engage in international agreements, thereby helping establish the complete legitimacy of the weather forecast services. These two meetings confirmed that the state was the guarantor and regulator of meteorological activities, not private observatories. Furthermore, they showed that the languages of cartography and statistics could serve as tools to producing weather forecasts and storm warnings—announcements which could also be used by states to reinforce their processes of national building.

However, in doing so states were taking sides. For the acceptance of the idea that the weather could be probabilistically anticipated from the isobaric forms on synoptic charts and from statistical methods opened the door to increasing state control over the forecast process—which in practice would mean the centralization and bureaucratization of national weather services. For if isobaric maps and statistics became the basis for a public service aimed at foreseeing the course of atmospheric disturbances, would not be there then a risk that this same service could be transformed into an administrative–statistical bureau, chained to routine tasks of image-production and digestion of observational data? As will be shown in Chapter 7, this belief in the power of isobaric maps and statistical methods to foretell the probable weather changes would contribute to the primacy of public service over research in national meteorological institutes.

But before going through the next chapters, let us summarize the situation. By the early 1870s synoptic charts and statistics had become the principle tools used by states to produce storm warnings and weather forecasts, providing them with hope science could anticipate the course of atmospheric disturbances as probabilities, rather than as infallible predictions or prophecies. The pressure, or baric, field would determine weather conditions during the passage of storms. Each state would take control over its national meteorological space. No event testifies more to the importance of isobaric maps and statistical science in encouraging these hopes than their backing at the meetings in Leipzig and Vienna, where they served as the basis for the consolidation of national weather services.

References

Achbari, Azadeh, 2017. *Rulers of the Winds. How Academics Came to Dominate the Science of the Weather, 1830–1870*. Amsterdam: Vrije Universiteit.

Achbari, Azadeh, and Lunteren, Frans van, 2016. 'Dutch Skies, Global Laws: The British Creation of "Buys Ballot's Law"'. *Historical Studies in the Natural Sciences*, 46 (1), 1–43.

Anderson, Katharine, 2005. *Predicting the Weather: Victorians and the Science of Meteorology*. Chicago: The University of Chicago Press.

Arntzen, Jon Gunnar, 2005. *Bergensere i tusen år: fra Olav Kyrre til Sissel Kyrkjebø*. Oslo: Kunnskapsforlaget.

Boeckh, Richard, 1863. *Die geschichtliche Entwickelung der amtlichen Statistik des Preussischen Staates eine Festgabe für den internationalen statistischen Congress in Berlin*. Berlin: Königliche geheime Ober-Hofbuchdruckerei.

Boguslawski, Georg von, 1874. *Bericht über Wetter-Telegraphie und Sturmwarnungen, abgestattet an den Meteorologischen Congress in Wien von dem dafür auf der Leipziger Conferenz ernannten Comité*. Berlin: Mittler.

Boumans, Marcel, 2015. *Science Outside the Laboratory: Measurement in Field Science and Economics*. Oxford: Oxford University Press.

Brian, Eric, 1989. 'Observation sur les origines et sur les activités du Congrès international de statistique (1853–1876)'. *Bulletin de l'Institut international de statistique* (47e session), 121–38.

Brown, Samuel, 1868. 'Report on the Sixth International Statistical Congress, Held at Florence, From 29th September to 5th October, 1867'. *Journal of the Statistical Society of London*, 31 (1), 11–24.

Bruhns, Carl, Jelinek, Carl, and Wild, Heinrich, 1872. 'Einladung zu einer im August d. J. in Leipzig abzuhaltenden Meteorologen – Versammlung'. *Zeitschrift der für Meteorologie*, 7 (12), 193–96.

Buchan, Alexander, 1865. 'Examination of the Storms of Wind which Occurred in Europe during October, November and December, 1863'. *Transactions of the Royal Society of Edinburgh*, 24, 191–205.

Buchan, Alexander, 1867. *A Handy Book of Meteorology*. Edinburgh and London: William Blackwood and Sons, 1867; 2nd ed. in 1868.

Buchan, Alexander, 1869a. 'On the Storms Which Passed over the United States between the 13th and 22nd March 1859, with Remarks on Storms which Occurred at the Same Time in the North Atlantic, Europe and Western Asia'. *Journal of the Scottish Meteorological Society*, 2, 198–214.

Buchan, Alexander, 1869b. 'The Mean Pressure of the Atmosphere and the Prevailing Winds of the Globe for the Months and for the Year'. *Proceedings of the Royal Society of Edinburgh*, 6, 303–07; Part II, *Transactions of the Royal Society of Edinburgh*, 25, 575–637.

Buchan, Alexander, and Mitchell, Arthur, 1875–1878. *On the Influence of Weather on Mortality from Different Diseases and at Different Ages*. Edinburgh and London: William Blackwood and Sons.

Burstyn, L. Harold, 1970. 'Buys Ballot, Christoph Hendrik Diederik'. In Charles Coulston Gillispie ed., *Dictionary of Scientific Biography*. New York: Charles Scribner's Sons, 2, 628.

Burton, James M.C., 1990. 'Meteorology and the Public Health Movement in London during the Late Nineteenth Century'. *Weather*, 45, 300–07.

Burton, Jim, 2008. 'Scott, Robert Henry (1833–1916)'. In Oxford University Press ed., *Dictionary of National Biography*. Oxford: Oxford University Press, 2004; online ed., Jan 2008.

Buys Ballot, Christoph Hendrik Diederik, 1847. *Les changements périodiques de température, dépendant de la nature du soleil et de la lune, mise en rapport avec le pronostic du temps, déduits d'observations néerlandaises de 1729 à 1846*. Utrecht: Kemink.

Buys Ballot, Christoph Hendrik Diederik, 1848. 'Iets over de meteorologische waarnemingen aan het observatorium te Utrecht' (On meteorological observations at the Observatory at Utrecht). *Algemeene Konst- en Letterbode*, 379–84.

Buys Ballot, Christoph Hendrik Diederik, 1850a. 'Die periodischen ... Änderungen der Temperatur etc.'. *Die Fortschritte der Physik im Jahre 1847*, 3, 623–29.

Buys Ballot, Christoph Hendrik Diederik, 1850b. 'On the Great Importance of Deviations from the Mean State of the Atmosphere for the Science of Meteorology'. *London, Edinburgh and Dublin Philosophical Magazine*, 37 (247), 42–49.

Buys Ballot, Christoph Hendrik Diederik, 1851. *Uitkomsten der Meteorologische Waarnemingen Gedaan in 1849 en 1850 te Utrecht en op Eenige Andere Plaatsen in Nederland* (Results of Meteorological Observations in 1849 and 1850 at Utrecht and at Some Other Locations in The Netherlands). Utrecht: Kemink.

Buys Ballot, Christoph Hendrik Diederik, 1854. 'Erläuterung einer Graphischen Methode zur gleichzeitigen Darstellung der Witterungs-Erscheinungen an vielen Orten, und Aufforderung an die Beobachter, das Sammeln an vielen Orten zu erleichtern'. *Poggendorf's Annalen der Physik und Chemie*, 4, 559–76.

Buys Ballot, Christoph Hendrik Diederik, 1857. 'Note sur les rapports de l'intensité et de la direction du vent avec les écarts simultanées du baromètre'. *CR*, 45, 765–68.

Buys Ballot, Christoph Hendrik Diederik, 1860. *Eenige regelen voor aanstaande weersveranderingen in Nederland*. Utrecht: Kemink & Zoon.

Buys Ballot, Christoph Hendrik Diederik, 1864. 'On the System of Forecasting the Weather Pursued in Holland'. In *Report of the Thirty-Third Meeting of the British Association for the Advancement of Science*. London: John Murray, 20–21.

Buys Ballot, Christoph Hendrik Diederik, 1865. 'On Meteorological Observations as Made in Holland'. *Civil Engineer and Architect's Journal*, 28, 245–46.

Buys Ballot, Christoph Hendrik Diederik, 1871. 'Der Barometer an Bord'. *Hansa. Zeitschrift für Seewesen*, 8, 197–98.

Buys Ballot, Christoph Hendrik Diederik, 1872. *Suggestions on a Uniform System of Meteorological Observations*. Utrecht: Printing Office "The Industry".

Buys Ballot, Christoph Hendrik Diederik, 1873. *A Sequel to the Suggestions on a Uniform System of Meteorological Observations*. Utrecht: Printing Office "The Industry".

C.A., 1904. 'Meteorology in Austria', *Monthly Weather Review*, 32 (6), 277–78.

Cannegieter, Hendrik Gerrit, 1963. 'The History of the International Meteorological Organization, 1872–1951'. *Annalen der Meteorologie*, 1, 7–280.

Cantoni, Jean, 1867. 'Organisation des stations météorologiques et formation d'une carte diurne de l'Europe'. *Congrès International de Statistique à Florence. Programme de la sixième session du 20 septembre au 5 octobre 1867*. Florence: Impr. de G. Barbèra, 31–33.

Dandeker, Christopher, 1990. *Surveillance, Power and Modernity. Bureaucracy and Discipline from 1700 to the Present Day*. Cambridge: Polity Press.

Desrosières, Alain, 1992. *The Politics of Large Numbers. A History of Statistical Reasoning*. Cambridge, MA: Harvard University Press.

Desrosières, Alain, 1994. 'Le territoire et la localité. Deux langages statistiques'. *Politix*, 7 (25), 46–58.

Desrosières, Alain, 2002. 'Adolphe Quetelet'. *Courrier des statistiques*, 104, 3–8.

Dettwiller, Jacques, 1982. 'La loi de Buys Ballot'. *La Météorologie*, 28, 61–68.

Dove, Heinrich Wilhelm, 1872. 'Ueber den Meteorologen-Congress zu Wien'. *Zeitschrift der Österreichischen Gesellschaft für Meteorologie in Wien*, 7, 297–98.

Dove, Heinrich Wilhelm, 1873. 'Geh. Regierungsrath H. W. Dove to Director C. Bruhns, Leipzig'. In *Report of the Proceedings of the Meteorological Conference at Leipzig. Protocols and Appendices*. London: Georges E. Eyre & William Spottiswoode, 45–46.

Dufour, Louis, 1947. *Notes pour servir à l'histoire de la météorologie en Belgique*. Uccle: Institut Royal Météorologique de Belgique.

Dufour, Louis, and Deprise, Pierre, 1981. 'Histoire de la météorologie en Belgique'. *Ciel et Terre*, 107 (1), 5–18.

Engel, Ernest, 1863. *Compte-rendu général des travaux du Congrès international de statistique dans ses séances tenues à Bruxelles, 1853, Paris, 1855, Vienne, 1857, et Londres, 1860*. Berlin: Imprimerie Royale.

Everdingen, Ewoud van, 1953. *C.H.D. Buys Ballot 1817–1890*. 's-Gravenhage: D.A. Daamen.

Eyler, John M., 1979. *Victorian Social Medicine. The Ideas and Methods of William Farr*. Baltimore, MD: Johns Hopkins Press.

Fedorov, Evgeny Konstantinovich, and Böhme, V., 1972. 'The Leipzig Meteorological Conference, 1872'. *World Meteorological Organization, Bulletin*, 21 (3), 136–41.

Fleming, James Rodger, 1997. 'Les réseaux d'observation météorologiques en Angleterre, en France, en Allemagne, en Russie et aux Etats-Unis d'Amérique: analyse et comparaison'. *Organisation Météorologique Mondiale, Bulletin*, 46 (3), 270–80.

Frängsmyr, Tore, 1989. *Science in Sweden: The Royal Swedish Academy of Sciences, 1739–1989*. Canton, MA: Science History Publications.

Giddens, Anthony, 1985. *The Nation State and Violence*. Berkeley: University of California Press.

Glaisher, James, 1855. *Report on the Meteorology of London and Its Relation to the Epidemic of Cholera*. London: HMSO.

Glaisher, James, 1859–1860. 'Determination of the Mean Pressure of the Atmosphere on Every Day in the Year as Deduced from all the Barometrical Observations Taken at the Royal Observatory, Greenwich, from the Year 1841 to 1858'. *Report of the British Meteorological Society*. London, 11–31.

Guldberg, Cato Maximilien, and Mohn, Henrik, 1876–1880. *Études sur les mouvements de l'atmosphère*. Christiania: Brøgger.

Halford, Pauline, 2004. *Storm Warning. The Origins of the Weather Forecast*. Stroud: Sutton Publishing.

Hammerl, Christa, Lenhardt, Wolfgang, Steinacker, Reinhold, and Steinhauser, Peter eds., 2001. *Die Zentralanstalt für Meteorologie und Geodynamik 1851–2001. 150 Jahre Meteorologie und Geophysik in Österreich*. Graz: Leykam.

Herschel, John F.W., 1831. *A Preliminary Discourse on the Study of Natural Philosophy*. London: Longman, Rees, Orme, Brown and Green.

Herschel, John F.W., 1867. 'On Barometric Waves, February 5th 1867'. *Proceedings of the Literary and Philosophical Society of Manchester*, 6, 91–93.

Higgs, Edward, 2004. *Life, Death and Statistics: Civil Registration, Censuses and the Work of the General Register Office, 1836–1952*. Hatfield: Local Population Studies.

Hildebrandsson, Hugo H., and Teisserenc de Bort, Léon Philippe, 1898. *Les bases de la météorologie dynamique*. Paris: Gauthier-Villars.

Hildebrandsson, Hugo, and Hellmann, Gustav, 1909. *Codex of Resolutions Adopted at International Meteorological Meetings 1872–1907*. London: HMSO.

Hunt, John L., 1978. 'James Glaisher, FRS (1809–1903)'. *Weather*, 33 (7), 242–51.

Hunt, John L., 1996. 'James Glaisher (1809–1903). Astronomer, Meteorologist and Pioneer of Weather Forecasting: "A Venturesome Victorian"'. *Quarterly Journal of the Royal Astronomical Society*, 37, 315–47.

Iafrate, Luigi, 2001. 'Verso un Ufficio meteorologico centrale anche in Italia? Dai primi fermenti organizzativi alla sua istituzione governativa: Regio Decreto n° 3534 del 26 novembre 1876'. In Domenico Vento ed., *Presenze scientifiche illustri al Collegio Romano: Celebrazioni del 125° anno di istituzione dell'Ufficio Centrale di Ecologia Agraria*. Roma: MIPAF-UCEA, 50–58.

Jelinek, Carl, 1850. *Beiträge zur construction selbstregistrirender meteorologischer apparate*. Wien.

Jelinek, Carl, 1867. 'Ueber die Stürme des November und December, 1866'. *Sitzungsberichte der Kaiserlichen Akademie der Wissenschaften. Mathematisch-Naturwissenschaftliche Classe. Abt. 1, Mineralogie, Botanik, Zoologie, Anatomie, Geologie und Paläontologie*, 55, 369–400.

J.P., March 1950. 'Scrap Album of Alexander Buchan'. *Weather*, 119–22.

Khrgian, Aleksandr Khristoforovich, 1970. *Meteorology: A Historical Survey*, 2nd ed. Jerusalem: Israel Program for Scientific Translations.

Klein, Frank H., 1863. *The Foretelling of the Weather in Connection with Meteorological Observations by F.H. Klein together with a Description of the Telegraphic Warning System Introduced in The Netherlands, June, 1860, as proposed by the Director of the Royal Netherlands Meteorological Institute Professor Dr. Buys-Ballot*. London: Benjamin Pardon. Translated from the original Dutch by A. Adriani.

Körber, Hans-Günther, 1993. *Die Geschichte des Meteorologischen Observatoriums Potsdam*. Offenbach am Main: Selbstverlag des Deutschen Wetterdienstes.

Kutzbach, Gisela, 1979. *The Thermal Theory of Cyclones: A History of Meteorological Thought in the Nineteenth Century*. Boston, MA: American Meteorological Society.

Landsberg, Helmut E., 1955. 'Weather "Normals" and Normal Weather'. *Weekly Weather and Crop Bulletin*, January 31, 7–8.

Maestri, Pietro, 1868. *Compte-rendu des travaux de la vie session du Congrès international de statistique réuni à Florence les 29, 30 septembre*. Florence: Impr. de G. Barbèra.

Marié-Davy, Hippolyte-Edme, 1868. 'Rapport sur une mission météorologique en Hongrie, en Turquie, en Grèce et en Italie'. *Archives des Missions Scientifiques et Littéraires*, 5, 457–68.

Marriott, William, 1903. 'James Glaisher F.R.S., 1809–1903'. *Quarterly Journal of the Royal Meteorological Society*, 19, 115–21.

Marriott, William, 1904. 'Some Account of the Meteorological Work of the Late James Glaisher, F.R.S.' *Quarterly Journal of the Royal Meteorological Society*, 30 (129), 1–28.

Meteorologia italiana, pubblicata per cura del Ministero di Agricultura, Industria e Commercio, Direzione di Statistica, Anni 1865 (Marzo a Dicembre), 1866, 1867, 1868.

'Meteorological Conference at Leipzig during August 1872'. *Nature*, 28 August 1873, 341–43.

Mitchell, Arthur, 1866. 'Weather of May'. *Good Words*, 116, 336–44.

Mitchell, Arthur et al., 1908. 'Contributions towards a Memorial Notice of Alexander Buchan, M.A., L.L.D., F.R.S.' *Journal of the Scottish Meteorological Society*, 14, 101–18.

Mohn, Henrik, 1868a. *Stormen den 7de til 8de februar 1868.* Christiania: H. Tonsbergs bogtrykkeri.

Mohn, Henrik, 1868b. *Stormenes love.* Christiania: P.T. Mallings bogtrykkeri.

Mohn, Henrik, 1870. *Det Norske Meteorologiske Instituts Storm-Atlas—Atlas des Tempêtes de l'Institut météorologique de Norvège.* Christiania: Bentzen.

Mohn, Henrik, 1872. *Om Vin og Vejr.* Christiania: Mailings Bogtrykrike.

Nash, William C., 1903. 'James Glaisher FRS'. *Proceedings of the Royal Meteorological Society,* 329, 129–32.

N.N., 'Professor Mohn og det norske meteorologiske institute'. *Naturen,* April–May 1985.

Pedersen, Olaf, 1981. 'Mohn, Henrik'. In Charles Coulston Gillispie ed., *Dictionary of Scientific Biography.* New York: Charles Scribner's Sons, 9, 442–43.

Prévost, Jean-Guy, and Beaud, Jean-Pierre, 2012. *Statistics, Public Debate and the State, 1800–1945: A Social, Political and Intellectual History of Numbers.* London: Pickering & Chatto.

Quételet, Adolphe, 1860. 'Sur le Congrès International de Statistique, tenu à Londres le 16 Juillet 1860'. *Bulletin de la Commission Centrale de Statistique de Belgique,* 9, 1–31.

Quételet, Adolphe, 1873. *Congrès international de statistique. Sessions de Bruxelles (1853), Paris (1855), Vienne (1857), Londres (1860), Berlin (1863), Florence (1867), La Haye (1869), et St-Pétersbourg (1872).* Bruxelles: F. Hayez.

Renou, Émilien. 1868. 'Rapport sur les résultats d'une mission météorologique en Allemagne et en Suisse'. *Archives des Missions Scientifiques et Littéraires,* 5, 449–55.

Report of a Committee Appointed to Consider Certain Questions Relating to the Meteorological Department of the Board of Trade. London: Eyre & Spottiswoode, 1866 (also PP, 1866).

Report of the Meteorological Committee of the Royal Society, For the Year ending 31st December 1868. London: Eyre and Spottiswoode, 1869, 18.

Report of the Proceedings of the Meteorological Congress at Vienna. Protocols and Appendices. London: G.E. Eyre & W. Spottiswoode for H.M. Stationery Office, 1874.

Report Weather Telegraphy and Storm Warnings, Presented to the Meteorological Congress at Vienna, by a Committee Appointed at the Leipzig Conference. London: Her Majesty's Stationery Office, 1874.

'Résumé des procès-verbaux des séances de la Commission *Centrale de Statistique.* N° 189. Du 7 février 1851'. *Bulletin de la Commission centrale de statistique,* 5 (1853), 1–105.

'Robert Henry Scott, F. R. S. 1833–1916' [Obituary]. *Quarterly Journal of the Royal Meteorological Society,* 42 (1916), 301–04.

Rosenberger, Ferdinand, 1884. *Die Geschichte der Physik.* Leipzig: F. Vieweg.

Roy, Marjory G. ed., 2007. *The Life, Achievement and Legacy of Alexander Buchan MA LLD FRSE FRS (1829–1907): Secretary of the Scottish Meteorological Society from 1860 to 1907.* Edinburgh: Royal Meteorological Society.

Royal Netherlands Meteorological Institute, 1954, *Koninklijk Nederlands Meteorologisch Instituut, 1854–1954.* Den Haag: Staatsdrukkerij- en Uitgeverijbedrijf.

Schoder, Hugo von, 1874. 'Appendix 1. To the Protocol of the Fourth Meeting'. In *Report of the Proceedings of the Meteorological Congress at Vienna. Protocols and*

Appendices. London: G.E. Eyre & W. Spottiswoode for H.M. Stationery Office, 48–49.

Schröder, Wilfried, and Wiederkehr, Karl Heinrich, 1992. 'Georg von Neumayer (1826–1909) und die internationale Entwicklung der Geophysik'. *Gesnerus, Swiss Journal of the History of Medicine and Sciences*, 49, 45–62.

Scott, Robert Henry, 1869. *Report of an Inquiry into the Connexion between Strong Winds and Barometrical Differences: Presented to the Committee of the Meteorological Office.* London: Eyre & Spottiswoode.

Scott, Robert H., 1873. *Report of the Proceedings of the Meteorological Conference at Leipzig: Protocols and Appendices.* London: E. Stanford for H.M. Stationery Off. Translated from the official report, appendix in *Zeitschrift der für Meteorologie*, 7 (24).

Shaw, W. Napier, 1907. 'Dr. Alexander Buchan, F.R.S.' *Nature*, 76 (1960), 83–84.

Sheynin, Oscar Boris, 1984b. 'On the History of the Statistical Method in Meteorology'. *Archive for History of Exact Sciences*, 31 (1), 53–95.

Silberman, Bernard, 1993. *Cages of Reason: The Rise of the Rational State in France, Japan, The United States, and Great Britain.* Chicago: The University of Chicago Press.

Solutions arretées dans la session du Congrès International de Statistique rénui à Florence en 1867 et recommandées par la junte organisatrice aux différents états. Florence: G. Barbèra, 1868.

Stark, James, 1862. 'On the Fallacy of the Present Mode of Estimating the Mean Temperature in England'. *Proceedings of the Royal Society of Edinburgh*, 4, 264–65.

Statistics of the United States: (Including Mortality, Property, etc.) in 1860, vol. 4. Washington, DC: Government Printing Office, 1866.

Steinhauser, Ferdinand, 1974. 'Jelinek, Carl'. In Historischen Kommission bei der Bayerischen Akademie der Wissenschaften ed., *Neue Deutsche Biographie.* Berlin: Duncker & Humblot, 10, 389.

Stevenson, Thomas, June 1867. *On Ascertaining the Intensity of Storms by the Calculation of Barometric Gradients.* Edinburgh: [s.n.] Extract from Paper Read at the General Meeting of the Scottish Meteorological Society.

Stok, J.P. van der, 1899. 'Levensbericht C.H.D. Buys Ballot'. *Jaarboek der Koninklijke Nederlandsche Akademie van Wetenschappen*, 59–100.

Supplement to 25th Annual Report of the Registrar General of Births, Marriages and Deaths for England and Wales (PP, 1865, XII).

Szreter, Simon, 1991. 'The GRO and the Public Health Movement in Britain, 1837–1914'. *The Society for the Social History of Medicine*, 4, 435–63.

'The Meteorological Congress at Vienna'. *Nature*, 7 May 1874, 17–18.

Tripe, John Williams, 1862. 'On the Medical Meteorology of the Metropolis during the Years 1859, 1860 and 1861'. *Proceedings of the British Meteorological Society*, 1, 186–96.

Van Berkel, Klaas, 1999. 'Part One: The Legacy of Stevin. A Chronological Narrative'. In Klaas Van Berkel, Albert Van Helden, and Lodewijk Palm, eds., *The History of Science in the Netherlands: Survey, Themes and Reference.* Leiden: Brill, 116–17.

Walker, J. Malcolm, 2012. *History of the Meteorological Office.* Cambridge: Cambridge University Press.

Weber, Max, 1978. *Economy and Society*, vol. II. Berkeley: University of California Press.

Westergaard, Harald, 1932. *Contributions to the History of Statistics*. London: King.

Wolfenstein, Gabriel K., 2007. 'Recounting the Nation: The General Register Office and Victorian Bureaucracies'. *Centaurus*, 49, 261–88.

Worboys, Michael, 2000. *Spreading Germs: Disease Theories and Medical Practice in Britain, 1865–1900*. Cambridge: Cambridge University Press.

6 The hegemony of the Eulerian approach and the beginning of its end

Introduction

On 15 February 1906, the professor of the U.S. Weather Bureau Alfred J. Henry delivered a lecture on 'Weather forecasting from synoptic charts' at the Franklin Institute of Pennsylvania.[1] Well aware of the rising interest in weather forecasting experienced as part of the rise of national meteorological institutes and the civilian control of the Bureau's operations, by then under the Department of Agriculture, Henry outlined the Bureau's (and world-wide) modus operandi in this field.[2] The practice for weather forecasting, he argued, rested not on any theoretical understanding of the atmosphere or on the knowledge about the physical principles governing the formation and motion of storms. Rather, it was essentially based on empirical reasoning and its close association with weather maps, an art aimed at accurately knowing the distribution of atmospheric pressure, more than any other element. This practice, he pointed out, was deeply rooted throughout the entire American territory. Ten years later, a report of the U.S. Weather Bureau stated that they continued 'to make all forecasts of every character as to future weather conditions solely on the basis of synoptic weather maps'.[3]

For Henry, modern weather forecasting was based on 'two well established facts': the 'general eastward drift of the atmosphere' in midlatitudes in the Northern Hemisphere (as anticipated by Buys Ballot's law); and the close relation between 'the weather and the distribution of atmospheric pressure'. These two facts should be regarded as the ground work of all scientific weather forecasting. To prepare a synoptic chart—the centrepiece of forecasting—there were several steps to be followed. First, 'simultaneous

---

1 Author of numerous papers relating to weather forecasting, hydrology, and climatology, Alfred Henry had worked as chief of the meteorological records division of the U.S. Weather Bureau, before becoming professor of meteorology. He was also one of the staff of official forecasters.

2 In 1891, the U.S. Weather Bureau became a civilian agency, initiating a policy of decentralization and promotion of special services such as research projects and a systematic program in aerology. See Whitnah (1961, pp. 61–114).

3 Quoted by Nebeker (1995, p. 38).

meteorological observations' were carried out 'over as large an area as possible'. Barometric readings were then reduced to sea level. The next step required that observations be telegraphed immediately to a central point. Here recourse to cipher codes became necessary—the code was a simple arrangement of vowels and consonants in a series of words so as to be easily translatable at first sight. As the weather reports arrived in the Bureau, they were translated into numbers and symbols, and transcribed on a set of blank maps of the United States. The next step, he stated, was to draw lines of equal pressure and equal temperature, since by this means 'the eye quickly perceives the salient features of the chart and the changes that have taken place'.[4] The chart was completed by marking the words 'High' and 'Low' in the centres of highest and lowest pressure, respectively.

Over the course of the last decades of the nineteenth century, according to Henry, the U.S. Weather Bureau worked tirelessly to turn weather forecasting into a visual graphic matter. Relying on a regional network of observers and a central team of draughtsmen and bureaucrats, the Bureau's forecasters located the areas of high and low pressure on the maps. Circular and oval curves of 'Highs' and 'Lows' extended over, and moved through, the vast state territories, creating the impression that the eastward movement of these shifting centres controlled to a great extent the weather. At the same time, a range of empirical and statistical rules was used to catalogue and memorize isobaric patterns and weather types, reminding the lay public that the character of the weather was closely related to the distribution of atmospheric pressure. For Henry and the Bureau's forecasters it was evident that the problem of weather prediction would be 'solved to a great extent whenever the course and rate of motion of the highs and lows could be accurately forecast'.[5] In their own way, Henry and the forecasters transformed the narrative of the practice of weather forecasting into a visual graphic image, a picture of the weather that was subject to the dictates and whims of the baric field.[6]

Henry's account shows both the potential and limitations of the synoptic charts of isobars as a tool for short-range weather forecasting. On one hand, they were powerful graphic tools for simplifying the complexity of phenomena, which until then had been expressed by verbal and written descriptions. These new tools allowed the empirical forecaster to assimilate a vast amount of information at a glance. Moreover, as visual objects they could be filed in a way that was easily accessible to meteorologists so that one could compare archived weather maps with patterns similar to current ones in order to forecast the weather. These qualities therefore made them

---

4 Henry (1906, pp. 298–302).

5 Ibid., pp. 305–06.

6 It should be noted that Henry's account was not the only one describing the art of weather forecasting at the U.S. Weather Bureau. For other accounts on the preparation and dissemination of weather forecasts at that time, see Gaster (1896, pp. 212–28); and 'How a Weather Map Is Made'. *Scientific American*, 1900, 82 (3), 38.

seem like ideal devices for identifying certain 'weather types' and 'isobaric forms'. Indeed, as I will show, this was the principal goal behind Ralph Abercromby's method of configurations of isobars—the attribution of a definite weather type to a given form of isobars. It was meant to give a geometric form to the abstract idea of the weather type, to inspire forecasters to think that past analogous synoptic situations could be repeated in the near future.

At the same time, however, such synoptic maps as the U.S. Weather Bureau's and Abercromby's forms showed important limitations, lacking both the authority and theoretical principles to serve as the basis for the rational solutions that would be sought from the 1900s onwards. Based almost exclusively on surface observations and bereft of any reference to a physical structure, they were viewed by their critics as idealized representations of a synoptic situation that could arise rather than as mirror images of weather systems underway. Professor William Blasius, for example, described the synoptic chart at the time as a 'vague, incoherent representation' of what was taking place in the atmosphere.[7] Forecasters generally made little effort to tackle the complex processes of hydrodynamics and thermodynamics involved in atmospheric phenomena, resorting instead to crude techniques of extrapolation, empirical rules (sometimes), and guesswork based on intuition. For all these reasons, then, it is not surprising to learn that with the turn of the century a number of physicists—among which are Napier Shaw, Max Margules, Felix Exner, and Vilhelm Bjerknes—began to seek methods of weather prediction on the basis of mechanical and physical theories.

By the beginning of the twentieth century the hegemonic role played in weather forecasting by the Eulerian approach and by the synoptic map as a reflection of weather systems would start to disappear. This chapter charts the evolution of this hegemony, showing how the Eulerian approach reached its apex with the search for weather types and the identification of isobaric configurations in the 1880s and 1890s, and how it began to decline progressively in the 1900s with the first attempts to seek solutions based on the principles of theoretical physics. I begin with a detailed examination of the search for statistical regularities in weather patterns as well as the use of empirical rules for methodical forecasting. I then show how these regularities and the old weather lore led Abercromby to ascribe several forms of isobars to certain weather patterns that were assumed to take place together with them. Abercromby's isobaric geometry, I contend, was the epitome of the use of the Eulerian approach in weather forecasting. In fact, it provided the foundation for a method of weather forecasting that would become a reference for generations of forecasters. The second part of the chapter focuses on the principal critical voices against this isobaric geometry, and the Eulerian approach in general, which arguably founded the scientific weather forecasting. A series of rational solutions were proposed: ones capable of predicting the weather on the basis of hydrodynamic and thermodynamic

---

7 Quoted by Khrgian (1970, p. 175).

theories rather than by sheer empiricism. These rational solutions brought the development of new forecasting tools such as air trajectories, paths, and fronts, especially by Vilhelm Bjerknes and the Bergen school of meteorology. All this marked the beginning of the end of the Eulerian hegemony.

Looking for statistical regularities: weather types

The basic strategy used to construct maps of future weather did not change substantially across the last quarter of the nineteenth century. This strategy can be displayed as a two-stage process: first, forecasters projected the ongoing movement of the barometric depression along its supposed storm track; then, they assessed the effects of this depression (intensity, gradient, etc.) on local meteorological elements such as temperature, wind, and precipitation. However, the techniques used for the assessment of these effects did change. By the 1870s public demand for increasingly accurate information about weather forecast led state forecasters to assess these effects by statistical rules, graphic models, or even simple algorithms, which, in turn, limited the theoretical focus of their atmospheric studies. By the end of the century it would go even further, triggering studies on weather types. At the U.S. Weather Bureau these studies gained special impetus with the publication of a classification and index of weather maps as an aid in weather forecasting.[8]

Armed with these assessment techniques, as opposed to cartographic or predictive techniques, meteorologists could now begin to provide general guidance, helping forecasters to match the projected pressure conditions to given typical situations. They often resorted to verbal algorithms in the form of conditional *if-then* statements, assigning probability ranges to a linear sequence of events. Readers and the general public could thus 'see' and recognize the sequence of atmospheric phenomena that were rarely 'seeable' and recognizable in synoptic charts. Typical of such efforts was a 1869 paper by Scott on the past and present of the work of the Meteorological Office, which formulated a series of probabilistic rules as the basis for forecasting. Scott made statements such as:

> if we take the area from Valencia to Helder, and from Nairn to Rochefort, we find that whenever the difference of barometrical readings between any two stations is 0.6 in. on any morning, the chance is 7:3 that there will be a storm before next morning.[9]

Like many other meteorologists, Scott's probabilistic rules were then accompanied with synoptic charts that provided the general public with a clear, but misleading image of the relationship between barometric readings and the chance of rain or storm.

8 Brown (1901).
9 Scott (1869, p. 340)—quoted by Nebeker (1995, p. 44).

As years passed and storm studies were published, meteorologists began to seek statistical regularities in order to define a 'path type' in cyclone movements. In 1876 a careful statistical study of a large number of cyclones by an English reverend, Clement Ley, for example, revealed that the track of a cyclone is at right angles to the direction of greatest barometric gradient.[10] Ley was an amateur meteorologist, a sharp observer of clouds, and a staunch supporter of the thermal theory of cyclones. He had just published a book on *The Laws of the Winds Prevailing in Western Europe* (1872), and later would propose an interesting three-dimensional structure model of cyclones.[11] In addition to searching for regular patterns within the barometric data sets, meteorologists also examined cyclone movement with regard to its isotherm direction, identifying when and how certain cyclones followed the same type of path. For example, in 1879 Russian professor Peter Ivanovich Brounov found that the path of a cyclone made an average angle of 28° with isotherms.[12] These rules were then often translated into cartographic form, giving readers a bird's eye view of the synoptic processes taking place. In the maps of hydrographer Iosif Bernardovich Shpindler from the Main Physical Observatory in St. Petersburg, for example, one could see the advance of a series of cyclones: two or three cyclones followed the same path which seemed to suggest that certain types of synoptic processes were taking place, though it remained unclear what these processes were.[13]

In addition to providing a type of cyclone path these statistical analyses also helped meteorologists identify weather types associated with the most common storm tracks and with certain periods of the year. In 1872, for example, a young man named Wladimir Köppen, also from St. Petersburg Observatory and who had just completed his dissertation on the relationship between temperature and plant growth, wrote about certain weather types, deduced 'according to the laws of probability'.[14] Instances of non-periodic variations of weather were interpreted as manifestations of probabilistic sequences and expressed by conditional statements, introducing in the minds of meteorologists the tendency of each weather type to persist. Köppen held, for example, that if it had rained in Brussels for ten days, then on the following day there would be rain in four cases out of five.[15]

---

10 Ley (1876, 1877). See also Khrgian (1970, p. 177).
11 For Ley's synoptic-statistical investigations on storm surface observations in the 1870s and his conviction on the need for observations from the atmosphere's upper strata, see Kutzbach (1979, pp. 120–23, 128–32, 238). As regards his life and work, see Kington (1999, 166–72); Kutzbach (1979, p. 238); 'Obituary Notice of the Rev. W. Clement Ley'. *Quarterly Journal of the Royal Meteorological Society*, 23 (1897), 103–05.
12 Khrgian (1970, p. 177). Known as one of the founders of agrometeorology, Brounov presented his doctoral thesis on 'Forward Movement of Cyclones and Anti-Cyclones in Europe and primarily in Russia' (original in Russian, 1882).
13 Khrgian (1970, p. 177).
14 Köppen (1872, pp. 369–75) dealt with the sequence of the non-periodic variations of weather, investigated according to the laws of probability.
15 According to Scott (1873, p. 379)—quoted by Nebeker (1995, p. 44). See also Köppen (1873).

In the same vein, he found that the longer the duration of a weather type the greater the probability this weather will persist in time. Likewise, by comparing a large number of daily synoptic charts, Köppen found that certain conformations of baric areas reappeared constantly; in fact, in 1874 he grouped two years of barometric observations at St. Petersburg into six classes according to the pressure distribution—a work that pioneered the pressure-field climatology.[16]

Despite these findings, however, meteorologists continued to doubt the overall reliability of statistical studies of weather patterns, seeing them as probabilistic rules that could be useful provided that they offered general guidance rather than particular predictions to forecasters. Willis Moore, for example—director of the U.S. Weather Bureau from 1895 to 1913 and who always had a strong appreciation of synoptic charts, including the commercial weather maps—stated in 1898 that 'no exact rule' in regard to low-pressure areas could 'be laid down'. For him 'empirical reasoning, and intimate association with the charts', equipped 'the successful forecaster for his important functions'.[17] This did not prevent Moore from proposing five local forecasting guidelines while serving as the U.S. Weather Bureau forecaster for Milwaukee, Wisconsin, in 1892. The second guideline reads as follows: 'A Low from the northwest that reaches western Minnesota and western Iowa without precipitation or clouds will pass over Wisconsin as a dry Low, unless the isobars are closer than five eights of an inch'.[18]

The search for statistical regularities in weather patterns also carried an analytical value, leading meteorologists to identify weather types by examining

---

16 Köppen (1874). After being appointed chief of the new meteorological research division of the Deutsche Seewarte in Hamburg in 1879, Köppen undertook systematic studies of the climate along similar lines already worked in 1874. He published his first maps of climatic zones in 1884. See Wegener-Koppen (1955), Greene (2008, pp. 152–55). For Köppen's climate classification, see Rubel and Kottek (2011, pp. 361–62); Greene (2008, pp. 153–54).

17 Moore (1898, p. 13)—quoted by Nebeker (1995, p. 40).

18 These guidelines were presented in his book *The New Air World: The Science of Meteorology Simplified* (Boston, MA: Little, Brown, 1922), p. 154. After leaving the U.S. Weather Bureau, Moore became professor of practical meteorology at George Washington University. Before we accept as fact that Moore renounced any forecast rule, we should hear what he said on this regard in 1916. In that year, Moore became partner with W.F. Carothers, a Houston attorney and long-range forecaster, who had posited a correlation between solar activity (i.e. the 'heat rifts' that burst through the photosphere) and the motion of equatorial air masses. As Moore wrote in *Washington Herald* on 12 March 1916: 'having located a heat rift on the sun [...] we may look with certainty for its return at the end of twenty-five days and for the cool weather conditions that in each case will follow in due time'. Thus, 'while not perfect as to the exact time of arrival of weather changes', this long-range forecast method would still be more accurate than the U.S. Weather Bureau's short-term forecasts. Moore's embrace to long-range forecast was not lacking in irony. As a U.S. Weather Bureau's official remarked, Moore was 'for many years, an ardent protagonist against the very principle which he now endorses'. For further details on Moore as a long-range forecaster and for sources of quotations, see Pietruska (2017, pp. 148–53).

analogous atmospheric situations that took place in the past. As meteorology responded to probabilities, the degree of reliability that a forecast could reach depended on the number of similar cases in the past that meteorologists had been able to gather, classify, and discuss. Here barometric pressure tended to play a decisive role in characterizing the atmospheric state, leading meteorologists to search through a large number of past isobaric maps for a weather map that closely resembled the present pressure distribution. An example of this practice can be found in Paul Garrigou-Lagrange, the secretary of the departmental meteorological commission of Haute-Vienne (one of the most active commissions in France) and the founder of the Observatory of Limoges in 1885. Garrigou-Lagrange pointed out that the general state of the atmosphere in a given place and time was a function of the atmospheric state in the preceding period.[19] To prove this, he looked through two hundred weather maps published by the US Signal Office's *International Bulletin* from 1875 to 1884. After determining the number of lunations, or lunar months, during those years and dividing each lunation into eight periods, he was able to identify eight types of atmospheric situations in the Northern Hemisphere (Figure 6.1).[20] A certain number of weather types were thus determined, within which some major common traits were recognizable[21]; as a result, forecasting became a comparative exercise: one had to find the concordance between these traits and the main features of the present weather, and assume that the weather would change as it did on previous occasions.

The search for statistical regularities could also be addressed to produce catalogues of storms and weather maps, so that one could discover in past records types of weather identical in pattern and evolution to those of recent days. Until then, the classification of weather types was part of the personal experience of forecasters rather than a source of information systematically compiled and preserved in print form for ready reference. The reason for this is that the areas covered by telegraphic reports were too limited to permit an acceptable classification of weather types. There records also required great effort and time to prepare.[22]

19 Garrigou-Lagrange (1893, p. 167). Garrigou-Lagrange was a self-taught meteorologist, who invented the device known as *anémobare*, a seasonal table aimed at forecasting the weather from wind and pressure. See Galliot (1995, pp. 19–20), Saumande (1990).

20 In 1892, Garrigou-Lagrange read before the *Société météorologique de France* a study on the pressure distribution on the surface of Europe in winter and its relationship with the moon's motion in declination.

21 For example, the first weather type said:

> Type 1 is characterized by the presence of a high pressure area on the northeast of Europe, which is more or less directly related from the east and southeast to the high pressure areas of Siberia. At the same time, the high pressure system is often over the Azores Islands; between the two, a zone of lower pressures extends from N-W to S-E, sometimes cut by a ridge N-E, S-W.

>   Garrigou-Lagrange (1893, p. 175)

22 Efforts to systematically compile and classify weather phenomena had focused on storm tracks, especially by Frank Hagar Bigelow, a professor of meteorology in the U.S. Weather

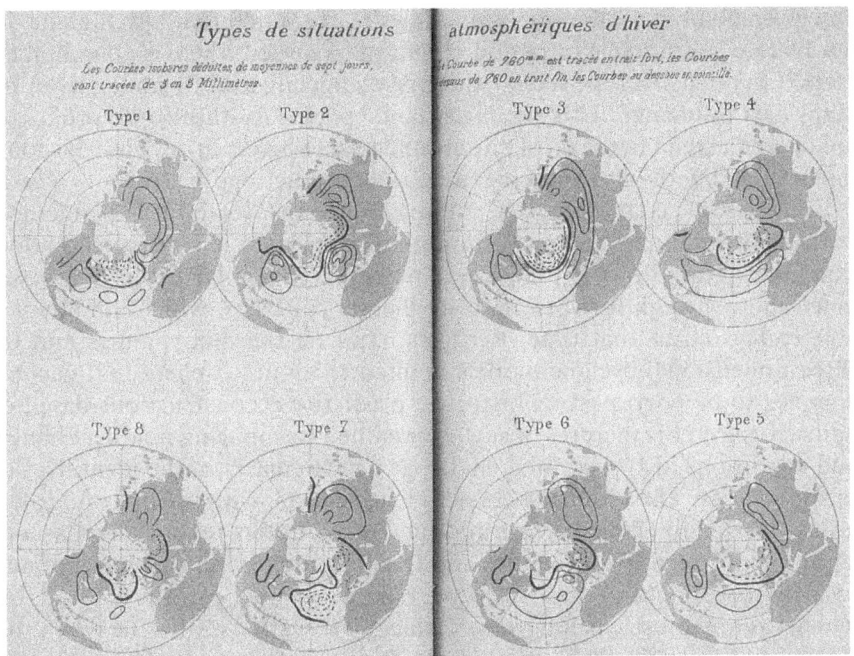

Figure 6.1 Series of maps of mean isobars for the different weeks of the winter season, covering a large part of the Northern Hemisphere. Each map represents a type of atmospheric situation.

Source: Garrigou-Lagrange (1893, pp. 176–77).

In the late 1890s, however, things began to change. Meteorologists constantly cited the efforts of the U.S. Weather Bureau's forecasters as critical to the classification and indexing of weather maps—a task that has been historically known as 'weather typing'. Chief Washington forecaster Edward Garriott, for example, was engaged in compiling and charting storm tracks of the Northern Hemisphere, and began to make short-term forecasts by the aid of classified weather maps. With further plans afoot to produce new catalogues, Garriott confidently asserted that these studies appeared to indicate 'recurring periods during which similar weather conditions and changes' could be found, especially over the middle latitudes of the Northern Hemisphere.[23] Similarly, Indianapolis observer W.V. Brown formed a file of twenty-two hundred weather maps, and proposed a classification and

---

Bureau from 1891 to 1910, who published a work on *Storms, Storm Tracks, and Weather Forecasting* (Weather Bureau Bulletin, no. 20, 1897). Bigelow derived—often concisely and with no information about how to plot such tracks—storm-track maps from an examination of 1,133 storms recorded from 1884 to 1893. For a critical review of Bigelow's storm-track maps, see Monmonier (1999, pp. 10–13). See also Henry (1924, pp. 165–66).

23 Garriott (1901, p. 549).

an index of maps that, he claimed, enabled one to find 'a map presenting any given combination of highs and lows'.[24] He stated that although he did not believe in any cycle or periodicity in the weather, 'there is abundant evidence' that when a type recurs 'there is a tendency toward similarity in the maps next following'. This similarity could persist for three or four days, or could even extend to the most minute and exact degree; in any case, it should 'be reckoned with in the making of weather predictions'.[25]

Despite some meteorologists' enthusiasm for the weather typing method, both theoretical and technical limitations prevented its widespread adoption. To begin with, the practice of recognizing similarities in maps was much more straightforward than its theoretical basis, which nobody was able to lay out at that time. Recent studies on thermodynamics and the three-dimensional cyclone models pointed to an atmosphere that was too complex to be portrayed by historical predictive reconstructions based on daily cartographic pictures of surface weather. Comparing weather changes and conditions also presented challenges, particularly in the identification of similarities. The visual process of identification, with virtually no calculation, tended to lead to errors and imagine fictitious realities. Apparently similar maps regarding pressure and temperature distribution did not necessarily correspond to the same type. Finally, weather records covered limited areas and mostly belonged to surface observations, with the result that storm tracks such as those by Frank H. Bigelow often seemed to be simplistic. For all these reasons meteorologists themselves tended to see catalogues of weather maps as more funds of information than guidance for short-term forecasts.

Empirical rules for methodical forecasting

The statistical search for weather types described earlier was but one side of the larger quest to systematically and methodically predict the weather; for any forecast method was only as valuable as its ability to generate systematic predictions. Without the patient search for statistical and empirical regularities in the behaviour and evolution of weather, no algorithm, rule, or graphic model would have been able to provide the necessary guarantee to make the forecasters' storm warnings believable. In fact, forecasters claimed their authority in part through the belief that the predictions and warnings were the result of a process of methodical empirical observation—a process that demanded the search for regularities in synoptic weather maps and graphs. In places unconnected with central meteorological institutes,

---

24 Brown (1901, p. 547). Brown was professor at Depauw University and a voluntary observer at the U.S. Weather Bureau's office in Indianapolis.
25 Brown (1901, p. 547): 'It seems not unreasonable [...] to expect that like conditions would produce like effects, and the maps of the next succeeding day would, in each instance, also show resemblances'.

however, this fact presented a problem, since local observations—usually by instruments and of the appearance of the sky—constituted the single basis for making forecasts. In those places, there was no room or opportunity for the study of daily synoptic charts.

A physical notion was therefore needed that could be of significant value for making local weather forecast. Local forecasters would discover that notion in the form of the baric field, which had proved to be a reference field in the forecasts made from an Eulerian approach by central meteorological institutes. Over the course of the last quarter of the nineteenth century the fundamental problem of weather forecasting would become the problem of the evolution and progress of barometric depressions. In the 1880s and 1890s the baric field would become the primary reference framework used by local meteorologists to derive the empirical rules that enabled one to arrive at a forecast, while Buys Ballot's law (or its extension, Buchan's rules) would determine wind direction and would become a valuable tool to ascertain the relationship between pressure and winds.

This interest in producing local forecasts did not arise only within the amateur meteorological communities, but also among official forecasters and collaborators of regional and provincial meteorological establishments. This was the case for Gabriel Guilbert, a Norman self-taught meteorologist, who as early as 1886 proposed a method for weather forecasting while working as secretary of the meteorological commission of Calvados Department in Caen, France.[26] Guilbert's proposal, outlined at the Meteorological Society of France in 1887 and read at the 1894 Congress of the *Association française pour l'avancement des Sciences* in Caen, was in many ways quite unusual.[27] On one hand, it included a set of practical rules derived from surface observations of barometer and clouds (or cloud sequences), which enabled him to deduce weather changes for the following day. On the other hand, by combining cloud observations and the application of these rules one could foretell 'the *inception* and the *dissolution* of storms',[28] either in Calvados or all over Europe, without needing to be serving at a meteorological bureau or receive weather telegrams from European stations.

As a meteorologist of the department bathed by the coasts of the English Channel with important merchant traffic, the question of how best to guarantee maritime safety was of special interest to Guilbert and his department. Troubled by the *Bureau Central Météologique*'s inability to foretell weather changes from isobaric maps—which led him to exclaim that

---

26 Guilbert's method was first advanced in the *Annuaire de la Société météorologique de France,* April-May 1887. See Guilbert (1887a, 1887b, 1887c). On Guilbert's career, see Khrgian (1970, pp. 199–200); *Notice sur les travaux scientifiques de M. Gabriel Guilbert.* Paris: Gauthier-Villars, 1920; 'Guilbert (Gabriel)'. *Enciclopedia universal ilustrada europeo-americana.* Barcelona: Espasa-Calpe, 1925, 27 (1), 231.

27 Guilbert (1887a, pp. 162–67; 1895, pp. 377–84).

28 Fassig (1907, p. 211).

'official meteorology had already given up on the solution of the forecasting problem'—Guilbert defined a new physical concept, the *normal wind*, on which his short-term forecasts would be based. Instead of placing emphasis on the principle of barometric gradient, an indicator of the degree of approach of storms for 'official meteorology', forecasters should focus on the normal wind, the wind whose force is directly proportional to the barometric gradient. In practice, the normal wind was the result of multiplying the barometric gradient by two on the Beaufort scale.[29] With that in mind, Guilbert calculated a scale of winds at a given time and place and showed them in a table, distinguishing between normal values and abnormal ones—either by 'excess or by deficiency'. Next, he applied a set of rules to determine the evolution of disturbances. As a general rule, while a wind abnormal by excess brought a rise of barometric pressure, a wind abnormal by deficiency brought a fall of pressure and therefore the birth of a storm.[30]

Guilbert's method for weather forecast moved away from the focus on the isobaric map as an object of analysis, refocusing it on the baric field by applying a set of twenty-five rules. Or, to express it in another way, Guilbert was more interested in predicting the evolution of centres of high and low pressure than in plotting their geometrical forms. Indeed, Guilbert's rules could be reduced to two basic principles in which the baric field was the cornerstone: (1) if the wind in a low-pressure area is weaker than normal, its pressure will rise (and the converse); and (2) a barometric minimum moves in the direction in which winds are weaker than normal.[31]

Yet at the same time, Guilbert's methodology also represented in some ways the recognition of a failure, lacking both the scientific basis and believability to serve as the driving force for the reorientation of the synoptic forecast then at its height. Based on empirical judgements, intuitions, and 'an incommunicable personal experience' (as he himself avowed), his rules could be viewed more as the outcome of an art rather than a science.[32] Guilbert, for example, had the intuition of an equilibrium between wind speed and barometric gradient, deviations from which would determine cyclone motion. He held that no depression could subsist unless there be a complete equilibrium between 'the force of the wind which it causes and the gradient which it forms'. To produce this equilibrium, he argued, there should be equality between two forces that are in constant struggle in every

---

29 Guilbert (1907, p. 211): 'Thus, on a scale of 0 to 9, a *light* wind (force 2) is *normal* for a gradient of 1 mm. per geographical degree of 111 km.; a *moderate* wind (force 4) is normal for a gradient of 2 mm.', etc.

30 Guilbert (1907, pp. 211–12).

31 For these two principles, see Khrgian (1970, p. 200).

32 Guilbert (1907, p. 212) recognized this fact in concluding his article on 'The Principles of Forecasting the Weather': [Following our method], 'the art of weather forecasting, empirical up to the present time, without strict rules, and based upon an incommunicable personal experience, will then become scientifically established'.

depression: 'the gradient represents the centrifugal force, the wind the centripetal force'. If at any point of a cyclone one of them predominates, then there is 'a change in the form of the cyclonic whirl'.[33] In this issue, however, Guilbert made little effort to tackle the complex questions about the storm thermal energy and dynamics examined in previous chapters. Nor did he make any attempt to explain the effects of the deviating force of the earth's rotation, leaving the reader with the impression that physical principles and theoretical concepts played no role in practical forecasting.[34]

Despite all these limitations, however, Guilbert's rules met with a widespread acceptance among meteorological circles, to the extent that his method won the first prize at a weather-forecasting international contest held in Liege in 1905.[35] In 1909, he published a compilation of his rules and ideas in his work *Nouvelle méthode de prévision du temps*. The then-director of Deutsche Seewarte, Köppen, supported his method in the meeting of the German Meteorological Society held in Hamburg in 1908, asserting that in the typical case of a moving depression its winds follow Guilbert's rules.[36] The founder of the Swedish storm warning system and then head of the Stockholm Meteorological Institute, Nils Gustaf Ekholm, disseminated these rules in an article published in *Meteorologische Zeitschrift* in 1907.[37] Finally, Guilbert's rules also found their echo into many of the weather-forecasting handbooks published in different countries across the world, including some regional agricultural weather services in Europe.[38]

Part of the explanation of the acceptance of Guilbert's rules was the non-theoretical character of the method and its seemingly high degree of success in short-range weather forecasts. In fact, there were both operational and social aspects of it that increased its acceptance and interest among meteorologists. To begin with, Guilbert's rules were exceedingly adaptable to a broad spectrum of local conditions, providing an empirical model for all those amateur meteorologists who were unfamiliar with physical theories of the atmosphere. They proved to be particularly helpful for those who could be called 'intermediary meteorologists' from peripheral services and who acted as a bridge between local amateur observers and the state meteorologists of central institutes, promoting local predictive projects as a result

33 Guilbert (1907, p. 211).
34 Examples of criticisms and objections to Guilbert's predictive rules can be found in Brunhes (1909, pp. 402–05); Reuter (1954, pp. 76–77).
35 Organized by the Belgian Astronomical Society, the jury was composed of 'six well-known meteorologists'. See Fassig (1907, p. 210).
36 For their degree of acceptance (and rejection) among European meteorological circles, see Brunhes (1909, p. 394).
37 Ekholm (1907, pp. 326–28). Other meteorologists followed suit, e.g. Bernard Brunhes (1909), Manuel Iranzo (1912).
38 For example, Manuel Iranzo (1912), the head of the *Servicio Meteorológico de la Federación de Levante* in Spain, adopted Guilbert's method in 1912. See Polop (2003, p. 70), Anduaga (2012, pp. 239–41).

of the cooperation between them. When, for example, the Meteorological Service of the Levantine Agrarian Federation in Spain became interested in Mediterranean weather forecast in 1912, its director Manuel Iranzo devised a table of weather types based on Guilbert's rules.[39] To reach this point, Iranzo had examined thousands of bulletins from the central bureaus of Madrid and Paris, and organized a regional pluviometric network for local data collection. The departmental meteorological commission of L'Hérault in France too adopted this system, when trying to make forecasts for the French Mediterranean coast.[40]

But the explanation of their acceptance can be also found in the degree of success attained by the application of the rules. In fact, Guilbert took core empirical beliefs about the importance of pressure and wind and projected them through empirical rules as elements for weather prediction. Perhaps the most fundamental of these beliefs was the idea that the relationship between pressure and winds determined the movement of storms. These ideas were subjected to statistical scrutiny. In 1904, French meteorologist Louis Besson attempted to determine what percentage of success would be obtained in rain forecast by taking the mean of statistical probabilities furnished by the barometric height and the wind direction.[41] The result was unequivocal: he applied this process to 1,516 cases and obtained 70% of correct forecasts, while if forecasts had been based on wind direction alone only 67% would have been correct. Moreover, had four new elements been added (using six instead of two elements), the percentage of success would not have increased (69%).[42] Given the weight of these two elements in the success of weather forecasts, it is small wonder that Guilbert's method aroused interest and receptivity despite its limitations.

Guilbert was not the only promoter of a series of projects that would seek to systematize weather forecasting. The U.S. Signal Corps, for example, was another important promoter of projects resulting in the proposal of rules aimed at atmospheric prediction. Here perhaps the most emblematic case was John P. Finley (known as tornado forecaster), a lieutenant of the Signal Corps, who had been recruited by Cleveland Abbe in 1877 for the Signal Service's weather bureau, where he wrote impressive reports on tornadoes.

---

39 In this, Guilbert's contemporary disciples were but following *au pied de la lettre* and accepting with relish what Guilbert strongly argued, namely that

> by studying analogies between certain types of baric situations, and by comparing isobar charts for approximately a half century, it might be possible to find a certain number of analogous, and perhaps even identical, situations. Clearly, in this way certain types could be distinguished, and then described, after which forecasting rules could be formulated.
>
> Guilbert (1909, p. 30)—as translated by Khrgian (1970, p. 200)

40 Moye (1911); Iranzo (1915, p. 270); Polop (2003, p. 141); Anduaga (2012, p. 241).
41 A former assistant at the Observatory of Tour Saint-Jacques, Besson was assistant director of the Physical and Meteorological Service in the City of Paris, where he published many studies on the Parisian climate. See Grisollet (1944, pp. 99–100).
42 Besson (1904, pp. 93–96).

A Bachelor of Science degree at Michigan State Agricultural and Mechanical College, Finley tried to put into practice much of the knowledge acquired there about the effects of weather on agriculture. Indeed, his chance came only after the late chief signal officer Albert J. Myer was replaced by General William B. Hazen, who established a research unit (the 'Study Room') in 1881, to which Finley would be assigned.[43]

Finley, a sharp observer, showed special ability to combine personal observations, statistical data, and historical records on tornadoes. After being assigned to be an assistant in the Study Room, he resumed his work on the collection of tornado reports begun in Philadelphia, and soon wrote a report, entitled 'Character of Six Hundred Tornadoes', that would become 'the most comprehensive study of the climatology of tornadoes to that date'.[44] This accomplishment convinced him that a feasible system of tornado forecast could be devised which would be predicated on certain empirical rules. A tornado-studies project with such aims was favourably viewed by the chief signal officer. In 1884, Finley organized a network of tornado reporters— that exceeded 2,400 observers in its best years—from whom he obtained statistical data on tornadoes.[45] Strict in his instructions, his particular merit lay in his ability to combine his own observations and statistical data with the aim of deducing a set of rules from which tornadoes' appearance could be foretold—a view that few meteorologists shared at that time.[46]

Finley's rules, advanced first in an 1884 article in *Science* and then redefined in an 1886 paper in the *Journal of the Franklin Institute*, were in many ways a reflect of an era. On one hand, they were essentially in line with the Eulerian reference framework that was defended in the meetings of Leipzig and Vienna and that was predicated on the emphasis on the field concept rather than on air trajectories or air masses. In particular, the baric field occupied a privileged position. The first rule of the two of the six rules dealing with this field claimed the existence of a portion of an area of low pressure (called 'the dangerous octant') within which 'the conditions for the development of tornadoes are most favorable'. The second rule was even more precise:

> The study of the relation of tornado regions to the form of barometric depressions seems to show that tornadoes are more frequent when the major axis of the barometric troughs trends north and south, or northeast and southwest, than when it trends east and west.[47]

43 Galway (1985, p. 1389); Cox (2002, p. 105); Whitnah (1961, pp. 39–40).
44 Bradford (2001, p. 36). Finley's work was published as a Signal Service Professional Paper. Finley (1884b).
45 For further details on Finley's network of reporters and the vicissitudes of his tornado-studies project, see Galway (1985, pp. 1389–91); Bradford (2001, pp. 36–37).
46 Interesting for its testimony to the difficulties and resistance which Finley's views encountered even within the Signal Service is the type of public expression used for official alerts of tornadoes: 'Conditions are favorable for severe local storms this afternoon'. The use of the word 'tornado' was forbidden by the Signal Corps in public releases. Whitnah (1961, p. 40).
47 Finley (1884c, pp. 767–68).

Yet at the same time Finley's rules also represented the application of statistical and synoptic techniques to the problem of forecasting atmospheric phenomena, moving from the early empirical rules of thumb to a synoptic-statistical prediction. In fact, in his 1886 paper he added a set of features of surface weather maps that should be taken into consideration to satisfactorily predict the appearance of a tornado. The first of them, for example, noted the following variables: 'Barometric trough, region, ratio of axes, pressure, and departure from normal'.[48] Although Finley's tornado-studies project was at first based on the collection of statistical and empirical data, in the course of his methodology he sought to combine surface map plots and climatological data.

However, Finley's efforts were not successful in the medium and long term. Affected by internecine strife—the Signal Corps was the centre of a civilian-military struggle for control of weather service—and overwhelmed by criticisms, his tornado-studies project was not capable of systematizing or normalizing tornado predictions. The new chief signal officer, General Adolphus W. Greely, closed down the Study Room in 1887 and transferred Finley to the Records Division, where his project survived only nominally as a section. Rather than forecasting tornadoes, Finley had to routinely compile tornado data and list the number of reporters.[49] Problems existed with the percentage of correct predictions as well, as Finley developed a method of counting whereby tornado occurrences were treated on a par with non-occurrences. Thus, he published in the *American Meteorological Journal* a table listing the total number of predictions for three months in 1884, with a per cent rate of success between 94 and 98 that included the two possibilities.[50] This counting method was very much questioned. G.K. Gilbert, for example, referred to a 'serious fallacy' in Finley's 'percentages of verification', for this method created situations in which tornado occurrences were highly rare and non-occurrences were too frequent.[51] According to Gilbert, using only actual occurrences to recalculate the success rate, the figure dropped to 23%.[52] As a result of all these issues, Finley's project was simply not capable of ensuring the reliability and credibility the Signal Corps' officers needed in order to justify the sponsoring of tornado forecasts before the bar of public opinion. As the chief signal officer stated in 1887, 'more harm would be done by the prediction of a tornado than from the tornado itself'.[53]

---

48 Finley (1886, pp. 255–56).
49 Bradford (2001, pp. 39–44).
50 Finley (1884a, p. 86).
51 Gilbert (1884, p. 166).
52 Murphy (1996) holds that, despite these criticisms, Finley's method marked the beginning of a conceptual development in the methods-oriented and practices-oriented discussions in the field of forecast verification.
53 Quoted by Cox (2002, p. 106). For the original report, see U.S. Army Signal Corps, *Annual Report of the Chief Signal Officer of the Army to the Secretary of War for the Year 1887*. Washington, DC: Government Printing Office, 1887, 21–22, on p. 22.

The epitome of the Eulerian approach:
Abercromby's isobaric geometry

We are now approaching the epitome of the use of the Eulerian approach in weather forecasting: Ralph Abercromby's isobaric geometry. Next the main argument of this chapter is developed, viz. that during the latter part of the nineteenth century, isobaric geometry was the foundation of a predictive method that would become the basis of weather forecasting for subsequent generations of meteorologists. This method was predicated on the attribution of weather types to certain geometric configurations, rather than on physical theories. This isobaric-geometric view grew to dominate the conceptions and practices of most forecasters in the years before and after the First World War.

From the mid-1880s, meteorologists in such different communities as Great Britain, the Central Europe, and the United States embraced Abercromby's isobaric-geometric method as well as the cyclone model associated with it. Such is the case with the memorable treatise *Lehrbuch der Meteorologie*, published by Julius von Hann in 1901. A cyclone model similar to Abercromby's appears intact in the first four editions of the work (from 1901 to 1926), being widely used in Europe until 1930.[54] For its part, the Meteorological Office also adopted this model—instead of Clement Ley's three-dimensional cyclone model—which Abercromby described in his book *Principles of Forecasting by Means of Weather Charts* in 1885. According to British historian James Burton, this book 'was to exert a massive influence on the development of weather forecasting within Britain. Extensive verbatim extracts from it were still being quoted as late as 1940'.[55]

Ralph Abercromby's isobaric-geometric method is worth examining closely, because it shows the steadfastness and fervour with which this idealized method was defended as well as its significant influence on the routine practice of weather forecast.

Ralph was the youngest son of the third Lord Abercromby of Tullibody.[56] A descendant of a long line of intrepid soldiers, he followed their steps, serving as lieutenant in a battalion quartered in Canada, until his continuing ill health obliged him to abandon the military career. In 1869, he was elected Fellow of the Scottish Meteorological Society, the same society that five years earlier had awarded a prize to his first essay, 'A scientific explanation of the popular weather prognostics of Scotland', sponsored by instrument maker Louis Casella.[57] At that time he started to become aware of the utility

---

54 See, e.g., Hann (1901, pp. 509, 545–47). For further details, see Bergeron (1981, p. 456).

55 Burton (1988, p. 146): 'The remarkable influence of Abercromby is surprising. He had no pretensions as a scientist and his interest in meteorology was that of the enthusiastic amateur'. See also Walker (2012, p. 124).

56 On Abercromby's life and work, see: Scott (1897, p. 55); Wood (2001).

57 Wood (2001, p. 102).

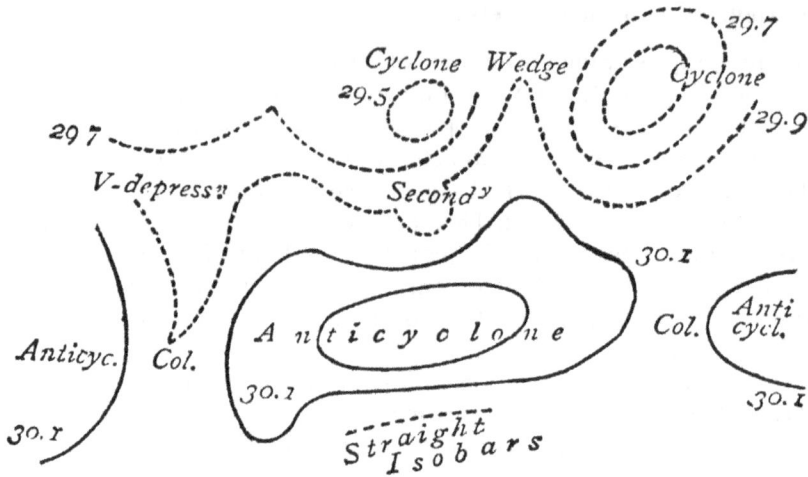

Figure 6.2 Abercromby's seven fundamental configurations of isobars.
Source: Abercromby (1887, p. 25).

of synoptic maps, and, in a note published in *Nature* in 1877, entitled 'Is meteorology a science?', he described the expectations created: 'Synoptic meteorology shows that the world is covered with shifting cyclones and anti-cyclones'. One of our greatest problems is how to 'determine their future courses and changes', given the position of the same at any instant. The answer lies in 'synoptic methods': 'mean values or harmonic series can never much advance meteorology'.[58]

Abercromby soon became the main standard bearer of isobaric geometry in Britain. In 1883, he began to serve on the Meteorological Council, the body that run the Meteorological Office, submitting a draft on 'weather prognostics', which was accepted for publication. His *Principles of Forecasting by Means of Weather Charts* was the first book accepted by the Council that had been written by non-Office staff personnel. Although other books such as those by FitzRoy and Ley had described predictive methods based on barometric and cloud observations, Abercromby's first showed how, by means of synoptic charts, isobars and pressure patterns could be used to prognosticate the weather. Thus, he linked seven kinds of isobaric shapes with certain weather patterns that were assumed to take place together with them. In addition to *straight* isobars, he identified the circular or oval isobars (prefigured by Buchan) encircling *cyclones* and *anticyclones*, the converging isobars of high-pressure *wedges* and *V-shaped depressions* (troughs of low pressure), the *cols* (saddles of low pressure), and the *secondary* cyclones (Figure 6.2). The problem of

58 Ralph Abercromby, 'Is Meteorology a Science?' *Nature*, 1877, 15, 510.

forecasting was then reduced to the ability to prognosticate the incidence and movement of these shapes.

However geometric these shapes could be, in many aspects Abercromby's method has a 'populist' spirit. His modus operandi is a strange hybrid of Buchan's synoptic method and the ideas of the old weather lore. A self-confessed and self-taught amateur, Abercromby was a positivist in his attitude towards constructing cyclone models. His cyclone diagrams constitute a merger of modern synoptic maps and observations regarding weather wisdom.

His cyclone model, for instance, is predicated on isobaric geometry and empirical evidence. The cyclone represented in Figure 6.3 shows weather conditions in various parts of the same: muggy weather in the front, cool weather in the rear, with diverse cloud forms bordering its course. Moreover, it combines scientific nomenclature (such as 'cirro-stratus' and 'strato-cumulus') with popular vocabulary ('restless animals', 'corns', etc.) and more descriptive terms (such as 'driving rain' or 'dirty sky'). Accordingly, the outcome was a highly idealized model; he interpreted, for example, that no sudden change occurred along the lines of confluence (as can be seen in Figure 6.3, the rear confluence appears as a trough line of clearing

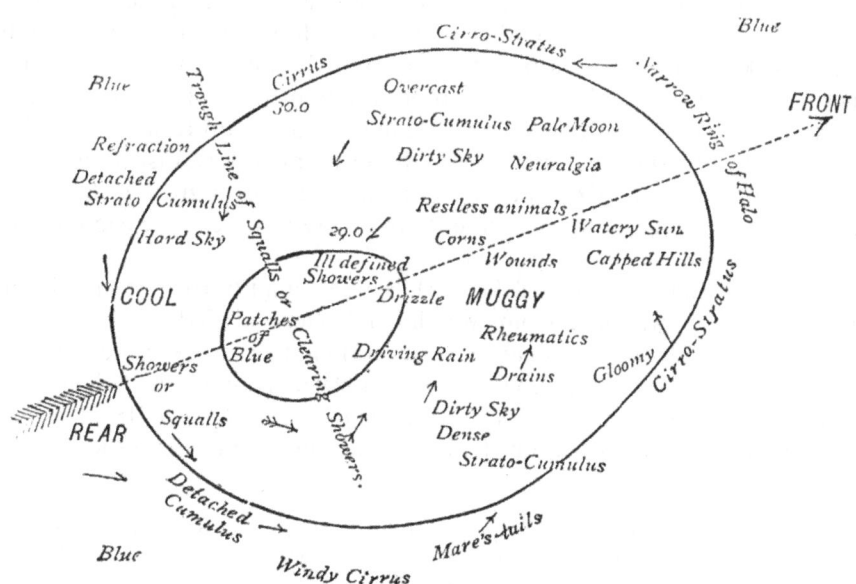

Figure 6.3 Abercromby's model of weather patterns in a cyclone. Note that the line of barometer change (or trough line) is perpendicular to the path of the cyclone centre, and coincides with the line of clearing showers. In actual fact, the rear-confluence line hardly ever has the orientation traced by Abercromby.

Source: Abercromby (1885, p. 16).

showers or squalls).[59] The assumption that cyclone models are basically idealizations, that all are constructed from hypothesis, was by no means novel. What is remarkable in the present case is that Abercromby's diagrams do not aim at representing any physical structure, but at relating old weather wisdom to low-pressure areas. The question of how perfectly geometric isobaric shapes can match weather lore becomes the challenge of the art of forecasting.

That Abercromby championed the isobaric geometrization of weather patterns and changes was evident in a series of works published by him in the mid-1880s. In one of these works, published in *Nature* in 1883, he asserted that 'nearly all our weather is of the cyclonic or anti-cyclonic type, and is entirely dependent upon the form and closeness of the isobars'. Thus, in storms the velocity of the wind 'depends upon the closeness of the isobars': 'the closer the isobars the greater is the difference in pressure', and therefore 'the stronger the wind'.[60] Deeply ingrained in his method is the notion of weather typing, a task that consists in classifying 'isobars by their shape', and then grouping 'certain phenomena of wind and weather round these forms'. Abercromby had no qualm about confessing that 'the great advantage of this method is that it avoids any theoretical considerations'.[61]

From the mid-1880s onwards, isobaric geometry became the basic technique for forecasting the weather in Britain. However, in the early 1880s the most known method was not Abercromby's isobaric technique, but a method that was based on cloud observations instead of the old weather wisdom as well as on the statistical evaluation of synoptic situations. The author of this method was reverend William Clement Ley, inspector for the Meteorological Council of stations in England. Interestingly, Ley developed a flow cyclone model, the forerunner to the current three-dimensional models of frontal low-pressure systems.

Ley was at that time one of the world's greatest authorities in clouds and nephology.[62] A graduate from Magdalen College at Oxford, he was ordained in 1863, working as curate in several chapels in Herefordshire and Leicestershire.[63] While serving as assistant cleric, he wrote his widely read book on *The Laws of the Winds Prevailing in Western Europe* (1872), a summary of his cloud investigations to date.[64] He concluded from his long series

59 Ludlam (1967, p. 29): 'Abercromby recognized sudden changes of barometer, temperature and wind as characteristic of all showers, and therefore to be expected along the "line of clearing showers", but he considered them to be insignificant details'.

60 Ralph Abercromby, 'Weather Prognostics and Weather Types'. *Nature*, 28, August 2, 1883, p. 331.

61 Quoted in Ludlam (1967, p. 28).

62 Nephology is the branch of meteorology that studies clouds and their formation.

63 As regards Clement Ley's biography, see Kington (1999); Kutzbach (1979, p. 238). See also the obituary in *Quarterly Journal of the Royal Meteorological Society*, 1897, 23, 103–05.

64 For Ley's investigations on clouds, see Hamblyn (2017, pp. 45, 63, 88, 106–08); Kutzbach (1979, pp. 128–32).

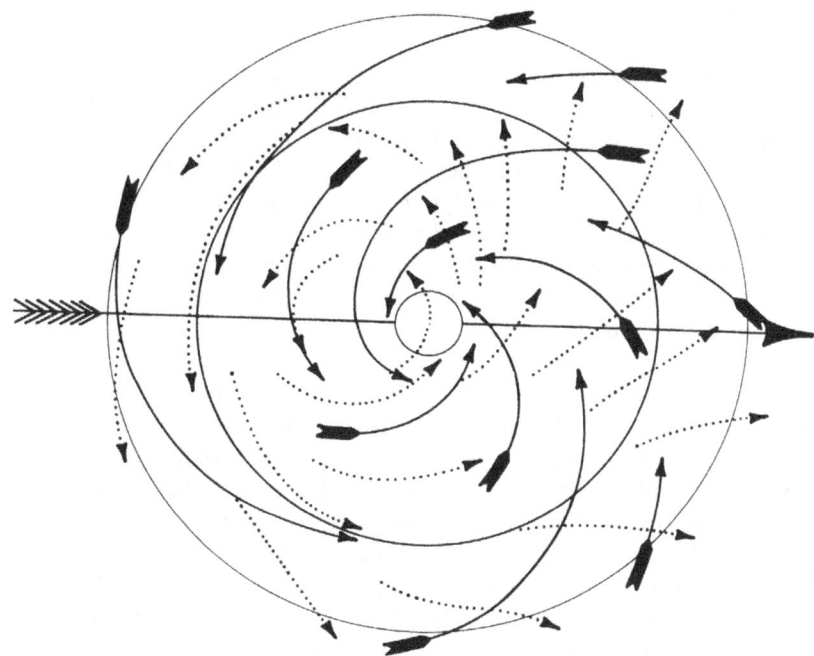

Figure 6.4 Diagram by Ley (1877) showing the relation between the directions of upper winds (dashed arrows) and surface winds (solid arrows) in cyclones. The small circle in the centre represents the lowest pressure area, whereas inner and outer circles demarcate area of cloud observation. The straight heavy arrow shows the trajectory of the cyclone system as a whole.
Source: Ley (1877).

of storm observations that it was absolutely necessary to examine weather conditions not only at surface level but also in the upper atmosphere—a view that few of his colleagues shared at the time. He also claimed that the cyclone axis would lean backwards—an idea that, although being technically correct, was also severely criticized.[65] It is somewhat ironic that a man who adumbrated these facts discarded the 'older theories of primary (polar and equatorial) currents' of Dove and FitzRoy, and embraced Espy's thermal theory of cyclones, the theory that cyclone development depended on the release of latent heat during condensation.[66] Be that as it may, the fact is that in the following years Ley read two papers to the Royal Meteorological Society in which he showed a diagram representing the mean directions of upper and surface winds round a mature progressive cyclone (Figure 6.4).[67]

65 Ley (1872; 1877, p. 445; 1879a).
66 Ley regarded a barometric depression as a thermal phenomenon. On this question, see Kutzbach (1979, pp. 120–23).
67 Ley (1877, 1879b). For further details, see Kington (1999, p. 168).

This is how Ley described the distribution and motion of clouds at the head of a mature cyclonic weather system:

> In the front of a large depression at its extreme exterior there commonly stretches a great bank of cirrostratus [...] The outside edge of the bank is pretty definite, and its outline in most cases is in rough correspondence with the contour of the advancing isobars. The movement of the upper current which carries the outlying parts of this elevated cloud-bank is often nearly tangential to the edge of the cloud-bank, and nearly opposite to the direction of the wind, which is presently about to spring up at the earth's surface, and in nearly all cases it makes a greater angle than 90" with this wind.[68]

In contrast to Abercromby's model, Ley's cyclone model was a result of the semi-Lagrangian method of synoptic analysis. From the physical point of view of atmospheric dynamics, his model presented undeniable advantages. On one hand, it afforded the essence of the three-dimensional structure of frontal low-pressure systems as were known after the First World War, with the studies of Bergen school's meteorologists. In this regard, Ley's model was a precursor of flow cyclone models published in the 1930s. As Bergen's physicist Tor Bergeron asserted, 'the main things missing in Ley's otherwise excellent model were the front concept and the developability'.[69] On the other hand, it provided 'for the first time a flow model for a level above the surface'.[70] No one had ever expressed the distribution and motion of high clouds in a cyclone by round isobars.

As an official inspector of meteorological stations, Ley was well acquainted with the methods and limitations of local forecasters. In his 1878 lecture on clouds and weather signs, he held that the form and movement of cirrus clouds were directly linked to the relationship between wind and pressure distribution.[71] This meant that the movement of cirrus clouds was essential to make weather forecasts. Ley gathered these ideas in a revision work on FitzRoy's *Barometer Manual*, entrusted by the Council. After some revision, the Council approved his draft, and Ley's book was published in 1880 as *Aids to the Study and Forecast of Weather*. Ley had clear ideas on this issue. For him, observations of the movements of upper clouds gave indications of the position and advance of cyclonic systems. But it had hitherto been impossible to train observers in this difficult art. Inasmuch as observations of the form, direction, and speed of upper clouds facilitated weather forecasting, he wrote in 1883, they could be telegraphed to central offices by

---

68 Ley (1880, p. 31).
69 Bergeron (1981, p. 454).
70 Ludlam (1967, p. 28).
71 The lecture was reprinted in *Modern Meteorology. A Series of 6 Lectures Delivered under the Auspices of The Meteorological Society in 1878*. London: E. Stanford, 102–36.

specific codes.[72] Ley thought that cloud observations from strategic enclaves in the North Atlantic to anticipate the arrival of storms, such as Valentia and Brest, could complement the synoptic data collected by the then existing storm warning service. Ley's proposals, and in particular his manual on weather forecasting, had to compete with Abercromby's book, *Principles of Forecasting*, which was to be also published by the Council in 1885.[73]

Ley's ideas gained little acceptance, however, and there are several reasons for this. First, the budget constraints for weather forecast service led the Council to support viable and simpler models. Indeed, Ley's model required records from mountain stations as well as the costly training of staff in cloud observations. Second, Ley had suggested that the axis of the extratropical depressions would be tilted backwards at 89° to the vertical if his cyclone model was right. But some considered this idea was ludicrous and unsustainable.[74] Finally, Ley's new concepts were never used on a regular basis in the daily synoptic maps; rather, they were regarded as local ingenious ideas with few repercussions on the large-scale synoptic practice.

As a consequence of the simplicity and visualization of isobaric geometry, the use of charts representing Abercromby's isobaric configurations in weather forecasting became widespread in Britain. In assessing the march of meteorology in the past decades, English meteorologist Napier Shaw expressed in 1934 his astonishment when he joined the Meteorological Council in 1897 and found the lack of interest in forecasting among its members.[75] The daily forecasts were still produced and issued by 'the same forecasters using the same methods as in the 1870s'.[76] A more detailed and insightful argument of this stagnation of weather study and the lack of information about the atmospheric structure in the vertical is that formulated by Frederic Gaster. This meteorologist of the Office's Forecast Division delivered a lecture before the Royal Meteorology Society in 1896 on 'Weather forecasts and storm warnings: how they are prepared and disseminated'. Here, the isobaric-geometric approach appears as a generalized, if imperfect method reflecting the accommodation of meteorologists to graphic models of weather behaviour. Forecasting was a two-dimensional problem in which cyclonic and anticyclonic systems, gradients, lows and highs, and

72 Codes such as that adopted by the Meteorological Office to transmit weather observations, consisting in a series of five-figure groups. See Clement Ley, 1883. *Note on a Proposed Scheme for Observations of the Upper Clouds* (printed in Leicester for private circulation)—quoted by Kington (1999, p. 167).

73 In the Council's view, Abercromby's first draft was incompatible with the text written by Ley in 1880. Consequently, they decided to publish the two works as separate texts. See Burton (1988, pp. 145–46); Walker (2012, pp. 123–24).

74 For the Deputy Director of the Meteorological Office, 'this was piling Pelion upon Ossa. They did not fit; the [cyclone] structure toppled over'. Quoted by Ludlam (1967, p. 28).

75 Shaw (1934a, p. 112).

76 Burton (1988, p. 190). 'The techniques of weather forecasting in 1905 had not advanced since Abercromby's day' (Ibid).

isobaric shapes were the main players. It was evident, however, that 'we still require some means of ascertaining the movements in the higher regions of the atmosphere, and to accomplish this various means have been tried'—here Gaster mentioned the failed attempts with ballooning, the guns producing clouds of dense smoke, and mountain observatories.[77]

Shaw, Gaster, and the Meteorological Office's forecasters were not the only ones that highlighted the utility of Abercromby's isobaric configurations. These geometric shapes also proved an effective tool for the U.S. Weather Bureau's storm-track strategy. Similarly, influential treatises such as Hann's *Lehrbuch der Meteorologie* spread Abercromby's ideas in Central Europe. His ideas were regarded as valuable not so much because they explained the cyclone dynamics and atmospheric circulation, but because they introduced an isobaric classification in weather forecasting.

Eloquent in this regard is the influential work on *Weather Forecasting in the United States*, written by Alfred J. Henry and other co-workers from the U.S. Weather Bureau.[78] The prime object of this book was to aid the Bureau's employees to fit themselves for the forecast work. In addition to present an account of the methods and practices of weather forecasters, Henry broached the thorny issue of terminology. He deemed as *standard* the classification of isobaric shapes outlined by Abercromby, and since it had not hitherto been published for the information of employees, he gave a detailed account of the same.[79] He added a glossary of technical words and phrases at the end of the book because, he stated, 'there is still a need of qualifying terms to describe more accurately the characteristics of individual Highs and Lows'.[80]

The acceptance of Abercromby's classification as standard had more than just merely terminological implications, since its isobaric configurations also served to reinforce the primacy of the baric field over other elements. In his aforementioned outline of the historical development of synoptic meteorology, Tor Bergeron analyzed the pervasive influence of Abercromby's ideas on even the work of 'non-Eulerian' meteorologists. Bergeron lamented that 'Shaw (who otherwise did so much for the advancement of Meteorology) in all editions of his *Forecasting Weather* (1911, 1923 and 1940)' had taken 'over sixteen pages, literally and within quotation marks, from Abercromby's *Principles* to form the substance' of one of his main chapters—viz. that dealing with the relation of temperature and weather to barometric pressure.[81] Bergeron had doubts as to whether

---

77 Gaster (1896, pp. 224–25).
78 For the origins of this work, see Henry (1930, p. 4).
79 Henry et al. (1916, pp. 74–75).
80 Ibid., pp. 75, 349 for the glossary.
81 Bergeron (1981, p. 453). It would be more accurate to say that although many British meteorologists did indeed see the limitations of Abercromby's method, it continued casting a very long shadow. For example, C.K.M. Douglas, 'the greatest British synoptic weather

'in any other natural science an exposition from 1885 could still serve as a canon fifty-five years later'.[82]

The importance that some influential textbooks and manuals attached to Abercromby's classification, meteorologists' strong predisposition to convert forecasting into an exercise of sheer isobaric geometry, and their endeavour to combine statistical science with synoptic techniques—often to the detriment of the study of air trajectories and the physics of the atmosphere—in order to systematize and automatize the predictive practice all point to the hegemony of the Eulerian approach in weather forecasting. As I will show in the next chapter, the centralization and bureaucratization of national meteorological institutes not only facilitated but also nourished the maintenance of this hegemony.

Critics of the Eulerian approach

If one examines textbooks and manuals on storm and weather forecasting from the 1870s to the 1900s, one finds a high percentage of works that provide synoptic and statistical methods for such purposes. More remarkable, however, is the almost total lack of critical voices against the Eulerian approach within the meteorologists' community. Normally, this criticism focuses on the empirical nature and accuracy of predictions, i.e. on scepticism about forecasting itself rather than the approach adopted. Yet although the critical eye in most works is limited in part to issues of accuracy and terminology (e.g. the use of 'probabilities' instead of 'forecasts'), the very existence of 'anti-Eulerian' voices denotes the hegemony of a forecasting method that prioritized the pressure field over air trajectories.

---

forecaster of all time' according to Walker (2012, p. 320), described this influence in his article on 'The evolution of 20th-century forecasting in the British Isles':

> The main sources of information about the pre-1914 period are Shaw's *Forecasting Weather,* published in 1911, and the *Daily Weather Reports* of that time, which contain forecasts for the 24 hours from noon [...] The empirical methods described by Shaw are broadly the same as those in Abercrombie's *Weather,* which is highly praised. There is no doubt that Abercrombie did much for synoptic meteorology, and he certainly helped to give it the status of a science, though not of an exact science. See Shaw (1911)
> See: Douglas (1952, p. 2)

82 Bergeron (1981, p. 453). Although Shaw distanced himself from Abercromby's Eulerian approach, introducing the analysis of air trajectories and paving the way to the concept of fronts (see his 1906 work *The Life History of Surface Air Currents*, in collaboration with R.G.K. Lempfert), he does not seem to have fully untied. In a book published in 1934, he depicted the primacy of isobars as follows:

> The whole drama of the weather comes to be associated with the ground-plan of the isobars shown on the map [...] It is not unfair to describe the Victorian meteorologist's idea of the play as wind and weather accompanying a barometer-solo [...] Proper names were assigned to recognisable groups of isobars which were regarded as having a definite role to play in the drama [...] The cyclonic depression and anticyclone were the chief, and have remained the principal actors on the stage.
> Shaw (1934b, p. 229)

Though sporadic at first, criticisms against the dominion of the baric field and the geometrization of forecast space began to emerge from amateur weather scientists and physicists, rather than from forecasters. This is the case, for example, with William Blasius, a German professor of natural sciences in Hanover who migrated to America following the political revolutions of 1848 and who had to subsist precariously by giving lectures at Cambridge, Massachusetts, under the protection of naturalist Louis Agassiz.[83] In 1875, he published his book on *Storms: Their Nature, Classification and Laws*, in which he deprecated the tendency among meteorologists to think that the character of storms was largely determined by the atmospheric pressure distribution. In his view, the problem was more profound: 'I am convinced that the existing theories of the nature and laws of changes of weather are intrinsically erroneous'.[84] The key lay in the physical processes of the atmosphere:

> Twenty years ago I in vain exhausted every effort to obtain a hearing for what I believe to be the truth, and I have patiently awaited the development of the science [...] The consequence of the method of observation generally pursued is that an area of barometric depression is considered the storm itself, and the *cause* of the movement of air-currents, while I am certain that the storm is the conflict of air-currents of different temperatures, and that the barometric depression the *effect* of their movement. And, in addition, the most important elements in the life of storms––heat, their originator, and the clouds, their embodiments–– are those to which least attention has been paid.[85]

Here, Blasius was criticizing the determinism of the baric field relative to the nature of storms and the Eulerian approach that had been applied from the mid-1850s to the problem of storm forecasting.[86]

The acceptance of the Eulerian approach was closely connected for Blasius to the routinization of the forecasting task, to the purely empirical, one-sided, and standardized study of baric field and its changes. In 1887, in a paper read before the American Philosophical Society, Blasius critically evaluated the methods and results of the U.S. Signal Service Bureau. Where the Bureau's forecasters saw the proper application of synoptic techniques and empirical rules, Blasius saw the neglect of physical laws. In his view, the Bureau 'has devoted its chief efforts to *prediction* and *signaling*, while the study of nature and its laws has received but scant attention'.[87] In particular, he called atten-

83 Ludlum (1970, p. 154); Ficker (1927); Sandlin (2013, p. 124).

84 Blasius (1875, p. 5).

85 Ibid. Also quoted by Bergeron (1981, pp. 443–44).

86 The rebutting of the determining role of the baric field, represented in synoptic maps by isobars, is surely the single most common issue expressed by the 'anti-Eulerian' critical voices. So Blasius—as quoted by Khrgian (1970, p. 175): a synoptic chart gives just 'an incoherent, vague representation of what is happening'.

87 Blasius (1887, p. 179).

tion to the neglect of 'cloud forms'—the most important element together with heat in the life of storms.[88] At this point, Blasius linked forecasters' prioritization of baric field to the routinization of forecasting task.

> Upon receipt of [telegraphic] reports at Washington, they are written down upon blank maps, on which the respective stations are marked. The points of lowest equal barometric pressure are then united by a line called the Isobar which usually form an ellipse. In the same way the stations of five or ten points of higher pressure are joined by a line. The field enclosed by these somewhat concentric Isobars is the "area of low pressure," or the "cyclonic storm" so called. From the results of the next reading the next position of the area of low pressure is ascertained in the same way, and by comparing the distance traveled with the time occupied, the probable position for some hours ahead is calculated, and predictions are issued. The labor is thus of a routine character.[89]

Blasius's view is somewhat provocative, and his ruminations likely generated reflections among weather scientists about the practice of forecasts reliant on the *yoke* of the baric field for reasons of operational economy and routine. For if one finds meteorologists tending to forecast weather and storms on the basis of personal judgement, experience, and routine techniques, often acting mechanically or subconsciously, then it is reasonable to think that those scientists being acquainted with the physical processes in the atmosphere could also have doubts about the art of prediction, especially when its methods did not conform to physical theories. Such is the case with Henrik Mohn, who had proposed in the 1870s an empirical cyclone model in which thermal processes played a key role in the propagation of cyclones. His conformity with the Eulerian approach used in synoptic weather forecasting, however, did not prevent him from expressing doubts about this practice:

> The weather of each time depends essentially on the distribution of the atmospheric pressure, on which the wind is primarily dependent, and, secondly, almost all other meteorological elements. But the determination of changes in atmospheric pressure is a task [...] which science has

---

88 Twelve years earlier, in his book on storms, Blasius (1875, p. 6) had pointed to the position of the clouds surrounding storms as a key element to elucidate the formation and motion of storms:

> Instead of the modern invention of the "barometric gradient," from which so much has been hoped and so little realized, I have introduced the *plane of meeting* [...] It is defined in its position to the storm by the clouds, and has a meaning and an intimacy with the life of storms which make it more comprehensible and useful than the gradient.

One thus meets once again (cf. the earlier notes) Blasius's insistent demand that meteorologists' views conform with dynamic views of the atmosphere.

89 Blasius (1887, p. 180).

not yet solved. The determination in advance of the weather, therefore, remains for the most part an experience-based opinion, which relies more on practical view than on the knowledge of the laws on atmospheric movements.[90]

Another remarkable case of questioning the forecast practice, understood this time as the result of a shift from inductive to deductive reasoning, involved American meteorologist Cleveland Abbe. A graduate in science from Harvard in 1864 and trained at the Observatory of Pulkovo, Russia, Abbe became director of the U.S. Weather Service in 1870 after having directed the Astronomical Observatory in Cincinnati.[91] In the new era opened after the Civil War, Abbe had organized a national reporting system in which Army's observers and forecasters used purely empirical techniques without demur, possibly because of their widespread use in Europe. In fact, Abbe himself issued the first official weather forecast in 1871, without questioning the empirical nature of forecasting process.[92] At that time, his main concern was precision in forecast language.[93]

However, with the proliferation of empirical rules and the growing tendency to seek statistical patterns, Abbe's view began to change; he began to demand physical or dynamical insights to forecast the weather. In a report entitled *Preparatory Studies for Deductive Methods in Storm and Weather Predictions*, which the already experienced Abbe published in 1890, he left behind early empirical approaches to firmly aver that 'meteorology is essentially the application of hydrodynamics and thermodynamics to the atmosphere'—thus began his report.[94] With this in mind, his aim was to provide 'an approximately complete rational system in meteorology, a deductive or philosophical method, in which the pertinent physical laws' would be 'followed out to their conclusions'; by doing so, the methods of meteorology would become 'comparable in accuracy and perhaps in elegance' to the analysis developed in exact sciences.[95] The differences with respect to his 1871 stance could not be more evident: while then his main concern had been precision in language, now they were the dynamics of storms, horizontal

---

90 As quoted by Mohn in his book *Grundzüge der Meteorologie*, 3. Aufl. Berlin, 1883, p. 341. Translation mine from German original.

91 Humphreys (1919, pp. 472–73); Henry (1917, pp. 62–63); Abbe (1955, pp. 40–128); Fleming (1990, pp. 150–02); Raines (2011, pp. 50–51).

92 By 1872, Abbe regularly sent hundreds of daily maps and bulletins from the U.S. Weather Service and in return he received data from foreign observatories.

93 For example, Abbe ordered that the first sentence of a 'probability' issued by the Weather Service must include a verb in a future tense (e.g. 'the barometer will rise'). For his concerns in empirical rules and precise expressions, see, for example, C. Abbe, 'Rules for the Preparation of Synopses and Probabilities'. *Annual Report of the Chief Signal Officer*, Government Printing Office, 1873, 379–83.

94 Abbe (1890a, p. 7).

95 Ibid.

air flow, cloud motion, viscosity of air, vortex motion, and radiation, among many other issues.[96] Five years later, in an article on the needs of meteorology published in *Science*, he did not hesitate to state it was necessary to pass from the empiric to philosophical methods:

> Meteorologists can never be satisfied until they have a deeper insight into the mechanics of the atmosphere. Something more is needed than the most perfect organization for observing, reporting and publishing the latest news from the atmosphere. [...] We must have a deductive treatise on the laws governing the atmosphere as complete and rigorous as the *Celestial Mechanics* of Laplace, and this will necessarily be a treatise on the application to the atmosphere of the general laws of force, or what is technically known as the dynamics and thermo-dynamics of gases and vapors.
>
> [...] Further progress in meteorology demands a laboratory and the consecration of the physicist and the mathematician to this science.[97]

As the years passed, it became increasingly clear that Abbe's reservation about the use of graphic and statistical methods for weather forecasting was growing. In 1901, Abbe made a stout defence of the mathematically driven deductive reasoning in his article 'The physical basis of long-range weather forecasts'. A synthesis of his theoretical studies, it is especially remarkable for two reasons, first for anticipating Bergen meteorologists' view with his defence of the methods of analytical mechanics, and second because it went virtually unnoticed.[98] In this paper, Abbe described how meteorologists tended to observe weather phenomena, and match them to past events to seek for a correlation with said phenomena. This practice, he acknowledged, is permissible for short-range forecasting but not for long-term prediction. Abbe argued against the long-term forecasts based on solely empirical methods, namely, on 'generalizations based upon observations' into which 'physical theories have as yet entered in only a superficial manner if at all'. These generalizations are 'quite elementary in character'.[99] For that reason, the most appropriate methods for attacking the problem of forecasting, he said, 'are not those of types and averages, but those of analytical mechanics'. 'Our problems cannot be resolved by merely examining charts of the

96 Similar concerns can be found in other reports written by Abbe (1890b, 1893) in those years: 'Recent progress in dynamic meteorology', issued in the 1888 report of the Smithsonian Institution; 'The needs of meteorology', published in the *American Meteorological Journal*, 1893, 9, 560–62.

97 Abbe (1895, pp. 181–82).

98 Abbe's 1901 paper, which has been barely cited—Saltzman (1967, p. 590) and Nebeker (1995, p. 51) are an exception—has received special attention only recently, by Willis and Hooke (2006, pp. 322–24), Pudykiewicz and Brunet (2008, pp. 431–32), and Majdik, Platt, and Meister (2011).

99 Abbe (1901, p. 551).

globe, charts of the diurnal insolation', or maps of temperature, moisture, and pressure; rather, they should be solved by mathematical, theorem-based reasoning:

> It is only after the problems have been fully stated analytically and alge-braically that we are likely to succeed in devising auxiliary geometrical and graphical methods of solving them that may eventually be made the basis of a system easily handled by practical experts.[100]

In this regard, he propounded the integration of the fundamental equations of physics and fluid mechanics, barely known by most forecasters, as a solution for long-term weather forecasting. This task was a major challenge for one had to formulate the basic equations of motion, continuity, and thermodynamics (mass and energy). Abbe tackled it in several steps: he first displayed the equations of motion, providing the necessary background for each equation; he then sought for the solutions in terms of spherical harmonics; and finally he explored numerical and graphical methods. However, before solving these equations, a challenge he deemed technically difficult, one must present general theorems containing the principles of atmospheric mechanics, he said, and these theorems should include assumable conditions. He mentioned, for example, that hitherto the problem of the general circulation had been oversimplified: the globe was assumed to be 'without friction' and 'of uniform perfect spherical shape'; and the atmosphere, 'dry' and 'cloudless'. As a result, the thermodynamic and hydrodynamic aspects of the problem were 'violently separated from each other'.[101] He also pointed out the need for upper-atmospheric data.[102] In all these issues, Abbe

---

100 Ibid., p. 554.
101 Ibid., p. 552.
102 In an essay on the obstacles to the progress of meteorology, read before the Franklin Institute in 1911, Abbe stated:

> From the annual means of [records] we may deduce several interesting coincidences and relations, correspondences and periodicities during past years, but every attempt at accurate forecasts of temperature, either daily, monthly or annual, based on these simple statistics turns out to be elusive.

Next, he added:

> At first forecasts of storms and weather were an application of our el ementary statistical knowledge of general climatology, were not based on our knowledge of mechanics, because we know so little of this latter science as applied to the atmosphere. We believe that the fundamental laws of forces must be still better understood than now, before we have any right to expect detailed predictions.

In fact, 'we may always obtain sufficient observational material for [daily weather maps], but how can accurate predictions be made from the most perfect graphic maps and statistical figures?' Abbe (1912, pp. 57, 63). One thus meets once again Abbe's insistent demand that weather forecasts must be fully consistent with knowledge of mechanics.

was just anticipating the mechanical and physical solutions to weather forecasting which were soon to be championed by Vilhelm Bjerknes and his colleagues at the Bergen school.[103]

My review of this set of critical voices from the meteorological community at the end of the nineteenth century indicates that the statistical and isobaric-geometric methods of weather forecasting did not meet the requirements of physical theory, and that the questioning of forecast practice was fuelled by advances in the hydrodynamic and thermodynamic theories of the atmosphere. Next, I show subsequent attacks against these methods, which became the first serious cracks in the hegemony of the Eulerian approach to weather forecasting.

The beginning of the end

Abbe's proposals never led to any actual forecasting attempt. Even so, Abbe was not the only meteorologist who sought to develop without success methods of weather prediction on the basis of mechanical and physical theories. Max Margules and Felix Exner in Vienna, and N. Shaw and Rudolph G.K. Lempfert in England, also sought rational physical solutions in the 1900s. However, the most successful was Vilhelm Bjerknes, who conceived weather prediction as a mathematically well-defined physical problem. The value of all these studies lies not so much in the greater or lesser success of their forecasting attempts as in creating serious cracks in the dominance of the Eulerian approach.

In the late 1890s, Max Margules was one of the most promising theoreticians of dynamical meteorology. Born in Ukraine, he had studied mathematics and physics at the universities of Vienna and Berlin, where he was a pupil of Ludwig Boltzmann.[104] He worked as a lecturer in physics at Vienna, before joining the Central Institute for Meteorology and Geomagnetism as an assistant, where he served from 1882 to 1906, conducting studies on electrodynamics, hydrodynamics, and physical chemistry of gases. He first focused on physical-chemical investigations. Yet he was soon interested in atmospheric dynamics and storm energetics, in part because of the stimulus of Julius von Hann, who urged him to study the role played by both theory and observation in meteorology.[105] In 1893, he studied the diurnal and semi-diurnal oscillations in barometric pressure due to solar effect,

---

103 For a comparison study of Abbe's 1901 paper and an article published by Bjerknes on the physical basis of long-range weather forecasts, which will be discussed in the next section, and which has been widely lauded as the initiator of a new era in meteorology, see Willis and Hooke (2006, pp. 322–24).

104 On Margules's personal life, see Gold (1920); Kutzbach (1974, pp. 107–08; 1979, pp. 239–40).

105 In the 1890s, Margules organized a network of meteorological stations around Vienna with the aim of studying pressure and wind variations during the passage of storms.

analyzing the problem on the basis of Laplace's tidal theory. He found two types of solutions to tidal equations, identifying them as what is now known as 'inertia-gravity waves' and 'rotational waves'.[106]

Twelve years later, Margules turned his attention to the issue of the source of energy of storms.[107] He introduced the concept of 'available potential energy' in his search for the source of kinetic energy within storms. He concretely showed that the potential energy originating from horizontal temperature differences within storms is, in itself, sufficient to explain the motion of winds. Thus, the storm's kinetic energy would be derived from the release of internal and gravitational potential energy during a redistribution of air masses of different temperatures (cold and warm air) in unstable equilibrium. By recovering and sharpening the concept of air masses, Margules's studies introduced modifications in the thermal theory of cyclones and adumbrated essential features of the polar front theory which would emerge around 1920.

As both a physicist and a meteorologist, Margules considered the possibility of predicting theoretically the weather in a paper published in 1904 in tribute to the memory of Ludwig Boltzmann.[108] His aim was to predict pressure changes by using the continuity equation. As is well known, the barometric pressure is the weight of air in the column above a point. The pressure changes only if there is an air flow within or outside the column. According to Margules, calculating the pressure change becomes an arduous task for there is usually flow through the column and the net change is given by the balance between influx and outflow, a tiny difference between two large numbers. Hence, to calculate pressure changes, one should know the winds with a very high degree of precision. He concluded that any attempt to forecast pressure changes by the mass conservation principle was notoriously error-prone.[109] According to his biographers, Margules would have stated that weather forecasting is 'immoral and damaging to the character of a meteorologist'.[110]

Paradoxically, Margules's failed attempts at forecasting pressure changes by the application of physical principles encouraged his colleagues in Vienna to make meteorology (and particularly, the much-disdained forecasting) an exact science. Thanks to Margules's work—but also to that of Julius Hann and Viktor Conrad in climatology—by 1900 Vienna had become a

106 Margules (1893).
107 Margules (1905). In 1906, Margules derived a formula for the slope of inclination of the surface of discontinuity between two air masses. This formula was applied by Jacob Bjerknes and the Bergen group to their cyclone model. Kutzbach (1974, p. 108; 1979, pp. 147–49, 186–94; 207–18).
108 Margules (1904).
109 The difficulty identified by Margules in 1904 is nowadays that of the balance in initial conditions. For an insightful analysis of this problem and Margules's paper, see Lynch (2001; 2003, pp. 186–90). See also Pudykiewicz and Brunet (2008, pp. 432–33).
110 Quoted by Kutzbach (1974, p. 240). See also Fortak (2001).

reference centre for meteorological research as well as the promoter of one of the field's leading journals, the *Meteorologische Zeitschrift*.[111] A young assistant from the Central Institute, Felix Exner, would follow Margules's steps.[112]

In 1902, Exner attempted to forecast weather patterns from physical laws on a large geographic scale. Instead of using the full set of hydrodynamic and thermodynamic equations, however, he opted for a system of equations reduced to the base essentials. He assumed a series of simplifying conditions, ignoring key issues such as heat transfer, friction, and the vertical motion of air. This allowed him to derive a first-order linear differential equation with which he was able to forecast changes in barometric pressure. Next, he examined a number of observations from European stations and realized that his forecasts coincided with them 70%–90% of the cases. He concluded that his prediction model could 'be of value for the prognosis of a pressure distribution'.[113] Here Exner resorted to statistical, rather than physical, reasoning to justify his approach: what determined the value of a predictive model was not the empirical veracity of its simplifying assumptions, but its ability to forecast the weather changes.

In 1908, Exner moderated his initial optimism when he published an article on his dynamical model in the *Meteorologische Zeitschrift*. As in 1902 paper, he adopted a deductive approach to a problem that could be defined as a geophysical flow system. He first identified a given feature of the weather, then constructed a highly simplified model, and finally examined its role in the flow evolution. This time the chosen feature was the increase in the surface pressure over an area affected by the passage of a polar air mass. Although the calculated pressure changes and the observed changes agreed reasonably well, he admitted his model was a 'crude approximation' to the solution of weather forecasting problem.[114] But despite the limited applicability of his method, Exner would lay some foundations for the numerical weather prediction that was to be developed by Lewis Fry Richardson in the 1920s.[115]

Behind Exner's predictive model was the idea that there were free and independent air masses whose trajectories were worthy of study and special attention. The idea that meteorologists should study airflow trajectories was not obvious at that time. The main argument used by its promoters was that the models based on the air-mass concept were more physical, concrete, and productive in the long run, though they were not as appealing

111 On the Vienna Central Institute and its leading position in the fields of dynamical meteorology and climatology, see the papers edited by Hammerl, Lenhardt, Steinacker, and Steinhauser (2001), esp. Davies (2001, pp. 301–12); and Coen (2007, esp. pp. 282–92; 2010).

112 Lauscher and Skoda (1981, pp. 94–102); Volkert (2008, pp. 425–28).

113 Quoted by Coen (2007, p. 285).

114 Exner (1908, p. 62). See also Coen (2007, p. 285).

115 Exner's numerical method of weather prediction is summarized in his textbook on *Dynamische Meteorologie* Exner (first ed., 1917).

as the isobaric-geometric methods. This was in part because the former were more complicated to handle graphically and mathematically, but also because they required much of the energy was channelled toward research rather than public service. It is precisely in the early 1900s that we find at the British Meteorological Office an atmosphere receptive to the idea that the Office's duty was to stimulate physical investigations on the atmosphere. This is the new scenario created after the appointment of Napier Shaw as the Office's director, who instilled in its staff a research ethos.[116] It was necessary to promote physical research because predictive success depended on it and the Office's prestige depended on predictive success.

The most rigorous study of air trajectories used to show the flow accompanying cyclones was published in this context in 1906. Shaw and the Office's scientific assistant Rudolf G.K. Lempfert traced such trajectories in an official paper, *The Life History of Surface Air Currents*.[117] They calculated the surface air trajectories associated with low-pressure systems, and explained the physical processes responsible for rainfall. As is shown in Figure 6.5, they developed a flow model in which the confluence lines that separate the two air currents (warm and cool) are associated to certain rain-belts (shaded areas). Reflecting their identification with the tradition of the Lagrangian approach (as adopted by Dove, FitzRoy, and Jinman), Shaw and Lempfert chose air trajectories over the dominant practice of isobaric patterns.[118]

The end results of this development of a flow model based on air trajectories were mixed. On one hand, they showed the Bergen meteorologists the way towards air-mass analysis and anticipated features of the front concept. They suggested, for example, that the changes in the wind direction in a cyclonic depression showed discontinuities. Thus, when winds changed from one persistent direction to another, this transition was abrupt. This fact would be recognized by the Bergen meteorologist as 'the first clear statement of the dissymmetry of the depression'.[119] In short, they transformed air trajectories into something that was analyzable and representable—that through the use of flow charts could be seen and measured, and studied their changes with time.

Yet at the same time, the results of all these seminal papers often failed to attract the immediate interest of forecasters and most meteorologists. Numerical prediction and flow models remained a theoretical, intellectual

---

116 Before and after his appointment in 1905, Shaw promoted research projects such as the study of the upper atmosphere by kites and the distribution of fog and rain in the London area. For Shaw's impact on the Meteorological Office, see Walker (2012, pp. 142–50).

117 *The Life History of Surface Air Currents. A Study of the Surface Trajectories of Moving Air*. London: Darling, 1906. Lempfert drew the air trajectories and all the charts in the work.

118 This point is especially stressed by Ludlam (1967, p. 31): Shaw and Lempfert 'had the merit of redirecting attention to the flow – rather than the pressure–pattern'.

119 Gold (1945, p. 215).

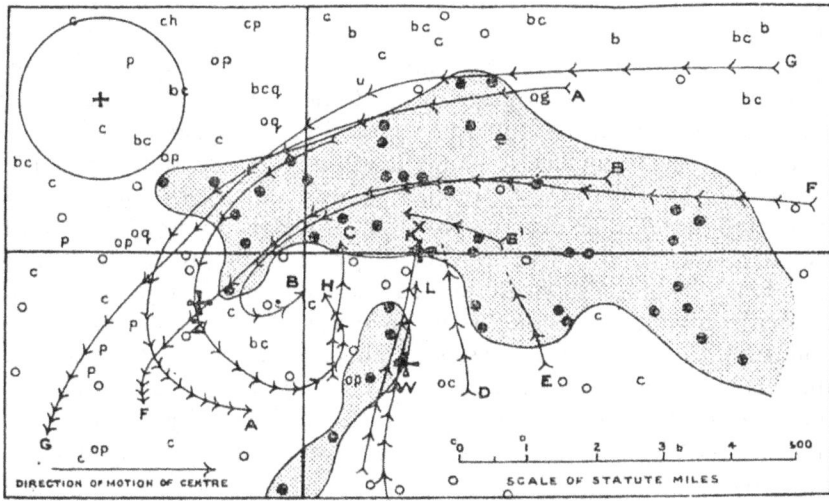

Figure 6.5 Trajectories of parcels of air relative to the centre of a travelling storm (date: 11–13 November 1901). Precipitation is distributed in two areas (shaded regions): a broad rain-belt ahead of a forward-confluence line, and a narrow rain-belt at a rear-confluence line. Streamlines or continuous curves with arrowheads represent the motion of the storm centre for two hours intervals; small letters show weather conditions, e.g. *n* (no rain), *o* (overcast), etc.

Source: Shaw and Lempfert (1906).

curiosity whose immediate application was constantly being called into question by forecasters and the Eulerian methods that accompanied them. As such, they were viewed as contributions alien to the world of forecasters since the authors of these papers—those who considered weather forecasting as a problem of physics—were not among those who prepared predictions. But they also aroused the mistrust of forecasters, because the authors themselves put into question the utility or applicability of their models. Nowhere did this fact become more obvious than in Shaw's flow model, whose value was dismissed by Shaw himself in regarding it as a superficial representation of a cyclone, especially after obtaining the results of balloon soundings which seemed to point to the existence of symmetrical flows.[120]

However, other academic meteorologists were to challenge seriously isobaric geometry and the Eulerian approach to forecasting as a whole. Certainly the most convincing and successful among them was Norwegian physicist Vilhelm Bjerknes, who in 1904 promoted a rational program for

120 Ludlam (1967, p. 32); Volkert (1999, p. 16).

scientific weather prediction. His program, based on the new physical hydrodynamics, would lay the foundations of the ensuing Leipzig and Bergen schools of meteorology.

The son of a professor of mathematics who had formulated a hydrodynamic analogy of the electromagnetic field, Bjerknes followed in his father's footsteps, studying physics at the University of Christiania.[121] He then conducted research on electromagnetic waves in Bonn, under the supervision of Heinrich Hertz. After completing his doctorate with a dissertation on electrodynamics, he became professor of mechanics and mathematical physics in Stockholm, where he devoted himself to the study of geophysical hydrodynamics. In 1898, he developed a circulation theorem as a result of his lengthy investigations on classical hydrodynamics, and soon sought its application to atmospheric physics.[122] This theorem had a twofold importance, as on one hand it showed the way circulation evolves in atmospheric fluids and, on the other hand, it symbolized a shift in Bjerknes's research line from the electromagnetic theory and pure hydrodynamics to atmospheric dynamics.[123]

The transition of these research lines brought with it the rapprochement between hydrodynamics and thermodynamics. During his Stockholm time from 1893 to 1907, Bjerknes worked and often discussed his ideas with his colleague Nils Ekholm, who was studying the role of warm and cold currents of air for the propagation of cyclones. Ekholm had already distinguished the difference between barotropic and baroclinic conditions, which inspired him to applying his circulation theorem to the atmosphere. Bjerknes realized that while the motion of classical ideal fluids is homogeneous and incompressible (barotropic conditions), the motion of real fluids depends on differences of density due to changes of temperature and humidity (baroclinic conditions). Or, to put it another way, while in a barotropic atmosphere the density of fluids is only a function of the barometric pressure (this ideal condition is given when the flow is dry and laminar), in a baroclinic atmosphere this density is determined by factors such as solar thermal energy, the Earth's rotation, and the latent energy absorbed and released by phase changes. By applying his circulation theorem to the atmosphere, Bjerknes

121 On Vilhelm Bjerknes, see Devik, Bergeron, and Godske (1963); Sverdrup (1951); Pihl (1981); Volkert (2005, pp. 357–58); and Jewell (2017).
122 Bjerknes (1898). As a result of these investigations, from 1900 to 1902 Bjerknes published his two-volume work *Vorlesungen über hydrodynamische Fernkräfte nach C. A. Bjerknes' theorie*. Leipzig: J.A. Barth.
123 Bjerknes extended Helmholtz's theorem on the conservation of the vorticity in incompressible fluids. The result was a theorem that gives the rate of variation of the vorticity for a compressible fluid. This theorem was formulated in a paper published by the *Proceedings of the Royal Swedish Academy of Sciences*, entitled 'On a Fundamental Theorem of Hydrodynamics and Its Applications Particularly to the Mechanics of the Atmosphere and the World's Oceans' (translated from German original). See Thorpe, Volkert, and Ziemianski (2003); Kutzbach (1979, pp. 159–71).

not only extended Lord Kelvin's theorem on fluid motions from barotropic to baroclinic conditions, but also built a bridge between hydrodynamics and thermodynamics.[124]

A few years later, Bjerknes launched an ambitious program for the prediction of the weather by using the equations of hydrodynamics and thermodynamics.[125] In 1904, he published a landmark article in *Meteorologische Zeitschrift*, entitled 'Das Problem der Wettervorhersage, betrachtet vom Standpunkte der Mechanik und der Physik' (Weather Forecasting as a Problem in Mechanics and Physics). In its first paragraphs, he established two 'necessary and sufficient conditions for the rational solution' of weather forecasting problem:

1   A sufficiently accurate knowledge of the state of the atmosphere at the initial time and
2   A sufficiently accurate knowledge of the laws according to which one state of the atmosphere develops from the other.[126]

Bjerknes called these two diagnosis and prognosis, respectively. In his view, the diagnosis required adequate observations if one aimed at ascertaining the three-dimensional state of the atmosphere. He hoped that the time would soon come when, with the aid of upper air and marine observations, a complete diagnosis of the state of the atmosphere would be available. The prognosis involved solving the differential equations that express the evolution of the seven variables describing the atmosphere: pressure, temperature, humidity, density, and the three components of the wind velocity. These fundamental equations were seven: the continuity equation, the three hydrodynamic equations of motion, the equation of state, and the first and second laws of thermodynamics on the conservation of energy. However, he soon realized that they could not be easily solved by exact analytical integrations; the computing work was simply too arduous. He then changed strategy. He would first draw several charts that represent the distribution of the seven variables and the initial state of the atmosphere. Simplified graphical methods could be used to draw additional charts showing the state of the atmosphere in the very short term (say, a few hours). He would repeat this

124 According to Kelvin's theorem, vortex motions in incompressible fluids are conserved, and the density in a heterogeneous fluid depends only on pressure. Friedman (1989, p. 19); Gramelsberger (2009, pp. 669–70).
125 In an article published in *Meteorologische Zeitschrift* in 1902, Bjerknes (1902, p. 108) advanced his plans:

> Each task of theoretical mechanics is, when it is placed in direct form, a prognostic, just like the most well-known task of practical meteorology. The goal is to predict the dynamic and physical condition of the atmosphere at a later time, if at an earlier given time, this condition is well known.
>
> —as translated by Fleming (2016, pp. 23–24)

126 Bjerknes (1904, p. 1).

technique until the expected length of prediction was attained. Although his method only worked for very simple, idealized flows, he was optimistic:

> It may be possible some day, perhaps, to utilize a method of this kind as the basis for a daily practical weather service. But however that may be, the fundamental scientific study of atmospheric processes sooner or later has to follow a method based upon the laws of mechanics and physics.[127]

Or, as Bjerknes even suggested, one could always resort to mechanical integration of the differential equations, either through graphical methods or through numerical computation, when they were not analytically integrable. Simply put, Bjerknes's program made all the empiricist methods of weather forecasting vanish, replacing them with a precise, *deterministic* model of scientific prediction.

Bjerknes's program for the scientific prediction of the weather brought the illusion of determinism many physicists had long desired. Yet beneath this veneer of determinism certain nuances should be recognized, especially in the way in which the accurateness of knowledge was understood. Most of these physicists would have embraced Laplace's famous maxim of universal determinism—that the laws of mechanics could accurately predict the future state, position, and motion of all mass particles, provided that the mechanical problem was previously reduced to stating their present state, position, and motion. However, the 'sufficiently accurate knowledge' of the natural laws— which formed the bedrock of Bjerknes's criteria—was not matched with the Laplacian ideal, the determination of future states by a perfectly precise observation of nature. Bjerknes was well aware of the difficulties of measuring with precision the initial state of the atmosphere over a region; hence he demanded calculations of 'sufficient accuracy' rather than perfect knowledge of the laws. Bjerknes took, therefore, a quasi-deterministic, neo-Laplacian view in addressing the prediction of future atmospheric states.[128]

As a result of the interest aroused by his program, Bjerknes was invited to lecture at Columbia University in 1905. He delivered a series of lectures on 'Fields of force', and subsequently he received an invitation from Washington. The issue of the program for scientific prediction was naturally of great interest to the U.S. Weather Bureau, and, under its auspices, he gave a lecture on 'Weather forecasting as a problem in mechanics and physics'.[129] Accompanied by the charts constructed by his student Johan W. Sand-

---

127 Ibid., p. 4.
128 The perception that Bjerknes developed a strictly deterministic model of weather was widespread among some meteorologists at the time, even among historians such as Coen (2007, pp. 290–01), who holds that his manner 'struck his Austrian colleagues as *dogmatic*', and, therefore, contrary to the liberal Austrian tradition of probabilistic reasoning.
129 For a summary of this lecture, delivered at the Carnegie Institution because the U.S. Weather Bureau had not means to house large lectures, see Fleming (2016, pp. 29–31).

ström, and by an audience eager to learn about his plans—'if at any time a lecture by me has been a success, it happened this time', he later admitted—Bjerknes displayed a new way of representing the weather.[130] Instead of tracing isobars and wind patterns on a synoptic map (which was in essence the Eulerian view, focused on air moving past fixed coordinates), forecasters would follow the *trajectories* of parcels of air, the paths of particles moving through the airflow. Here, paths were the true graphical representations of the fields of force on which atmospheric motions depended. Bjerknes described them in almost four-dimensional terms: while streamlines, or lines of airflow, represented air motion in space for a given fixed time, trajectories were a function of air motion in space and time. Streamlines provided a static picture of horizontal air movement (i.e. wind velocities over a region); in contrast, trajectories, Bjerknes argued, would be curves tracing successive locations of an air parcel.

The concept of an air trajectory, first advanced by Shaw and Lempfert in their 1906 work *The Life History of Surface Air Currents*, was in many regards revolutionary.[131] In fact, it essentially broke with a long practical tradition in weather forecasting that ran through Le Verrier and Marié-Davy and that was predicated on the study of motion fields with respect to a fixed reference frame. As Shaw and Lempfert noted in their memoir, by applying the method devised by them one could follow 'the actual path described by an isolated portion of air moving along the surface'.[132]

Bjerknes's new method of synoptic representation borrowed Shaw and Lempfert's concept of air trajectory and focused on the air motion as an object of analysis. Galton had already constructed streamlines by drawing on very limited surface data in his 1863 work *Meteorographica*. For Bjerknes, however, it was not sufficient to represent the two-dimensional wind field by a number of arrow-headed lines, but rather a continuous function amenable to graphic algebra and differentiation was needed. Thus, he proposed two continuous representations: streamlines depicting lines of flow and curves of equal wind intensity (now called *isotaches*). Indeed, Bjerknes began to produce these types of representations with the publication of his two volumes of *Dynamic Meteorology and Hydrography* (1910/11), in particular the second part, *Kinematics*, which was written with his students Theodor Hesselberg and Olaf Devik.[133]

130 Quoted by Devik, Bergeron, and Godske (1963, p. 14).
131 Shaw and Lempfert (1906). This work is a continuation of Shaw's study on air trajectories in storms presented in 1903 in his article 'The Meteorological Aspects of the Storm of February 26–27, 1903'. *Quarterly Journal of the Royal Meteorological Society*, 29, 233–58. According to Kutzbach (1979, p. 180), quoting in turn A. Defant, it was Köppen who first expressed the idea to analyze air trajectories in cyclones in 1877.
132 Shaw and Lempfert (1906; reprint. 1955, p. 16).
133 The first part, *Statics*, was written with his assistant and collaborator at the Carnegie Institution, J.W. Sandström. Although published in 1910, the manuscript was submitted in 1907. See: Bjerknes and Sandström (1910). Bjerknes had already traced streamlines and

Again, Bjerknes's methodology also represented the most developed application of computational and synoptic techniques to the problem of weather forecasting, paving the way to the concept of a cold front which was so central to the Bergen school of meteorology. Although his atmospheric kinematics owed much to Shaw and Lempfert's concept of air trajectory and Helmholtz's hydrodynamics, at the heart of his program was an attempt to provide practical rules for weather forecasts.[134] His analyses on the charts of motion, such as the one showed in Figure 6.6 depicting weather conditions for the ground level over the United States, aimed to do that. The streamlines, drawn by continuous curves with arrowheads, enable the wind direction to be derived by interpolation at any point of the chart. Isotachs, traced by round and oval curves, represent the wind speed. The spiralling of streamlines around Minnesota indicates the existence of a cyclonic centre. Moreover, the figure shows certain singularities and lines of convergence where the streamlines come together. Areas of convergence denote areas of marked vertical motion and probable precipitation. In short, streamlines would become graphic tools to analyze hydrodynamic processes and introduce new concepts. As geophysicist Arnt Eliassen noted in his review of Bjerknes's studies of atmospheric motions, 'the confluence lines are undoubtedly the seeds of the Bergen school fronts'.[135]

Bjerknes's forecasting program found great (though not universal) appeal within meteorological circles, to the extent that the Carnegie Institution awarded him a grant which enabled him to form and maintain his own research team wherever he stayed (including his son Jacob Bjerknes). In 1912, he accepted the position of professor and director of the new Geophysical Institute in Leipzig—a step that would subsequently lead to the development of the Leipzig school of meteorology.[136] His three Carnegie assistants also pursued a successful career: Sandström and Hesselberg became directors of the Swedish and Norwegian Meteorological Institutes, respectively, and Devik became a world-leading hydrologist. Finally, Bjerknes's program

---

curves of intensity of the air motion (for 28 November 1905) in a paper on 'Synoptical representation of atmospheric motions' published in 1910.

134 For the influence of Helmholtz's hydrodynamic studies on Bjerknes's contributions to dynamical meteorology and weather prediction, including the importance of surface discontinuities in the atmosphere, see Darrigol (2008, p. 177).

135 Eliassen (1999, p. 7).

136 In Leipzig, Bjerknes and his team laid the foundations for addressing the issue of weather prediction on the basis of physical theories by the study of the three-dimensional structure of cyclones. Their aerological (or upper-atmospheric) observations and their calculations of the future states of the atmosphere by hydrodynamic and thermodynamic equations paved the way to the accomplishments achieved in Bergen. Comparing the breakthroughs in the two places, Bjerknes (1938, p. 60) noted: 'However insignificant the contribution of the Leipzig School to the final success in Bergen may appear, this contribution is not only great but was *absolutely necessary for the final success*'—as translated by Kutzbach (1979, p. 208).

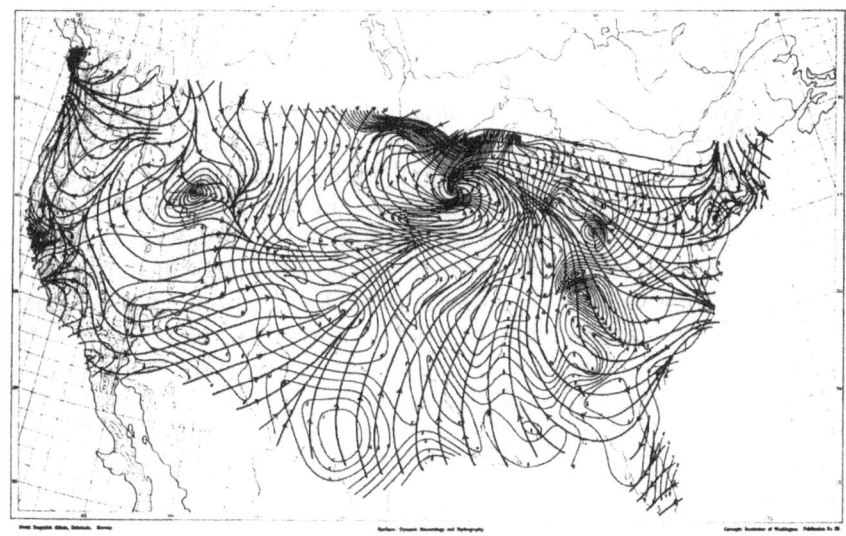

Figure 6.6 Streamlines and curves of equal wind intensity (isotachs) for the ground
   level over the United States on 28 November 1905.
Source: Bjerknes et al. (1911, plate 38).

also sowed seeds in many of the conceptions and investigations developed
by the members of the Bergen school of meteorology, including his son
Jacob's finding that the origin of cyclones was associated with sloping fronts
separating different masses of cold and warm air.[137]

   But even regardless of the recognition of Bergen meteorologists' contribution to dynamic meteorology, there is one characteristic of Bjerknes's
legacy that sets it completely apart from the Eulerian representations of
storm systems. His models opened up new possibilities for dealing with
the issue of the three-dimensional structure of storm systems. It is no coincidence that his models were advanced just at the time when the exploration of the free upper atmosphere started. Upper air direct and indirect
observations are found in abundance in Germany, France, and the United
States at the turn of the century. German meteorologist Richard Assmann
organized systematic observations from manned balloons in the 1890s.
Unmanned paper balloons and ballonsondes with automatic recording instruments were used by Gustave Hermite, Georges Besançon, and Léon
Teisserenc de Bort in France. Kites techniques were tried by A. Lawrence

---

137 The first frontal cyclone model was proposed by Jacob Bjerknes, who drew on Margules's
   generalization of Helmholtz's formula for the surface of discontinuity. See Ludlam
   (1967, pp. 32–38), Khrgian (1970, pp. 215–16), Kutzbach (1979, pp. 206–18), Jewell (1981,
   pp. 824–30), Friedman (1978, 1989), Volkert (1999), Monmonier (1999, pp. 57–69).

Rotch at the Blue Hill Observatory, Massachusetts.[138] The global study of the upper air (or 'aerology', a term coined by Köppen in 1906) provided field data for the reconstruction of the three-dimensional structure of the cold-front model.[139]

In brief, following Bjerknes's peculiar journey through the fields of fluid motion, we began with the circulation theorem, outlined his weather forecasting program, and ended with streamlines and trajectories. Bjerknes built a bridge between classical hydrodynamics and classical thermodynamics, and incorporated the atmosphere and the oceans into the domains of these two disciplines. More importantly for our purposes, he identified the streamline as a fundamental tool depicting lines of flow, and inaugurated a powerful approach—a Lagrangian approach—to weather prediction in which trajectories, paths, and fronts became the new forecasting tools. Together with Shaw, Margules, and Exner, he marked the beginning of the end of the Eulerian hegemony.

Conclusion

The basic tool used by national meteorological institutes in forecast practice was the synoptic weather map. During its first five or six decades, both the expertise and the empirical experience gained in the use of synoptic maps virtually constituted the sole basis of weather forecasting.

The handling of weather maps took many forms. In the small sample showed in this chapter, several kinds of handling are involved. In order of appearance, they imply the use of verbal algorithms (Scott), the search for statistical regularities in cyclonic path types (Ley, Brounov, Shpindler), weather types associated with storm tracks (Köppen), local forecasting guidelines (Moore), the identification and classification of weather types (Garrigou-Lagrange, Garriott, Brown), and empirical rules for methodical forecasting (Guilbert, Finley). Although their techniques, aims, and objects differed considerably, all of them shared the same key principle: the weather *travels* through the synoptic maps. Yet the similarities do not end there; each of them is predicated on the assumption that the character of the weather is largely determined by the atmospheric pressure distribution.

The epitome of the use of this Eulerian approach to weather forecasting was certainly Abercromby's configurations of isobars. Abercromby's isobaric geometry was no instance of sporadic happy thought arisen after FitzRoy's death, but instead was the product of a mindset and approach

---

138 For a review of these achievements, see Labitzke and Loon (1999, Chapter 1) and Hergesell (1927, pp. 73–80). See also Anduaga (2000).

139 For the emergence of aerology as a research program in Imperial Germany as a consequence of trans-imperial networks and field work, as well as Köppen's leadership in this program, see the suggestive article of Wille (2017).

that were rooted at the Meteorological Office as to the dominant role of the baric field in weather changes. As was shown, the Office adopted his model instead of Ley's three-dimensional cyclone model, and his isobaric forms had a great influence on the practice of weather forecasting in Britain. His success was lasting but misleading and unproductive. During next forty years, forecast methodology remained stagnant, trapped by its own shortcomings and limitations—the lack of upper air observations and its detachment about physical theories are only two examples. As meteorologist Richard J. Reed rightly stated, 'the inherent limitations of the empirical method ruled out the hope of significant further improvement in forecast skill'. By 1920, 'forecasting had reached a dead end'.[140]

Bjerknes's program for scientific weather forecasting marked a turning point in the history of meteorological prediction, proving to be the transitional moment when the empirical two-dimensional thinking about the travelling centres of high and low pressure was challenged by one anchored in mathematico-physical science. As I showed, the attempts to seek methods based on hydrodynamic and thermodynamic theories were not unknown before 1904. Here the names of Shaw, Lempfert, Margules, and Exner must be mentioned. However, for different reasons their impact on immediate forecast practice was slight. In contrast to this, Bjerknes's investigations confirmed for many meteorologists the principle that the path or trajectory of parcels of air should be the object par excellence of analysis, not the baric field. Furthermore, he taught them that paths, streamlines, and fronts could serve as tools for ascertaining the motion and behaviour of cyclonic systems—tools which could be also used to scientifically predict weather changes.

Nevertheless, in this turning point the states played their part. In fact, it was the role played by different official institutes in Leipzig, Vienna, and London in promoting theoretical investigations on the atmosphere that established the legitimacy of the new scientific weather prediction—note that Bjerknes was not alien to the world of forecasters, and was able to implement his ideas in the meteorological institutes in Norway and Sweden. Now, if national meteorological institutes will permit the implementation of scientific forecast methods from the 1920s, then why not they followed and permitted the same path in earlier decades by combining the public service with the promotion of physical investigations? The answer lies largely in governments' interests. As is shown in the next chapter, governments, through sponsorship of central meteorological institutes, would help to not only shape a forecast practice based on mass production of weather maps but also define the character and direction of the research itself.

---

140 As stated by Reed (1977, p. 391) in his Bjerknes Memorial Lecture, entitled 'The Development and Status of Modern Weather Prediction'.

References

Abbe, Cleveland, 1873. 'Rules for the Preparation of Synopses and Probabilities'. *Annual Report of the Chief Signal Officer.* Washington: Government Printing Office, 379–83.

Abbe, Cleveland, 1890a. *Preparatory Studies for Deductive Methods in Storm and Weather Predictions.* Washington, DC: Government Printing Office, Annual Report of the Chief Signal Officer for 1889. Appendix 15.

Abbe, Cleveland, 1890b. 'Recent Progress in Dynamic Meteorology'. *Annual Report: 1888*, Smithsonian Institution, 355–424.

Abbe, Cleveland, 1893. 'The Needs of Meteorology'. *American Meteorological Journal*, 9, 560–62.

Abbe, Cleveland, 1895. 'The Needs of Meteorology'. *Science*, 1 (7), 181–82.

Abbe, Cleveland, 1901. 'The Physical Basis of Long-Range Weather Forecasts'. *Monthly Weather Review*, 29, 551–61.

Abbe, Cleveland, 1912. 'The Obstacles to the Progress of Meteorology'. *Journal of the Franklin Institute*, 173, 55–71.

Abbe, Truman, 1955. *Professor Abbe and the Isobars. The Story of Cleveland Abbe, America's First Weatherman.* New York: Vantage Press.

Abercromby, Ralph, 1877. 'Is Meteorology a Science?' *Nature*, 15, 510.

Abercromby, Ralph, 1883. 'Weather Prognostics and Weather Types'. *Nature*, 28, 330–34.

Abercromby, Ralph, 1885. *Principles of Forecasting by Means of Weather Charts.* London: Printed for H.M. Stationery Off.

Abercromby, Ralph, 1887. *Weather: A Popular Exposition on the Nature of Weather Changes from Day to Day.* New York: Appleton.

Anduaga, Aitor, 2000. *La Aerología o el estudio de las altas capas de la atmósfera en España en el primer tercio del siglo XX.* Madrid: Instituto Nacional de Meteorología.

Anduaga, Aitor, 2012. *Meteorología, ideología y sociedad en la España contemporánea.* Madrid: Consejo Superior de Investigaciones Científicas/Agencia Estatal de Meteorología (AEMET).

Bergeron, Tor, 1981. 'Synoptic Meteorology: An Historical Review'. *Pure and Applied Geophysics*, 119, 443–73.

Besson, Louis, 1904. 'Attempts at Methodical Forecasting of the Weather'. *Annuaire de la Société Météorologique de France*, 52, 93–97.

Bigelow, Frank Hagar, 1897. *Storms, Storm Tracks, and Weather Forecasting.* Washington, DC: Weather Bureau (Weather Bureau Bulletin, no. 20).

Bjerknes, Vilhelm, 1898. 'Über einen hydrodynamischen Fundamentalsatz und seine Anwendung besonders auf die Mechanik der Atmosphäre und des Weltmeeres'. *Kongliga Svenska Vetenskaps-Akademiens Handlingar*, 31, 1–35.

Bjerknes, Vilhelm, 1900–02. *Vorlesungen über hydrodynamische Fernkräfte nach C.A. Bjerknes' theorie.* Leipzig: J.A. Barth.

Bjerknes, Vilhelm, 1902. 'Cirkulation relativ zu der Erde'. *Meteorologische Zeitschrift*, 19, 97–108.

Bjerknes, Vilhelm, 1903. 'The Meteorological Aspects of the Storm of February 26–27'. *Quarterly Journal of the Royal Meteorological Society*, 29, 233–58.

Bjerknes, Vilhelm, 1904. 'Das Problem der Wettervorhersage, betrachtet vom Standpunkte der Mechanik und der Physik'. *Meteorologische Zeitschrift*, 21, 1–7.

Bjerknes, Vilhelm, 1910. 'Synoptical Representation of Atmospheric Motions'. *Quarterly Journal of the Royal Meteorological Society*, 36 (155), 267–86.

Bjerknes, Vilhelm, 1938. 'Leipzig-Bergen'. *Zeitschrift für Geophysik*, 14, 50–62.

Bjerknes, Vilhelm, Hesselberg, Theodor, and Devik, Olaf, 1911. *Dynamic Meteorology and Hydrography. Part II, Kinematics*. Washington, DC: Carnegie Institution.

Bjerknes, Vilhelm, and Sandström, Johan W., 1910. *Dynamic Meteorology and Hydrography. Part I, Statics*. Washington, DC: Carnegie Institution.

Blasius, William, 1875. *Storms: Their Nature, Classification and Laws. With the Means of Predicting Them by Their Embodiments: The Clouds*. Philadelphia, PA: Porter & Coates.

Blasius, William, 1887. 'The Signal Service Bureau. Its Methods and Results'. *Proceedings of the American Philosophical Society*, 24 (125), 179–83.

Bradford, Marlene, 2001. *Scanning the Skies. A History of Tornado Forecasting*. Norman: University of Oklahoma Press.

Brown, W.V., 1901. 'A Proposed Classification and Index of Weather Maps as an Aid in Weather Forecasting'. *Monthly Weather Review*, 29 (12), 547–48.

Brunhes, Bernard, 1909. 'L'évolution des dépressions barométriques et les règles de prévision de M. Guilbert'. *Revue générale des sciences pures et appliquées*, 20, 393–406.

Burton, James, 1988. 'The History of the British Meteorological Office to 1905'. PhD diss., The Open University.

Coen, Deborah R., 2007. *Vienna in the Age of Uncertainty. Science, Liberalism, and Private Life*. Chicago: The University of Chicago Press.

Coen, Deborah R., 2010. 'Climate and Circulation in Imperial Austria'. *The Journal of Modern History*, 82, 839–75.

Cox, John D., 2002. *Storm Watchers: The Turbulent History of Weather Prediction from Franklin's Kite to El Niño*. Hoboken, NJ: John Wiley & Sons, Inc.

Darrigol, Olivier, 2008. *Worlds of Flow. A History of Hydrodynamics from the Bernoullis to Prandtl*. Oxford: Oxford University Press.

Davies, Huw C., 2001. 'Vienna and the Founding of Dynamical Meteorology'. In Christa Hammerl, Wolfgang Lenhardt, Reinhold Steinacker, and Peter Steinhauser eds., *Die Zentralanstalt für Meteorologie und Geodynamik 1851–2001. 150 Jahre Meteorologie und Geophysik in Österreich*. Graz: Leykam, 301–12.

Devik, Olaf, Bergeron, Tor, and Godske, Carl Ludvig, 1963. 'In Memory of Vilhelm Bjerknes on the 100th Anniversary of his Birth'. *Geofysiske Publikasjoner*, 24, 6–37.

Douglas, Charles Kenneth MacKinnon, 1952. 'The Evolution of 20th-Century Forecasting in the British Isles'. *Quarterly Journal of the Royal Meteorological Society*, 78 (335), 1–21.

Ekholm, Nils Gustaf, 1907. 'Die Wetterregeln des Herrn G. Guilbert'. *Meteorologische Zeitschrift*, 24, 326–28.

Eliassen, Arnt, 1999. 'Vilhelm Bjerknes's Early Studies of Atmospheric Motions and Their Connection with the Cyclone Model of the Bergen School'. In Melvyn A. Shapiro and Sigbjorn Gronas eds., *The Life Cycles of Extratropical Cyclones*. Boston, MA: American Meteorological Society, 5–14.

Exner, Felix M., 1908. 'Über eine erste Annäherung zur Vorausberechnung synoptischer Wetterkarten'. *Meteorologische Zeitschrift*, 25, 57–67. Translated from German as: *A First Approach towards Calculating Synoptic Forecast Charts*. Dublin: Met' Eireann, 1995, with a Biographical Note on Exner by Lisa Shields and an Introduction by Peter Lynch.

Exner, Felix M., 1917. *Dynamische Meteorologie*. Leipzig. B. D. Teubner.

Fassig, Oliver L., 1907. 'Guilbert's Rules for Weather Prediction'. *Monthly Weather Review*, 35 (5), 210–11.

Ficker, H. Von, 1927. 'Das meteorologische System von Wilhelm Blasius'. *Sitzungsberichte der Preussischen Akademie der Wissenschaften*, 33, 248–67.

Finley, John P., 1884a. 'Tornado Predictions'. *American Meteorological Journal*, 1, 85–88.

Finley, John P., 1884b. *Character of Six Hundred Tornadoes*. U.S. Signal Service, Professional Papers No. 7.

Finley, John P., 1884c. 'Intelligence from American Scientific Stations'. *Science*, 3, 766–68.

Finley, John P., 1886. 'Tornado Study: Its Past, Present, and Future'. *Journal of the Franklin Institute*, 71, 241–62.

Fleming, James Rodger, 1990. *Meteorology in America, 1800–1870*. Baltimore: Johns Hopkins University Press.

Fleming, James Rodger, 2016. *Inventing Atmospheric Science. Bjerknes, Rossby, Wexler, and the Foundations of Modern Meteorology*. Cambridge, MA: The MIT Press.

Fortak, Heinz C., 2001. 'Felix Maria Exner und die österreichische Schule der Meteorologie'. In Christa Hammerl, Wolfgang Lenhardt, Reinhold Steinacker and Peter Steinhauser eds., *Die Zentralanstalt für Meteorologie und Geodynamik 1851–2001. 150 Jahre Meteorologie und Geophysik in Österreich*. Graz: Leykam, 354–86.

Friedman, Robert Marc, 1978. *Vilhelm Bjerknes and the Bergen School of Meteorology, 1918–1923: A Study of the Economic and Military Foundations for the Transformation of Atmospheric Science*. Baltimore, MD: Johns Hopkins University Press.

Friedman, Robert Marc, 1989. *Appropriating the Weather: Vilhelm Bjerknes and the Construction of a Modern Meteorology*. Ithaca, NY: Cornell University Press.

Galliot, Michel, 1995. '1880–1910: L'âge d'or de la Météorologie à Limoges'. *La Météorologie*, special issue, 17–21.

Galway, Joseph G., 1985. 'J.P. Finley: The First Severe Storms Forecaster'. *Bulletin of the American Meteorological Society*, 66 (11), 1389–95, 1506–10.

Garrigou-Lagrange, Paul, 1893. 'Sur la prévision du temps et sur l'enchaînement des situations atmosphériques'. *Annuaire de la Société Météorologique de France*, 41, 166–93.

Garriott, Edward B., 1901. 'Classified Weather Types'. *Monthly Weather Review*, 29, 548–49.

Gaster, Frederic, 1896. 'Weather Forecasts and Storm Warnings: How they are Prepared and Disseminated'. *Quarterly Journal of the Royal Meteorological Society*, 22 (99), 212–28.

Gilbert, G.K., 1884. 'Finley's Tornado Predictions'. *American Meteorological Journal*, 1, 166–72.

Gold, Ernest, 1920. 'Dr. Max Margules [Obituary]'. *Nature*, 106, 286–87.

Gold, Ernest, 1945. 'William Napier Shaw, 1854–1945'. *Obituary Notices of Fellows of the Royal Society*, 5, 203–30.

Gramelsberger, Gabriele, 2009. 'Conceiving Meteorology as the Exact Science of the Atmosphere: Vilhelm Bjerknes's Paper of 1904 as a Milestone'. *Meteorologische Zeitschrift*, 18, 669–73.

Greene, Mott T., 2008. 'Köppen, Wladimir Peter'. In Noretta Koertge ed., *New Dictionary of Scientific Biography*. Detroit: Scribner, 4, 152–55.

Grisollet, H., 1944. 'Louis Besson (1872–1944)'. *L'Astronomie*, 58–59, 99–101.

Guilbert, Gabriel, 1887a. 'De la prévision des nuages et des successions nuageuses'. *Annuaire de la Société météorologique de France*, 35, 162–67.

Guilbert, Gabriel, 1887b. 'Etude sur les dépressions secondaires du golfe de Gênes et observations relatives à leur prévision'. *Annuaire de la Société météorologique de France*, 35, 328–35.

Guilbert, Gabriel, 1887c. *Études météorologiques. De la prévision des nuages et des tempêtes*. Caen: Impr. de V.A. Domin. (Extracts from the *Annuaire de la Société météorologique de France*, April–May, 1887).

Guilbert, Gabriel, 1895. 'De la prévision du temps par l'observation simultanée du baromètre et des nuages'. *Compte Rendu, Congrès de Caen, Association française pour l'avancement des sciences*. Paris, 377–84.

Guilbert, Gabriel, 1907. 'Principles of Forecasting the Weather'. *Monthly Weather Review*, 34 (372), 211–12.

Guilbert, Gabriel, 1909. *Nouvelle méthode de prévision du temps*. Paris: Gauthier-Villars.

Guilbert, Gabriel, 1925. *Enciclopedia universal ilustrada europeo-americana*. Barcelona: Espasa-Calpe, 27 (1), 231.

Hamblyn, Richard, 2017. *Clouds: Nature and Culture*. London: Reaktion Books Ltd.

Hammerl, Christa, Lenhardt, Wolfgang, Steinacker, Reinhold, and Steinhauser, Peter eds., 2001. *Die Zentralanstalt für Meteorologie und Geodynamik 1851–2001. 150 Jahre Meteorologie und Geophysik in Österreich*. Graz: Leykam.

Hann, Julius von, 1901. *Lehrbuch der Meteorologie*. Leipzig: C.H. Tauchnitz.

Henry, Alfred Judson, 1906. 'Weather Forecasting from Synoptic Charts'. *Journal of the Franklin Institute*, 162, 297–316.

Henry, Alfred Judson, 1917. 'Memoir of Cleveland Abbe'. *Annals of the Association of American Geographers*, 7 (1), 61–67.

Henry, Alfred Judson, 1924. 'Frank Hagar Bigelow, 1851–1924'. *Monthly Weather Review*, 52, 165–66.

Henry, Alfred Judson, 1930. *Weather Forecasting from Synoptic Charts*. Washington, DC: United States Department of Agriculture.

Henry, Alfred Judson, et al., 1916. *Weather Forecasting in the United States*. Washington, DC: Government Printing Office.

Hergesell, Hugo von, 1927. 'The Development of Aerology. A Retrospect and a Glance into the Future'. *Quarterly Journal of the Royal Meteorological Society*, 53, 73–80.

'How a Weather Map Is Made'. *Scientific American*, 1900, 82 (3), 38.

Humphreys, William Jackson, 1919. 'Biographical Memoir Cleveland Abbe, 1838–1916'. *Biographical Memoirs*, Vol. VIII. Washington, DC: National Academy of Sciences, 469–508.

Iranzo, Manuel, 1912. *Service météorologique de la Fédération Agraire du Levant de l'Espagne. Les règles du Guilbert et leur application*. Montpeller: Société Astronomique Flammarion.

Iranzo, Manuel, 1915. 'Reproducción de la ponencia en el tema Meteorología agrícola, presentado al Congreso internacional de agricultura de Gante'. In *Asociación Española para el Progreso de las Ciencias. Congreso de Madrid*. Madrid: Imprenta de Eduardo Arias, 265–72.

Jewell, Ralph, 1981. 'The Bergen School of Meteorology: The Cradle of Modern Weather–Forecasting'. *Bulletin of the American Meteorological Society*, 62, 824–30.

Jewell, Ralph, 2017. *The Weather's Face: Features of Science in the Story of Vilhelm Bjerknes and the Bergen School of Meteorology*. Bergen: Fagbokforlaget.

Khrgian, Aleksandr Khristoforovich, 1970. *Meteorology: A Historical Survey*, 2nd ed. Jerusalem: Israel Program for Scientific Translations.

Kington, John, 1999. 'Meteorologist's Profile – William Clement Ley'. *Weather*, 54 (6), 166–72.

Köppen, Wladimir, 1872. 'Die Aufeinanderfolge der unperiodischen Witterungserscheinungen nach der Wahrscheinlichkeitsrechnung untersucht'. *Zeitschrift der Osterreichischen Gesellschaft für Meteorologie*, 7, 369–75 (Originally reproduced in *Repertorium für Meteorologie*, 1872, 2, 189–238).

Köppen, Wladimir, 1873. 'Uber Mehrjährige Perioden der Witterung, Insbesondere Über die 11jährige Periode der Temperatur'. *Zeitschrift der Osterreichischen Gesellschaft für Meteorologie*, 8, 241–48, 141–50.

Köppen, Wladimir, 1874. 'Über die Abhängigkeit des klimatischen Charakters der Winde von ihrem Ursprünge'. *Repertorium für Meteorologie*, 4 (4), 1–15.

Kutzbach, Gisela, 1974. 'Margules, Max'. In Charles Coulston Gillispie ed., *Dictionary of Scientific Biography*. New York: Charles Scribner's Sons, 9, 107–08.

Kutzbach, Gisela, 1979. *The Thermal Theory of Cyclones: A History of Meteorological Thought in the Nineteenth Century*. Boston, MA: American Meteorological Society.

Labitzke, Karin G., and Loon, Harry, 1999. *The Stratosphere: Phenomena, History, and Relevance*. Berlin: Springer.

Lauscher, Friedrich, and Skoda, Georg, 1981. 'Zum Gedenken an Felix M. Exner'. *Wetter und Leben*, 33, 94–102.

Ley, William Clement, 1872. *Laws of the Winds Prevailing in Western Europe*. London: Stanford.

Ley, William Clement, 1876. 'Results of an Inquiry into the Inclination of Winds towards the Lower Isobars'. *Journal of the Scottish Meteorological Society*, 4, 66–72.

Ley, William Clement, 1877. 'The Relation between the Upper and Under Currents of the Atmosphere around Areas of Barometric Depressions'. *Quarterly Journal of the Royal Meteorological Society*, 3 (24), 437–48.

Ley, William Clement, 1879a. 'On the Inclination of the Axis of Cyclones'. *Quarterly Journal of the Royal Meteorological Society*, 5, 167–77.

Ley, William Clement, 1879b. 'Clouds and Weather Signs'. In *Modern Meteorology: A Series of 6 Lectures Delivered under the Auspices of The Meteorological Society in 1878*. London: E. Stanford, 102–36.

Ley, William Clement, 1880. *Aids to the Study and Forecast of Weather*. London: Her Majesty's Stationery Office.

Ley, William Clement, 1883. *Note on a Proposed Scheme for Observations of the Upper Clouds*. Leicester, for private circulation.

Ludlam, Frank Henry, 1967. *The Cyclone Problem: A History of Models of the Cyclonic Storm*. London: Imperial College of Science and Technology.

Ludlum, David MacWilliams, 1970. *Early American Tornadoes, 1586–1870*. Boston, MA: American Meteorological Society.

Lynch, Peter, 2001. *Max Margules and his Tendency Equation*. Dublin: Met Éireann. It includes a translation of Margules' 1904 paper: 'On the Relationship between Barometric Variations and the Continuity Equation'.

Lynch, Peter, 2003. 'Margules' Tendency Equation and Richardson's Forecast'. *Weather*, 58, 186–92.

Majdik, Zoltan P., Platt, Carrie Anne, and Meister, Mark, 2011. 'Calculating the Weather: Deductive Reasoning and Disciplinary *Telos* in Cleveland Abbe's Rhetorical Transformation of Meteorology'. *Quarterly Journal of Speech*, 97 (1), 74–99.

Margules, Max, 1893. 'Luftbewegungen in einer rotierenden Sphäroidschale'. *Sitzungsberichte der Kaiserliche Akademie der Wissenschaften in Wien*, II A, 102, 11–56.

Margules, Max, 1904. 'Über die Beziehung zwischen Barometerschwankungen und Kontinuitätsgleichung'. In Stefan Meyer ed., *Festschrift Ludwig Boltzmann gewidmet zum sechzigsten geburtstage 20. februar 1904*. Leipzig: J.A. Barth, 585–89.

Margules, Max, 1905. 'Uber die Energie der Stürme'. *Jahrbuch der Zentralanstalt für Meteorologie and Erdmagnetismus*, 40, 1–26. English translation by Cleveland Abbe, 1910. *The Mechanics of the Earth's Atmosphere*. Washington, DC: Smithsonian Institution, 553–95.

Mohn, Henrik, 1883. *Grundzüge der Meteorologie, 3. Aufl*. Berlin: Reimer.

Monmonier, Mark, 1999. *Air Apparent. How Meteorologists Learned to Map, Predict, and Dramatize Weather*. Chicago: The University of Chicago Press.

Moore, Willis Luther, 1898–99. 'Weather Forecasting: Some Facts Historical, Practical, and Theoretical'. *Bulletin* No. 25. Washington, DC: U.S. Department of Agriculture.

Moore, Willis Luther, 1922. *The New Air World: The Science of Meteorology Simplified*. Boston, MA: Little, Brown.

Moye, Marcel, 1911. *Les types du temps dans le bassin de la Méditerranée occidentale*. Montpellier: Roumégous Déhan.

Murphy, Allan H., 1996. 'The Finley Affair: A Signal Event in the History of Forecast Verification'. *Weather and Forecasting*, 11 (1), 3–20.

Nebeker, Frederik, 1995. *Calculating the Weather. Meteorology in the 20th Century*. New York: Academic Press.

Notice sur les travaux scientifiques de M. Gabriel Guilbert. Paris: Gauthier-Villars, 1920.

'Obituary Notice of the Rev. W. Clement Ley'. *Quarterly Journal of the Royal Meteorological Society*, 23 (1897), 103–05.

Pietruska, Jamie L., 2017. *Looking Forward: Prediction and Uncertainty in Modern America*. Chicago: The University of Chicago Press, 2017.

Pihl, Mogens, 1981. 'Bjerknes, Vilhelm Frimann Koren'. In Charles Coulston Gillispie ed., *Dictionary of Scientific Biography*. New York: Charles Scribner's Sons, 2, 167–69.

Polop, Josep Antoni, 2003. *Manuel Iranzo Benedito. Un pioner de la meteorologia valenciana*. Valencia: Universitat de València.

Pudykiewicz, Janusz, and Brunet, Gilbert, 2008. 'The First Hundred Years of Numerical Weather Prediction'. In Mohamed Gad-ElHak ed., *Large-Scale Disasters: Prediction, Control, and Mitigation*. Cambridge: Cambridge University Press, 427–46.

Raines, Rebecca Robbins, 2011. *Getting the Message through a Branch History of the U.S. Army Signal Corps*. Washington, DC: Center of Military History, United States Army.

Reed, Richard J., 1977. 'The Development and Status of Modern Weather Prediction'. *Bulletin of the American Meteorological Society*, 58, 390–99.

Reuter, Heinz, 1954. *Methoden und Probleme der Wettervorhersage*. Vienna: Springer.

Rubel, Franz, and Kottek, Markus, 2011. 'Comments On: "The Thermal Zones of the Earth" by Wladimir Köppen (1884)'. *Meteorologische Zeitschrift*, 20 (3), 361–65.

Saltzman, Barry, 1967. 'Meteorology: A Brief History'. In Rhodes W. Fairbridge ed., *Encyclopedia of Atmospheric Sciences and Astrogeology*. New York: Reinhold, 583–91.

Sandlin, Lee, 2013. *Storm Kings: The Untold History of America's First Tornado Chasers*. New York: Pantheon Books.

Saumande, Pierre, 1990. 'Paul Garrigou-Lagrange; un curieux homme, un esprit curieux'. *Bulletin de la Société archéologique et historique du Limousin*, 118, 157–66.

Scott, Robert Henry, 1869. 'On the Work of the Meteorological Office, Past and Present'. In Stanley Keith Runcorn ed., *The Earth Sciences*. London: Applied Science Publishers, 1971, 1, 333–45.

Scott, Robert Henry, 1873. 'On Recent Progress in Weather Knowledge'. In Stanley Keith Runcorn ed., *The Earth Sciences*. London: Applied Science Publishers, 1971, 1, 378–83.

Scott, Robert Henry, 1897. 'Hon. Ralph Abercromby'. *Nature*, 57, 55.

Shaw, W. Napier, 1903. 'The Meteorological Aspects of the Storm of February 26–27, 1903'. *Quarterly Journal of the Royal Meteorological Society*, 29, 233–58.

Shaw, William Napier, 1911. *Forecasting Weather*. London: Constable.

Shaw, William Napier, 1934a. 'The March of Meteorology: Random Collections'. *Quarterly Journal of the Royal Meteorological Society*, 60 (254), 101–20.

Shaw, William Napier, 1934b. *The Drama of Weather*. Cambridge: Cambridge University Press.

Shaw, William Napier, and Lempfert, Rudolf Gustav Karl, 1906. *The Life History of Surface Air Currents: A Study of the Surface Trajectories of Moving Air*. London: Darling.

Sverdrup, Harald U., 1951. 'Vilhelm Bjerknes in memoriam'. *Tellus*, 3, 217–21.

Thorpe, Alan J., Volkert, Hans, and Ziemianski, Michal J., 2003. 'The Bjerknes' Circulation Theorem: A Historical Perspective'. *Bulletin of the American Meteorological Society*, 84 (4), 471–80.

U.S. Army Signal Corps, 1887. *Annual Report of the Chief Signal Officer of the Army to the Secretary of War for the Year 1887*. Washington, DC: Government Printing Office, 21–22.

Volkert, Hans, 1999. 'Components of the Norwegian Cyclone Model: Observations and Theoretical Ideas in Europe Prior to 1920'. In Melvyn A. Shapiro and Sigbjorn Gronas eds., *The Life Cycles of Extratropical Cyclones*. Boston, MA: American Meteorological Society, 15–28.

Volkert, Hans, 2005. 'Vilhelm Bjerknes' Vision for Scientific Weather Prediction'. In Helge Drange et al. eds., *The Nordic Seas: An Integrated Perspective. Oceanography, Climatology, Biogeochemistry, and Modeling*. Washington, DC: American Geophysical Union, 357–66.

Volkert, Hans, 2008. 'Exner-Ewarten, Felix Maria Von'. In Noretta Koertge ed., *New Dictionary of Scientific Biography*, vol. 2. Detroit: Charles Scribner's Sons / Thomson Gale, 425–28.

Walker, J. Malcolm, 2012. *History of the Meteorological Office*. Cambridge: Cambridge University Press.

Wegener-Koppen, Else, 1955. 'Wladimir Koppen. Ein Gelehrtenleben fur die Meteorologie'. In Hans Walter Frickhinger ed., *Grosse Naturforscher*. Stuttgart: Wissenschaftliche Verlagsgesellschaft.

Whitnah, Donald R., 1961. *A History of the United States Weather Bureau*. Urbana: University of the Illinois Press.

Wille, Robert-Jan, 2017. 'Colonizing the Free Atmosphere: Wladimir Köppen's 'Aerology', the German Maritime Observatory, and the Emergence of a Trans-Imperial Network of Weather Balloons and Kites, 1873–1906'. *History of Meteorology*, 8, 95–123.

Willis, Edmund P., and Hooke, William H., 2006. 'Cleveland Abbe and American Meteorology, 1871–1901'. *Bulletin of the American Meteorological Society*, 87 (3), 315–26.

Wood, Mick, 2001. 'Meteorologist's Profile – Ralph Abercromby'. *Weather*, 56, 102–06.

7 Behind weather forecasting

National interests and the primacy of public service over research

Introduction

Today, like yesterday, the weather map is the tool and symbol par excellence of the national weather forecasting agencies: it represents their anima and their purposes. Interests related not only to commerce and agriculture, but also to domestic surveillance and social control, make us direct our attention to the national control contexts of synoptic meteorology. Under the impulse and pulse of many Western governments, the period from the establishment of these agencies to the First World War forms an epoch in which 'physical concepts and theoretical concepts played little, if any, role in practical weather prediction'.[1] Especially regarding the role assigned to research, forecasting was developed as an art rather than as a science. This, at least, is what could be deduced from the character and purpose of the national weather agencies, particularly those in the United States, Britain, France, and Italy. As I will show in this chapter, the science of meteorology underwent a qualitative change in its shapes, aims, and character, focusing primarily on providing public service in their pursuit of national surveillance and control through ever more advanced mapping technologies.

The conclusion brought forward here may seem strange to those historians and meteorologists who have viewed weather prediction as a sustained and continuous scientific development, a *continuum* of science and art. Then and now, many meteorologists regarded the proliferation of national weather agencies and their synoptic maps as a good in and of itself, a commendable step towards scientific prediction devoid of political connotations. Further, they always supposed that public service and research were hand in hand.

However, those in these decades who allocated funds and pulled the strings of the national weather agencies assumed, at least tacitly, what will be discussed in this chapter. The weather map soon emerged as a powerful tool. As a U.S. War Department circular stated as early as 1871, justifying the use of synoptic weather maps instead of the traditional display in tabular form:

> The graphic charts are of additional value, from the fact that it is often possible to trace upon them, in lines, the progress of storms, or the

---

1 According to meteorologist Richard Reed (1977, p. 391).

change of meteoric condition (as the movement of an area of high or low barometer) from report to report, and thus, by considering the past, and by applying the laws and generalizations reasonably well established, to estimate more easily the 'Probability' of the future.[2]

On the other hand, those who knew the complexity and pressure of the process of image-production were well aware of how important a central office was to meet the increasing demands of weather mapping. As the journal *Nature* stated in an article printed shortly after the British Meteorological Office resumed daily charts in 1872, the method of preparation of charts was 'the fruit of much thought': the printing process went from the pantograph 'which reduces the original drawing' of charts, to the engraving device that could rapidly etch into metal, to the templates which would 'ensure uniformity' in the typefaces of characters.[3] No observatory or institution other than a national agency had the necessary technology to daily produce weather maps in industrial quantities.[4]

Certainly, those directing the forecasting divisions and sections of the national weather agencies kept synoptic weather maps clearly in mind, and were quite convinced about their efficiency and pertinence. In fact, as a U.S. Weather Bureau official held in 1898, 'the daily weather map is the one publication around which the structure of our Bureau is being erected'.[5] By then he could boast that the Bureau had an annual budget of more than one million dollars and a paid staff of 1,400 workers.[6] Behind all this laid the dramatic increase in the appropriation of public funds to those agencies in their beginnings.

However, as striking as their initial impulse was a common circumstance that has gone rather unnoticed among historians of science: the stagnation in public funds and research in those agencies in the 1880s and 1890s. In this chapter, I aim to clarify that question, by suggesting that there was an *imbalance*—i.e. an unequal apportionment of efforts and resources—between the provision of public service and the pursuit of research. For the meteorologists who suffered its consequences, like Tor Bergeron from the Norwegian Bergen School, there was no doubt how and why it all happened: 'the meteorological institutes of the world—some of them with the main task of forecasting weather—were rapidly transformed into administrative–statistical bureaus, mainly loaded with the burden of ruling administratively and digesting statistically the observational data'.[7]

2 Chief Signal Officer, War Department, 1871, *The Practical Use of Meteorological Reports and Weather Maps*. Washington, DC: Government Printing Office, 8–9.
3 'The Times Weather Chart'. *Nature*, 15 April 1875, 11, 473–74, on p. 474.
4 This point has been suggested by Anderson (2006, p. 77).
5 The officer was A.F. Sims, from the station of Albany, New York. See Beals and Sims (1899, p. 76)—quoted by Monmonier (1988, p. 17).
6 Monmonier (1988, p. 17).
7 Bergeron (1981, p. 448).

The commitment to public service and the subordination of research, as well as the subsequent bureaucratization and stagnation that characterized national weather agencies between 1870 and 1914, were reflected to a large extent in the public funds allocated to those agencies in Britain, the United States, and France. As is shown in Figure 7.1, in FY '64 (just before FitzRoy's death) the total British Meteorological Department budget was similar to that in FY '54; and in FY '71, the newborn U.S. Weather Bureau had a symbolic budget of $15,000—in neither case there was special appropriation for research. By roughly 1880, these budgets increased by a factor of more than three for the Meteorological Office (FY '77) and by a factor of thirteen for the U.S. Weather Bureau (FY '83) in constant currency, while the Office budget for research represented only 6.6% of the total budget and the Bureau continued without specific appropriation for research.[8] Over the following three decades total Office budgets virtually came to a standstill, rising only 1.1 times in FY '00 and FY '09, in comparison with FY '77, while total Bureau budgets smoothly oscillated down and up vis-à-vis FY '83 (0.96 times in FY '96 and 1.4 times in FY '08, in constant dollars). A similar pattern can be observed in France, where there was a strong initial boost in fund allocation (prompted by the establishment of the Bureau Central Météorologique (BCM) in 1878), reaching a peak in 1882, and a virtual standstill for almost three decades (Figure 7.2). These numbers reflect a strong commitment to weather forecasting and national control on the part of those governments, as well as a lack of progress in research and funds in subsequent decades.

As I will show below, government patronage was important not simply because it helped to shape weather forecasting by financing the process of image-production, but also because it reshaped the very character and direction of the meteorological research itself.

Funds for synoptic forecasting

In the times of the establishment of national weather institutes each government responsible for these agencies made assessments of the significance of weather forecasting for their nations' present and future. Each arrived at the same conclusion: this kind of forecast and warning, derived from isobaric maps, was of the utmost importance for commerce and agriculture. In the United States, the Congressional act of 10 June 1872 provided funds for 'expenses of storm signals announcing the probable approach and force

---

8 See, for example, the statements of amounts appropriated for the support of the U.S. Army Signal Service for FY '84 and FY '86, in: 'Report of Officer in charge of the Property and Disbursing Division', *ARCSO*, 1884, 50–58, on pp. 57–58; and *ARCSO*, 1886, 22–25.

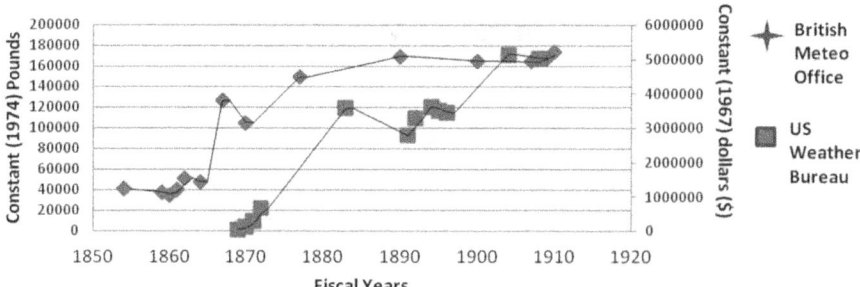

NATIONAL BUDGET IN METEOROLOGY IN GREAT BRITAIN & USA

Figure 7.1 Budgets in the national meteorological institutes in the United States and United Kingdom, 1850–1910.

The figure is folded so as to clearly show the increment of budget in the national meteorological institutes in the United States and United Kingdom. In the United States, budgetary appropriations for the National Weather Service were provided by a military institution, the US Signal Corps, from FY '71 to FY' 91. In 1891, the U.S. Weather Bureau became a civilian agency under the Department of Agriculture. General W.B. Hazen's and other 146 officers' judgement that 'a civilian bureau controlling the weather service would increase expenditures by approximately $350,000 per year' is not withstanding [S. Misc. Doc. No. 82, 938 (1886), quoted by Whitnah (1961, p. 54)]; the change to civilian status implied no appreciable change in appropriations, except the reduction and end of the annual deficits occurring from FY '84 to FY '91.

Formed as the Meteorological Department of the Board of Trade with a specifically maritime purpose, the appropriations for the Meteorological Office were provided by both the Board of Trade and the Admiralty during FY '1854 – FY '1865. By FY '67, the Meteorological Committee of the Royal Society of London controlled the Meteorological Office's budget and expenditure. Under the Royal Society, its employees were no longer considered public servants.

Sources: As regards the U.S. Weather Bureau, for FY 1869–70 to FY 1872–73, Miller (1931b, pp. 69–70), and Fuller (1990, p. 4); for FY '71, FY'83, FY'92, and FY1908, see 'Table 1: Appropriations for selected years of Weather Bureau operations', in Whitnah (1961, p. 21); for FY in the 1890s, Whitnah (1961, p. 62, 75); for FY 1904–05, Khrgian (1970, p. 120). As regards the Meteorological Office, from FY 1856 to FY '59, see *Report of the Meteorological Department* for those years, and Dry (2009, p. 36); for FY '57, *Parliamentary Papers*, 20 (1857), 283–372; for FY '54, FY '60, FY '61, FY '62, FY '64, FY '67, FY '70, FY '77, FY '90, FY 1900, FY '07, FY '09, and FY '10, see Walker (2012, pp. 22, 40, 41, 47, 51, 73, 113, 119, 122, 123, 151, 155, 169, and 170).

In the figure, the above-mentioned U.S. Weather Bureau's monetary data have been translated into constant dollars according to the Consumer Price Index (CPI) of federal government as estimated by Ethel D. Hoover (CPI) for the 1851–90 period and by Albert Rees (Cost of Living Index) for the 1890–1912 period, and included in Table 121 from *Handbook of Labor Statistics* (Washington, DC: US Department of Labor, 1973), 287. Taking 1967 as base year, the multipliers used for determining constant dollars are:

1870	1871	1872	1883	1891	1892	1894	1895	1896	1904	1908
2.6	2.7	2.7	3.6	3.7	3.7	3.8	4	4	3.7	3.7

Again, the Meteorological Office's monetary data have been converted from nominal British pound values to real pound values by using the study *Consumer Price Inflation since 1750* by Jim O'Donoghue, Louise Goulding and Grahame Allen (London, Economic Trends No. 604, 2004), p. 43. Taking 1974 as base year, the multipliers are:

1854	1859	1860	1861	1862	1864	1865	1867	1870	1877	1883	1890	1900
9.8	11.1	10.7	10.5	10.7	11.2	11.1	9.9	10.5	10.3	10.7	11.3	10.8

1907	1909	1910
10.6	10.5	10.4

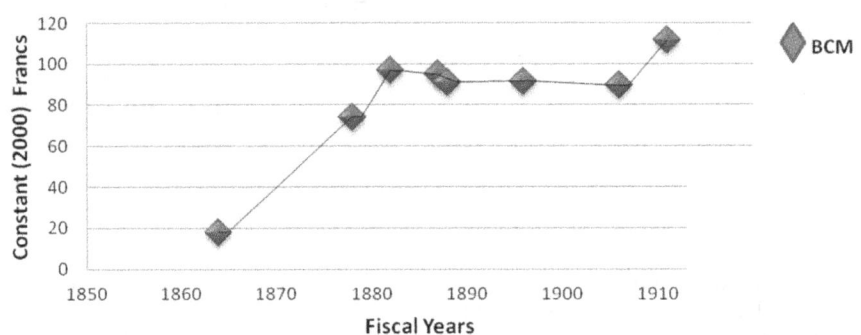

Figure 7.2 Budgets in the *Bureau Central de Météorologie* in France, 1864–1911.

Sources: As regards the Bureau Central de Météorologie, for FY '78, FY '82, FY '87, FY '88, FY '96, FY 1906 and FY 1911, see Fierro (1991, p. 197). The 1864 figure is the additional credit allocated by the French government to the Paris Observatory's budget for the establishment of forecasting service. Locher (2008, p. 122).

In the figure, the Bureau Central de Météorologie's monetary figures have been converted into real francs after adjusting nominal values to account for inflation. The conversion was carried out according to the Consumer Price Index (CPI) estimated by Jordi Maluquer de Motes (2010, p. 207), Appendix 1, for the 1840–2009 period, with 2000 as base year. For the complexity of the price indexes in France in this period, see Jean Claude Toutain, 'L'imbroglio des indices de prix français du XIXe siècle'. *Économies et Sociétés*, série HEQ, 1997, 1 (11), 137–97. The multipliers are:

1864	1878	1882	1887	1888	1896	1906	1911
48×10^{-5}	53×10^{-5}	53×10^{-5}	50×10^{-5}	50×10^{-5}	45×10^{-5}	48×10^{-5}	55×10^{-5}

of storms' for 'the benefit of commerce and agriculture'.[9] General A.J. Myer named this service 'The Division of Telegrams and Reports for the Benefit of Commerce' (extended to include 'Agriculture' in 1877). In France, the Minister of Public Instruction founded the BCM by the Decree of 14 May 1878, with its two first tasks being 'the study of atmospheric movements and meteorological forecasts to seaports and agriculture'.[10] In Britain, the circular of 30 November 1867, dispatched by the Board of Trade to all ports

9 Quoted by Weber (1922, p. 9). In the same vein, Congressman H.E. Paine introduced on 2 February 1870, a Joint Congressional Resolution requiring the Secretary of War to give 'notice on the northern (Great) lakes and on the seacoast' by telegraph and marine signals, 'of the approach and force of storms'. Grice (1991, p. 14); Hughes (1994, p. 26).
10 The decree was signed by the Minister Agénor Bardoux. Fierro (1991, p. 199). A similar concern was expressed earlier by a governmental commission reporting on the meteorological service at the Paris Observatory: 'the study of weather is making rapid progress and the forecasts rival those of England'. Vice Admiral Fourichon, 'Report to the Ministry of Public Instruction', 4 January 1868, p. 6. AN, carton F 17, 3719.

and fishing stations, stated that the Meteorological Office was prepared to freely transmit by telegraph 'immediate intelligence of all storms of whose existence it may have received information'.[11]

The big storms and gales not only awakened but also tremendously re-inforced a conviction, already firmly established in the minds of FitzRoy, Le Verrier, and other leaders prior to the foundation of national weather institutes—that these tragedies could be avoided by forecasting their arriv-als. The storm of November 1854 during the Crimean War and the sinking of the Royal Charter in 1859 were only two examples of forces for mobiliza-tion of large-scale government-driven ventures. Following the succession of shipwrecks on the Great Lakes in the late 1860s, voices were raised advocat-ing an independent meteorological department within the US government.[12] Analyzing these developments, historian A.H. Dupree held that, in terms of the economic interests of the country, these kinds of ventures 'so extensively supported by the federal government told almost entirely in favour of com-merce'. In his view, 'a counterdemand for scientific aid to agriculture' would have been 'the natural response to such a one-sided program'. This would have been true especially in the aftermath of those shipwrecks, when 'pres-sure for a comprehensive forecasting service became particularly strong'.[13]

When meteorologists created the isobaric techniques of mapping in the early 1870s, fears towards warnings and forecasts began to be dispelled: 'the need for immediate and practical results', as Dupree reminds us in dealing with the U.S. Signal Service, placed meteorology on the governments' po-litical agenda. Immediately the funds allocated by Western governments to meteorology multiplied in a few years, reaching hitherto the highest figures. Thereafter new divisions of weather forecasting were established through-out these Western nations. Although the weather forecasting by isobaric-synoptic methods in the style of the Paris Observatory was much discussed and even rejected in Britain, storm warnings became progressively known—especially by the Americans—first as 'probabilities', then as 'indications', and finally as 'forecasts'.[14]

---

11 In the words of the Meteorological Committee. The circular was issued by Thomas Henry Farrer on behalf of the Board of Trade on 30 November 1867. Walker (2012, p. 86). Agricultural meteorology was one of the first issues considered by the Meteorological Council in 1877, as a result of which a forecast service for agriculturists was established within the Meteorological Office. 'Minutes of the Meteorological Council 1877–1878', p. 11. National Meteorological Library, Exeter.

12 The ferocity of nature is reflected in figures: storms sank or damaged 1,164 vessels in 1868 and 1,914 vessels in 1869, killing 321 and 209 people, respectively, on the Great Lakes. Hughes (1994, p. 26).

13 Dupree (1957, p. 110, 187). According to Dupree, 'the government was the obvious agent to undertake this work', and 'the need for forecasts was pressing more economic groups all the time'.

14 It was not until 1889 that the term 'forecast', introduced by FitzRoy, was officially adopted by the U.S. Weather Bureau. Halford (2004, p. 248).

To no insignificant extent, these state commitments gained strength in the late 1870s. In barely a decade, the U.S. Weather Bureau doubled its full-time personnel, multiplied its budget (in current dollars) by a factor of 27.5, and increased more than fivefold the number of meteorological stations.[15] A similar pattern followed the BCM: it was established with a staff seven times as large as the Paris Observatory had for meteorological activities in the early 1870s, a budget almost five times as large, and almost twice as many rain gauges.[16] The British Meteorological Office, whose budget and structure followed that which the Galton Report had proposed for the re-organization of the Board of Trade's Meteorological Department in 1867, employed a qualified staff of ten (as opposed to four before that year) and performed around 1,650,000 sets of observations (as opposed to 500,000), annually expending, on average, 20,000 pounds more than that year.[17]

Notwithstanding the growing commitment of Western governments and the extraordinary increase in the budget, staff, and structure of the national weather agencies in the 1870s, there was very little progress in the following three decades. In general, both budgets and staffs remained practically flat, almost at a standstill, and there was barely any research budget or stimuli for innovation. As Dupree states with regard to the U.S. Weather Bureau, 'the need for immediate results' was so dominant that 'from the beginning its routine operations of gathering data and forecasting swallowed up both funds and energy'.[18]

Thus began the era of the centralization and bureaucratization of national weather agencies based upon the statistical digestion of data. As the Norwegian Bergen School's meteorologists allowed many years later, 'the meteorological institutes of the world'—among whose objectives included weather forecasting—were 'rapidly transformed into administrative-statistical bureaus', often overloaded with administrative tasks, and burdened with the statistical digestion of observational data and the great deal of calculation.[19]

The U.S. Weather Bureau

On 8 November 1870, the US Army Signal Office issued the first storm warning in America, and soon after printed its first isobaric maps, with a

15 Whitnah (1961, p. 21).
16 Fierro (1991, pp. 201–02).
17 Walker (2012, pp. 61, 75); Burton (1988, p. 109).
18 Dupree (1957, p. 189). In his article on the methods and results of the Signal Service Bureau, one of the most mordant against this body, William Blasius (1887, p. 179) stated that the Bureau 'devoted its chief efforts to *prediction* and *signaling*, while the study of nature and its laws' received 'but scant attention'. If one should assess its results, he must allow that 'it is found to be utterly disproportionate to the means at its disposal, even in the matter of prediction. Its methods must therefore be at fault'.
19 Bergeron (1981, p. 448).

budget of $15,000 for FY '71.[20] By 1887, the Signal Office had reached an annual production of 175,387 maps issued, or a daily average of 480. The U.S. Weather Bureau had an annual budget of almost a million dollars.[21] Post-war inflation ended with the economic stringency caused by the Civil War. For America, the 'nationalization' of meteorological service meant an unprecedented concentration of resources in a single central institution. It brought the U.S. Weather Bureau booming growth in the 1870s, growth that then stagnated amongst considerable doubts and uncertainty in the 1880s and 1890s. State appropriation increased twentyfold the meteorological maximum budget level of the Smithsonian Institution.[22] Examined more carefully, however, this growth seems to follow three very different trends: Between 1870 and 1883 state appropriations for weather service increased 2,750% (in constant dollars); in the next two decades annual budgetary appropriation showed no growth at all; and in the 1900s this support increased 143%.[23]

The first and most important surge came with the expansion of telegraphic lines in the 1870s. Professor Cleveland Abbe admitted that the main reason for the choice of Signal Service in 1870 was the high number of government telegraphic lines.[24] Forecasts were distributed by telegraph to weather and railroad stations. By 1883, the Congressional budget for the U.S. Weather Bureau had multiplied by 27.5 and the number of stations had quintupled (from 25 in 1870 to 132 in 1886). Early reluctance to use forecasts soon disappeared. The *New York Herald* published the government daily reports from 1871 onwards. In its editorial of 11 August 1877, the *New York Tribune* averred that the Signal Service had become a leading institution.[25] Already by 1877, according to the *New York Daily Tribune*, the United States led the

20 These early isobaric maps were drawn by the first forecaster and assistant to the Chief Signal Officer, Increase A. Lapham. See *ARCSO* (Washington, DC: U.S.G.P.O., 1871), 7, 167–72, and fifteen charts. Miller (1931a, p. 4); Weber (1922, pp. 3–4); Glassford (1891, pp. 163–64).
21 Whitnah (1961, p. 28).
22 The Smithsonian meteorological budget peaked in 1860 at $4,431.97. Fleming (1988, p. 187).
23 The scope of the sudden spurt in the state support to storm warning service following the establishment of the Weather Bureau can be inferred from the letter that Spencer F. Baird wrote to Joseph Henry, the first Smithsonian Secretary. Although he did not mention—as Fleming (1990, p. 164) rightly points out—the appropriation for weather service under the Signal Office ($25,000), he listed the Congressional budget hearings for 1870, including $10,000 to the Museum, $10,000 for the Smithsonian building, $25,000 for Ferdinand Hayden's western geological survey, $50,000 for Arctic Expedition, $29,000 for solar eclipse observations by the Coast Survey, $50,000 for the Naval Observatory, and $12,000 for John W. Powell's expedition to Colorado. Note that the Signal Office's amount grew by a factor of five in only two years. See Smithsonian Institution Archives, Acquisition Records. Baird Papers, Baird to Henry, 11 July 1870.
24 S. Misc. Doc. No. 82, 98 (1886)—quoted in Whitnah (1961, p. 53).
25 *New York Tribune*, 11 August 1876, p. 4.

service of public weather information, together with France, Great Britain, and Holland.[26]

The 1870–82 period was, nevertheless, one of exceptional growth. However, this was not the case the first decade after the U.S. Weather Bureau was moved under the Department of Agriculture and put under the administration of General William B. Hazen. Between 1883 and 1892, when the Bureau achieved its civilian status, appropriations for the weather service were gradually reduced, causing cutbacks in public spending as well as in the number of stations and full-time personnel.[27] Reductions were especially conspicuous during the military-service days, in which annual deficits were accumulated. But even when the first civilian Chief, Mark W. Harrington, took office in 1891, he had to conduct the Bureau functions at an average of $400,000 less annually than indicated by military officers, to the point that FY '96 was the lowest appropriation of the 1890s.[28]

Several factors undermined the confidence placed by Congress on the U.S. Weather Bureau. First, a fiscal scandal initiated by the officer of the Signal Service, Captain Henry W. Howgate, provoked petitions to Congress for a thorough investigation of said service. Howgate was indicted for the use of fraudulent vouchers and embezzlement of $90,000 and subsequently arrested by federal authorities.[29] Second, the struggle for control and the eventual transfer of the National Weather Service into a civilian body caused serious controversy in the sessions of Congress during the 1880–87 era. Among other questions, the dispute involved the question of what degree of autonomy the Signal Office should have as a body of the Army.[30] And third, there was abundant evidence of the lack of efficiency in the Signal Office, as well as of discontent, jealously, and rivalry between its two factions, military and civilian. Indeed, a joint Congressional commission was formed in 1884 to investigate the efficiency of the Signal Corps and other

---

26 *New York Daily Tribune*, 26 January 1877, p. 4.
27 The number of regular stations ranged from 132 in 1886 to 161 in 1893, 146 in 1896, and 159 in 1899. As regards the figures of full-time personnel of the Weather Bureau, they varied from 485 in 1880 to 465 in 1890 and 546 in 1900.
28 In 1885, the $241,000 allocated to U.S. Weather Bureau (excluding salaries) differed considerably in their shares of the expenditures in public services and research activities. Under the items for 'observation and reporting of storms', the percentages assigned included: telegraphic reports (56.4%); rent, offices, civilian observers, heat, and ice (16.5%); maps and bulletins (10.3%); storm signals (lanterns, flags, and observers) (4.1%); observations at river gauges (4.1%); cotton-belt reports (2.9%); instruments (2.2%); lifesaving stations (2.2%); and instrument shelters (0.8%). An additional $245,158 was given to cover salaries. S. Misc. Doc. No. 82, 273–75, 280, 355–59—quoted by Whitnah (1961, pp. 40–41).
29 Grice (1991, p. 6); Whitnah (1961, pp. 46–48); Popkin (1967, p. 64).
30 Several episodes of this dispute are described in *New York Tribune*, 17 August 1881, p. 5; and 21 November 1881, p. 4.

agencies.[31] The hearings taken before the so-called Allison Commission pointed to charges such as mistreatment, poor training, and no incentive for advancement. The final report of the commission, released on 8 June 1886, was an indictment of the model of mixed military and civilian regime in the weather service, and urged measures to promote a change.[32]

The U.S. Weather Bureau had to wait for its demanded transformation into a civilian service to make its way through the implementation of more and more public services and strategic research projects. From 1895 to 1913, the expansion of special services such as long-range forecasting, river and flood services, and cold-wave and hurricane warnings coincided with an ever increased investment in research, including the fields of climatology, aerology, solar radiation, and cloud studies. However, despite this expansion and the notable increase in the number of stations (from 146 in 1896 to 197 in 1915) and full-time personnel (465 in 1890 to 792 in 1910), the number of constant dollars allocated by the federal government to U.S. Weather Bureau barely grew from 1895 to late 1920s.[33] All rose considerably from 1932 onwards, and particularly steeply in those years when aviation meteorology was in the greatest demand.[34]

Figure 7.3 has been drawn to show the sudden spurt in federal funds following the establishment of U.S. Weather Bureau and the subsequent relative stagnation from 1883 to the 1920s. This evidence would refute the view of the supposed sustained (and even exponential) growth of the government funding for the National Weather Service as maintained by most of early studies on meteorology in America. With the fillip provided by the isobaric map-based forecasting in Europe, the scarce resources of the Army's Signal Service were no obstacle for the injection of enormous funds. In the three years following the foundation of U.S. Weather Bureau, the total of appropriations increased fivefold in a period in which U.S. War Department expenditures sharply went down. Even in 1892, after the change from military to civilian status, the U.S. Weather Bureau received nearly 40% of the Department of Agriculture's total budget.[35] By then, weather agencies

---

31 Other agencies under federal investigation were the US Geological Survey, the Coast and Geodetic Survey, and the Navy Hydrographic Office. Notwithstanding this, the Commission listed the Signal Service, and particularly the struggle for control between civilian and military authorities, first among its subjects of discussion. See S. Misc. Doc. No. 82 (1886); Dupree (1957, pp. 190–92).

32 Allison (1886). For the Allison Commission and its proposal for a department of science, see Dupree (1957, pp. 215–31). For a comprehensive description of Congressional comments, see Whitnah (1961, pp. 50–58).

33 Whitnah (1961, p. 21).

34 Dines (1918); Whitnah (1961, pp. 167–240).

35 MacMahon and Millett (1939, pp. 384–87). Cf. the Smithsonian's appropriation for meteorology between 1849 and 1874, which represented only 6% of the total Smithsonian budget. Fleming (1988, p. 187).

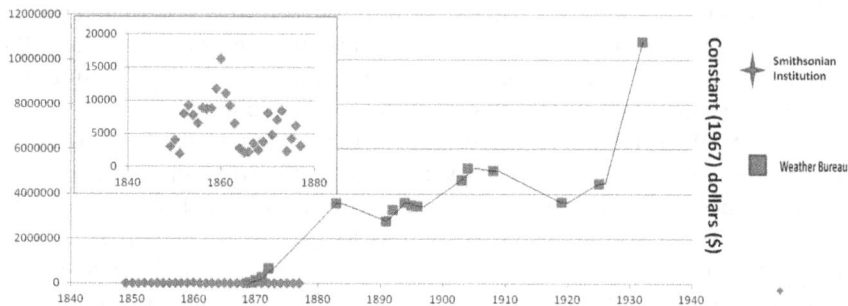

Figure 7.3 Increments of meteorology budget in the Smithsonian Institution (1849–
    1877) and the U.S. Weather Bureau (1870–1932).
The figure shows the increment of meteorology budget in the Smithsonian Institution
    (1849–1877) and the U.S. Weather Bureau (1870–1932). Data for Smithsonian meteor-
    ology budget are those given by Fleming (1988, p. 189). Supplementary funds were pro-
    vided by other agencies such as the Navy Department, the Patent Office (from 1855
    to 1861), and the Department of Agriculture (1863). Again, National Weather Service
    figures are those mentioned in Figure 1. Moreover, for FY 1919, FY'25, and FY'32, see
    'Table 1: Appropriations for selected years of Weather Bureau operations', in Whit-
    nah (1961, p. 21). The prolongation of these figures forwards to 1932 is intended to be
    merely illustrative. Earlier studies do not identify the budgetary decline slowing to near-
    stagnation from 1883 to late 1890s and the feature of uneven high plateau characteriz-
    ing the 1883–1925 period. They do however defend a sustained (and even exponential)
    growth of the federal government funding for the National Weather Service from its
    inception on the basis that they show state appropriations in nominal rather than con-
    stant or non-inflationary dollars.
Whitnah (1961, p. 21) gives the following figures for appropriations for U.S. Weather Bureau
    operations in thousands of current dollars:

1883	1892	1908	1919	1925	1932
993	889	1,361	1,912	2,352	4,497

However, when converted to constant dollars (taking 1967 as base year), the figures show a
    whole different scenario:

1883	1892	1908	1919	1925	1932
3,576	3,292	5,038	3,634	4,469	10,794

Smithsonian meteorology budget figures, here reproduced in two graphs to display
    their growth on two different scales, clearly reflect the budgetary differences between
    them, as well as the budgetary fluctuations and constraints in the Smithsonian period
    (1849–1877).

in the United States had become greatly dependent on federal government
support, as never before in history.

The British Meteorological Office

The first International Maritime Conference of Brussels in 1853 had so interested the British government with Maury's plan to collect oceanic observations and tabulate results that the First Lord of the Admiralty, James
Graham, announced in the House of Commons the establishment of a
meteorological department that would proceed with the modus operandi

agreed there.[36] It could be so, however, only if the government was able to persuade a non-military institution, like the Board of Trade, to accept military assignments. The Admiralty, by tradition more open to this kind of mission, needed to save every penny of its budget and its strength for the war declared by Britain on Russia in Crimea. Finally, the intermediary solution whereby the Board of Trade and the Admiralty would cover about 75% and 25% of budget, respectively, was accepted in 1854.[37] Although the House of Commons endorsed the new department looking to marine statistics, there were some who saw a window open to forecasting. The Member for Carlow, John Ball, 'anticipated that in a few years, notwithstanding the variable climate of the country, we might know in this metropolis the condition of the weather twenty four hours beforehand'.[38]

Through the first five years after its foundation, in spite of organizing the supply of instruments to ships and despite the appeal made by the Board of Trade asking for cooperation of persons to collect simultaneous observations, its director Robert FitzRoy had rather modest resources. Meteorology budgets remained flat until 1859 and neither the Board of Trade nor the Admiralty showed intent to raise them. Indeed, the Admiralty's appropriation remained constant in nominal terms from the 1854 to 1863 period. Even the sinking of the steamer Royal Charter in October 1859 because of a colossal storm, though it led to the British Association for the Advancement of Science's (BAAS) authorization of FitzRoy's prediction system and subsequently to his first telegraphed storm warning, brought no more favourable response.

36 The name 'Meteorological Department of the Board of Trade' appears in the first four reports of this body (1855, 1857, 1858, and 1862). The name 'Meteorological Office' was not made official until some time after FitzRoy's death, and following its divorce from the Board of Trade. For the steps leading to the establishment of the Meteorological Department, see Burton (1986, pp. 150–52; 1988, pp. 24–31); and Lewis (1981, pp. 221–27).

37 Halford (2004, pp. 88–100) and Walker (2012, pp. 19–38) discuss background, foundation, and early activities of the Meteorological Department. In addition to the primary sources published by Lewis and Burton, the following reports and documents by the Meteorological Office staff are valuable: *Report of the Meteorological Office of the Board of Trade* (1855, 1857, 1858, 1862); *Report of the Meteorologic Office of the Board of Trade* 1863, 1864); Kew, NAUK, BJ 3/61, BJ 7/2, BT 3/50; 'Draft memorandum, 5 November 1853', by Robert FitzRoy, Kew, NAUK, BJ 7/113.

38 Quoted in Walker (2012, p. 22). In 1848, Irish glaciologist and politician John Ball, reading the paper entitled 'On rendering the electric telegraph subservient to meteorological research', in the meeting of the BAAS held in Swansea, had expressed his conviction on the potential of telegraphy for weather forecasting. He had distinguished the meteorological cases studied a posteriori (comparing already known results with current observations) from the cases studied a priori (thanks to telegraphy, which allowed receiving information in a short space of time). 'We should in almost all cases be enabled to calculate on the state of the weather for twenty-four hours in advance'—quoted by Ball (1848, p. 12). How predicting considerations such as those from Ball, if did not stay behind, at the very least had an influence on government support of the meteorological department is something that has gone virtually unnoticed.

Although being foundational members, the Royal Society, and some members of the Board of Trade contributed to this lack of support. Two basic considerations centred the core of their criticism. First, many of them thought that FitzRoy had gone far beyond his original program, which was—according to them—to collect data for investigation of weather laws and transmit storm warnings, rather than provide daily forecasts. Second, there were serious doubts as to the scientific principles on which FitzRoy's method of daily forecasting rested. The strength of these fears is indicated by appropriations, which experienced important ups and downs from 1859 to 1865. In FY 1864, a year before FitzRoy's death, the government appropriation was almost 10% lower than in FY 1863 (from £4,800 to £4,270).[39]

When, however, in January 1867 the Meteorological Committee of the Royal Society was constituted as a consequence of the inquiry led by Francis Galton into FitzRoy's meteorological activities, the pursuit of control and power became manifest. Then the disparity between present and past became alarming. The Meteorological Department embodied both fears and ambitions of a scientific lobby concentrated in London and in the Royal Society. Immediately, the government funds allocated for the Department for that fiscal year almost tripled, reaching £12,800, a sum that was within the £12,950 recommended by the Galton Report.[40] Although the abolition of FitzRoy's storm warning service was much discussed, very quickly the Royal Society, and the British government in general, opted for money, power, and control.[41]

Thus began the era of the centralized and unified enterprise of meteorology based upon the funding of the state. As with the U.S. Weather Bureau, these figures would refute the view that 'from its inception, the Meteorological Department provided an example of reasonably generous government funding', as held by most of studies on the history of meteorology in Britain.[42]

---

39 Walker (2012, p. 51).

40 Both the Royal Society's archives and the Galton Report make clear the extent to which the Meteorological Department attracted the bulk of appropriations. According to the estimates for expenditure agreed by the Meteorological Committee on 21 January 1867, costs amounted to £12,800, with £9,900 for the Department and an additional £2,900 for equipment, made up of £1,700 'to complete the instruments of the eight observatories' and £1,200 for 'buildings and alterations at Kew'; the costs estimated by the Galton Report totalled £12,950, composed of £10,450 for the routine work of the Meteorological Department (including £3,000 for weather telegraphy and storm warnings, £3,200 for ocean statistics—namely, £1,700 for discussion of results, and £1,500 for instruments—and £4,250 for maintenance of six stations), and £2,500 for buildings at Kew. 'Report of a Committee Appointed to Consider Certain Questions Relating to the Meteorological Department of the Board of Trade' [Galton Report]. pp. 1866 (3646), LXV. 329, p. 40; Burton (1988, pp. 77–79); and Walker (2012, p. 73).

41 Petitions for the restitution of the storm warning service were even presented to parliament. As a result, it was resumed in a somewhat modified form in 1868.

42 Anderson (2005, p. 283). Walker (2012, p. 170) shows a graph of the Meteorological Office's annual expenditure from 1900 to 1939. In addition to giving an image of continuous growth in government funding (from 1918 onwards), Walker's graph is also misleading because monetary data have not been translated into constant pounds.

With this renewed support, the Royal Society sought to promote observational precision and a research-oriented meteorology disengaged from any obligation to publicize weather warnings. Yet this initial stance not only met considerable resistance from the public audiences following Fitzroy's forecasts but also ran counter to the accepted belief of public funding—i.e. that meteorology had to have utility and immediate application for the nation. This was especially true in the periphery, in places like Scotland, where the utilitarian concerns and local pressures from domestic merchant navy captains and the Scottish Meteorological Society emphatically opposed the cancellation of storm warnings, while in London the scientific elite strove to control the central office and Kew Observatory for their own interests.[43]

Likewise, there was considerable rivalry that might more properly be called nationalist politics. This derived from the production by the U.S. Weather Bureau, of international weather charts combining European and American coastal observations from 1874. Yet, what particularly hurt British pride came in 1877, when the *New York Herald* began to transmit storm warnings to the United Kingdom through the Atlantic cable. Such nationalist political considerations moved the Meteorological Committee and specially its successor, the Meteorological Council, to project a comprehensive forecasting service so close to social concerns as to be politically realizable. This question was far from negligible, for the Treasury's officials every year insisted that the Meteorological Office vote was only experimental, and under no conditions could its staff be considered as civil servants.[44] The Treasury was nearly always reluctant to sanction financial provision for meteorological projects about which it had scant knowledge.

It was just such circumstances which made forecasting support from the British government particularly congenial to the advances in storm warning in America, when a few years earlier the same government had cancelled FitzRoy's storm warning service, following the Galton Report. The more settled outlook of the members of the Meteorological Council sought to increase forecasting costs through the new synoptic techniques based on isobaric maps and far from the empiricism that had aroused Galton's repudiation, while reducing costs for oceanic statistics. The expenditure of telegraphy and storm warnings doubled between 1864 and the 1870–75 period, and increased 13% between the periods of 1870–75 and 1898–1903, while the

43 The significance of Scottish concerns is clearly expressed in an account on the gale of 3 October 1860, written by the Astronomer Royal, Charles Piazzi Smyth. In addition to reflect the tense relations between Edinburgh and London, Smyth (1871, pp. 85–86) criticized the Royal Society's decision to apply the standards of 'exact science of the highest order' to forecasting: 'High science is one thing, and storm warnings so completely another [and] it is not fair to measure its use and right to existence by a test derived from anything else of so entirely different a nature'. For a discussion of this view, Anderson (2005, pp. 238–42).

44 In 1867, the Office was removed from the Civil Service, and its costs were to be funded by a Treasury grant-in-aid until 1920. Brunt (1951, p. 118).

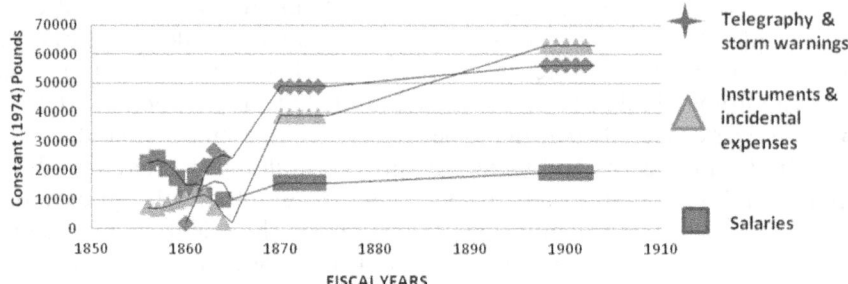

The British Meteorological Office's expenditure charged to Parliamentary vote

FISCAL YEARS

Figure 7.4 The British Meteorological Office's expenditure charged to parliamentary vote. It shows the annual expenditure charged against grant in
the periods of 1856–64, 1870–75, and 1898–1904, for 'telegraphy and
storm warning', 'instruments and incidental expenses', and 'salaries',
respectively.

Expenditures are in nominal or non-inflationary British pounds (taking 1974 as base year).
The figures of expenditure for 1856–65 are drawn from the Galton Report (1866, appendix 18,
p. 39); for 1870–75 and 1898–1904, the figures express average annual expenditures, and they
are drawn from, respectively, the 'Report of the Treasury Committee appointed to inquire into
the conditions and mode of administration of the annual grant in aid of meteorological observations'. PP, 1877, XXXIII, 731–957, on p. 872; and the 'Report of the Committee appointed to
inquire into the administration by the Meteorological Council, of the existing Parliamentary
Grant', P.P. 1904, Cd. 2123, XVIII, p. 838. They can be also found in Burton (1986, p. 333).

expenditure of ocean statistics was reduced by 36% from 1867 to the 1870–75
period (Figure 7.4).[45]

What was true of promotion of weather maps and storm warnings was
truer still of their diffusion and circulation. Although the first printed
weather maps appeared in the *Bulletin* of the Meteorological Office on 23
March 1873, they were not disseminated in national press until late 1870s,
thanks to the even more rapidly growing funds allocated by the British Parliament.[46] While in 1876 the Office arranged with the owners of *The Times*
for an evening service of telegrams and a special weather map, the cost of
which (about £2,500 a year) was borne by the newspaper, in 1879 the Office's
forecasts were placed at the disposal of all newspapers *free of charge* for

45 As regards the expenditure of ocean statistics, the percentage (36%) is calculated from
the figures provided by the Galton Report (1866, p. 40) and the expenditure of 'ocean meteorology' charged against grant 1870–75, as given by the 'Report of the Treasury Committee appointed to inquire into the conditions and mode of administration of the annual
grant in aid of meteorological observations'. pp. 1877, XXXIII, 731–957, on p. 872.
46 Research was usually left to universities and meteorological societies. For the 'special
researches' commissioned by the Council in the 1880s and 1890s, see Walker (2012,
pp. 119–22).

their morning and afternoon editions.[47] By 1880, the parliamentary appropriation to the Office was increased enough so as to enable the Office to offer a free evening forecasting service to all newspapers.[48]

Faced with important funds, neither weather forecasting nor storm warning was evocative enough to convince the Meteorological Council to support research programs, or stirring enough to generate scientific illusion and opportunities within the Office. Although the Council provided funding for research after 1877, the work was carried out on a contract basis that remained outside the walls of the Office.[49] Research had no place within the Office until Napier Shaw replaced Robert H. Scott in the direction in 1900.[50] But despite all this, funding kept dropping for the Office. Analyzing the allocation of public money to meteorology, historian James Burton points to reasons of state:

> That money was always available for meteorology in such relatively full measure is, *prima facie*, surprising. Clearly the internal Treasury feeling was hostile but approval seems to have come from higher levels.[51]

When one recalls how meteorologists stated that, with this money, they drew weather maps without knowing what laws governing the atmosphere, or how these laws affected the weather whose maps they drew, it follows then that forecasting work was much closer to art than science, to an administrative task than research.[52]

---

47 In 1873, Galton devised a method for reproducing weather charts so that they could easily printed by newspapers. This idea was adopted by *The Times* in April 1875, when it started publishing regularly weather maps. See Lempfert (1913, p. 179).

48 The substantial growth in the annual grant for forecasting service was evidently the result only of a well thought-out strategy, not of circumstantial decisions, aimed at promoting forecasts' dissemination. Only a few months earlier, in November 1878, the Treasury refused to grant extra funds for ocean data, arguing that the Meteorological Council should 'economize in other directions'. Unlike weather forecasting, which had both a public opinion and dimension, ocean statistics affected the remote seafarer. Burton (1986, p. 153). For the role played by newspapers in disseminating the reports and maps prepared by the Office, see Burton (1986, pp. 149–52).

49 Harrington (1894, pp. 330–31).

50 5.2% of the average annual expenditure charged against grant in the 1898–1903 period was for 'special research'. 'Report of the Committee appointed to inquire into the administration by the Meteorological Council, of the existing Parliamentary Grant', pp. 1904, Cd. 2123, XVIII, p. 838.

51 Burton (1986, p. 216).

52 In examining the obstacles to the progress of meteorology, Cleveland Abbe bemoaned the ignorance of the physical laws in the atmosphere: 'How can accurate predictions be made from the most perfect graphic maps and statistical figures? We may indeed estimate the probabilities'. But 'if one depends only on the daily weather maps and knows little or nothing of nature's laws, he finds himself in a maze of limitations'. Abbe (1912, p. 63, 65). Faced with so much darkness, he urged the meteorologist to follow the motto: *Nil desperandum. Labor omnia vincit* (Man must never despair). Abbe (1912, p. 55).

In the 1880s and even well into the 1890s, voices were raised in concern over this lack of scientific initiative and the despondency and discontent of the Office's staff. These were not only predisposed and familiar voices like Frank Thomas Bullen's, a clerk from 1882 to 1899, who, according to his biographer, 'thoroughly disliked his experience in the Office, where he encountered the petty tyranny of senior clerks over their junior, the demand for absolute conformity, the denial of any initiative, and the total lack of any variety'.[53] Even a world authority on meteorology and the inspiring leader of the new Office from 1900, Napier Shaw, so often given to temperance and rationalization, confessed in 1934 that he had had 'resentment at the indifference of some, and the hostility of others, during the later years of the nineteenth century'.[54] He seemed 'distressed by the hostility in high quarters to any effort at progress'.[55] The fact was widespread: the Office's and university physicists had lost the active interest in atmospheric problems shown by their early predecessors.

Perhaps the most striking instance of the stagnation and bureaucratization of government-sponsored activities at the Meteorological Office can be found in the section of income and expenditure. Thus, in the ten years following Shaw's assumption of office, the novelty feature was that the Office generated income for services rendered and instruments supplied. In fact, the Office received £3,397 in 1908–09, as opposed to £1,811 in 1903–04.[56] Yet, however considerable the income generated for institutional services was, the parliamentary appropriation to the Office remained practically unchanged—roughly 0.7%—in non-inflationary pounds in the 1900–09 period.[57] The insignificant increase in the Office's budget resulted not from any intrinsic quantitative commitment to basic research, but rather by specific circumstances related to technical developments.[58] By 1909, Shaw lamented that 'no application for an increase had yet been made by the Meteorological Committee' during their term in office.[59]

The Bureau Central Météorologique

Through the 1850s, the only significant recipient of funds for the study of meteorology in France was the Paris Observatory, whose mission was de facto predominantly astronomical. The Ministry of Public Instruction

53 Kennerley (1996)—quoted by Walker (2012, p. 117).
54 Brunt (1956, p. 197). Napier Shaw to David Brunt, 7 April 1934.
55 Brunt (1956, p. 197).
56 Walker (2012, pp. 170–71).
57 In current pounds, the sum increased from £15,300 in 1900 to £15,500 in 1907 and £15,900 in 1909. Walker (2012, p. 169).
58 In 1907, there was a slight increase due to a subscription of £240 for telegrams from Iceland; and in 1909, because of the cost of radiotelegraphic reports from Atlantic liners.
59 Walker (2012, p. 170).

provided almost all funds for the so-called astronomical establishments, though almost two thirds of state appropriations were allocated to the Bureau des Longitudes, an academic and military institution responsible for the *Connaissance des temps* including annual astronomical charts for seafarers.[60] The institutional and organic relationship between the Bureau and the Observatory was broken at the beginning of the Second Empire, when Le Verrier replaced François Arago in 1854 after the death of the distinguished astronomer. By then, Le Verrier complained that, of 125,000 francs funded by the Ministry, only 45,000 francs were for the 'constant and extraordinary expenses' of the Observatory, while '80,000 francs were absorbed by the sinecures of the Bureau des Longitudes'.[61] But although the Paris Observatory's budget grew 44% from 1855 to 1869, its share in the promotion of meteorological activities remained very modest.[62] With a small fraction of its budget going into the *Bulletin International* and the storm warning telegraphic service, its mission remained mainly astronomical.

The Ministry of Public Instruction, by tradition more open to foster learned culture and from 1863 determined with the new minister Victor Duruy to support Le Verrier's forecasting service, took the largest steps to monopolize the state's ties with the Observatory and meteorology.[63] Le Verrier, who had managed to involve telegraph clerks in routine meteorological work in 1856, flirted in a balanced manner with the *Direction générale de l'administration des télégraphes* and the Ministry of Public Instruction, while having tense relations with the Ministries of Marine and War and with private societies.[64] As early as Marié-Davy published his first forecasts in the *Bulletin International,* Duruy requested the French government to allocate an additional credit of 37,500 francs to the Observatory's budget

60 For the relations between the Paris Observatory and the Bureau des Longitudes: Flammarion (1866, p. 3); Bigourdan (1932); and Locher (2008, pp. 56–61).

61 Quoted by Fierro (1991, pp. 111–12).

62 Counts for FY '55 are taken from Jourdain (1867, pp. 227–29), and those for FY '69 are taken from *Journal Général de l'Instruction Publique [et des Cultes],* 1868, 38, 397, 410–11, and 425. Drawing on these figures and the budgets allocated to other three key institutions (College de France, Museum d'Histoire Naturelle, and Institut de France), Fox (2012, p. 119) has underlined the Ministry of Public Instruction's conception of 'a broadly based but unified structure for learned culture' in which the elites of these institutions 'would preside over the efforts not only of professors in the faculties of science and letters but also the members of *sociétes savantes* and even the humblest *curés* or teachers in primary schools'.

63 On the early attempts by the Ministry of Public Instruction to organize meteorological observations in the *écoles normales* in the mid-1850s, see Le Verrier (1868, pp. 62–63); Noël-Waldteufel (1995, pp. 11–15); and Locher (2008, pp. 74–76). See also Radau (1868).

64 For negotiations between Le Verrier and the Compte de Vougy, the director of the *Administration des télégraphes* (a body under the aegis of the Ministère de l'Intérieur) from February 1855 to April 1856, see Locher (2008, pp. 41–44). See also 'Circulaire du 20 avril 1856'. *Ministère de l'Intérieur, Direction des lignes télégraphiques, Lois et règlements,* t. 3, année 1856. Paris: Imprimerie nationale, 1860.

for the forecasting service. The money was used to repair meteorological instruments and recruit three assistants for Marié-Davy.[65] Thus, Le Verrier and Duruy could feel that in a new scenario of tolerance with forecasting, as it appeared to be, they would control the situation. Yet, when Le Verrier devised a plan to study the climate of France by establishing observers in the *écoles normales primaires* in 1864, he soon realized that the government appropriation of 250 francs to each school to purchase instruments was clearly insufficient.[66]

The establishment and centralization of services (storm warning, *écoles normales*, and marine observations) required a special sort of operation: additional income through private societies. By 1865, in stark contrast with 1860, meteorological work already occupied so large a fraction of the Observatory's routine tasks as to render state appropriation for meteorology clearly insufficient. The Observatory simply had not enough resources and observers to meet its growing needs. To Le Verrier, this scarcity represented not only a serious setback but also something reflecting the limited engagements of the state and the amateurs. Then, in April 1864, just as the mentioned new services were knocking at the door, Le Verrier founded a *société savante*, the *Association pour l'avancement de l'astronomie et de la météorologie*, with 'the sole purpose of financing meteorological studies'.[67] In less than a year, the *Association scientifique de France* (as it was known later)[68] had 3,500 members and had gathered 21,000 francs to support scientific studies, mainly in meteorology and astronomy.[69] By 1867, the figure amounted to nearly 75,000 francs, an amount close to half the Observatory's annual budget.[70] Between 1864 and 1869, the total sum invested solely in meteorology reached about 60,000 francs,[71] which helped the Observatory produce important publications such as the *Bulletin hebdomadaire* and several atlases on storms and the atmosphere, as well as to materialize different meteorological projects.[72] This was especially true in certain provinces,

---

65 The new assistants were Georges Rayet, Émile Fron, and Léon Sonrel. Locher (2008, p. 122).

66 Le Verrier (1868, p. 49). Although the plan originated from the director of the *école normale* of Vesoul, Le Verrier and Duruy were the ones who brought the climate project to fruition. By July 1865, Le Verrier could assert that 'the vast majority of *écoles normales* had the necessary instruments' and that 'nearly all had been already verified by the Paris Observatoire'. Le Verrier (1865, p. 142). See also Le Verrier to Duruy, 3 February 1864, 29 July 1864. AN F17, 3728; Mascart (1891, pp. 170–76); Noël-Waldteufel (1995, pp. 11–15); and Locher (2008, pp. 74–77).

67 Le Verrier to Minister of Public Instruction, 10 February 1872. AN, F17, 17198.

68 For this little studied Association see Moigno (1867, pp. 353–54); and Davis (1990, pp. 84–91). On French savant societies in general see Chaline (1998).

69 Fox (2012, p. 117).

70 Aubin (2003, p. 92).

71 Locher (2008, p. 123); Lequeux (2013, p. 108).

72 That the Association had encouraged, through grants and subscriptions (ten francs per year), 'marine observations, observers of storms and hail, useful works for agriculture

where there was a strong commitment from their chambers of commerce (interested in drawing on Marié-Davy's forecasts for local needs)[73]; and in some political spheres, for about two hundred deputies appeared among the members of the Association.

The 'official meteorology' in France did not wait for the Paris Observatory to make its way through the complex institutional labyrinth. Yet, it had less skilled personnel and far fewer economic resources than Le Verrier's bastion for meteorological work. The *commissions départamentales de météorologie*, created at the request of Le Verrier himself in 1864 to take charge of studying storms, had the favourable disposition of prefects and took steps to serve as research sponsors.[74] Moreover, their reports were directly delivered to the Ministry of Public Instruction. But even so, their budgets were always modest. The accounts of one of the most successful and long-lived departmental weather commissions, that of Gironde, show its staff participating in the preparation of several general atlases, while its annual allotments for meteorological work remained flat at 500 francs from 1866 to 1880 and at 2,500 francs from 1881 to 1902.[75]

Even more striking is the case of the Observatory of Montsouris, the institution inspired by the *Société météorologique de France*, the only centre that escaped from Le Verrier's authority, providing some services belonging to the Paris Observatory during the tumultuous and confused years of 1870–73.[76] The City of Paris, having agreed with representatives of the Ministry of Public Instruction the cession of a building, approved an annual appropriation of 60,000 francs in 1870 to establish the new observatory. This sum proved to be around one third of the Paris Observatory's annual budget.[77]

---

and the study of little known places' was in fact admitted by Le Verrier (1868, p. 50). The publications financed include: *Atlas des orages de l'année 1865*. Paris: Charles Chauvin, 1866; *Atlas météorologique de l'Observatoire impérial: Année 1866*. Paris: Charles Chauvin, 1867; *Atlas des mouvements généraux de l'atmosphère*. Paris: Charles Chauvin, 1864–65; and *Bulletin hebdomadaire de l'Association pour l'avancement de l'astronomie et de la météorologie (puis de l'Association scientifique de France)*. Paris, 1865–87.

73 The list of the chambers of commerce joint to the Association includes those of Bordeaux, Boulogne, Caen, Cherbourg, Dunkerque, Honfleur, Le Havre, and Morlaix. For the commitment of these chambers and others to the Observatory's service of storm warning for seaports, see Le Verrier (1868, pp. 29–36).

74 Lequeux (2013, p. 291, 294).

75 Rayet (1903)—quoted by Doublet (1912, p. 173). For the extraordinary activity developed by the departmental weather commissions of Allier and Rhône, see Fierro (1991, pp. 196–97). For the organization and work of departmental commissions (after the fixation of their attributions in 1873), see Grandidier (1882, pp. 532–43), and Fierro (1991, pp. 206–08).

76 One of these services transferred to Montsouris was the international meteorological service, entrusted to Marié-Davy from 1872 to 1873. Grisollet (1950, pp. 136–37); Fierro (1991, pp. 212–13).

77 The supremacy of the Paris Observatory was evinced by numerous facts and figures. The City of Paris subsidized the meteorological observatory of Saint-Jacques (created in 1885), granting 2,000 francs in 1891 and 3,000 francs in 1892. Fierro (1991, p. 214).

To perhaps Le Verrier's most feared foe, astronomer Charles-Eugène Delaunay, who succeeded him in Paris from 1870 to 1872, the Observatory's supremacy was offensive: 'the [Paris] Observatory has been given the luxury of a general staff, to the detriment of an army of workers'.[78] Not in vain, Le Verrier's annual salary quadrupled the titular astronomer's wage; and was much higher (20%) than his counterpart in England.[79]

With the establishment of the BCM in 1878, state financing changed. There was a growing conviction among the various ministerial representatives that the three new services of the BCM (weather warnings, climatology, and general meteorology) required more economic and human resources than the Observatory had hitherto had.[80] Thus, in 1878 the real (constant) funds allocated by the French government for the BCM were four times greater than those granted to the Observatory for the forecasting service in 1864.[81] This spectacular increase in the state appropriation to meteorology, though deriving immediately from the quantitative growth of qualified personnel, resulted mediately from the increased attention to weather forecasting (rather than to climatology), with advances in synoptic mapping that followed the international meteorological meetings in Vienna, Utrecht, and London.

The growing importance of weather warnings—with state funds and for maritime and agricultural purposes, sent to seaports and farmers—was directly related to the creation of the BCM and the emancipation of meteorology from astronomy. It is no coincidence that in 1875 Le Verrier established at the Observatory a warning service for seaports replacing the recently eliminated forecasting service of the Ministry of Marine; or that in 1876, he founded

---

The Marseille Observatory, a branch of the Paris Observatory, and inaugurated in 1864, received a grant of 15,000 francs annually from the City of Marseille. Lequeux (2013, p. 122).

78 BOP, MS 1060–61. For the relationship between Delaunay's directorship and the Observatory of Montsouris, see Davis (1984, pp. 377–79).

79 The Decree of 3 April 1868, stipulating the regulations for meteorological and astronomical activities at the Paris Observatory, fixed the following annual salaries: 1,200–1,900 francs for auxiliaries; 2,000 francs for assistant astronomers; and 7,000–8,000 francs for titular astronomers. 5,500 francs was the salary of the Marseille Observatory's director. The official documents do not mention Le Verrier's remuneration, but it should reach 30,000 francs annually, according to Lequeux (2013, p. 176). BOP, Documents divers sur l'Observatoire de Paris, 1854–1872, cote 3567 (4), folder V.

80 The Conseil or body sanctioning the BCM budgets and other technical issues was made up by representatives of the Ministries of Agriculture and Commerce, Public Works, War, Marine, Foreign Affairs, Interior, and Public Instruction. For the establishment and role of the BCM, see Fierro (1991, pp. 199–206).

81 Fierro (1991, p. 201). In 1875, the meteorological service of the Paris Observatory received the 30,000 francs promised in 1874 by the Ministry of Public Instruction, thanks to which the international meteorological service begin functioning. Lequeux (2013, p. 294). When Delaunay assumed the position of director in 1870, he requested 220,000 francs for reorganizing the Observatory, with *only* 25,000 francs being for meteorological activities. Eight years later, the BCM's budget was 140,000 francs. BOP, MS 1060–61.

an agricultural weather warning service that was so successful that the state obtained additional incomes enabling the emancipation of meteorology: as early as 1878, 1,660 receivers paid forty francs annually for receiving weather warnings.[82] As the Commission appointed by the Minister of Public Instruction stated in its report on the reform of meteorological organization in 1878:

> such an important service [...] should not continue being a dependency of the Paris Observatory [...] It is apparent that there is no scientific relationship between astronomy and meteorology. Insofar as the science of meteorology has not its official representation, absolutely independent, nothing will be achieved.[83]

Nevertheless, despite the close link between the French government's increasing support to weather forecasting and the spectacular growth of meteorological funding in the beginning, through the following three decades the state financing came to a standstill. The BCM's budgets remained virtually flat, on a high plateau, compared to the Paris Observatory's funds for meteorology. A total of 182,500 francs (in 1882), equivalent to 96.7 francs (in 2000), reached the peak. Thus, in 1906 real funds allocated by the state for the BCM budget were some 7.5% smaller than those expended in 1882, rather than the double or triple to be expected from the upward trend in other scientific centres (Figure 7.2).[84]

The impression of stagnation in the official meteorology in France is not only substantiated by the BCM budgets. Staff as well as its ageing also corroborated it. In his book *Histoire de la meteorology*, Alfred Fierro summarized the BCM's intrinsic characteristics like this: 'a stagnant budget, a stable but aged direction: Mascart quits at the age of 70 years and leaves his place to a sexagenarian Angot [...] the BCM's image barely evokes dynamism'. He considered moreover that, 'on the whole, situation changed little between 1878 and 1914 [...] The nature of the work undertaken barely changed until the eve of World War I'.[85] In fact, the number of qualified staff

---

82 The figures are given by Fierro (1991, p. 196). For Le Verrier's second term as director of the Paris Observatory, see Davis (1984, pp. 379–81).
83 Quoted by Fierro (1991, p. 197).
84 The BCM's annual budgets in 1878, 1882, 1887, 1888, 1896, 1906, and 1911 are provided by Fierro (1991, p. 197). In real francs, the budgets of the Muséum d'histoire naturelle, Collège de France, and Institud de France grew (in round numbers) 44%, 60%, and 18%, respectively, between 1855 and 1869. See Jourdain (1867, pp. 227–29), *Journal Général de l'Instruction Publique [et des Cultes]*, 1868, 38, 397, 410–11, and 425; and Fox (2012, p. 322). For incomes and growth trends and rates in academic institutions in France in the second half of the nineteenth century see Weisz (1983, pp. 56–60, 228–33); and Prost (1968, pp. 356–60, 372).
85 Quoted by Fierro (1991, p. 201, 204). By 1914, the number of the requests for weather records and information stemming from the rapid expansion of aviation had doubled in less than a year.

fixed by the Decree of 14 May 1878 for the BCM was roughly equal to the number of skilled employees stipulated by the Decree of 5 August 1904—about twenty-four—while the volume of weather data and records remitted by the increasingly more numerous climatologic stations—30 in 1879, 143 in 1889, 201 in 1899—considerably increased.[86] The stagnation of the state public financing during this period, quite similar to the patterns given in other counties, run in parallel with, and largely as a result of, the centralization and bureaucratization of meteorological activities.

State sponsorship and the direction of meteorology

All the earlier considerations lead to a question of great significance: what *direction* was there in the progress of science, and therefore what kind of meteorological science developed from government patronage? This burgeoning state awareness of the atmospheric environment, did it lead to a service-oriented meteorology or a research-oriented meteorology? In this regard, we are too accustomed to reading in the foundational texts of national weather agencies the achievement of ambitious aims such as the 'meteorological and climatic characterization on a sound scientific basis'—as says the Circular of 14 January 1865 establishing the *Ufficio Centrale Meteorologico* in Italy—and too accustomed to their utilitarian rhetoric—'if only to render services of utility', as the same circular states, to sectors in the country such as 'agriculture and merchant marine', as well as 'demographic studies and public hygiene'.[87]

More pretentious perhaps is the list of objectives pursued by A. Bardoux, the French Minister of Public Instruction, when he founded the BCM. Both the statements of this minister and those earlier made by the Italian Minister of Agriculture, Industry, and Commerce and their counterparts in Britain and the United States, that one of their prime aims was the understanding of atmospheric laws, are not too convincing. Here is what Bardoux asserted in the foundational decree of 14 May 1878:

> The Bureau central météorologique [...] comprises the study of atmospheric movements, meteorological forecasts to seaports and agriculture, organization of meteorological observatories and regional our departmental commissions, publication of their works and research on meteorology or climatology.[88]

---

86 Doublet (1912, p. 121).
87 Quoted by Iafrate (2001, p. 52). See also Eredia (1914, p. 1006). In general, governmental authorities resorted to utilitarian and elusive appeals in their *usus loquendi*. Fox (2012, p. 117) reminds us that Le Verrier answered his many critics (like Victor Regnault and Jean-Baptiste Biot) 'with a smothering rhetoric of utility (for rather vaguely defined ends in agriculture, shipping, and warfare)'.
88 Quoted by Fierro (1991, p. 199). See also 'Le Bureau Central Météorologique'. *La Revue scientifique de la France et de l'étranger*, 1878, 21, 1124.

While meteorology and climatology had already set their course of development as research fields before the establishment of national weather services, the greater part of what we identify as predictive meteorology owed its very existence to the progress of those services. Hence particular attention should be paid to their scientific production rather than their foundational texts, if one wishes to find out what kind of meteorological science developed from government patronage.

The variety and disparity of scientific production in national weather agencies do not make this task any easier. Although objectively measuring the quality of an agency is virtually impossible, counting their publications is also not without risks. An agency may well be, by itself, much more productive in the space of a few years than several other agencies in a decade. In addition, a work may well be, in itself, immeasurably more significant than dozens of ordinary works. All this makes it difficult to quantitatively compare the individual productivity of several agencies. Notwithstanding these difficulties, it does seem feasible to quantitatively estimate the publications and fields of interest of a community of scientists. In order to provide a characterization of this production and direction of interest, I have undertaken a statistical survey of the publications on meteorology printed by three national weather agencies in the second half of the nineteenth century.

This survey draws on the subject classification introduced by Oliver L. Fassig in his monumental bibliography of meteorology. Fassig's work was issued by the U.S. Army Signal Corps in four volumes between 1889 and 1891.[89] Although his bibliography did not include all the publications printed by the agencies selected, it did clearly define all the subjects addressed by these national agencies. His work was regarded as 'the most ambitious and intensive bibliographic project ever undertaken in meteorology'.[90] The number of items gathered by this work (over 16,000) was extremely large and it comprised subjects on temperature, moisture, winds, and storms 'from the beginning of printing to 1889'. All the works published by the national agencies can be included in any of the numerous divisions and subdivisions of his subject classification.

The publications on meteorology printed by the national agencies selected are in part shown in Figure 7.5. The categories drawn on Fassig's subject classification are those of physics and mechanics of the atmosphere, weather prediction, and observations. While the physics of the atmosphere had already been established as a research field long before the 1850s— especially in areas such as temperature, moisture, pressure, and electrical

---

89 Fassig (1889–91). In addition to having been printed in small quantities, the original volumes were lithographed on acidic paper and the few copies preserved are largely in poor condition. A new edition of this nineteen-century bibliography was prepared by Fleming and Goodman (1994). For an account of this edition and Fassig's work, see Fleming (2009).

90 *Meteorological Abstracts and Bibliography*, 1955, 6, 85.

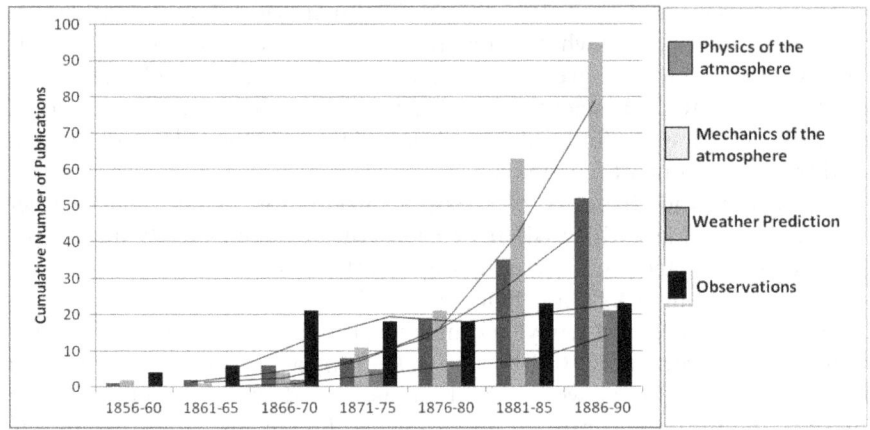

Figure 7.5 Cumulative number of publications printed by the national weather agencies in the United States, Britain, and Italy, in the areas of the physics and mechanics of the atmosphere, weather prediction, and observations, 1856–90.

The publication count is based on the following sources: For the United States, Oliver L. Fassig, 1892. *Extract no. 11, from Annual Report of the Chief Signal Officer, 1891: Report of Mr. Oliver L. Fassig, Bibliographer and Librarian.* Washington: Govt. Print. Off.; for Italy, *Elenco delle pubblicazioni del R. Ufficio centrale di meteorologia e geodinamica dal 1860 al 1910.* Roma: Tipografia Nazionale di G. Bertero, 1911; for Britain, compiled by the author based on the works published by the British Meteorological Office (initially Meteorological Department).

The selection of divisions is based on Oliver L. Fassig, 1889–91. *Bibliography of Meteorology. A Classed Catalogue of the Printed Literature of Meteorology from the Origin of Printing to the Close of 1881; with a Supplement to the Close of 1887, and an Author Index.* Washington City: Signal Office, 4 vols. (pt. 1. Temperature; pt. 2. Moisture; pt. 3. Winds; pt. 4. Storms).

phenomena—the greater and rapidly growing part of the state meteorological production corresponds to the mechanics of the atmosphere. This part owed its very existence to the development of synoptic techniques. The national weather institutes in the United States, Britain, and Italy, with their staffs of forecasters, devoted much of their time and resources to producing daily isobaric maps—a feature that is not reflected in Figure 7.5, for the division of weather prediction included monographs and case studies. The extremely rapid growth of publications on the mechanics of the atmosphere would not have been possible without synoptic boom.

The pleiad of research fields and techniques comprising the mechanics of the atmosphere included all the studies on atmospheric motion, notably general atmospheric circulation, winds and storms (including tornadoes, waterspouts, and storm waves), together with the newer geometrical and graphical methods for describing, charting, and monitoring the areas of low and high pressure, as well as the cyclonic laws derived from those methods.[91] Even so, studies on the dynamics and thermodynamics of the atmosphere were still a tiny minority within this category. The spectacular rise of publications on the mechanics of the atmosphere not only coincided with the irruption of national weather agencies but also sharply contrasted with the steadier and more regular pattern showed by the publications on meteorological observations before and during the period of the survey.

If only to help us to understand the reorientation of the science of meteorology, it is worth considering briefly the predictive involvement of the most productive national weather service. The US Army Signal Office moved towards domestic surveillance and social control after the Civil War. It drew on the military experience accumulated as an intelligence gathering agency to maintain a national storm warning and telegraphic weather service. 'Meteorology provided the rationale and funding for a wide-ranging set of military activities involving telegraphy and surveillance'.[92] The Signal Office never demobilized: what had been defined as a special military unit had now the peacetime mission of gathering intelligence on both military and natural threats. The Signal Service's activities were concentrated on weather forecasting and warning, by far the largest in the Service. Studies on the mechanics of the atmosphere experienced an unparallel growth. Publications on meteorological organization and methods proliferated in parallel with, and largely in consequence of, the establishment of weather stations and international cooperation (see Figure 7.6).[93]

The activities in weather prediction were centralized in the forecasting office in Washington, by far the largest of the Signal Service, and wholly controlled by military authorities until 1891, when it became part of the U.S. Weather Bureau. The central office prepared maps three times a day from synchronized observations received from all the local weather stations. These stations also usually prepared weather charts, though not all of them received three daily reports from all localities. Mapping absorbed most of the resources in the central office: in fact, it published the *Daily Weather Bulletin*, *Weekly Weather Chronicle*, and the *Monthly Weather Review*, which were available for sale to the public. General A.J. Myer estimated that 'at

---

91 The range of topics is that gathered by Fassig (1889–91) in the volumes 2 (Winds) and 4 (Storms) of his *Bibliography of Meteorology*. For methods of forecasting used in the United States at that time, Russell (1895, esp. pp. 141–97).

92 Fleming (2000, p. 331). On the contrary, Raines (2001, p. 46) frames the assignment of weather duties to the Signal Corps within the Army's 'essentially civil functions'—i.e. 'for the welfare of the nation'.

93 For a table showing the number of weather stations in operation from 1870 to 1888, see *Report of the Chief Signal Officer for 1888*. Washington, DC, 1889, 171. Darter (1942, pp. i–xxxvi) gives valuable indication of the breadth and character of the weather service activities; so also does the often cited work by Abbe (1895a) portraying the arduous and tortuous way of research at the Signal Service. Some secondary works that deal with the Signal Office's weather duties are Hawes (1966) and Whitnah (1961, pp. 22–60).

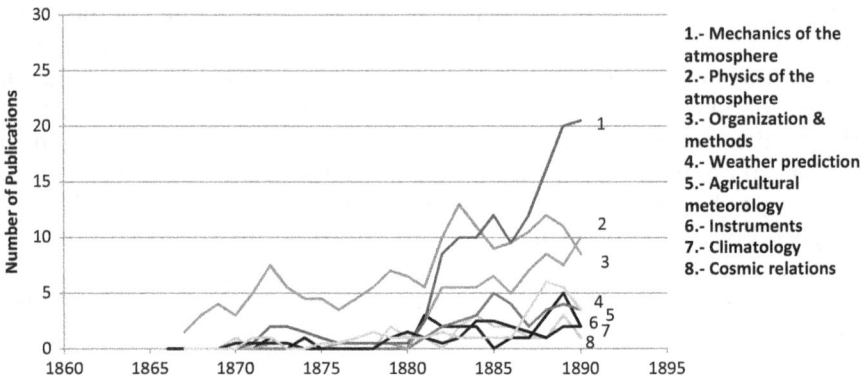

Figure 7.6 Publications by the US Signal Service's Weather Bureau, 1861–91.
All curves are based on year-by-year data, and are drawn through data points that are indicated by bullets of different form and colour. The selection of divisions is based on Oliver L. Fassig, 1889–91. *Bibliography of Meteorology*. Washington City: Signal Office.

Source: Oliver L. Fassig, 1892. *Extract no. 11, from Annual Report of the Chief Signal Officer, 1891: Report of Mr. Oliver L. Fassig, Bibliographer and Librarian*. Washington, DC: Government Printing Office.

least one third of American households received the Signal Corps' weather information in some form'.[94]

A striking instance of the paramount importance of weather maps appears in a report written in 1903 by Willis Moore, chief of the U.S. Weather Bureau from 1895 to 1913. He launched an appeal for funds on the premise that 'the weather maps' afforded 'the only effective means possessed by the Weather Bureau for promptly placing before the public its daily observations and summaries'. For this reason, he 'urgently recommended' the 'improvement and extension of maps'.[95] His appeal was a consequence of the gradual decentralization initiated when the U.S. Weather Bureau authorized local forecasters to operate in New York, San Francisco, and other localities. Between 1887 and 1903, the circulation of the locally produced station weather maps multiplied by a factor of 166, rising from near 150 in 1887 to 8,800 in 1893, and 25,000 in 1903.[96]

More broadly and vigorously than any other institution, the Signal Service pushed international meteorological cooperation as a basis for the centralized data recording on a global scale. On the initiative of General Myer and

94 Quoted by Raines (2001, p. 49).
95 The quoted phrases are from Willis L. Moore, *Report of the Chief of the Weather Bureau for 1903*. Washington, DC: Government Printing Office, 1903, 627.
96 These figures were estimated by Monmonier (1999, p. 53, 252).

with the endorsement of the First International Congress of Meteorology (Vienna, 1873), it organized in 1875 what might well be regarded as the first *Bulletin of International Simultaneous Observations*, extending to the Northern Hemisphere.[97] This daily *Bulletin* was supplemented in 1878 by a daily graphic synoptic chart, called the *International Weather Map*, and, some time later, by monthly international charts, displaying the monthly storm tracks, isobars, isotherms, and prevailing winds in the Northern Hemisphere.[98] Both bulletin production and mapping work demanded a hierarchical service-type organization—military and voluntary observers who were subordinated to a chief meteorologist, as well as a central receiving staff—and resources oriented to image-production. Not in vain, of the dozen sections into which the Signal Office was divided, over half was closely related to mapping work and prediction—'general correspondence and records', 'telegraph room', 'printing and lithographing room', '*International Bulletin*', 'instrument room', 'map room', 'computation room', and 'library'.[99]

In the production and diffusion of meteorology cartography in Europe, an especially important role was played by the British Meteorological Office.[100] In addition to the statistical or climatological branch processing data from both the continuous recording observatories and voluntary observers, the Office developed the telegraphic branch, in which weather reports and storm warnings were prepared. The daily weather report was the main occupation of the Office's forecasters; it consisted of a large sheet of royal quarto size and included two weather charts prepared for 8am, which was daily supplied to *The Times* newspaper and other means.[101] In dividing the British Isles into eleven districts, with forecasts for each region, these maps afforded an image of national unity and internal homogeneity. In 1878 and at the suggestion of agricultural authorities, the Office began producing weekly weather reports including temperature, sunshine and rain averages, and weather maps for the whole of Europe.[102]

97 Hazen (1884, pp. 19–21); 'Daily Bulletin of International Meteorological Observations'. *Symons' Monthly Meteorological Magazine*, 1879, 14, 50–52; Harrington (1894, p. 335).

98 Anderson (2006, pp. 83–85) draws on the example of this *Bulletin* to emphasize the ever-increasing global scale of meteorology and the ever-decreasing importance of local observations and individual observers over the last decades of the nineteenth century.

99 The rest of the sections was 'property room', 'artisan's room', 'station room', and 'the study room'. Raines (2011, p. 54).

100 Not all the national meteorological agencies encouraged weather mapping. In the report read at the International Meteorological Congress held in Chicago in 1893, R.H. Scott bemoaned that one of the oldest meteorological bulletins in Europe, the *Bulletin Meteorologique du Nord* (including observations from Denmark, Norway, and Sweden), contained no map. Scott (1894, p. 7).

101 As Monmonier (1999, p. 273) notes, although the *Times* weather chart is supposedly the first regular newspaper weather map, the London trade newspaper *Shipping Gazette* had been publishing daily wind charts drawn at the Meteorological Office since 1871. See also 'The '*Times*' Weather Chart'. *Nature*, 1875, 11, 473–74.

102 For the Meteorological Office and weather chart production and dissemination, see Scott (1875) and Gaster (1896). See also the report on the publication of daily weather

In contrast to the Meteorological Office, the bulk of the French BCM's mapping work was of an international nature, the number of dispatches received at the Bureau from outside France and Algeria being around two-thirds of the total.[103] This high presence of worldwide data at the BCM's warning service was essentially owing to the legacy left by Le Verrier—the *Bulletin Météorologique International*—who by the mid-1850s had already realized that forecasting was becoming an issue of international interest and rivalry. Émile Fron, the chief of the warning service, set both mapping program and its budget for the BCM, for as the Minister of Public Instruction Jules Ferry decreed in 1880, said *Bulletin* must be distributed among all the *écoles normales* in order to stimulate meteorological observations. From international dispatches, Fron and his assistants drew five maps in the morning and three in the afternoon, aimed at graphically representing physical conditions in the atmosphere.[104]

In the 1890s nearly all the national weather institutes were highly involved in some forecasting and mapping work. Then and through the 1900s a certain balance was maintained: forecasting and mapping work absorbed most of the resources at the central offices and district centres, whereas climatological work required continued attention by observatories and stations of second and third order. Moreover, virtually all the synoptic maps had a dominant feature: isolines.[105] Later, with the new techniques of the Bergen school, weather maps were inundated with warm fronts, cold fronts, and air masses. With these innovations, the statement by Cleveland Abbe that the Northern Hemisphere map forced meteorologists to give 'prime attention to the fundamental mechanics of the atmosphere' was much more realizable and plausible.[106]

In brief, national weather institutes afford an illustrative example of *scientific bureaucratization*. Their governmental sponsors not only oriented meteorology towards public service but also constrained research. As Abbe, one of the leading promoters of meteorological research, stated about the position of this science in America, the 'daily routine of observing and

---

maps and bulletins read by Scott (1894) read at the International Meteorological Congress in Chicago.
103 Fierro (1991, pp. 203–04).
104 In 1896, Alfred Angot adopted a new model of map for the BCM's *Bulletin International*. See Angot (1896, 1898). For French meteorological cartography and the *Bulletin International*, see Hildebrandsson and Teisserenc de Bort (1898, pp. 134–36). See also Grandidier (1882, pp. 532–43) and Monmonier (1999, p. 44). In her otherwise well-documented study on 'mapping meteorology', Anderson (2006, pp. 81–82) seems to suggest that the publication of international maps was conducted by Niels Hoffmeyer from 1873 to 1876 and Georg von Neumayer from 1880 onwards, and at no time by the BCM.
105 In 1894, Harrington (1894, pp. 330–35) compared copies of weather maps for the dates January 1–5, 1892, remitted to the Weather Bureau library from meteorological services all around the world. Though various and varied in size, colour, and symbols, nearly all maps contained isobars and/or isotherms.
106 Abbe (1914, p. 37).

forecasting is not in itself research'.[107] As I will show below, only a small part of efforts and resources were devoted to physical research.

Imbalance between public service and research

After seeing the determination with which governments oriented meteorology towards public service, it is worthwhile to return to the other realm of activity and examine in detail the scope and character of the research that resulted in the national weather agencies in the decades before 1914. Already in 1892, physicist Arthur Schuster anticipated in part the results of our enquiry when he wondered, before the BAAS: 'Why not relieve our numerous National Weather Bureaus from their previous onerous statistical work and turn their best men into the study of the physics and mechanics of the earth's atmosphere?' 'Why not stop all observing and recording and the daily predictions, for five years, and devote ourselves to *study*?'.[108] Since governmental patronage of weather service scientists was intended to solve utilitarian questions of immediate concern, what should be expected from the relationship between service and research? Further, what will be the scope and character of such research?

In the advance of research-oriented meteorology, an especially important role was played by Abbe and the US Army Signal Office's chief, W.B. Hazen. Unlike his predecessor General Myer, who prioritized the practical and the immediate, Hazen established the Scientific and Study Division in 1881, where Abbe organized the so-called 'study room', and signed a cooperation agreement with Fort Myer, where members of the study room began training and technically instructing new recruits for the Office.[109] Abbe and his four assistants (known as 'computers') formed the central group that conducted research projects and 'innumerable matters bearing on the instruments and work of the stations'.[110] But in 1886, the study room would be abolished by order of the Secretary of War. According to Abbe, its real contribution to the service was that it awakened in the Office the 'necessity of introducing a higher grade of scientific training and thought and experimental laboratory work'.[111]

---

107 Abbe (1902, p. 843). Despite his efforts to introduce research programs at the Weather Bureau, he had to admit that results were limited: 'I do not expect great benefits from large central institutes, clubs or societies; the present desideratum is the production of a large number of able investigators'. Abbe (1902, p. 838).

108 Quoted by Abbe (1912, pp. 62–63).

109 U.S. Congress, Senate, *Testimony before the Joint Commission...*,49th Cong., 1st Sess., Senate, Misc. Doc. No. 82, Washington, DC: Government Printing Office, 1886, pp. 19, 91–94, 248, 302–05, 475, 530, 950, 959, 986–89, 1050, 1059. On the importance and nature of the study room, Ibid., pp. 987–89; *ARCSO*, 1871, pp. 12,14; 1872, pp. 86–87; 1879, p. 195.

110 Abbe (1895a, p. 241). The works carried on in the study room are fully enumerated in Abbes' appendices to the *ARCSO* from 1882 to 1886.

111 Ibid., p. 242. A new series for professional papers, known as *Signal Service Notes*, was introduced by Hazen. Moreover, he resorted to several consulting specialists and the National Academy of Sciences, who acted as advisory agents. Raines (2011, pp. 55–56).

The veiled allusion to an element of constraint in the scope of the research under governmental patronage can be deduced from the words of Abbe himself, in his review on the meteorological work of the U.S. Signal Service:[112]

> The study of the daily weather map and methods of forecasting was promptly undertaken by the study room, in order to prepare the way for the researches into storms and atmospheric motions that were to be their ultimate work. This, however, I am sorry to say, was within a year discontinued by order of Gen. Hazen, so that from that time forward until the engagement of Prof. Ferrel, there remained no systematic effort on the part of the Service to attack the great problems of dynamic meteorology. Our time, however, was abundantly occupied with duties of an eminently practical nature.[113]

This very suggestion of certain constraint was confirmed by Abbe in a forthright description of the needs of meteorology, addressed to his fellow physicists in the journal *Science* in 1895:

> Further progress in meteorology demands a laboratory and the consecration of the physicist and the mathematician to this science. Something like this was started [with our] 'Study Room,' but it was ruled out by the report of a committee of Congress, and since that day meteorology has more than ever looked to the universities for its higher development. The applications of climatology to geology, physiography, hygiene, irrigation and other matters have been developed, but meteorology itself [...] still remains to be provided for.[114]

The view expressed by Abbe on the 'no systematic' research efforts, the reference to 'duties of an eminently practical nature', with pernicious consequences for investigations on dynamic meteorology, and the explicit allusion to the interference of a committee of the US Congress in the closure of the study room all suggest not only a utilitarian attitude concerning research but also a subordination of this to the public service.

Inasmuch as domestic surveillance and social control underlay the state support of meteorological research, pressures for utilitarian ends, public service, and limitation of the scope and character of research were common. In this period there was barely any basic research or studies on the dynamics and thermodynamics of the atmosphere. Research was simply a *handmaiden*

---

112 Ibid., p. 241.
113 William Ferrel joint the Signal Office's civilian scientific staff, where he compiled a textbook of meteorology, published as part 2 of the 1885 *ARCSO*, for classroom use at Fort Myer.
114 Abbe (1895b, p. 182).

of services, an activity in which a considerable part of the responsibility for deciding what scope and character it should have fell upon the military authorities and ruling politicians. This fact was clearly denounced by Abbe, when he saw that the establishment of the study room was a way to free itself of government restrictions:

> [At that time], it began to be realized that the Signal Service was no longer able to accept meteorological and physical science as it came from the hands of outside physicists, but that it must itself provide for advanced scientific investigation.[115]

The interventionist implications of this research policy subjugated to public service were not always well perceived in the Signal Office. Yet, the military pressure to control the accuracy of forecasts and research budgets became increasingly important.[116] The members of the Allison commission, formed in 1884 to investigate the role of science in the federal government, recommended not only the closure of the study room and the reduction of appropriation to the Signal Office but also the conferral of full powers to the Secretary of War:

> That the Weather Bureau shall be organized as a civil establishment to promote meteorological investigations, and shall be under the direction of the Secretary of War [SW]; that in organizing the Weather Bureau the Secretary of War shall assign the work to several appropriate divisions, not exceeding ten; [that the SW shall] submit to Congress at its next session a detailed estimate of the number, classification, and salaries of the employees required to conduct economically and efficiently the office work of the Bureau; [and that the SW] shall prescribe [the] rules and regulations [under which] the observations [made by auxiliary state stations] may be received by the Weather Bureau and used in making up statistics and prognostics and reports of the weather, the expense of transmission and printing the same, and of necessary instruments for observations.[117]

It goes without saying that such a control of the U.S. Weather Bureau could easily lead, on one hand, to push investigations on operational and instrumental developments rather than applied research, and, on the other hand, to engage meteorologists in the provision of routine services such as data collection and statistical recording.[118]

---

115 Abbe (1895a, p. 242).
116 For further details of this pressure, see Whitnah (1961, pp. 46–60, 74–81).
117 House of Representatives, serial set. 2443, 49th Cong., 1st sess., 1886, pp. 63–64.
118 In her work on the history of the U.S. Army Signal Corps, Raines (2011, p. 75) highlights the prevalence, at the Weather Bureau, of the practical and the statistical over the

The interventionist character of this research policy was also present after the U.S. Weather Bureau became a civilian agency under the Department of Agriculture. In the new civilian era Major H.H.C. Dunwoody, who was in charge of the Forecast Division and was also Executive Officer, had no qualms about reducing the appropriations for the year 1895 by $95,000.[119] This should come as no surprise, as Dunwoody—at one time the best forecaster at the Bureau— had been indeed 'one of the major opponents to Signal Service expenditures for research in the 1880's'.[120] Still, the situation changed for the worse, when he forced research professors to pass a series of forecast examinations, which caused serious frictions between members of the Bureau and the Department of Agriculture. Abbe was very disturbed by the fact that Dunwoody claimed that neither he nor Charles F. Marvin had attained the standard established during the practice periods.[121] With like preoccupation and with an eye on Dunwoody's attitude, the Chief of the Bureau, Mark W. Harrington, exclaimed: 'we do not need military control of the Weather Bureau'.[122]

Frank H. Bigelow's attempts for expanding the traditional work of the U.S. Weather Bureau illustrate well the potentialities and constraints of physical research in said bureau. Graduated from Harvard in 1873, he entered the Episcopal ministry, before becoming a professor of mathematics, first at Racine College and from 1891 at the U.S. Weather Bureau.[123] As an expert on solar physics and its relation with long-range seasonal forecasting, he made a passionate defence of the building of the Mount Weather Research Observatory. The reasons why the government should build such

---

applied: 'Although the Signal Office had included a study room for several years, its work apparently consisted more of compiling statistical data than preparing original reports'. The annual reports of the Chief Signal Officer provide a faithful portrait of the scientific research conducted under Abbe's direction: of the nearly seventy-seven separate investigations carried out in the period 1871–80 and included in the annual reports (twenty of which can be attributed to Abbe), the vast majority of them deals with climatology, instrumentation, and case studies of severe weather. See Willis and Hooke (2004, p. 50). Furthermore, of thirty-six works published by members of the study room in the reports of the Chief Signal Officer (years 1884 and 1886), half of them was about tornadoes and instruments (in equal percentages). See 'Report of the Assistant in charge of the study room', *ARCSO*, 1884, 59–67, on p. 66; 1886, 208–12, on p. 210.

119 Whitnah (1961, p. 79).
120 Ibid. Dunwoody was opposed to the possibility of producing accurate long-range forecasts, and regarded as mere guesswork the forecasts made in the period from thirty-six to forty-eight hours. U.S. Weather Bureau, *Report of the Third Annual Meeting of the American Association of State Weather Services, Co-operating with the Weather Bureau, U.S. Department of Agriculture*. Washington, DC: Weather Bureau, 1894 (*Bulletin*, 1894, 14), 18.
121 The bitter feud between the U.S. Weather Bureau's Chief Mark W. Harrington and the Secretary of Agriculture J. Sterling Morton led US President Grover Cleveland to remove Harrington from office.
122 Harrington to Morton, 3 January 1894. Secretary of Agriculture Correspondence— quoted by Whitnah (1961, p. 79).
123 Wiley (1924, p. 423); Henry (1924, pp. 165–66); Kutzbach (1979, pp. 226–27).

a first-class observatory, he said in a convention of U.S. Weather Bureau officials, 'is that we may not fall behind other nations', on one hand, and 'the hope of practical advantages in forecasting to be derived' from advances in cosmical physics, on the other hand.[124] With this hope, he planned the details of a research laboratory adapted to the future needs of American meteorology, which would be able to make long-range forecasts by studying the synchronism between solar and terrestrial actions.[125] Yet, his hopes and expectations did not come to fruition. Finding scientists sufficiently expert to research in these subjects was not easy. As historian Donald R. Whitnah acknowledged many years later, 'slow progress and waning enthusiasm on the part of colleagues led Bigelow to leave the Weather Bureau in 1910 and continue his studies in Argentina'.[126]

Bigelow's claim for a new practical organization of meteorology underlay his abandonment of the Bureau. This was clearly stated by one of his closest colleagues at the Bureau, Alfred J. Henry, who, in addition to say that Mount Weather only fulfilled in part his ambitions, pointed to certain intellectual incomprehension by the Bureau's authorities:[127]

> Owing to the highly mathematical and often obscure character of his papers the leading officials of the Weather Bureau found it difficult to follow the force of his arguments or concur in the integrity of his conclusions. This led [him] to a sort of isolation. Indeed, discouragement at the outlook probably was an important factor in the termination of his connection with the bureau. [Before he left the post], the idea of entirely revamping meteorology had taken possession of him, at first as a mild sort of obsession, which later became the impelling object of his existence.[128]

Similarly to what was said regarding the U.S. Weather Bureau, little progress was also made towards a research-based meteorology in the British

---

124 Bigelow (1904, p. 14).
125 By 1904, the year in which he read this proposal in the Third Convention of Weather Bureau officials held at Peoria, Bigelow was a world authority on cosmical meteorology: author of over thirty-three publications, he had conducted researches on solar physics, terrestrial magnetism, eclipses, and terrestrial meteorology, including a book on *Storms, Storm Tracks, and Weather Forecasting* (Washington, DC: Government Printing Office, 1897), together with Willis L. Moore.
126 Whitnah (1961, p. 106). According to Wiley (1924, p. 423), he 'accepted a call from Argentina to organize the weather service of that country'.
127 Henry (1924, p. 166).
128 Bigelow's demands for reform included meteorological theory. In a series of papers printed between 1902 and 1906 dealing with the thermodynamics and dynamics of the atmosphere, he sharply criticized the theories of atmospheric circulation held by Ferrel and other authors. Instead, he propounded a 'counter current' theory of cyclones and anticyclones, which placed the focus on the boundaries of air currents. See Bigelow (1902, 1903, 1906); and Kutzbach (1979, pp. 173–80, 213–14).

Meteorological Office. In 1877, the Treasury released the report on an inquiry into meteorological work, pressed by public complaints about the expenditure of said Office. The inquiry intended to answer key questions, such as: how far have statistics 'led to the discovery or confirmation of any meteorological laws?'; or, 'how far have the principles on which storm warnings are given, been justified by results?' And last but not least,

> how far the appropriation of a large sum of public money in aid of meteorology is justified, bearing in mind the fact that it is not the policy of Government in this country to give direct assistance to the study of any science?.[129]

The Committee, formed by representatives from the government and scientific institutions in equal shares, recommended in its famous 1877 report both the transfer of ocean studies to the Admiralty and the fact that a new paid Council—different from the Royal Society Meteorological Committee—be formed with the overall control of appointments and allocation of funds.[130] In addition, it recommended that a grant be allocated to 'original investigations' or 'special researches', which would be administered by the new Council. For this purpose, it assigned a sum of £1,000 /annum.[131]

However much liberty the Council had to administrate these funds—and even, as the report suggested, 'to select the investigators and fix their remuneration'—the important fact is that the Treasury Committee set the direction of research. The identification of specific items of research and the engagement of appropriate researchers should be effected by a contractual arrangement with an agency outside the very Office, an arrangement that favoured the external contractor to the detriment of the Office's internal research workers.

In a thorough work on 'The history of the British Meteorological Office to 1905', J. Burton points out that the Committee never intended that research be carried out primarily within the Office. The strong orientation of special researches towards operational and technical developments is clear in the list of projects suggested by the Council: of the ten proposals made, four of them were concerned with instrumentation and observational methods, three with synoptic meteorology and one with atmospheric electricity,

---

129 The question added a proviso: '[the study of any science] except with a view to the more immediate application of scientific theories to practical purposes in which the public rather than individuals have a direct interest'. *Report of the Treasury Committee* (1877, p. 732).

130 The Committee consisted of two members of the House of Commons, one of the Treasury, one of the Board of Trade, one of the Royal Society, two of the Meteorological Committee, and one of the Scottish Meteorological Society.

131 Burton (1988, pp. 135–36); Anderson (2005, pp. 143–44).

almost all of them going to centres outside the Office.[132] Ironically, the only basic research related to the harmonic analysis of meteorological variables, consisting of the identification of certain coefficients from a purpose-built machine, was abandoned because of lack of satisfactory results. Apart from these projects, the Office's staff also conducted some research, mostly on their own initiative, with similar orientation: the vast majority of over thirty papers published by them in the *Quarterly Journal of the Royal Meteorological Society* were simply case studies of weather events and statistical analysis of observations.[133] In any case, 'it was ironic', as Burton says so compellingly,[134]

> that the apparent victory of the pure (Galton) faction led to a situation where research had effectively no place within the Office until [Napier] Shaw took over in 1900. Funding for a research programme was, in fact, provided by the Meteorological Council after 1877, but the work was done on a contract basis that remained outside the ambience of the Office, and so any resultant stimulation was lost on the staff who worked there.[135]

More forcefully and broadly than any other meteorologist, physicist Napier Shaw pushed research on the physics of the atmosphere as a basis for national weather forecasting service.[136] In an unpublished memorandum that circulated among the Council members, Shaw proposed in 1900 what might well be regarded as the first seedbed for a 'special scientific staff' at the Office—as Galton called it[137]—with the formation of 'special inspectors' who would undertake physical investigations.[138] This proposal was supplemented by a series of research topics suggested in a second memorandum— this time he did submitted it officially to the Council—including studies on

132 For a detailed description of the program for research and its results, see Burton (1988, pp. 141–49).

133 Walker (2012, p. 117). In the same vein, he points out that the majority of around twenty papers published by the members of the Council, particularly George Stokes and Richard Strachey, in the two decades that followed its formation, 'simply described instruments or discussed observations'.

134 Burton (1988, p. 219).

135 Walker (2012, p. 117): 'Those who expected the change in management of the Office to bring about renewed dynamism were surely disappointed. Inspiration and innovation were as lacking in the 1880s and 1890s as they had been before the Council assumed control'.

136 Napier Shaw was director of the Meteorological Office from 1905 to 1920, and a member of the Meteorological Council from 1897. For biographical notes see Gold (1945); Brunt (1946); Kutzbach (1979, pp. 243–44); Burton (2004); Walker (2012, pp. 127–32).

137 Galton to Shaw, 10 June 1900, together with N. Shaw, 'Memorandum on Meteorological Office staffing', n.d. [circa May 1900], William Napier Shaw Papers, Cambridge University Library.

138 In March 1902, Shaw's idea took form when a Scientific Assistant was appointed to deal with the issues exclusively related to research.

atmospheric electricity, depressions, earth temperatures, streamlines, and radiation from clouds.[139] Furthermore, he organized biweekly evening discussions of foreign papers at the Office, which served as a reference forum for the presentation of work in progress in those topics. With these initiatives, Shaw aimed to turn what was certainly an external contractual activity into something intrinsic to, and typical of, the Office. Still, the Office did not achieve a high level of research commitment.[140]

To an experimental physicist like Shaw, who made significant efforts to promote investigations and seminars on the physics of the atmosphere, but who also bore the old burden of administrative and procedural inertia, the unbalance between research and service was both noticeable and regrettable. Meteorology, he would write in his seminal *Manual of Meteorology*, might have fared better 'if meteorologists had not been compelled to issue a series of 50,000 sets of forecasts, only more or less correct'. In that case, 'they might have given their attention to more purely scientific aspects'.[141] Shaw, like Abbe and many other meteorologists, was too aware that within the routine application of a forecasting service there was always an overload of mapping work and paperwork derived from operating procedures.

Conclusion

During the five decades spanning the 1860s and 1900s, weather prediction—above all in the form of short-range forecasts and storm warnings—flourished in national meteorological institutes. This flowering was largely due to the funds provided by central governments—through civil bodies generally related to ministries of Public Instruction, Agriculture, and Trade—and to the peculiar mode of allocating it: as money awarded to basically administrative–statistical bureaus for the mass production of weather maps and the assimilation of observational data.

If all these state patronages led to promote a service-oriented meteorology rather than a research-oriented meteorology, this was not solely due to the limited training of state forecasters or the rise of image-production technologies. Rather, in post-revolutionary Europe the science of meteorology had become so deeply permeated and affected by 'reasons of state' that the most emblematic manifestation of this discipline (i.e. weather forecast) always had a marked political character linked to national interests.

---

139 'Minutes of the Meteorological Council, 1901–1902', pp. 34–37, 42–44. National Meteorological Library, Exeter. See also Burton (1988, pp. 188–89).
140 The failure of the Meteorological Office to put research effort on a par with service provision is emphasized by Burton (1988, p. 219): 'Regrettably, and despite Shaw's efforts, a high level of research commitment did not become enshrined within the Office structure until after the Second World War'.
141 Shaw (1926, p. 9).

Seen through its historical development, the practice of weather forecasting appears to be strongly conditioned by a state pressure that tended to ensure there was 'correspondence' between the predictive methods financed and the social demands of meteorological prediction, understood in their broadest sense—namely, not only for the benefit of commerce and agriculture but also for domestic surveillance and national control. However, it must be noted that 'conditioned' does not mean 'determined'. In practice, the pressure was made possible by the strategies followed by a group of political and scientific authorities (in the case of the United States and France) or by an elite (in Britain), on the basis of their own national interests, and an ever-increasing demand for weather forecasts within the societies of those countries. On paper, other science policies would have been possible in national meteorological institutes if other members of their staffs had been able to adopt bolder and more ambitious aims, aspiring to the realization of a weather prediction on mathematico-physical bases.

Throughout this chapter, I have taken for granted a remarkable circumstance that characterized the nineteenth-century national weather institutes: the embodiment of scientific bureaucratization, especially in those fields centring on weather forecasting. The bearers of that bureaucratization were indeed seasoned scientists, but their science was largely subordinated to and geared towards public service. By contrast, the scientism to be found among the precursors of the new scientific weather prediction (both synoptic and numerical), especially among the physicists associated with the Bergen and Vienna schools of meteorology, subordinated the public service to the results of physical research. Their response to the growing bureaucratization and stagnation of national weather institutes was to stimulate the theoretical treatment of atmospheric motions on the basis of the laws of hydrodynamics and thermodynamics,[142] a claim and goal which scarcely a single one of the many state forecasters would have shared.[143] Thus if we ask what feature characterized the research promoted in such institutes as a result of their scientific bureaucratization, I think the answer must be: its subordination to public service.

---

142 Some physicists, such as the director of the *Preussische Meteorologische Institut* Wilhelm von Bezold (who also held the first professorship in meteorology in Germany), viewed the application of the laws of thermodynamics in the 1860s as the beginning of the transformation of the science of meteorology into a 'physics of the atmosphere'. Bezold (1890, p. 9). See also Kutzbach (1979, p. 46).

143 Cleveland Abbe's complain in 1890: 'Hitherto, the professional meteorologist has too frequently been only an observer, a statistician, an empiricist—rather than a mechanician, mathematician and physicist'. Abbe, untitled report, in: *Proceedings of the American Association for the Advancement of Science*, 39, 77. Among the members of the Weather Bureau's staff, Abbe was almost the only one who expressed himself publicly in such terms.

References

Abbe, Cleveland, 1895a. 'The Meteorological Work of the U.S. Signal Service, 1870 to 1891'. (U.S. Weather Bureau) *Bulletin*, 11, 232–85. Also published in: O.L. Fassig, ed., *Report of the International Meteorological Congress Held at Chicago, August 21–24, 1893. Part II*. Washington, DC: Weather Bureau, 232–84.

Abbe, Cleveland, 1895b. 'The Needs of Meteorology'. *Science*, 1 (7), 181–82.

Abbe, Cleveland, 1902. 'Meteorology and the Position of Science in America'. *The North American Review*, 174 (547), 833–44.

Abbe, Cleveland, 1912. 'The Obstacles to the Progress of Meteorology'. *Journal of the Franklin Institute*, 173, 55–71.

Abbe, Cleveland, 1914. 'The Weather Map on the Polar Projection'. *Monthly Weather Review*, 42, 36–38.

Allison, William Boyd, 1886. *In the Senate of the United States. June 8, 1886. Ordered to be printed. Mr. Allison, from the Joint Commission on the Signal Service, &c., submitted the following report: Report of the Joint Commission To Consider the Present Organizations of the Signal Service, Geological Survey, Coast and Geodetic Survey, and the Hydrographic Office of the Navy Department, with a view to secure greater efficiency and economy of administration of the public service in said bureaus*. Washington, DC: USGPO, 1886.

Anderson, Katharine, 2005. *Predicting the Weather: Victorians and the Science of Meteorology*. Chicago, IL: The University of Chicago Press.

Anderson, Katharine, 2006. 'Mapping Meteorology'. In James Rodger Fleming, Vladimir Jankovic, and Deborah R. Coen eds., *Intimate Universality. Local and Global Themes in the History of Weather and Climate*. Sagamore Beach, MA: Watson Publishing International, 69–91.

Angot, Alfred, 1896. 'La Nouvelle Carte du Bulletin International du Bureau Central Météorologique'. *Annales du Bureau Central Météorologique de France*, I (1898), B151–58.

Atlas des mouvements généraux de l'atmosphère. Paris: Charles Chauvin, 1864–1865.

Atlas des orages de l'année 1865. Paris: Charles Chauvin, 1866.

Atlas météorologique de l'Observatoire impérial: Année 1866. Paris: Charles Chauvin, 1867.

Aubin, David, 2003. 'The Fading Star of the Paris Observatory in the Nineteenth Century: Astronomers' Urban Culture of Circulation and Observation'. *Osiris*, 18, 79–100.

Ball, John, 1848. 'On Rendering the Electric Telegraph Subservient to Meteorological Research'. In *Report of the Annual Meeting/British Association for the Advancement of Science*. London, 12–13.

Beals, E.A., and Sims, A.F., 1899. 'Topic No. 7—Relations with the Press, Commercial Bodies, and Scientific Organizations. How Promoted'. *U.S. Weather Bureau Bulletin*, 24, Proceedings of the Convention of Weather Bureau Officials, held at Omaha, Nebr., 13–14 October 1898, 69–79.

Bergeron, Tor, 1981. 'Synoptic Meteorology: An Historical Review'. *Pure and Applied Geophysics*, 119, 443–73.

Bezold, Wilhelm von, 1890. 'Die neuere Witterungskunde und die Lehre von der Niederschlagsbildung'. *Himmel und Erde*, 2, 9–23, 65–71.

Bigelow, Frank H., 1902. 'Studies on the Statics and Kinematics of the Atmosphere in the United States'. *Monthly Weather Review*, 30, 13–19, 163–71, 80–87, 117–25, 250–58, 304–11, 347–54.

Bigelow, Frank H., 1903. 'The Mechanism of Countercurrents of Different Temperatures in Cyclones and Anticyclones'. *Monthly Weather Review*, 31, 72–84.

Bigelow, Frank H., 1904. 'The Mount Weather Research Observatory'. In *Proceedings of the Third Convention of Weather Bureau Officials Held at Peoria, III, September 20, 21, 22, 1904*. Washington, DC: Government Printing Office, 14–31.

Bigelow, Frank H., 1906. 'Studies on the Thermodynamics of the Atmosphere'. *Monthly Weather Review*, 34, 9–16, 74–78, 265–71, 307–15, 360–70, 470–78, 511–17, 562–72.

Bigelow, Frank H., and Moore, Willis L., 1897. *Storms, Storm Tracks, and Weather Forecasting*. Washington, DC: Government Printing Office.

Bigourdan, Guillaume, 1932. 'Le Bureau des longitudes entre la séparation de 1854 et la réforme de 1874'. *Annuaire du Bureau des longitudes*, A89–A117; 1933, A1–A91.

Blasius, William, 1887. 'The Signal Service Bureau. Its Methods and Results'. *Proceedings of the American Philosophical Society*, 24 (125), 179–83.

Brunt, David, 1946. 'Obituary Notice: Shaw, William Napier'. *Monthly Notices of the Royal Astronomical Society*, 106 (1), 35–37.

Brunt, David, 1951. 'A Hundred Years of Meteorology'. *The Advancement of Science*, 30, 114–24.

Brunt, David, 1956. 'The Centenary of the Meteorological Office: Retrospect and Prospect'. *Science Progress*, 44, 193–207.

Bulletin hebdomadaire de l'Association pour l'avancement de l'astronomie et de la météorologie (puis de l'Association scientifique de France). Paris, 1865–87.

Burton, James, 1986. 'Robert FitzRoy and the Early History of the Meteorological Office'. *British Journal for the History of Science*, 19, 147–76.

Burton, James, 1988. 'The History of the British Meteorological Office to 1905'. PhD diss., The Open University.

Burton, James, 2004. 'William Napier Shaw–Father of Modern Meteorology'. *Weather*, 59 (11), 307–08.

Chaline, Jean-Pierre, 1998. *Sociabilité et érudition, les sociétés savantes en France*. Paris: Comité des Travaux Historiques et Scientifiques.

Chief Signal Officer, War Department, 1871. *The Practical Use of Meteorological Reports and Weather Maps*. Washington, DC: Government Printing Office.

'Circulaire du 20 avril 1856'. *Ministère de l'Intérieur, Direction des lignes télégraphiques, Lois et règlements*, t. 3, année 1856. Paris: Imprimerie nationale, 1860.

'Daily Bulletin of International Meteorological Observations'. *Symons' Monthly Meteorological Magazine*, 1879, 14, 50–52.

Darter, Lewis J. Jr., 1942. 'Weather Service Activities of Federal Agencies Prior to 1891'. Introduction to L.J. Darter Jr., *List of Climatological Records in the National Archives*. Washington, DC: National Archives, 1942, i–xxxvi.

Davis, John L., 1984. 'Weather Forecasting and the Development of Meteorological Theory at the Paris Observatory, 1853–1878'. *Annals of Science*, 41 (4), 359–82.

Davis, John L., 1990. *Physics in France circa 1850–1914: Its National Organisation, Characteristics and Context*. Ph.D. diss., University of Kent.

Dines, W.H., 1918. 'Meteorology in Relation to Aeronautics'. *Scientific American Supplement*, 86, 351–52, 366–68.

Doublet, E., 1911–12. 'La Météorologie en France et en Allemagne'. *Revue Philomathique de Bordeaux et du Sud-Ouest*, 1911, 213–32, 250–67; 1912, 103–28, 169–83.

Dry, Sarah, 2009. 'Safety Networks: Fishery Barometers and the Outsourcing of Judgement at the Early Meteorological Department'. *The British Journal for the History of Science*, 42 (1), 35–56.

Dupree, Anderson Hunter, 1957. *Science in the Federal Government: A History of Policies and Activities to 1940.* New York: Harper Torchbooks.

Elenco delle pubblicazioni del R. Ufficio centrale di meteorologia e geodinamica dal 1860 al 1910. Roma: Tipografia Nazionale di G. Bertero, 1911.

Eredia, Filipo, 1914. 'L'organizzazione del servizio dei presagi del tempo in Italia'. *Rivista Meteorico-Agraria*, 35, 1001–48.

Fassig, Oliver L., 1889–91. *Bibliography of Meteorology. A Classed Catalogue of the Printed Literature of Meteorology from the Origin of Printing to the Close of 1881; with a Supplement to the Close of 1887, and an Author Index*, 4 vols. Washington, DC: Signal Office.

Fassig, Oliver L., 1892. *Extract no. 11, from Annual Report of the Chief Signal Officer, 1891: Report of Mr. Oliver L. Fassig, Bibliographer and Librarian.* Washington, DC: Government Printing Office.

Fierro, Alfred, 1991. *Histoire de la météorologie.* Paris: Éditions Denoël.

Flammarion, Camille, 1866. 'Bureau des longitudes et l'administration astronomique en France'. *Le Siècle*, 10 February 1866, 3.

Fleming, James Rodger, 1988. *Meteorology in America, 1814–1874: Theoretical, Observational, and Instrumental Horizons.* Ph.D., Princeton University.

Fleming, James Rodger, 1990. *Meteorology in America, 1800–1870.* Baltimore, MD: Johns Hopkins University Press.

Fleming, James Rodger, 2000. 'Storms, Strikes, and Surveillance: The U.S. Army Signal Office, 1861–1891'. *Historical Studies in the Physical and Biological Sciences*, 30, 315–32.

Fleming, James Rodger, 2009. 'The International Bibliography of Meteorology: Revisiting a Nineteenth-Century Classic'. *History of Meteorology*, 5, 126–37.

Fleming, James Rodger, and Goodman, Roy E., 1994. *International Bibliography of Meteorology: From the Beginning of Printing to 1889. Four Volumes in One: Temperature, Moisture, Winds, Storms.* Upland, PA: Diane Publishing Co.

Fox, Robert, 2012. *The Savant and the State: Science and Cultural Politics in Nineteenth-Century France.* Baltimore, MD: The Johns Hopkins University Press.

Fuller, John F., 1990. *Thor's Legions. Weather Support to the U.S. Air Force and Army, 1937–1987.* Boston, MA: American Meteorological Society.

Gaster, Frederic, 1896. 'Weather Forecasts and Storm Warnings: How They Are Prepared and Disseminated'. *Quarterly Journal of the Royal Meteorological Society*, 22, 212–28.

Glassford, W.A., 1891. 'Synoptical Sketch of the Progress of Meteorology in the United States'. *The American Meteorological Journal*, 9, 151–64. Reprinted for the *Annual Report of the Chief Signal Officer for 1891*, Appendix 8.

Gold, E., 1945. 'William Napier Shaw. 1854–1945'. *Obituary Notices of Fellows of the Royal Society of London*, 5, 202–30.

Grandidier, Alfred, 1882. *Exposition universelle international de 1878 à Paris. Rapport sur les cartes et les appareils de géographie et de cosmographie, sur les cartes géologiques et sur les ouvrages de météorologie et de statistique.* Paris: Imprimerie Nationale.

Grice, Gary K., 1991. *The Beginning of the National Weather Service: The Signal Service Years (1870–1891) as Viewed by Early Weather Pioneers.* Washington, DC: National Weather Service.

Grisollet, Henri, 1950. 'Histoire administrative et scientifique du Service d'études et de statistiques de la ville de Paris et des observatoires de Montsouris et de la Tour Saint-Jacques'. *La Météorologie*, 129–42.

Halford, Pauline, 2004. *Storm Warning. The Origins of the Weather Forecast.* Stroud: Sutton Publishing.

Handbook of Labor Statistics. Washington, DC: US Department of Labor, 1973.

Harrington, Mark W., 1894. 'History of the Weather Map'. *U.S. Weather Bureau Bulletin*, 11, 327–35. Reproduced in Oliver L. Fassig ed., *Report of the International Meteorological Congress, Held at Chicago, Ill., August 21–24, 1893. Part II.* Washington, DC: Weather Bureau, 1895, 327–35.

Hawes, Joseph M., 1966. 'The Signal Corps and Its Weather Service, 1870–1890'. *Military Affairs*, 30, 68–76.

Hazen, William Babcock, 1884. *History of the Signal Service, with Catalogue of Publications, Instruments, and Stations.* Washington: U.S. Signal Office.

Henry, Alfred J., 1924. 'Frank Hagar Bigelow, 1851–1924'. *Monthly Weather Review*, 52, 165–66.

Hildebrandsson, Hugo H., and Teisserenc de Bort, Léon Philippe, 1898. *Les bases de la météorologie dynamique.* Paris: Gauthier-Villars.

Hughes, Patrick, 1994. 'The Great Leap Forward'. *Weatherwise*, 47 (5), 22–27.

Iafrate, Luigi, 2001. 'Verso un Ufficio meteorologico centrale anche in Italia? Dai primi fermenti organizzativi alla sua istituzione governativa: Regio Decreto n° 3534 del 26 novembre 1876'. In *Presenze scientifiche illustri al Collegio Romano: Celebrazioni del 125° anno di istituzione dell'Ufficio Centrale di Ecologia Agraria.* Roma: MIPAF-UCEA, 50–58.

Jourdain, Charles, 1867. *Rapport sur l'organisation et les progrès de l'instruction publique.* Paris: Imprimerie Impériale.

Kennerley, Alston, 1996. 'Frank Thomas Bullen, 1857–1915: Whaling and Nonfiction Maritime Writing'. *The American Neptune*, 56, 353–70.

Khrgian, Aleksandr Khristoforovich, 1970. *Meteorology: A Historical Survey*, 2nd ed. Jerusalem: Israel Program for Scientific Translations.

Kutzbach, Gisela, 1979. *The Thermal Theory of Cyclones: A History of Meteorological Thought in the Nineteenth Century.* Boston, MA: American Meteorological Society.

'Le Bureau Central Météorologique'. *La Revue scientifique de la France et de l'étranger*, 1878, 21, 1124.

Le Verrier, Urbain-Jean-Joseph, 1865. 'Organisation de quelques entreprises météorologiques'. *Comptes Rendus de l'Académie des Sciences*, 60, 136–44.

Le Verrier, Urbain-Jean-Joseph, 1868. *Historique des entreprises météorologiques: 1864–1867, Observatoire impérial de Paris.* Paris: Gauthier-Villars.

Lempfert, R.G.K., 1913. 'British Weather Forecasts: Past and Present'. *Quarterly Journal of the Royal Meteorological Society*, 39, 173–84.

Lequeux, James, 2013. *Le Verrier – Magnificent and Detestable Astronomer.* New York: Springer. Edited and with an introduction by William Sheehan. Translated By Bernard Sheehan.

Lewis, R.P.W., 1981. 'The Founding of the Meteorological Office, 1854–55'. *Meteorological Magazine*, 110, 221–27.

Locher, Fabien, 2008. *Le savant et la tempête. Étudier l'atmosphère et prévoir le temps au XIX siècle.* Rennes: Presses Universitaires de Rennes.

MacMahon, Arthur W., and Millett, John D., 1939. *Federal Administrators: A Biographical Approach to the Problem of Departmental Management.* New York: Columbia University Press.

Maluquer de Motes, Jordi, 2010. 'El coste de la vida en Francia y España en perspectiva comparada (1840–2009)'. In Gérard Gayot and Gérard Chastagnaret eds.,

Los niveles de vida en España y Francia, siglos XVIII-XX: 'in memoriam' Gérard Gayot. Alicante: Publicaciones de la Universidad de Alicante, 189–212.

Mascart, Éleuthère Élie Nicolas, 1891. 'Rapport sur le service des observations météorologiques dans les écoles normales primaires'. *Revue pédagogique*, 18, 170–76.

Miller, Eric R., 1931a. 'The Evolution of Meteorological Institutions in the United States'. *Monthly Weather Review*, 59 (1), 1–6.

Miller, Eric R., 1931b. 'New Light on the Beginnings of the Weather Bureau from the Papers of Increase A. Lapham'. *Monthly Weather Review*, 59 (2), 65–70.

Moigno, François Napoléon Marie, 1867. 'Nouvelles scientifiques de la semaine: Association scientifique de France'. *Les Mondes*, 14, 353–54.

Monmonier, Mark, 1988. 'Telegraphy, Iconography, and the Weather Map: Cartographic Weather Reports by the United States Weather Bureau, 1870–1935'. *Imago Mundi*, 40 (1), 15–31.

Monmonier, Mark, 1999. *Air Apparent. How Meteorologists Learned to Map, Predict, and Dramatize Weather*. Chicago, IL: The University of Chicago Press.

Moore, Willis Luther, 1903. *Report of the Chief of the Weather Bureau for 1903*. Washington, DC: Government Printing Office.

Noël-Waldteufel, Marie-France, 1995. 'La mise en place des réseaux d'observation météorologique dans les écoles normales d'instituteurs'. *La météorologie*, special issue April, 11–15.

O'Donoghue, Jim, Goulding, Louise, and Allen, Grahame, 2004. *Consumer Price Inflation since 1750*. London: Office for National Statistics and Economic Trends No. 604.

Popkin, Roy, 1967. *The Environmental Science Services Administration. Including: The Coast and Geodetic Survey, the Weather Bureau, the Institute for Telecommunications Sciences and Aeronomy and Other Related Services*. New York: Praeger.

Prost, Antoine, 1968. *Histoire de l'enseignement en France, 1800–1967*. Paris: Librairie Armand Colin.

Radau, Rodolphe, 1868. 'L'observatoire de Paris depuis sa fondation'. *Revue de Deux Mondes*, 73, 751–52.

Raines, Rebecca Robbins, 2011. *Getting the Message through a Branch History of the U.S. Army Signal Corps*. Washington, DC: Center of Military History, United States Army.

Rayet, Georges, 1903. *Note historique sur la Commission météorologique de la Gironde et sur ses travaux*. Bordeaux: Imprimerie de G. Gounouilhou.

Reed, Richard J., 1977. 'The Development and Status of Modern Weather Prediction'. *Bulletin of the American Meteorological Society*, 58, 390–99.

Report of a Committee Appointed to Consider Certain Questions Relating to the Meteorological Department of the Board of Trade [Galton Report], vol. 65. London: Eyre & Spottiswoode, 1866. Also in *Parliamentary Papers*, 1866.

Report of Officer in Charge of the Property and Disbursing Division, ARCSO, 1884, 50–58; and *ARCSO*, 1886, 22–25.

Report of the Assistant in Charge of the Study Room, ARCSO, 1884, 59–67; and *ARCSO*, 1886, 208–12.

Report of the Chief Signal Officer for 1888. Washington, DC, 1889.

Report of the Committee Appointed to Inquire into the Administration by the Meteorological Council, of the Existing Parliamentary Grant, pp. 1904, Cd. 2123, XVIII. *Report of the Meteorologic Office of the Board of Trade* (1863, 1864).

Report of the Meteorological Office of the Board of Trade (1855, 1857, 1858, 1862).

Report of the Treasury Committee Appointed to Inquire into the Conditions and Mode of Administration of the Annual Grant in Aid of Meteorological Observations. 1877, XXXIII, 731–957.

Russell, Thomas, 1895. *Meteorology, Weather, and Methods of Forecasting; Description of Meteorological Instruments, and River Flood Predictions in the United States.* New York, London: Macmillan and Co.

Scott, Robert H., 1875. 'Weather Charts in Newspapers'. *Journal of the Society of Arts*, 23, 776–82.

Scott, Robert H., 1894. 'The Publication of Daily Weather Maps and Bulletins'. In Oliver L. Fassig ed., *Report of the International Meteorological Congress Held at Chicago, Ill., August 21–24, 1893.* Washington, DC: Weather Bureau, 1, 6–9.

Shaw, William Napier, 1926–31. *Manual of Meteorology*, 4 vols. Cambridge: Cambridge University Press.

Smyth, Charles Piazzi, 1871. 'Hyperborean Storm of the 2nd and 3rd October 1860'. *Edinburgh Astronomical Observations*, 13, T83–T141.

'The Times Weather Chart'. *Nature*, 15 April 1875, 11, 473–74.

Toutain, Jean Claude, 1997. 'L'imbroglio des indices de prix français du XIXe siècle'. *Économies et Sociétés*, série HEQ, 1 (11), 137–97.

U.S. Congress, Senate, 1886. *Testimony before the Joint Commission to Consider the Present Organizations of the Signal Service, Geological Survey, Coast and Geodetic Survey, and the Hydrographic Office of the Navy Department, with a View to Secure Greater Efficiency and Economy of Administration of the Public Service in Said Bureaus.* Washington, DC: Government Printing Office.

U.S. Weather Bureau, 1894. *Report of the Third Annual Meeting of the American Association of State Weather Services, Co-operating with the Weather Bureau, U.S. Department of Agriculture.* Washington, DC: Weather Bureau (*Bulletin*, 1892, 14).

Walker, J. Malcolm, 2012. *History of the Meteorological Office.* Cambridge: Cambridge University Press.

Weber, Gustavus A., 1922. *The Weather Bureau. Its History, Activities and Organization.* New York and London: D. Appleton and Company.

Weisz, George, 1983. *The Emergence of Modern Universities in France, 1863–1914.* Princeton, NJ: Princeton University Press.

Whitnah, Donald R., 1961. *A History of the United States Weather Bureau.* Urbana: University of the Illinois Press.

Wiley, Harvey W., 1924. 'Frank H. Bigelow'. *Science*, 59 (1532), 423.

Willis, Edmund P., and Hooke, William H., 2004. 'Cleveland Abbe and the Birth of the National Weather Service, 1870–1891'. *Proceedings of the International Commission on History of Meteorology*, 1, 48–54.

8 Meteorological cartography

Introduction

In the spring of 1863 the assistant of Le Verrier and head of meteorology section at the Paris Observatory, Hippolyte Marié-Davy, decided to develop a new synoptic method that could foretell the arrival of storms from meteorological charts. Well aware of the increasing interest in storm warnings as part of the surge of Napoleon III's nationalist spirit of defence and Le Verrier's reluctance towards forecasting, he noted the lack of progress in this endeavour. The key to storm forecast, he argued in a note read before the Academy of Sciences, lay not in the identification of *ondes atmosphériques* whose existence at first seemed consistent with the barometric data showed on the charts, and whose search was the practice theretofore followed at the Observatory.[1] Rather, it was a matter of dynamical isobaric geometry, a view far removed from Maury's statistical wind charts, which sought the most favourable conditions for navigation rather than the origin and course of storms.[2] The method, he announced in his note, consisted of identifying areas of low pressure—or coherent atmospheric patterns, also called *bourrasques*—from the isobars plotted on the maps. By the end of 1863, he had a successful technique, culminating with the first charts ever published in the Observatory's *Bulletin* and the launch of the weather forecasting service (see Figure 3.2 in Chapter 3).[3]

For Marié-Davy, the charts published in the *Bulletin* aimed to be more than just retrospective maps of past weather or pictures of the combined elements of the weather. Rather, in line with the Napoleonic security policy, they were seen as effective tools for the study of individual storms and their

1 The note was presented by Le Verrier on 17 August 1863. Marié-Davy (1863b, pp. 384–86). By April 1863, he had virtually dismissed the hypothesis of atmospheric waves: 'it is difficult to trace any isolated waves with distinct patterns of movement'. Marié-Davy (1863a, p. 233)—quoted by Locher (2009, p. 86).
2 On Maury's statistical charts and how they enabled sailing ships to considerably reduce the duration of journeys, see Lewis (1927, pp. 51–63, 83, 144) and Pinsel (1981, pp. 123–37).
3 The Paris Observatory's *Bulletin* of 16 September 1863 first included morning charts of isobars and winds from the days 7 and 10 September. See Locher (2008, pp. 114–15).

progress, which he thought would be foreseeable only by determining 'the state of the atmosphere over the largest possible extent' in synoptic charts.[4] To meet this ambitious goal, he proposed an ingenious method: first, by marking on a map each weather station and its pressure recorded and linking these points, he would create the map of isobars. As the drawing of isobars was made by interpolation among stations, these isobars would never intersect. Then, plotting 'courbes quasi-circulaires' instead of applying the linear rule 'point à point', he would draw isobars in the form of tight curves, which he defined as areas of low pressure.[5] Thus, forecasting seemed feasible. For as he noted with regard to a synoptic chart of Europe, 'the form of the isobarometric lines that cross the north of France and Holland suggests that the center of the phenomenon is over Scotland'.[6] The low-pressure system, as it deformed isobars, would certainly allow him to ascertain where its centre was and how it moved.

This fact was not trivial at all. Marié-Davy's maps were the centrepiece— and symbol par excellence—of what was to become the most common way of analyzing air motions in the atmosphere: the Eulerian approach. This approach was synoptic by its very nature for the observer focused on air movement as recorded at fixed points, or stations. The essence of synoptic representation was to produce a series of snapshots of the atmosphere (from a bird's eye view) at chosen intervals of time. The effect of this representation was twofold: on one hand, the observer had the impression that weather travelled throughout the map—in fact, in this displacement, the 'lows' seemed to be the true carriers of the bad weather; on the other hand, for him, it was the barometric pressure distribution, more than anything, that determined the character of weather—not in vain did the 'lows' embody the areas of low pressure.[7]

Over the course of the next three decades, the path initiated by the Paris Observatory was followed by others, and many national meteorological institutes began issuing daily weather maps. First to do so was the U.S. Weather Service in 1871, followed by the British (1872), the Russian (1873), the Danish and the Swedish (1874), the Belgian and the German (1876), plus

4 Marié-Davy (1866, p. 37).
5 *Courbes quasi-circulaires* or quasi-circular curves. Following this rule, the 760 mm Hg isobar should be located equidistantly between the weather stations whose barometric values are 757 and 763 mm, respectively. This question is discussed by Locher (2008, p. 115).
6 Marié-Davy (1863b, p. 385).
7 The dominance of the areas of low and high pressure was highlighted by U.S. Weather Bureau's professor, Alfred J. Henry, in a lecture given on 'weather forecasting from synoptic charts' in 1906. Henry (1906, p. 302): 'The terms "high pressure" [and] "low pressure" have a special significance in modern weather forecasting. [...] The character of the weather is closely related to the distribution of atmospheric pressure; if, therefore, the distribution of atmospheric pressure on any certain day can be accurately foreseen a forecast of the weather would follow quite easily'.

thirteen more by 1900.[8] As I will show below, although each used different techniques and scales, the various weather services gradually stitched various local data to create a picture of the state of the atmosphere over each nation-state. Heavy black lines (isobars) and coloured dotted lines (isotherms) ride the skies on a background of blue water and white land, creating an aerial overview of the national—and sometimes international—meteorological space. At the same time, a set of symbols and arrows representing winds and precipitations reflected the sutured scars of the interior of the country, reminding viewers of the importance of the local atmospheric conditions within the framework of national unity. Each in their own way, then, and adopting Humboldt's known sentence that maps 'speak to the senses without fatiguing the mind',[9] the services transformed the traditional tables into a picture of the nation that stretched 'as high and as far as the weather reaches'.

Synoptic charts show both the potentials and limitations encountered in the efforts to use graphic methods to solve the riddle of how the weather could be forecasted. On one hand, they were powerful instruments not only for elucidating the general natural laws governing atmospheric changes, but from the 1850s onwards, they went a step further with the advent of telegraphy, becoming a means to forecast the development of atmospheric disturbances. Their function, therefore, shifted and went from being a tool *for diagnosis* to a tool *for prognosis*. Moreover, as visual objects, they were inscribed in a language that overcame the cultural barriers derived from languages and verbal expression, and that was highly comprehensible to the general public. These attributes, therefore, made them seem like ideal tools to simplify the complexity of atmospheric phenomena and verbal description.[10] Indeed, this was the common goal behind the spirit of most weather charts and climatological maps—the creation of the official picture of a national meteorological space that theretofore did not exist. It was meant to provide a tangible and national form to the intangible and universal idea of the weather. As stated in a book on weather forecasting, 'once the symbolism is known, the [weather] map allows us to assimilate a vast amount of

---

8 Shaw (1926, p. 287) provides a list with dates of the first daily weather-maps published by several national meteorological services. In addition to the above-mentioned services, the list includes the services of Algeria (1877), Australia (1877), Austria (1877), India (1878), Italy (1880), Switzerland (1880), Portugal (1882), Japan (1883), Spain (1886), Holland (1890), Canada (1895), Greece (1896), Mexico (1899), Argentine (1902), Egypt (1902), Hungary (1905), China (1906), Romania (1908), Norway (1912), and Poland (1922).

9 Humboldt (1811, p. cxxxii)—quoted in Robinson and Wallis (1967, p. 122).

10 The Chief of the U.S. Weather Bureau, Willis Moore, went even further in 1898, maintaining implicitly, and to a certain extent explicitly, that any weather forecast that was graphically published was ipso facto understood by the mass public, regardless of the techniques with which it had been produced: 'There is hardly a daily paper that does not publish weather forecasts in a prominent place, and there is scarcely a reader who fails to note the predictions'. Moore (1898–1899, p. 12).

information at a single glance. To transmit the same information in writing, a long, hard-to-read, and tough-to-memorise text would be required'.[11]

At the same time, however, such synoptic maps as the Paris Observatory's can also be viewed as tools with important limitations. As data-derived images, they were produced by a process of routine tasks that formed part of the state administration in the same way that the dense data collections were being processed by national statistical bureaus. Like national censuses and statistical atlases, they became effective devices both for crafting images of the national space and for generating knowledge for control and security purposes. Nevertheless, they lacked both the completeness and believability to serve as the basis for the meteorological investigations then underway. Based almost exclusively on surface observations and alleged correlations between pressure and the weather, they constituted series of charts each showing separate elements rather than one comprehensive view of the atmospheric state useful for researchers. Hermann von Helmholtz and Theodor Reye, for example, developed their thermal and convective theories of storms in the 1870s without the benefit of synoptic charts.[12] Nor was there hardly any attempt made to show subtle nuance or mixing, with all the information integrated into a single central map. For all these reasons, then, it is no wonder that, as theorists themselves avowed, the daily weather maps were not crafted to uncover the hydrodynamic and thermodynamic laws governing atmospheric phenomena.[13]

By the beginning of the twentieth century the dissonance found in national weather maps between their potential as synoptic representations and their use for forecasting service rather than for research would begin to dissipate slowly. This chapter charts all this extensive development of more than six decades, first showing how synoptic weather maps evolved into authoritative representations capable of displaying not only the state of atmospheric conditions (i.e. for weather *diagnosis*) but also serving as operational tools for storm forecasts (i.e. for weather *prognosis*). Next it follows with a detailed examination of synoptic maps produced in the 1850s and 1860s, showing how these maps reflected the coexistence of two opposing approaches—the Eulerian and the Lagrangian—to weather cartography and the study of prediction in general. While the former emphasized the visual power of isolines

---

11 Javelle, Rochas, Pastre, Beaurepaire, and Jacomy (2000, p. 86). Translation mine from French original.
12 Kutzbach (1979, pp. 88–99).
13 As I quoted in the Introduction (p. 6), Vilhelm Bjerknes clearly revealed this fact when he characterized the role played by weather maps in the conceptualization of the theory of 'fronts' at the Bergen School: 'During 50 years meteorologists all over the world had looked at weather maps without discovering their most important features. I only gave the right kind of maps to the right young men, and then they soon discovered the wrinkles in the face of Weather'. Quoted in Bergeron, Devik, and Godske (1962, p. 18).

and the determining role of the baric field in weather conditions, the latter drew on symbols and paths to show the motion of air masses.

The second part of the chapter focuses on the primacy of the Eulerian line of thought over the Lagrangian approach in weather mapping from the early 1870s onwards. From then the isoline became the dominant feature of synoptic map. This primacy, I contend, not only resulted from procedural and operational issues such as standardization and image-production but was also stimulated by the expansion of statistical cartography. The rise of statistical graphics and bureaus, together with technological advances in map-making and printing, helped endow maps with scientific legitimacy. Then, I will show how the support of government agencies both reinforced the isolinear character of weather charts and geared them towards forecasting and storm warning rather than investigation. The last section deals with the relation between local observation, national expertise, and global knowledge. Instead of the supposed balance between the local and the global as often defended by historians, I contend that there was a *struggle* of forces expressed at three levels (local, national, and global), a struggle in which the global was shaped by a voluntarist internationalism that was subordinate to national interests.[14]

Pre-1851 weather maps: diagnosis and retrospection

With the advent of telegraphy the synoptic weather map progressively became the analytic tool for meteorological studies. National weather agencies never pioneered synoptic techniques, quite the contrary. Inasmuch as demands for maritime security became more pressing, they picked up the torch left by innovators in cartographic meteorology, such as Heinrich Wilhelm Brandes and Alexander von Humboldt, especially between the 1820s and 1840s. However, these innovators had developed the graphical displaying of

---

14 Studies on the development of 'scientific' maps and charts have recently experienced an increase in scale and number in the historiography of science. Map genealogy and thematic mapping have received attention from Konvitz (1987), Monmonier (1988, 1999), Palsky (1996, 1998), Robinson (1982), and Robinson and Pentchenik (1976). Useful studies on the genealogy of maps include Robinson (1971) and Robinson and Wallis (1967, 1982). This literature has established the milestones in the history of meteorological cartography, convincingly determining who were precursors and which were their influences, showing for example how the introduction of isothermal lines was inspired by the magnetic declination charts of Edmund Halley. However, this literature tends to show a somewhat teleological and progressivist development of meteorology mapping, and tends to ignore its shortcomings and constraints, as well as institutional and socioeconomic aspects related to this development. Anderson (2006) has analyzed the influence of the standard genealogies, questioning the supposed stability of the conventions of weather mapping transmitted by such genealogies. This chapter will address issues that have been ignored or have received little attention by the genealogist literature of meteorology mapping, and will readdress issues outlined in Anderson's study (for instance, the relations between local observations and global knowledge).

atmospheric conditions as a means to an exclusively physical end: weather *diagnosis*. As I will show below, for Humboldt, as for many physicists dealing with this issue, synoptic maps were not a means to foretell the course of atmospheric phenomena—they were an essential tool for ascertaining the laws of nature and its physical causes.

The physicists and meteorologists of the 1820s turned their attention to weather mapping as a result of different factors: their view of the weather as a spatial concept, their interest in causal laws of nature (rather than celestial influences or relations between weather, insects, and health), the resort to precision measurement to ensure basic standards and constants, and their emphasis on drawing lines (rather than on merely measuring and calculating) as a means for precision measurement. As I will show, these interests led them to create *retrospective* maps of weather—i.e. with reference to distant past events.

In the 1810s Brandes of the University of Breslau regarded weather data as worth mapping, and he did so on the basis of a retrospective reading of data. Thus, when he sent a letter to the *Annalen der Physik* in 1816, he stated that the plotting of graphic symbols 'would please and instruct the public more than would weather tables', and he wondered about their advantages, 'if one could draw maps of Europe according to the weather for all the 365 days of the year'.[15] Or, when three years later, in an article sent to the same journal, he announced the results of this initiative, asserting that he had identified two major storms—occurred on 6 and 12 March 1783—as centres of low pressure, ringed by isobars or 'lines in which the barometric pressures were equally low'.[16] Although his article contained no map, there seems to be no question that he mapped said storms by following through 365 weather maps for 1783. Or when in 1820, in his book on atmospheric conditions during 1783, *Beiträge zur Witterungskunde*, he investigated the relation between winds and air pressure—supposedly plotting their isolines and pressure centres—from data recorded by the Societas Meteorologica Palatina at Mannheim in that year.[17] In either case, Brandes introduced a

15 Brandes (1817). Quotation from an English translation of some extracts of Brandes's letter, by Schneider-Carius (1975, p. 179).
16 Brandes (1819). Quotation from Schneider-Carius (1975, p. 182). The lack of cartographical evidence caused a certain amount of controversy among meteorologists, historians, and scholars in general. Most authors accepted Brandes's reconstruction of weather maps for March 1783. For a comprehensive discussion of this issue, see Monmonier (1999, pp. 18–22). Two examples of misjudgement of Brandes's contribution are Khrgian (1970, p. 138), who pointed out that Brandes himself used the conditional in referring to the maps, and Scultetus (1943), who poorly reconstructed a chart drawn by Brandes concerning deviations from mean air pressure on 24 December 1821.
17 Brandes (1820) intuitively perceived the value of the synoptic method in meteorology: 'We have an almost incalculable series of weather observations, most of which, like a buried treasure, are without any use for science, for no one will take the trouble—to be sure, very great trouble—to derive from the thousands of observations appropriate

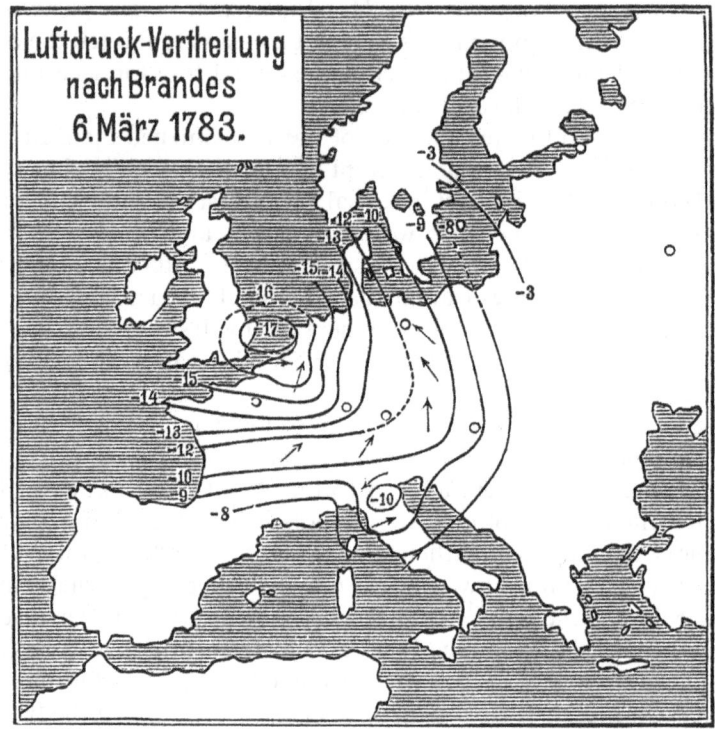

Figure 8.1 Reconstruction of the weather map for 6 March 1783 that Brandes supposedly drew from data recorded by the Societas Meteorologica Palatina
at Mannheim in that year.
Source: Trabert (1905, p. 65).

twofold innovation: on one hand, he represented atmospheric pressure with
isolines; and on the other hand, he juxtaposed air pressure and wind on the
same chart (Figure 8.1).[18]

However praiseworthy Brandes's innovation and his connection of weather
events with systems of low pressure may seem, it must be recognized that
his maps intended not so much to determine the causes of phenomena—or,
using a medical simile, to diagnose a disease from its signs and symptoms—
as to instruct public by retrospective weather analyses. Grappling as we

---

comparisons and thus to test whether we can get results from them'—translated by
Schneider-Carius (1975, p. 183). See also Robinson (1982, pp. 73–74), Kington (1988,
pp. 18–19).
18 On Brandes's two innovations, see Monmonier (1999, p. 23), and Robinson and Wallis
(1987, pp. 156–57, 223–24). For the use of daily weather maps retrospectively derived from
the Palatina data, and as a means to obtain information about atmospheric circulation
patterns associated with historical events in 1781, see Kington (1980, 1988).

are with the connection between synopsis and diagnosis, we should recall Edmund Halley's famous isogonic maps (1701), as well as his no less well-known chart of the winds of the tropical Atlantic (1688), deemed by historians as 'the parents of a most numerous progeny'.[19] Not in vain did both his isogons (or curved lines joining points of equal or constant magnetic declination) and his directional arrows (or symbols representing winds) have a markedly 'diagnostic' character: the former, in Halley's view, 'shew at one View all places where the Variation of the Compass is the same'[20]; and the latter 'shew at one view all the various Tracts and Courses of these Winds'.[21] In fact, it was these wind charts that led him to formulate in 1686 one of the earliest theories of the general circulation of the atmosphere, i.e. that the trade winds resulted from 'the action of the sun's beams upon the air and water', rather than from the air lagging behind the rotating Earth—as propounded by Edme Mariotte almost at the same time.[22]

Directly or indirectly, Halley's isogonic maps not only generated but also nourished the conviction, barely established in the minds of physicists and cartographers in the 1810s, that isolines—or, to be more precise, isometric lines portraying numerical distributions related to geographical facts—could be an analytical tool for weather diagnosis, as the values they represented could in fact exist at points on the earth.[23] Following the trail of Halley and independently from Brandes, German naturalist Alexander von Humboldt mapped temperature conditions on the earth in a study on the geographical distribution of plants in 1816. In this paper, he introduced the term isothermal parallel as a 'curve drawn through the points on the globe which receive an equal quantity of heat'.[24] Thus, his proposition that 'the lines of equal annual heat, or to coin a new word, the isotherms, do not run parallel to the equator but like the magnetic lines, they also cut the geographical latitudes at a different angle',[25] was but a

19 Chapman (1941, p. 122). Considered by cartographic specialist Norman Thrower (1969, p. 664) as one of the most important maps in the history of cartography, Halley's Atlantic chart is one of his five major contributions to thematic geo-cartography. The remaining legacy includes a map of wind patterns, a map of tides, a chart of the Channel between England and France, and a map of the shadow left by the Moon in a solar eclipse. See Monmonier (1999, pp. 23–8), Robinson (1982, pp. 46–50, 69–71, 84–6).
20 Quoted by Monmonier (1999, p. 24).
21 Halley (1686, p. 163).
22 Ibid., p. 165. For Halley's identification of solar heating as the cause of atmospheric motion and its differences with Mariotte's theory, see Burstyn (1966, pp. 171–75).
23 Robinson (1971, pp. 49–50) highlights this characteristic in distinguishing isometric lines from isopleths (or isolines portraying more complex distributions): their essential feature is that they show values than can exist on the earth's surface and atmosphere, irrespective of whether the mapped data were derived by single measurement, averaging of time series, or some other statistical manipulation.
24 Humboldt (1816, p. 229).
25 This definition of isotherm appeared in a later study, Humboldt (1817a). Quotation from an English translation of parts of Humboldt's work, by Schneider-Carius (1975, p. 169).

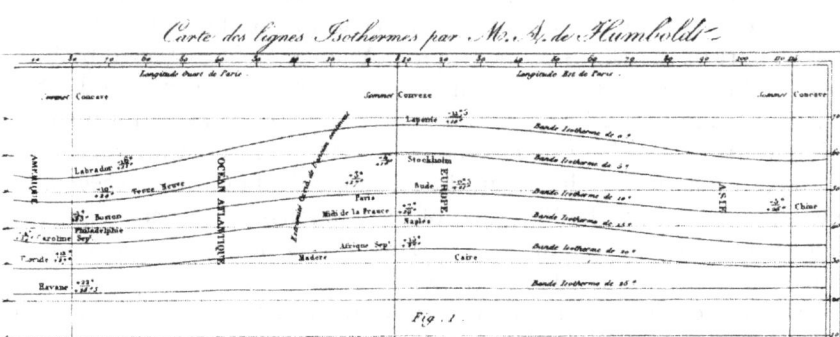

Figure 8.2 Alexander von Humboldt's map of average temperatures. The area portrayed ranges from the Equator to 70°N and from North America to China.
Source: Humboldt (1817b, pp. 112–13).

tacit acknowledgement of the influence of Halley's lines: isogons had an activating effect on isotherms.

It is noteworthy, however, that Humboldt's chief object was not to afford a compendium of temperatures at points on the earth but rather to deduce from his data the major laws governing the temperature distribution, and, thus, to challenge the prevailing view that climate depended only on latitude. He was not interested in small variations or coastlines, but in relations between his *isothermes* and the geographical latitude; this fact is clearly shown in the 1817 map of average temperatures (Figure 8.2), where his isotherms formed convex and concave curves diverging from the parallels of latitude.[26] Hence his map was extremely simple and intuitive: as he himself confessed, a map 'overcharged with signs, becomes confused, and loses its principal advantage, the power of conveying at once a great number of relations'.[27]

To a considerable extent, behind Humboldt's cartographic innovation lay his efforts to distance himself from earlier thermometric observers, like Louis Cotte, who unsuccessfully tried to correlate local mean temperatures with predetermined grids of geometrical latitude.[28] Insomuch as temperature, like magnetism, was the result of many simultaneously acting physical

26 Although a long discussion of this map is found in Humboldt (1817a), the map was indeed published in Humboldt (1817b), on a folded sheet between pages 112 and 113.

27 Quotation from an English translation of parts of Humboldt's 1817a paper, by Robinson and Wallis (1987, p. 120).

28 For the mapping projects of Louis Cotte and others in the 1770s and 1780s, as well as the growing interest in precision in instrumentation and measurement at that time, Feldman (1990, pp. 164–77), and Jordanova (1979, pp. 119–46). For Humboldt, these projects had far more to do with correlating numbers for a bureaucratic administration than with scientific cartography. See Dettelbach (1993, Chapter 3).

causes, the laws of climate, Humboldt argued, could not be established by gathering reports of mean temperatures and then averaging at short intervals of latitude—as his predecessors did. As he categorically asserted in his 1817 paper, temperature was an essentially physical phenomenon:

> Temperature and magnetism are not like those phenomena which, derived from a single cause or a central force, can be freed of the influence of disturbing causes by restricting attention to the mean results of a great number of observations, in which these foreign effects reciprocally counteract and destroy one another. The distribution of heat, like the declination and inclination of the compass needle or the intensity of magnetic force, is essentially conditioned by location, composition of the soil, by the proper ability of the earth's surface to radiate heat. One must beware of eliminating what is sought.[29]

The value of isolines as a tool for climate diagnosis became clear, at least symbolically, in the empirical nature that isotherms had for Humboldt. 'Before establishing the bases of a system', he stated in the same paper, 'one should gather facts, set numerical relations, and, [...] subject heat phenomena to the empirical laws, as Halley did with those of terrestrial magnetism'.[30] In the same way that Halley's Atlantic isogons were inexorably empirical, isotherms were inexorably the outcome of precision measurement: while the former reflected the departure of magnetic north from geographic north, the latter revealed the deviation of average temperature from the theoretical patterns of geographic latitude.[31]

Humboldt's isotherms stimulated a certain amount of similar cartographic works in physicists' and geographers' circles. German meteorologist and a professor of physics at the University of Halle, Ludwig Friedrich Kämtz showed in 1832 the distribution of annual mean temperatures over the globe from records of 145 stations (instead of just 58 stations used by Humboldt).[32] Dove followed in part his steps when, noticing the insufficiency of annual isotherms, first published isothermal charts of the globe for each month in 1852.[33] Such isotherms, in the two cases, represented records of averages encapsulating multiple (and not necessarily synchronized) observations, performed for climatologic purposes.

---

29 Humboldt (1817a, p. 469). Quotation from the English translation made by Dettelbach (1999, p. 482).

30 Ibid., p. 487.

31 This similarity in purpose is pointed out by Monmonier (1999, p. 25). For isotherms as expressions of precision measurement, Dettelbach (1999, pp. 481–90). See also Munzar (1967, p. 361), and Schneider-Carius (1975, pp. 164–79).

32 Kämtz (1832). Similar maps were constructed by meteorologist Wilhelm Mahlmann in 1836 and 1840. See Shaw (1926, p. 291).

33 Dove (1852). For the development of annual isotherm maps from Humboldt to Dove, see Meinardus (1899).

However, when assessing the influence of Humboldt's isothermal map and Brandes's isobars on their successors, the question arises of whether the isoline mapping awakened interest solely in the meteorological world or whether it extended to other circles such as geographers and cartographers. There is every indication that the geographical dictionaries and atlases on both sides of the Atlantic included isoline maps in their contents, and that the interpretation of atmospheric phenomena in these books was presented as an empirical fact, as a *fait accompli*. Although the beginning of the dissemination of Humboldt's isothermal map can be found in J.S.L. Gehler's *Physikalischen Worterbuch* in 1827,[34] isoline mapping was greatly stimulated by its inclusion in the seminal *Physikalischer Atlas* published by Heinrich Berghaus in several editions.[35] In what is considered by many as the first comprehensive physical atlas, the isarithmic technique was not only shown through Humboldt's isothermal curves but also applied to other phenomena such as barometric pressure, annual precipitation, and thunderstorms.[36]

The maps and especially the texts accompanying those illustrations depicted a reassuring portrayal of the strength of science to disclose natural laws and patterns of atmospheric phenomena. The remarks on what to map, the criteria whereby the atlas authors defined the different types of isolines, the decisions determining the display of averages—instead of minima and maxima—of atmospheric values all of these judgements contributed to disseminate the isarithmic technique and its isolines among the map users.[37]

For the purposes of my investigation the key question is not to what extent climatic maps were disseminated through physical atlases or who pioneered the various types of isolines. Rather, the crux of the matter is how and why synoptic map became a means to diagnose in retrospect the state of the

---

34 Robinson and Wallis (1987, p. 122).
35 The first edition, published in its entirety by Justus Perthes at Gotha in 1845, includes the map entitled 'Alexander von Humboldt's System der Isotherm-Kurven in Merkator's Projection'; in this map, the isotherms are black with two coloured bands: green for temperatures above 0°C, and green below.
36 For Berghaus's physical atlas and its distribution maps as both an established means of scientific argumentation and a self-contained representation of nature, see Camerini (1993). On the English version of *Physikalischer Atlas* and its dissemination in the Anglo-Saxon world see Engelmann (1964).
37 Let us see two examples. On 14 October 1841, Berghaus mentioned the analogy with Humboldt's isarithmic technique in depicting the distribution of precipitation in Europe: 'On the map in question, prepared in the winter of 1838–39, the graphical representation of Europe's rainfall has been attempted. By analogy with the principle of Physiko-Graphik established by A. von Humboldt in connection with heat distribution, the points which have equal amounts of rain were connected by lines which can be called isohyets. By this means a series of curves is produced'. On 25 April 1842, he extended the analogy to his map of thunderstorms: 'If one connects with lines the points which have an equal number of days with thunderstorms in the average year this provides a system of curves analogous to isotherms and isohyets'. The quoted phrases are taken from Robinson and Wallis (1987, p. 122) (authors' translation from the original in German).

weather. In this regard, the following distinction is fundamental: while the weather of the tropics (with warm temperatures, monsoon winds, and seasonal shifts in rain patterns) was essentially a matter of climate, the weather of the temperate latitudes was more subject to local geographical conditions and to variable and unforeseeable winds. And while for the study of the former a single map of records of averages proved extremely instructive, for the study of the latter a sequence of cartographic images was especially required. Their aims significantly differed.

The studies on storms in the middle latitudes by young American mathematician Elias Loomis exemplify better than any other the employment of synoptic maps made long after the occurrence of phenomena, with purposes of diagnosis related to cyclone theory. Familiar with the construction of magnetic charts and with the lines of equal magnetic dips, in 1841 this professor at Western Reserve College, in Cleveland, Ohio, read to the American Philosophical Society a paper on an 1836 storm experienced in the United States.[38] Loomis began his address by questioning the usefulness of climate maps for his purposes: since I am convinced, stated he,

> that meteorology is to be promoted, not so much by taking the mean of long-continued observations, as by studying the phenomena of particular storms developed over a widely extended country, I resolved to select some single storm of strongly marked characteristics, and trace its progress as extensively and minutely as possible.'[39]

Thus, on a single large map of the United States, he drew, instead of barometric averages, a series of lines joining the points where the barometer was at its lowest point, as well as arrows showing wind direction, so that one could visualize how the storm had progressed by viewing the progress of the lines from hour to hour on the map.[40]

Two years later, in a study on two winter storms of the year 1842, he presented before the same Society what would be regarded as the 'first printed examples of the *synoptic weather map*' (Figure 8.3), a map that integrated 'information for pressure, temperature, wind, sky conditions, and precipitation'.[41] This time, however, instead of lines of minimum barometric pressure, he drew lines of equal deviation from the normal average pressure for each place. Now the reasons why he opted for synoptic cartography techniques were much more explicit than two years earlier:

---

38 Loomis published his first magnetic chart with lines exhibiting the amount of magnetic dip as early as 1838. Newton (1895, pp. 218–20). See also Cox (2002, pp. 41–49).

39 Loomis (1841, p. 125).

40 Ibid., plate II. Moreover, the paper included two time-series graphs (Plate I) showing changes in barometric pressure and temperature at various places.

41 Monmonier (1999, p. 36). The paper contains thirteen hand-coloured and lithographed maps, which were drawn from reports provided by 131 observers.

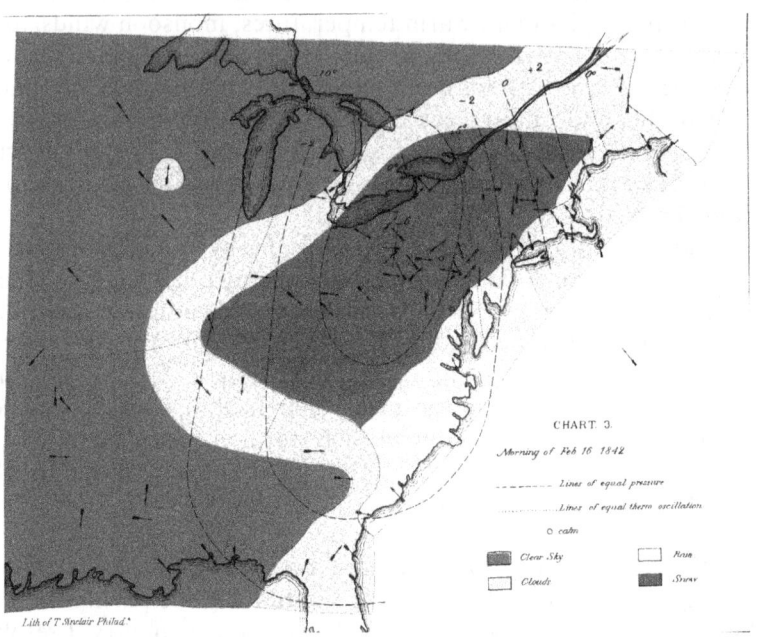

Figure 8.3 Synoptic chart, representing pressure, wind, temperature, and precipita-
tion distribution, by Loomis, from his article 'On Two Storms Which Were
Experienced throughout the United States, in the Month of February, 1842'.
Source: Loomis (1846, plate 3).

It appears to me, that if the course of investigation adopted with respect
to the two storms of February, 1842, were systematically pursued, we
should soon have some settled principles in meteorology. If we could
be furnished with two meteorological charts of the United States, daily,
for one year, charts showing the state of the barometer, thermometer,
winds, sky, etc., for every part of the country, it would settle for ever
the laws of storms. No false theory could stand against such an array of
testimony. Such a set of maps would be worth more than all which has
been hitherto done in meteorology.[42]

In other words, the function of Loomis's synoptic weather maps was eluci-
datory, not predictive. In his 1841 paper, Loomis did not intend to afford
a method for weather forecasting. Rather, he aimed to settle the celebrated
storms controversy between William C. Redfield and James P. Espy—i.e.
whether storm winds had a rotary motion according to which they blew
in circles anticlockwise around a centre (Redfield), or whether they had a

42 Loomis (1846, p. 183).

centripetal motion according to which the air followed an upward trajectory in the centre of the storm (Espy).[43] Reluctant to embrace one theory or another, Loomis provided an explanation of the storm's heaviest rain, showing in a diagram how moist air from the southeast was pushed upward as cold air advanced from the northwest (Figure 1.4).[44] In his 1846 paper, the maps for 1842 winter storms described winds that were neither completely rotational nor fully inward, but rather they showed both a 'tendency inward' and a 'disposition to circulate around the center'—which was in fact nothing more than the corroboration that Redfield's and Espy's theories were in part correct.[45]

The contributions to synoptic weather mapping by Loomis and his predecessors corroborate the first contention of this chapter: the synchronous weather maps made long after the occurrence of storms provided to scientists a powerful tool to elucidate the progress of storms and their physical laws. The protagonists themselves in the controversy were given not only to constructing weather charts but also to acknowledging their value to reveal the movement and evolution of storms. As Redfield noted in a study on three hurricanes of the American seas in 1846: 'The accounts [exhibited in the chart] serve to determine the route, character, extent, and progression of this memorable hurricane, and of the barometric depression and the expanded storm which preceded it'.[46] His study concluded with the hope that the recently invented electric telegraph could warn navigators against the arrival of storms.

In brief, on the threshold of telegraphy revolution the weather map was a diagnostic tool to identify a storm and determine its laws and progress from its signs and symptoms on the map.

From diagnosis to prognosis

The advent of electric telegraphy had important repercussions in all aspects of the science of meteorology, from organizational and practical to theoretical ones. It specially influenced the organization of observational systems, with an emphasis on simultaneity of observations and immediacy of data collection (as opposed to retrospective analysis). Regarding meteorological cartography, interest gradually shifted from the use of synoptic map as a tool to diagnose to its use for the prognosis of weather, discrete storms in particular.[47]

---

43 For a thorough account of this controversy, Fleming (1990, pp. 23–73). For the storm controversy and its repercussion on American science, De Young (1985, pp. 657–60). See also Kutzbach (1979, pp. 19–44), Cox (2002, pp. 27–39).

44 Loomis (1841, p. 159). For a detailed and critical discussion of Loomis's papers, see Kutzbach (1979, pp. 29–35).

45 Loomis (1846, p. 165).

46 Redfield (1846, p. 335).

47 Barboza (2012, pp. 188–89) makes clear the extent to which some historians tended to identify the studies on the general movement or the state of the atmosphere with weather forecasting studies. In her view, the scholarly discussions on the physical-mathematical

The telegraph industry was perhaps nowhere as revolutionary as in Britain and the United States. In all Western countries it represented a catalyst of communications, of the dissemination of scientific knowledge, and of some of the characteristic processes in the contemporary world—such as industrialization and urbanization.[48] In the United States and especially in Victorian Britain these sociocultural transformations were accompanied by important economic and scientific developments which led Britain to turn cable telegraphy into a successful ocean-spanning technology.[49] Economically, British capital and engineering expertise dominated the world cable industry until the 1920s. Scientifically, they set the pace of the progress of electrical science—a process that led her not only to establish the system of electrical units and standards used today (ohms, volts, and amps) but also helped telegraphy play a crucial role in the rise and consolidation of the field theory in electromagnetic physics.[50]

In the mid-1840s, Loomis and others had remarked the predictive potential of telegraphy but their comments had never moved from the rank of anecdote or insinuation. For example, when at the end of his 1846 paper, Loomis wondered whether it was not time to embark in a 'general meteorological crusade' and spread a well-arranged system of observations over the country, so that 'men would cease to ridicule the idea of our being able to predict an approaching storm'[51]; or when in his 1846 study on hurricanes, Redfield suggested that the approach of a gale could 'be made known by means of the electric telegraph'.[52] Yet from the opening of the first commercial telegraph lines in 1845 things changed. These lines enabled data exchange between observatories, making it possible to track weather changes across countries with regularity and immediacy. In addition, the telegraph was associated with the expansion of railways and press media, which contributed to the diffusion of practical knowledge.

These developments were noticed in Britain at three levels, academic, commercial, and political. In July 1849, a London newspaper, the *Daily*

---

laws governing the general movement of the atmosphere at that time referred to regular oscillations of the atmospheric pressure, rather than to weather changes or storms.

48 For the beginnings of the telegraphic news agencies and the methods of the early telegraphic information services, see Nalbach (2003, pp. 69 –71); for the links between the telegraphy and the press in America in the 1840s see Blondheim (1994, pp. 30–46).

49 For a clear account of how cable telegraphy shaped imperial affairs and relations in international politics, see Headrick (1991, pp. 11–27); for a study on how telegraphy and subsequent 'new technologies' provoked public reactions and affected social habits and customs in popular media and specialized engineering journal, see Marvin (1988, esp. pp. 3–8).

50 For an illuminating study of the impact of telegraphy on the physical sciences, with a special focus on Britain, see Hunt (1997, pp. 316–29). For an economic study on the role played by Britain on standardization and measurement conventions, see Velkar (2012, esp. pp. 29–94).

51 Loomis (1846, p. 183).

52 Redfield (1846, p. 334).

News, turned the vision of James Glaisher, the head of the Meteorological Department at Greenwich Observatory, into reality, and printed a bulletin of observations, taken the previous day at fifty railway stations in England and Scotland and dispatched by telegraph and rail to London.[53] Although the newspaper printed no map at that time, both Glaisher and the Astronomer-Royal George Airy drew on those reports to draw weather maps for their own use. A further step was taken in 1851 on the occasion of London's Great Exhibition, held in Hyde Park. There, a private firm, the Electric Telegraph Company, posted weather observations on a map of Great Britain, and printed lithographed versions for visitors.[54] This has been cited as the 'first same-day weather map' in the world.[55]

These developments were also contemporary with the first attempts in the telegraph era, promoted by Joseph Henry and the Smithsonian Institution between 1847 and 1858, to track storm path and changes by using the telegraph and the weather map.[56] The secretary and director of the newborn institution took *au pied de la lettre* Loomis's idea and similar suggestions made by James Espy, and in 1847 he proposed the Smithsonian's Board of Regents to organize a system of telegraphic observations. Henry well knew what weather networks were and how electromagnetic phenomena could affect the state of the atmosphere.[57] Given that population centres concentrated on the South and the West, he argued, the telegraph would 'furnish a ready means of warning the more northern and eastern observers to be on the watch for the first appearance of an advancing storm'.[58] His ideas about the potential of the telegraph as a warning tool were not preposterous, if we take into account the commercial context of the time.[59]

It is not easy to assess how comprehensively Henry mapped the Smithsonian's weather data. In the 1850s the prevailing perception in the Reports of the Board of Regents was that the Smithsonian had abundant human and technical resources but that Henry was not much interested in the arduous manual compilation of three weather maps a day. The few quantitative data

---

53 Marriott (1903, pp. 123–31). The first daily telegraphic report was printed by the *Daily News* on 31 August 1848. For further details about the Daily News' weather reports, see Monmonier (1999, p. 42, 248).

54 The maps were exhibited from 8 August to 11 October. Lempfert (1913, pp. 173–74).

55 Monmonier (1999, p. 42).

56 For a summary of the meteorological work at the Smithsonian Institution, see Langley (1895, pp. 216 –20) and Benjamin (1897, pp. 647–78).

57 While teaching at the Albany Academy in the 1820s, he collaborated with the regional weather network organized by Simeon DeWitt. See Fleming (1990, pp. 10–21), Reingold (1972, pp. 277–81). For Henry's little known contributions to the telegraph, see Hughes (1994, p. 23).

58 Henry (1854, p. 139)—quoted by Monmonier (1999, p. 39).

59 According to some sources, like Abbe (1897, pp. 142–43) and Miller (1931a, p. 3), in 1849 an Associated Press reporter in New York and later president of Atlantic Mutual Insurance, Alexander Jones, sent to Henry some experimental maps drawn from weather data for 19 and 20 July. This could be the first American weather map based on telegraphic reports.

that are available regarding the amount of observers and observations involved corroborate this perception. The annual reports of the Board inform that the network of observers considerably fluctuated in the 1850s, going from around 150 in 1850 and 210 in 1856 to around 500 in 1859 and 300 in 1863.[60] The report for 1858 features a technical staff of twelve to fifteen persons who were overwhelmed by 'the records of upwards of half a million separate observations, each requiring a reduction involving an arithmetical calculation'.[61] In his annual report for 1858, Henry himself admitted that 'a large number of maps have been constructed for the investigation of storms'.[62] Everything suggests, therefore, that the Smithsonian generated enough data for efficient mapping of storms.[63]

The most visible result of all this cartographic production was the celebrated giant map of the United States that the Smithsonian exhibited to visitors to the Institution in 1856, on which a set of coloured cards was hanged.[64] The cards indicated weather conditions by different colours and were suspended on iron pins inserted at points corresponding to telegraphic sources. Later these cards were replaced by coloured disks with eight holes which sported arrows to show wind direction.[65] Interestingly, it was the brilliant combination of a tool for communication and entertainment and a means for information dissemination that led the Smithsonian authorities to use the map in order to awaken public interest in the science of meteorology.[66]

The value of Henry and the Smithsonian's contributions to meteorological cartography must be put in perspective. They were the first to show the efficacy of weather mapping by using telegraphic data. However, the Civil War (1861–1865) undermined the number of observers and telegraphic lines and aborted Henry's plans to establish a storm warning service.[67] Many of the telegraph companies that formerly promised to render a service were no longer willing to take the leap after the war. This is no other than the

60 For the total number of meteorological observers reporting to the Smithsonian between 1849 and 1874, see Fleming (1990, p. 82).
61 *Annual Report of the Board of Regents of the Smithsonian Institution, for the Year 1857.* Washington, DC, 1858, pp. 27–28—quoted by Monmonier (1999, p. 40).
62 *Annual Report of the Board of Regents of the Smithsonian Institution, for the Year 1858.* Washington, DC, 1859, p. 84—quoted by Monmonier (1999, p. 40).
63 In his annual report for 1850, Henry (1854, p. 191) stated that the person responsible for meteorological correspondence, professor Edward Foreman, had drawn an outline map 'for representing the successive phases of the sky over the whole country, at different points of time'—quoted by Monmonier (1999, p. 40).
64 Benjamin (1897, pp. 658–59), Langley (1895, p. 219), True (1929, pp. 303–04).
65 Miller (1931a, p. 3). For further details of the map, see Fleming (1990, p. 143).
66 In the report for 1858 (p. 32), Henry wrote: 'This map is not only of interest to visitors in exhibiting the kind of weather which their friends at a distance are experiencing, but is also of importance in determining at a glance the probable changes which may soon be expected'—quoted by Benjamin (1897, p. 659).
67 In addition to the Civil War, there is the fact that a large fire destroyed part of the Smithsonian building in early 1865. See True (1929, pp. 36–37, 45, 128, 299–308).

final destiny of any venture marked by tragedy. However, throughout his long trajectory of almost two decades, Henry experienced a progressive but fundamental shift from the early years in which he envisaged the benefits that storm warnings would yield to the eastern states, to the post-war years in which he yearned for government-funded weather forecasting.[68]

At a time when the first national weather agencies were being founded, the shift in the conception of the synoptic map from diagnosis to prognosis, and within this concept the move from storm warning to general weather forecasting, or even the combination of the two elements, was important for the establishment of these agencies.

In this regard, it is worthwhile to examine the evolution of the one meteorological director who enjoyed the status of an exception, C.H.D. Buys Ballot. In fact, this case reveals that the acceptance of synoptic map as a tool for the analysis of weather patterns did not necessarily imply a conceptual evolution from diagnosis to prognosis. Then-professor of mathematics at the University of Utrecht had set up a meteorological observatory of his own in 1848.[69] Six years later, he would found the Nederlandsch Meteorologisch Instituut. In what was a clear expression of Buys Ballot's early intentions, which clearly reflect the processes of scientific emulation at that time, he first published some synoptic charts in the 1852 *Jaarboek* of the Institute— 'because of what I had seen at London's Great Exhibition in 1851', as he himself recognized.[70] On these charts, he plotted wind directions and the deviations from the mean values of temperature for several coastal stations of Western Europe (see Figures 5.1 and 8.4).[71] Given that the deviations below and above the mean were printed in different characters, the simultaneous distribution of temperature or pressure could be easily distinguished at a single glance.[72]

---

68 That the anticipation of weather changes—i.e. the act of prognosticating the course of atmospheric phenomena—was a fundamental aim for meteorologists was admitted by Henry in his annual report for 1865, in the aftermath of the Civil War: 'It has been aptly said that man is a meteorologist by nature. He is placed in such a state of dependence upon the atmospheric elements that to watch their vicissitudes and to endeavor to anticipate their changes become objects of paramount importance'. Nevertheless, implicit in the forecasting task was not necessarily the pursuit of the physical laws of the atmosphere, as Henry reminded us: 'There is, perhaps, no branch of science relative to which so many observations have been made and so many records accumulated, and yet from which so few general principles have been deduced'. See *Annual Report of the Board of Regents of the Smithsonian Institution, Showing the Operations, Expenditures, and Condition of the Institution for the Year 1865*. Washington, DC: Government Printing Office, 1866, p. 50. See also True (1929, p. 300).

69 Lunteren (1998, pp. 229–35), Achbari and Lunteren (2016, pp. 6–11).

70 Buys Ballot (1882, p. 9).

71 *Nederlandsch Meteorologisch Jaarboek*, 1852, 1. These charts were reproduced in J.C. Poggendorff's *Annalen der Physik und Chemie*, in Buys Ballot (1854, p. 564).

72 Hildebrandsson and Teisserenc de Bort (1898, pp. 88–92), Shaw (1926, pp. 303–04), Dettwiller (1982, pp. 66–68).

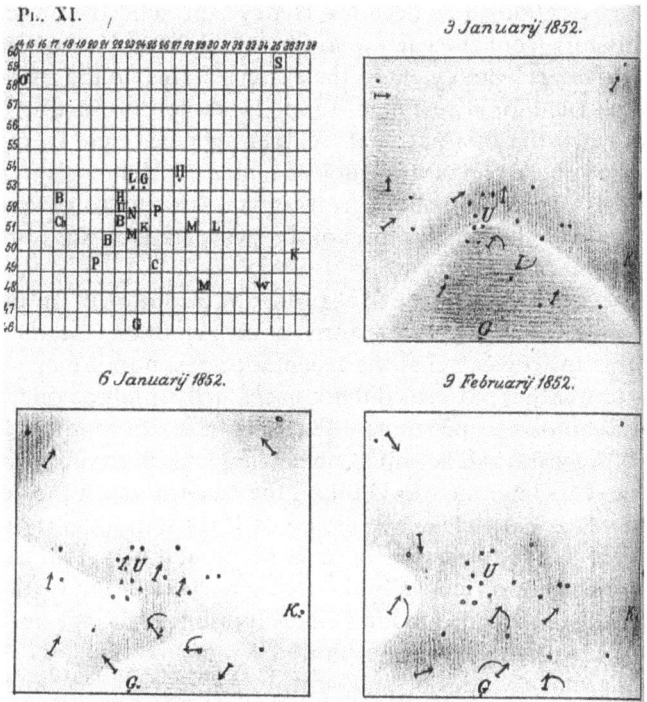

Figure 8.4 Early meteorological charts by Buys Ballot, January and February 1852. The position of stations is indicated by a letter. Wind direction is indicated by an arrow, showing its rotation throughout the day; wind strength by the length of the arrow. Temperature deviations from normals are indicated by hatches.

Source: Buys Ballot (1854, p. 635).

Buys Ballot's charts afforded a relationship of temperature with wind direction. Although their form was somewhat rudimentary, the message was clearly visible. Stations were marked either by initial letters or by dots; wind direction was indicated by arrows; and wind force by the length of the arrows. Moreover, he introduced shadings showing deviation of the temperature from the normal: the areas where temperature was above the normal, they were vertically striated; below the normal, strias were horizontal. As a mode of representing weather conditions, Buys Ballot's early charts embodied tools for diagnosis rather than for prognosis.[73]

---

73 Buys Ballot plotted the same kind of chart in his report on the cold in Europe from 25 to 28 December 1853, published in the *Comptes-rendues* of the Amsterdam Academy of Arts and Sciences. Hildebrandsson and Teisserenc de Bort (1898, p. 88).

Soon after, he went one step further. He later admitted that by 1855, he sought a sign that must precede strong winds.[74] Drawing on local observations during five successive years, he found that this sign was related not so much to the height of the barometer or its rise or fall, but rather to the differences in the barometric readings. In a brief note presented at the Amsterdam Academy in 1857 he summarized this local rule of thumb revealing an empirical regularity: 'The difference between simultaneous absolute readings or deviations from the normal reading', he stated, 'gives the most certain indication, even when one pays attention to places not too far apart'.[75] Thus, when this difference at the northern and southern stations in Holland was over five millimetres, then strong winds could be expected.

As early as October 1857, many years before the elevation of his eponymous wind rule to a global law, Buys Ballot discussed with the Inspector of the telegraphic service in the Netherlands the feasibility of his new system of storm warnings.[76] It is interesting to note that he aimed to provide storm warnings that would be based on his wind rule rather than on synoptic maps. This fact is not trivial; a significant shift took place for reasons of pragmatism and immediate utility in the arguments employed by Buys Ballot with respect to the scope of meteorological telegraphy. Where in 1852 he deemed his first charts as a means to diagnose weather conditions and thereby elucidate the laws of nature, five years later he regarded empirical rules as a means to prognosticate the course of storms. Not in vain was his warning system predicated on storms already observed as well as on storm predictions.[77] Such empirical approach was instrumental in persuading the Minister of the Interior to endorse Buys Ballot's proposal for the establishment of the Dutch telegraphic system of weather reporting and storm warning in 1860.[78]

Eulerian approach versus Lagrangian approach

In March 1858, Captain Robert FitzRoy first applied the word *synoptic* to meteorology.[79] Five years later, he defined the *synoptic chart* as the chart

---

74 Buys Ballot (1868, p. 19).
75 Buys Ballot (1857, p. 76)—quoted from the translation made by Achbari and Lunteren (2016, p. 13).
76 For the transformation of this local rule into the well-known Buys Ballot's law as a result, in part, of the favourable reports from the Royal Society of London and its subsequent spread all over the world, see Achbari and Lunteren (2016, esp. pp. 11–43); Achbari (2017, pp. 69–118).
77 As Halford (2004, p. 175) correctly points out, Buys Ballot's first warning preceded FitzRoy's by just eight months, 'making him the first to issue a storm notice based on weather prediction'.
78 For the 'Excerpt from Resolution by the Minister of the Interior, dated 21 May 1860', by J. Schröder, see Buys Ballot (1860, pp. 75–76).
79 From the Greek word *súnopsis* (view together). He used it in the instructions given to his assistant Thomas Henry Babington at the Board of Trade's Meteorological Department. Halford (2004, p. 123) points to a possible link to its usage in the Bible.

'being intended to express consecutive simultaneous states of the atmosphere'.[80] From then on, synoptic forecasting and the organization and consolidation of national weather agencies were hand in hand. In light of this, no one would say that synoptic weather forecasting was to follow two radically different lines of thought that coexisted throughout the 1860s.

In Chapters 3 and 4, I showed that in the 1850s and 1860s meteorologists adopted two opposing approaches to the study of weather forecasting: these coincided with the Eulerian and Lagrangian methods of fluid mechanics. The Eulerian method stressed the field concept—in particular, the baric field or the atmospheric pressure distribution—and its main object was to predict the changes of this field. It was assumed that the horizontal baric field basically determined weather conditions. In contrast, the Lagrangian method stressed the air-mass concept—or air currents—and its object was the arrangement of different air masses and their relationship with physical factors such as temperature, cloudiness, and, of course, air pressure. However, to what extent were these methodological differences reflected on weather maps? Or, in other words, how did an Eulerian weather analyst or forecaster represent the hegemony of baric field on a map? How did a Lagrangian analyst mirror the import of air masses?

The Lagrangian frame of reference for the study of atmospheric motion was adopted by FitzRoy at the British Board of Trade's Meteorological Department. Since its foundation in 1854, this department had been devoted to the preparation of charts for navigators. One of its first productions was the so-called 'wind star'. Modifying the nautical wind roses developed by Maury and others, FitzRoy devised a chart whose aim was to show at a glance the direction and strength of the winds navigators could find in a given region.[81] The wind star consisted of a series of lines radiating from a central circle, marked with latitude and longitude. While the lines represented wind directions, the size of the circles indicated the proportions of calms.[82] Between August 1855 and March 1856, the Department produced eleven wind charts.[83] FitzRoy and his staff gave priority to the collection and diffusion of marine wind data; this field was regarded as vital to the

80 The definition appears in his *Weather book*. FitzRoy (1863, p. 103). He depicted the synoptic charts as being 'as if an eye in space looked down on the *whole* North Atlantic *at one time* and afterwards took similar views (much more extensive than "*bird's eye*") at regular intervals of hours or days, so as to obtain sequences of synoptic conditions' (Ibid.).

81 For the influence of marine experts like A.B. Becher, William Marsden, and especially Isaac Greenwood on Maury's works and on FitzRoy himself, see Agnew (2004, esp. pp. 30–36). See also Pinsel (1981, pp. 123–36).

82 The wind stars used numbers to indicate wind force according to the Beaufort scale, and other notations for barometric pressure, sea temperature, sea depth, and magnetic variations. FitzRoy (1863, p. 414). His modus operandi is abridged in FitzRoy (1855, pp. 39–40).

83 FitzRoy (1863, pp. 413–18), Anderson (2005, pp. 189–90), Burton (1986, pp. 155–57), Halford (2004, pp. 98–99).

improvement of marine security at a time when many ships used sails and winds were the motor of trade and war.[84] Unlike Henry and others who constructed weather maps for the investigation of storms, Fitzroy and his staff shared a common interest in elucidating the course of air currents for a practical and nautical use.[85]

Prussian physicist H.W. Dove and his book *Das Gesetz der Stürme* (The Law of Storms) played a prominent role in persuading FitzRoy that storms originated from polar and tropical air currents rather than from thermal processes in atmospheric circulation. As early as Christmas 1856, FitzRoy wrote a memo addressed to the Board of Trade, in which he suggested a method for mapping a simultaneous picture of the atmosphere, and showed how to incorporate diverse states of wind and weather onto a sequence of charts.[86] Although his initial aim was to track the atmospheric waves, he soon realized that this concept presented difficulties to explain the nature of storms. In the months that followed, he pressed for a translation of Dove's seminal book into English, which was published by his Department in 1858.[87] By the time he prepared his official report for 1858 he had already come to the conclusion that the results of the synoptic charts, plotted by his assistant T.H. Babington between 1856 and 1857, provided graphical evidence of a struggle between two opposing wind currents.[88]

FitzRoy's storm warning system was a product of this scenery. As a consequence of the Royal Charter storm of 24 and 25 October 1859 that not only shipwrecked the modern iron vessel but also affected densely populated areas, he conducted an investigation into the catastrophe, coming to the conclusion that a proper warning system could have prevented it.[89] To this end, he resorted to his synoptic charts, and soon produced an obscure map displaying pressure, temperature, and wind data by means of thick, thin, and short straight lines, respectively (see Figure 4.4 in Chapter 4).[90] Thus, wind force was indicated by the length of wind lines. He also introduced several symbols to depict cloud cover and different forms of precipitation. His charts, together with a telegraphic network of coastal observers,

---

84 Burton (1986, p. 157): 'Before the work of Maury and FitzRoy there had been no firm meteorological or oceanographical data available over most of the world's trading routes'.

85 On the marine thrust in the early years of the Meteorological Department and its relationship with the project endorsed at Brussels Maritime Conference for devising a uniform system of meteorological observations at sea, see Walker (2012, pp. 19–32).

86 For further details of this memo, see Halford (2004, pp. 111–12).

87 For the first English translation, see *Third number of meteorological papers*. London: Board of Trade, 1858. The second translation was made by Robert H. Scott in 1862.

88 Halford (2004, p. 123), Cox (2002, p. 78).

89 FitzRoy (1861, pp. 39–44, 1863, pp. 298–329).

90 FitzRoy (1863, plates XII and XIII). It also includes two schematic charts showing polar and tropical air currents (plates VI and VII).

formed the basis of the storm warning service approved by the Board of Trade in June 1860.[91]

Perhaps the most important question here is that the double duty to respond to the needs of seafarers and fishing communities and to understand the nature and progress of storms led FitzRoy to a substantial shift in his modus operandi of graphical methods. In 1855, he plotted wind stars to inform navigators about the expected winds in their routes while two years later he used outline maps, wind-markers, and glasses with circles to obtain a global image of wind flow patterns. A decade later he unfalteringly advocated the importance of wind direction and expressed his reservations about isobaric charts: 'neither isobars, curves or any kind of wave or crest lines seem to show the direction of the wind'.[92] The proper interpretation of barometric changes was essential in his method, but the use of isobars was not. Instead of drawing isobars, he delineated atmospheric pressure by using lines plotted from west to east across a map, marked with latitude and longitude. The pressure at a given place was represented by the separation between the pressure line and the closest line of latitude.[93]

By early 1860s, FitzRoy omitted the construction of synoptic charts from the forecast process, so that his weather prediction method, based on rules of thumb, local weather-watching, and individual judgement, was regarded by his numerous critics as extremely empirical—i.e. as a 'heavy missile' in the mouths of scientists.[94]

FitzRoy's Lagrangian approach powerfully contrasts with the Eulerian frame of reference adopted by Urbain Le Verrier and his assistants for the study of atmospheric motion at the Paris Observatory. The history of these early maps can be reconstructed from the graphical demonstration that the French astronomer carried out before the Academy of Sciences on 31 December 1855.[95] One year earlier, on occasion of the great storm that had sunk part of the allied fleet in the Crimean War, the French government

91 Anderson (2005, pp. 109–14; 2006, pp. 72–74), Burton (1986, pp. 160–61), Dry (2009, pp. 46–48), Halford (2004, p. 148–54), Khrgian (1970, p. 145).

92 FitzRoy in his *Report of the Meteorological Department of the Board of Trade*. London, 1864, p. 23. 'An outline chart, with wind-markers, is useful; likewise a transparent horn, or a glass, with circles; but a certain amount of practice enables one to dispense with such assistance, and work out the questions mentally (like a chess-player who need not look at the board)', in FitzRoy (1863, p. 218).

93 In a report on 'British storms, illustrated with diagrams and charts', presented at the 30th meeting of the British Association for the Advancement of Science, held in Oxford in 1860, FitzRoy made clear the extent to which the lines of barometric pressure appropriately depicted the physical state of the atmosphere. After closely looking onto his diagrams, he concluded that 'while the atmosphere in the British islands varied in its pressure from time to time, such variation was not on a particular line, but extended over a large area'. FitzRoy (1861, p. 41).

94 This allegorical expression was used by Brunt (1951, p. 117).

95 Le Verrier (1855, pp. 1197–204).

had entrusted him to develop a storm warning service.[96] Le Verrier dele-
gated the task to his assistant Emmanuel Liais. In his attempt to persuade
academics of the viability of following storm tracks in real time, Le Verrier
displayed a map prepared by Liais from cartographic techniques devised by
Adolphe Quetelet some years before.[97] In his map (Figure 3.1 in Chapter 3),
Liais drew two types of lines: atmospheric waves, like sea waves, showing
the propagation of disturbance across Europe through the days preceding
and following the storm; and 'lines of transport of the waves', showing the
maxima and minima of atmospheric pressure on the crest of a wave, equivalent to isobars.[98]

The idea that the track of a storm could be plotted in a map was not obvious at that time. The main consequence that Le Verrier resorted to this
idea—i.e. the use of cartographical techniques—was that he persuaded the
government to endorse a telegraphic observation network. It also contributed to the publication of a daily weather bulletin with observations from
eighteen cities, which was known as the *Bulletin Météorologique International*. Although he had displayed the map of the Black Sea storm on an
ad hoc basis in 1855, the *Bulletin* did not include daily weather maps until
September 1863. Then the map-maker was not Le Verrier, but his assistant
Marié-Davy, who, basing on historical data printed in the *Bulletin*, marked
weather stations on a map of Europe, noting down in each point the pressure recorded as well as the wind direction with an arrow.[99] He then linked
those points with hypothetical isobars, with values varying at intervals of
five millimetres of mercury. The arrows had barbs showing wind intensity
by six categories (more barbs, stronger wind). The outcome of this procedure was a succession of isobaric curved lines that could not intersect on the
map (Figure 3.2 in Chapter 3).[100]

Although the pressure lines devised by Liais and Marié-Davy—the lines
of transport of the waves and isobars—served different purposes, their
methods shared the same characteristic: the baric field (or the pressure

---

96 For the damage caused by the Black Sea storm on 14 November 1854, see MacMunn
(1935, pp. 144–48).
97 Barboza (2012, pp. 180–84), Davis (1984, pp. 364–65), Lequeux (2013, pp. 278–81), Locher
(2008, pp. 39–41). For a modern reconstruction of the Black Sea storm, see Landsberg
(1954, p. 352).
98 The fact that Liais' *lignes de transport des ondes* were equivalent to isobars, rather than
to 'the amount of rise or fall of the barometer since the previous chart', as suggested by
Davis (1984, p. 365), has been stressed by Barboza (2012, pp. 183–84). Although in the
article published by Le Verrier (1855, p. 1201) in the *Comptes Rendus de l'Académie des
Sciences* Liais mentions the term isobarometric lines to refer to those lines defined by
the absolute value of pressure rather than by barometric oscillations, it does not seem he
plotted them in his 1855 map.
99 Locher (2008, pp. 114–15). See also Davis (1984, pp. 168–70).
100 Locher (2009, p. 86) points out that, despite its limitations, Marié-Davy's mapping method
gave 'considerable latitude to the draughtsman', especially when 'there were rather few
weather stations across Europe where atmospheric pressure could be recorded'.

distribution) was the indicator of the weather state and its impending change. Therefore, the problem of foreseeing the changes of the atmospheric disturbance was reduced to the examination of the deformations imprinted on the curves of equal pressure over the surface, as Marié-Davy admitted in a memoir addressed to Napoleon III.[101]

The similarities between these two mapping methods do not end there. Two more features are worth mentioning. First, each method sought to place the meteorological focus on an international framework. In the same way that natural catastrophes such as the Black Sea storm evinced the need for a global account of the motion of the atmosphere, mapping techniques too had to be adapted to the new global perspectives, showing the tracks of storms across Europe and discerning continental phenomena from local storms. Although local conditions, such as topography, forests, and the soil, could play a certain role in the formation of lower clouds, they thought that their effects on the general pattern of atmospheric circulation were of little magnitude.[102] Second, the two methods sought to give a dynamic impression of the movement of the atmosphere. Unlike FitzRoy's maps, representing a mess of lines of pressure and temperature, overloaded with symbols for cloud cover and precipitation, the morning maps printed by the *Bulletin* from 1863 configured a series of isobarometric images creating the impression of an animated sequence. Looking at the contours of the areas of low pressure (or *bourrasques*) traced on the map, the reader had the impression that one was observing individual, consistent, and articulated entities travelling through Europe.[103]

In comparing the two mapping methods, a familiar pattern emerges. FitzRoy argued for the importance of air current tracks in combination with other atmospheric variables; Le Verrier and his assistants defended the essential role of the isobaric field in the determination of weather conditions. The former focused on tracking air parcels along their trajectories while the latter on the changes of atmospheric variables fixed at a particular point in space. Both sides affirmed the superiority of their mapping methods and questioned the arguments of the other.

However, between these methods there were middle ways. Perhaps the most illustrative example is the semi-Lagrangian approach adopted by Francis Galton.[104] In 1863, he published his well-known book *Meteorographica*,

---

101 E. H. Marié-Davy, 'Mémoire adressé à Sa Majesté L'Empereur' (Paris, 1866). It was printed on 21 March 1868 (p. 6). Archives Nationales, F.[17] 3718.

102 Anderson (2006, p. 82) gives a full account of the little importance that local conditions had for French meteorologists, drawing on Hildebrandsson and Teisserenc de Bort's views in 1898.

103 According to Locher (2008, p. 115; 2009, p. 90), who has analyzed how Marié-Davy publicized the concept of *bourrasque* (or depression), 'the process was almost akin to the impression of movement produced by a flickerbook'.

104 For Galton's contributions to meteorological cartography, see Friendly (2008, pp. 514–16), Gillham (2001, pp. 140–51), Tickell (1993, pp. 54–61), Anderson (2006, pp. 78–80).

which includes a series of ninety-three weather maps for Europe for December 1861. *Meteorographica* represents the synthesis of his meteorological thought, in which, as a culmination of this thought, he introduces the concept of 'anti-cyclone' or a high-pressure area where winds spiral outwards, circling clockwise in the Northern Hemisphere and counterclockwise in the Southern Hemisphere.[105] Beyond his theoretical contribution to atmospheric circulation, however, his book constitutes a staunch defence of the value of the map as a scientific visual tool as well as the need for representing simultaneous observations from a wide variety of sources. As he stated in his preface,

> when lists of observations are printed in line and column, they are in too crude a state for employment in weather investigations; after their contents have been sorted into Charts, it becomes possible to comprehend them. But it requires meteorographic maps to make their meaning apparent at a glance.[106]

In this case, he added, maps can make visible a wide variety of information extracted from the simultaneous observations of hundreds of meteorologists.[107]

Galton's approach was essentially geometrical and visual, resting mostly on the representation of numerical and graphical summaries of complex and spatial data. It included several stages of abstraction. First, he produced weather glyph maps of Europe, which showed rain, cloud cover, and the direction and force of wind by a series of templates or stamps as well as identifying letters for wind direction (N, NW, etc.) (Figure 4.5). Next, he drew iconic three-dimensional maps, which showed the deviations of barometric pressure from an average by using red and black symbols to reflect the degrees of divergence (Figure 4.6). Finally, he devised schematic micromaps or miniature abstract panels, which displayed the states of barometric pressure, temperature, wind direction, and rain, on the morning, afternoon, and evening on each day of the month (Figure 4.8). One of these micromaps, in particular, showed streamlines representing the lines tangent to the velocity vector of wind. In this way, the reader was able to uncover the movements of air currents and observe the replacement of one air current by another from the analysis of schematic micromaps. Previously undetectable air movements in isobaric maps could thus be identified and given visual expression,

---

105 Galton (1863, p. 7).
106 Ibid., p. 3. This paragraph is also quoted by Anderson (2006, p. 78).
107 In July, 1861, Galton issued a circular, accompanied by a map and addressed to meteorologists throughout Europe, in which he invited them to send him meteorological data for the month of December 1861. The results of this initiative were published in his work *Meteorographica*. See Gillham (2001, pp. 141–42), Pearson (1914–30, vol. 2, pp. 37–38), Shaw (1926, p. 305).

while anticyclones (or high-pressure areas) could be first recognized, with their outward spiral winds in clockwise direction.

In sum, the 1850s and 1860s witnessed the coexistence of two opposing approaches to meteorological mapping and the study of weather forecasting in general. FitzRoy and Galton deemed it necessary to represent the combined elements involved in weather, emphasizing the need to use symbols and icons and reflect the movements of air masses in their representations, while de-emphasizing the visual power of the isolines. In contrast, Le Verrier and his assistants granted a hegemonic role to isobars for, according to them, the baric field determined weather conditions. As shown in Chapter 6, thenceforth the Eulerian line of thought clearly prevailed over the Lagrangian approach.

Before going further in the development of meteorological cartography, some notes on the transformation of statistical graphics are necessary if one wants to determine to what degree the hegemony of the isobaric synoptic mapping could be influenced by the nation-state building interests in which meteorologists were involved.

Transformation of statistical cartography

The consolidation of isoline weather maps was stimulated by the transformation and expansion of statistical cartography. After the 1860s, and under the influence of the French cartography, meteorologists in almost all parts in the world began to apply isobarometric techniques as a way to forecast weather changes. Since isolines visually represented individual numbers (either as a record of averages or a synoptic record), isobarometric techniques formed part of a larger frame, statistical cartography. In its broadest sense, statistical cartography included such forms as contour lines, isolines, and colour and gradient tints, aimed at distinguishing isometric lines, dots, circles, grades of value, and similar forms and figures.[108]

By the 1860s, a technique of statistical representation rapidly generated widespread interest in the Western world—one that was perceived as more objective to reflect social and administrative realities of the states and that could serve as a tool for the reordering of politics in the West.[109] During the ensuing three decades—a period than was called by historians as the

108 Significant works on the history of statistical cartography in the second half of the nineteenth century are Beniger and Robyn (1978, pp. 4–6); Friendly (2008, pp. 508–31), Friis (1974, pp. 132–40), Funkhouser (1937), Mood (1946, pp. 209–25).

109 In 1862, French engineer Charles Joseph Minard, who produced sixty-three graphic works from 1843 to 1869 (including thematic maps and statistical diagrams and charts), encapsulated the rise and flare of visual thinking in this way: 'The dominant principle which characterizes my graphic tables and my figurative maps is to make immediately appreciable to the eye, as much as possible, the proportion of numeric results [...] Not only do my maps speak, but even more, they count, they calculate by the eye'. Minard (1862, p. 2, 4), as translated into English by Friendly (2008, p. 512).

golden age of graphics—the graphic method awakened tremendous enthusiasm not only among statisticians and government and municipal authorities but also among the demographers and cartographers mapping the inward composition of nations.[110] Maps began to gain scientific authority in the official national censuses and act as a complement to statistical annexes in state publications. The implementation of graphic method also increased its ability as a tool for visual communication, enabling mappers to trace more definite delimitations of the frontiers between peoples and national entities by means of colour contrasts and gradient tints. Finally, the states themselves increasingly drew on statistical graphic for their official publications, marking this method as a genre endowed with scientific authority and legitimacy.[111]

Three main factors contributed to the transformation of statistical cartography in this period. The first was the growing sense of certainty provided by statistical science. As shown in foregoing sections, the migration of techniques from descriptive statistics to mathematical statistics effected by men like Humboldt and Quetelet had evinced the great potentialities of statistical cartography to social sciences. Likewise, it had reinforced demographers' confidence to graphically display and accurately measure nationality and its associated censuses. Consequently, isolines, shadings, and other graphical forms began to gradually represent the various systematic data being of interest to the state, from vital statistics to transportation routes and population densities, to mention a few. This trend could already be perceived in the early cartograms (or maps with diagrams) of Charles Joseph Minard[112] and the nomograms of Léon Lalanne,[113] and it would continue in the works of subsequent statisticians such as Galton and his studies on correlation and regression[114] or Thomas Thorburn and his numerous volumes of diagrams on subjects as diverse as agriculture and vitality.[115]

Paralleling the rise of statistical science was the increasing interest in the graphic representation of statistical data in the international arena. In this regard, the Universal Exhibitions held in London (1851, 1862), Paris (1855, 1867, 1878), and other locations became a showcase for displaying the value

---

110 The designation of the golden age of statistical graphics to the period 1860–1890 can be found, for example, in Funkhouser (1937, p. 330) and Friendly (2008).

111 For the rise of official statistical cartography in Central Europe and America, Mood (1946, pp. 209–25); Palsky (1996) for France; and Patriarca (1996, pp. 176–209) for Italy. See also Monmonier (1994, pp. 3–8).

112 Robinson (1967, pp. 98–100).

113 Lalanne devised diagrams to calculate how much land should be moved when building railways lines with the aim of making labor costs efficient. See Friendly (2008, p. 511), Hankins (1999, pp. 61–66).

114 Stigler (1986, pp. 294–99).

115 Thorburn's propensity to represent too many items on one graph led him to graphically draw as many as thirteen different fields using a single scale of numbers for all in his book on *Diagrams, Illustrative of Facts, Principles, & Theories*. Thorburn (1855, p. 28).

of graphical methods. The event in Paris in 1878, for example, gathered extensive displays of all types of graphic work.[116] A similar interest can be appreciated in the successive international statistical congresses, which became especially prominent at the Vienna Congress in 1857 and still persisted during the last congress in Budapest in 1876.[117] The attention given to graphic representation in these congresses was but an expression of the general concern about the method, its application, and its standardization. Discussions on the role graphic representations could play in popularizing knowledge gathered momentum at the Vienna meeting. An example of this interest is the constitution of a section on the 'General application of cartography and graphics for statistical purposes'.[118] This topic received special attention in the next congress, held in London in 1863, where in addition to creating three specific sessions on the subject—one of them on 'meteorological statistics'—there were clear calls to the use of graphical forms.[119] This is the case with British physician William A. Guy, who encouraged medical statisticians to use 'signs and letters for words and sentences, represent quantities by lines and spaces, and intensity by varying depths of shadow and colour'.[120]

The attendees at the 1869 Congress held at The Hague went a step further, as in addition to promote interesting discussions on the topic, led by Quetelet, Adolf Ficker, Ernest Engel, and other prominent statisticians, they adopted several resolutions on the standardization of cartographic methods and practice.[121] The first said that 'the Congress, considering the graphic method appropriate for teaching and popularizing statistical science, recommend[ed] that official statistical documents be accompanied by maps and diagrams'.[122] Moreover, the Congress urged the organizers of the next congress to prepare 'a memoir on the different graphic methods employed in

116 In a report on these displays, the director of the French *Bureau de la Statistique Graphique*, Émile Cheysson (1878, p. 324), described the success of statistical cartography in this way: 'A visitor to the exposition cannot but be impressed by the very extended use which almost all nations have made of graphic statistics for representing the principal elements of their intellectual, economic and social life'.

117 For a review on the development of the graphical method in the international statistical congresses, see Funkhouser (1937, pp. 310–18).

118 For instance, in this section, K. Schmutz proposed the use of shading over colouring in order to reduce the great costs generating the production of maps. For the section on 'Anwendung der Kartographie und der Graphik überhaupt auf Zwecke der Statistik', see Ficker (1857, pp. 131–34).

119 The sessions were: 'Statistical Methods and Signs'; 'Meteorological Statistics'; and 'Military and Naval Statistics'. See Farr (1861) and Friis (1974, p. 144).

120 Quoted by Farr (1861, p. 379).

121 An example of these discussions is J.M. Obreen, 'La méthode graphique'. *Congrès international de statistique à la Haye, septième session, 1869, Programme*. La Haye (1869, pp. 29–31).

122 The resolution was recommended by Georg von Mayr, an authority in the field of vital statistics in Germany.

statistics and on the proper means of rendering the graphic tables uniform and comparable'.[123]

Finally, contributing to this fusion of cartography and statistics was a major factor that helped endow maps with scientific legitimacy: the rise of state institutions, especially statistical bureaus. Here statistical bureaus and national meteorological agencies had close similarity: the expansion of these institutions propelled the collection and processing of huge amounts of data related to fields such as demography and climatology that subsequently served as the sources of a wide variety of graphic maps. Thus, the *Bureau de la Statistique Graphique*, created by the French Ministry of Public Works in 1878, was devised to prepare figurative charts and diagrams expressing statistical documents relating to all the economic facts, technical or financial, which could be of interest to the administration of public works.[124] As a result, it published the *Album de Statistique Graphique* from 1879 to 1897, whose volumes have been regarded as the 'finest specimens of French graphic work in the century'.[125] The statistical bureaus also commissioned the preparation of statistical atlases based on the results of censuses. The US Census Office first included maps in its results of decennial censuses in 1874, when the Superintendent Francis Amasa Walker began to publish the series of national statistical atlases.[126] In general, the Western industrial states—inasmuch as they led the efforts to produce accurate statistical graphics—played a crucial role in heightening and guaranteeing the scientific legitimacy of maps.

The state, in brief, was a key element in the scientific legitimacy of statistical graphics. Its government agencies not only sponsored official statistical atlases and maps, but also legitimized the product. The graphic method became an essential feature of official publications, and, therefore, a valid and universal language. As the then young American psychologist G. Stanley Hall wrote in a newspaper, after returning from a year's study in Europe in 1879, 'the graphic method is fast becoming the international language of science'.[127] As will be seen below, the rise of statistical representation had implications for meteorological cartography.

---

123 Congrès international de statistique à la Haye. *Compte rendu des travaux de la septième session.* La Haye, 1870, p. 71.
124 Palsky (1996, pp. 141–42).
125 Funkhouser (1937, p. 336).
126 This series was continued by more volumes from each of the 1880 and 1890 censuses, edited under the direction of Henry Gannett. Walker (1874). For the use of mosaic diagrams, bilateral histograms, and other graphic forms in the US census atlases, Friendly (2008, pp. 522–25).
127 After earning the first psychology doctorate awarded in America, Hall went to Europe, and worked at Wilhelm Wundt's Laboratory in Leipzig: 'In Germany, where it is most developed, it has revolutionized certain sciences by its unique logical method, [converting] the lecture room into a sort of theatre, where graphic charts are the scenery, changed daily with the theme'. Hall (1879, pp. 238–39)—quoted by Funkhouser (1937, p. 331).

State auspice and the character of weather maps

In the preceding paragraphs I have given some idea of the transformation of statistical cartography in the period 1860–1890. However, did this transformation affect meteorological cartography as a constituent part thereof? In this section, I will examine more closely the character of the weather maps published by national meteorological agencies under state patronage in the last three decades of the century. By the late 1870s most countries with telegraph networks had founded these agencies, especially through their military and agricultural services.[128] Since they were responsible for predicting and charting the weather, as well as for ensuring instrumental calibration and standardization, and since states had the duty to establish standards and the symbols to be used on weather charts, what should we expect the character of the state-sponsored weather maps to be?

The agreements achieved in the international meteorological meetings provide a good starting point for an answer to this question. The 1870s witnessed a process of scientific internationalism that led to the holding of several congresses and the foundation of the International Meteorological Organization (IMO). The resolutions and standards adopted in these forums served only as recommendations, for national representatives were at liberty to refuse them. Very often, issues such as political sovereignty and national identity carried more weight than international standards, as many historians remind us.[129] As a result, each state chose its particular balance between its own methods and international standards.[130] However, these national dynamics did not preclude the existence of common features in the weather maps—a fact that went virtually unnoticed for those same historians.[131]

The first International Meteorological Congress illustrates the concurrence of common patterns. In the spring of 1873, the Government of Austria invited the governments disposing weather services to a congress of directors to be held in Vienna. The Congress, presided over by Buys Ballot, was

---

128 Such was the case in Prussia (1847), Russia (1849), Great Britain (1854), Italy (1863), Norway (1866), United States (1870), Denmark (1872), India (1875), and France (1878), to mention but a few. For a table of development of weather services, see Kutzbach (1979, p. 12). For a survey more in line with meteorological observing systems than with national weather services as a basis for comparisons, see Fleming (1988, pp. 359–62, 1997, pp. 249–58).

129 The lack of unanimity in the use of metric system (vying with the British units) and the diversity in the conventions of maps are commonplace in the historiography of meteorology. See Cannegieter (1963, pp. 10–11), Khrgian (1970, pp. 133–34), Edwards (2006, p. 231), Anderson (2006, pp. 70–71).

130 On how national rivalries stymied the implementation of an international formal system of maritime meteorology in the maritime conferences of Brussels (1853) and London (1874), see Naylor (2015, pp. 92–95).

131 Although the emphasis on the points of divergence rather than convergence in the conventions of maps is one of her aims, Anderson (2006, p. 71) also points to some common patterns, which I shall deal with later.

attended by thirty-two representatives of twenty governments.[132] The five commission vice-chairmen presented an agenda with twenty-eight items, most of which were discussed at a preliminary conference held in Leipzig in 1872. 'Uniformity of procedure among different nations', they stated in the *Report of the Proceedings*, 'is indispensable', since 'meteorology can be prosecuted with success only when it is treated internationally'.[133] Although the questions discussed in the commissions mainly concerned organization and instruments, there was room for discussion of notations and graphic forms.[134] Of such questions, it is worth noting two examples. These served organizers both to formulate definitions on meteorological phenomena for general instruction and demonstrate the need for a uniformity of symbols in official weather charts and climatological tables.

The first of the questions on which the Congress adopted a resolution was the list of the symbols to be used internationally to designate types of precipitation and other phenomena on weather charts and in climatological tables. Detailed instructions were provided on the form and place they should have in publications. For example, the symbols for the character of precipitation should be inserted in the Remarks column, and the monthly *résumé* should give the sum of the days of rain and snow separately.[135] The symbols adopted—twenty-one in all—ranged from a black circle for rain and a star for snow to a black triangle for hail and three parallel lines for fog (see Figure 5.4 in Chapter 5).[136]

The second question concerned storm warnings, in particular the need for observers to provide barometric gradients in addition to the wind direction and force at the time of observation. The issue of barometric gradients was a matter of heated debate in the subcommittee on weather telegraphy and storm warnings.[137] Finally, an agreement was reached whereby 'the reduction of the barometer readings to the mean sea-level' would be 'admissible' for those stations which were not 'above 300 metres above the sea'.[138]

132 Most of the directors of meteorological institutes and services in Europe and North America attended. Only France and Prussia were absent.

133 *Report of the Proceedings of the Meteorological Congress at Vienna. Protocols and Appendices.* London: G.E. Eyre & W. Spottiswoode for Her Majesty's Stationery Office, 1874. Translated from the official report—reproduced in 'The Meteorological Congress at Vienna'. *Nature*, May 7, 1874, 17–18.

134 On the resolutions drawn at Vienna, Cannegieter (1963, pp. 10–11), Khrgian (1970, pp. 133–34), Walker (2012, pp. 110–11). See also Grahame (2000, p. 109).

135 'The Meteorological Congress at Vienna'. *Nature*, May 7, 1874, 17.

136 The table of figures can be found in *Report of the proceedings of the Meteorological congress at Vienna*, 1874, pp. 48–49. Schoder (1874, pp. 48–49).

137 The issue in question was whether barometric gradients should be referred to differences between readings at different stations, to the mean normal height of the barometer, or to 30 inches at sea level. For more details, see Chapter 5.

138 *Report on Weather Telegraphy and Storm Warnings, Presented to the Meteorological Congress at Vienna, by a Committee Appointed at the Leipzig Conference.* London: Her Majesty's Stationery Office, 1874, 13.

For greater heights the gradients should be referred to mean normal heights of the barometer. Interestingly, the reason for the adoption of this resolution had more to do with established cartographical practices than with observational considerations. In a report of the subcommittee its members admitted that 'this agreement was attained with special reference to the drawing of isobars for the purpose of Storm Warnings'—which was nothing less than an affirmation of the importance of isobars in weather mapping.[139]

The questions on cartographical construction and the conventions of maps were taken up again at the second International Meteorological Congress in 1879.[140] The forty delegates of eighteen countries met in Rome, divided into five subcommissions, and addressed synoptic mapping. To begin with, they recommended the adoption of the meridian of Greenwich as 'a starting point for the construction of synoptic meteorological charts'.[141] Of the eight subjects of general interest chosen by the subcommission on 'organization', two dealt with charts of the movements of storms and daily synoptic charts—the rest was about data collection and tables. They also posed the question: 'in what manner can we best organise and develop the construction and publication of synoptic charts embracing considerable portions of the globe?'[142] To this end, Niels Hoffmeyer, the director of the Danish Meteorological Institute, prepared a report on the construction of maps.

If these examples of synoptic notation and construction may be taken as representative of special interest—and the national delegates clearly regarded them to be so—then one may say that support by government agencies and state participation had in effect reinforced the isolinear character of weather charts in the interests of the uniformity of procedures. In other words, the increasing involvement of the states pushed the visually predominant element in large-scale weather maps to shift from observation points to isolines. This conclusion is consistent with what Hendrik Gerrit Cannegieter, chief of the IMO secretariat, stated in 1963 that the aforementioned agreements on synoptic notations and symbols 'remained practically unchanged for several generations'.[143]

In this regard, the states validated the isolinear weather map, transforming it into an object endowed with scientific legitimacy. In 1893, Mark Harrington, the Weather Bureau's chief and the founder and editor of the *American Meteorological Journal*, directed a petition to the eighteen countries known to be the most important promoters of government-sponsored

139 *Report on Weather Telegraphy and Storm Warnings,* 1874, 13.
140 'The Meteorological Congress at Rome'. *Nature,* May 15, 1879, 57–59.
141 *Report of the Proceedings of the Second International Meteorological Congress at Rome.* London: Her Majesty's Stationery Office, 1879, 11.
142 *Report of the Proceedings of the Second International Meteorological Congress,* 1879, 37.
143 Cannegieter (1963, p. 10).

weather mapping, requesting copies of synchronous weather maps for the dates 1–5 January 1892.[144] The copies were sent to the Weather Bureau Library where they were compared. As a result, Harrington wrote a paper entitled 'History of the weather map' in which he compared, among other things, the maps' content, size, and symbols.[145] These copies were of great value for Harrington: they served him both to determine to what extent the standard form of isoline graphics was reproduced in the weather maps of the official institutions, and to identify the points of convergence and divergence in the conventions of maps.[146]

In his comparative study, Harrington showed that national weather services unanimously opted for the drawing of isolines, devoting a map to isobars and winds and other to isotherms and precipitation.[147] Moreover, virtually all of them included forecasts or probabilities on their maps, though only two, France and the United States, predicted the incidence of floods.[148] What is more, all maps had a common stamp: a synopsis containing a statement of the prevailing weather conditions, with reference to changes in the last twenty-four hours. These statements were often tabulated. Similarities can be also found in issues of colour and synchronicity: practically all maps were in 'black lines on blue water and white land'[149]; and the maps covering only one country were made with synchronous observations (Table 8.1).

Inseparable from the orientation of weather maps was the so-called 'national issue'. In this sense too, as was acknowledged at the time, weather cartography was shaped by issues of national pride and sovereignty. Harrington noted that while the international symbols were usually employed on the European maps, the Australian and Japanese maps used the American symbols. He also highlighted the high graphic uniformity reached in the maps of one nation—the case of the United States was paradigmatic:

> In addition to the bi-daily weather map issued at Washington, 73 stations of the United States issue maps independently, each for a considerable area about the station, as a center. There were thus issued 8,830

144 The only exception to the mentioned date was for the Spanish maps, which dated back to 2 April 1893.

145 The paper was presented at the International Meteorological Congress held in Chicago in 1893. Harrington (1895, pp. 327–35).

146 As regards Mark Harrington and his leadership at the Weather Bureau, see Whitnah (1961, pp. 61–81), and Cox (2002, pp. 109–16).

147 In the case of the Belgian, British, French, and Saxon services the two maps were equal in size; the Italian issue included an additional map (three in all) of isobars.

148 The exception was the Algerian weather service.

149 Harrington (1895, p. 331). The German, Swedish, and Australian services used other colours.

Table 8.1 List of national weather agencies issuing weather maps in 1893

Country	Agency	Date of beginning	Scale (geography)	Publication	Agency's director
Algeria	Bureau Africain	08.4.1877	Europe (westward)	*Bulletin Météorolo gique de l'Algérie*	
Australasia[a]	Sydney Obser.	1877	Australia and New Zealand	*Weather Chart*	H.C. Russell
	Queensland Post & Telegraph	07.1877		*Weather Chart of Australasia*	C.L. Wragge
Austria	K.K. Centralans-talt für Met.	1877	Greece to Russia (westward)	*Telegraphischer Wetterbericht*	J. Hann
Bavaria	K.K. Met. Central Station	04.1881	Italy to Finland (westward)	*Wetterkarte und Wetterbericht*	C. Lang
Belgium	Observatoire Royale Brussels	25.9.1876	Bavaria to Scandinavia	*Bulletin Météorologique*	
France	Bureau Central Météorologique	16.9.1863	Europe and Algeria	*Bulletin Météorologique*	E. Mascart
Germany	Deutsche Seewarte	01.1.1876	Europe	*Wetterbericht*	G. Neumeyer
Great Britain	Meteorological Office	23.3.1872	Bavaria to Gulf of Bothnia	*Daily Weather Report*	R.H. Scott
India	Meteo. Reporter for India	13.10.1887	India	*Indian Daily Weather Report*	J. Eliot
Italy	Ufficio Centrale di Met. e di Geo.	01.9.1880	France to Austria (southward)	*Bolletino Meteo-rico giornaliero*	P. Tacchini
Japan	Central Meteo. Observatory	01.3.1883	Japan	*Tri-daily weather maps*	K. Kobayashi
Russia	Obs. Phys. Centrale St. Petersburg	12.5.1889	France to North Cape (eastward)	*Bulletin Météorologique*	H. Wild
Saxony	K. Sächs. Meteor. Institut	07.1878	Austria to Finland	*Wetterbericht*	P. Schreiber
Spain	Instituto Central Meteorológico	01.5.1893	Spain	*Boletin*	A. Arcimis
Sweden	Meteorologiska Central Anstalten		Finland to Saxony (westward)	*Synoptisk Karta*	R. Rubenson
Switzerland	Schweiz. Met. Centralanstalt	01.7.1880	Greece to Finland (westward)	*Wetterbericht*	R. Bilwiller
United States	Weather Bureau	01.1.1871	United States and southern Canada	*Weather Map*	M.W. Harrington

Source: Personal compilation based on the information provided by Mark W. Harrington (1895, p. 334, Table 1), in his paper 'History of the Weather Map', read at the Chicago International Meteorological Congress, 1895.

[a] The weather maps from Australasia were issued by the Queensland Post and Telegraph Department at Brisbane, Australia.

maps, daily, on June 1, 1893 [...] These maps are of uniform size, 750 by 325 millimeters. They are in black lines on white and blue, like the Washington map. The isobars, isotherms, wind, state of sky, and rain are given, with a brief tabular list of rainfall, weather, or river notes, on the right hand margin, and a forecast for a limited area for the next day in the lower left-hand corner. The rain areas are given on many, and on some are marked out areas of decided rise or fall of temperature. [150]

As with many of the official statistical charts mentioned in the previous section, a national dimension was strongly present in these weather maps.[151] By representing weather conditions by certain standards being uniform in the whole country, American forecasters achieved an image of internal homogeneity.[152] This process cannot be reduced to mere chauvinism; rather, it was the result of the search, by the Weather Bureau, for a balance between the international standards and its own cartographic techniques.[153]

Recently some detailed historical studies have in part confirmed the deductions of Harrington and Cannegieter. Comparing the early charts of FitzRoy and Galton with later weather maps, historian Katharine Anderson has drawn attention to the remarkable shift away of the focus of representation in meteorological cartography in the 1860s and 1870s—first and foremost in the United States, France, and Britain—showing how this focus moved from the importance of local conditions to the need for global visions of the atmosphere. She found a 'shift away from types of representation that summarized individual data, but retained the record of individual or local detail, towards types of representation that also summarized multiple observations, but subordinated them to broad patterns'.[154] Again, Frederik Nebeker identified the main broad pattern, when he claimed that 'it was above all the pattern of isobars to which forecasters paid attention'; in his view, forecasters' main task was 'to form a mental or physical picture of the pressure distribution for the coming day'.[155] However, they did not regard

---

150 Ibid., p. 333.
151 Six national weather services produced maps covering a single country. In the cases that maps covered more than one country, the map-producing country often occupied the central position on the map.
152 As was shown in Chapter 7, the case of Willis Moore, chief of the Weather Bureau from 1895 to 1913, illustrates well how he adapted the national weather map to a small locality while working as the local forecaster at the Milwaukee weather station in 1892. Drawing on the graphic and instrumental techniques used at the Washington central office, he devised several local forecasting guidelines from storm tracks drawn on weather maps. Moore (1922, pp. 153–55).
153 Monmonier (1999, p. 9) argues that, although the Weather Bureau offices produced their own weather snapshots, local forecasters were able to 'relate the national pattern to local conditions'.
154 Anderson (2006, p. 86).
155 Nebeker (1995, p. 37).

this shift as reflecting both a general statalization and bureaucratization of the purposes of weather services, and a mental posture fostered by the standardization of international conventions.[156]

The central governments' willingness to sponsor meteorological cartography was not limited to weather maps or short-term forecasting. It also extended to the climatic maps (both regional and national) reflecting the geographic distribution of the monthly and annual average values of climatic variables. What distinguished the states' involvement in this venture was not so much the types of charts produced or funded as the determination of governments to make daily weather mapping and image-production a routine task of state administration, so that maps were readily available to the public, either for forecasting purposes or to provide official climatic reports. To some degree, the mapping of national weather and climate resembled the graphic representation of national statistics: weather synoptic charts, climatic maps, and the statistical synthesis in thematic maps were a way of generating knowledge as an instrument of control.

Maps for forecasting rather than research

The national weather services in the United States, France, and Britain were not the only ones to make forecasts from isobar maps. The Deutsche Seewarte in Hamburg, detaching itself from Dove's local and Lagrangian method as well as from his stance against the use of these maps, introduced synoptic meteorology in 1875.[157] By 1890 almost all national services followed their lead. However, most of them constructed synoptic maps for forecasting and storm warning, not for research. As I will show, there was a common feature that characterized weather mapping at that time: its immediate usefulness.

In the 1890s and 1900s, forecasters could share with climatologists several tasks involving a great deal of calculation and often their cooperation was essential for data processing. Yet they had different aims and tempos. For example, forecasters, like climatologists, needed to reduce barometric readings to standard conditions, apply corrections, convert units if necessary, estimate the values measured indirectly, and master cipher codes to interpret the telegraphed observations. There was, however, a fundamental difference: while climatologists used the processed data for non-immediate studies, forecasters urgently needed them for short-term weather predictions.[158]

---

156 Here statalization does not refer to nationalization as opposed to privatization, but rather to the process of development of the state attributes in scientific services.

157 Its first director, Georg von Neumayer, hired Wladimir Köppen from the Central Physical Observatory in St. Petersburg to introduce synoptic meteorology. See Köppen (1921, pp. 289–92), and Lüdecke (2008, p. 41).

158 Calculation was especially required for the construction of auxiliary maps. Such was the case with humidity charts and temperature-change charts, prepared prior to forecast at the Weather Bureau. See Moore (1898–1899, p. 12).

The quickness of action or occurrence and its immediate usefulness affected procedural aspects. It was well known that forecasters preferred graphical procedures and international units—the tools they needed if they wanted to obtain predictions in just a few hours. Looking back on the Meteorological Office's procedural methods in the late 1900s, British meteorologist Ernest Gold stated that 'the need for common units' was felt 'even more acutely' by forecasters than by climatologists.[159] They all were aware that the adoption of a single system of units avoided all the tedious conversions. The use of numerical tables elicited similar feelings. Whereas for the climatologists numerical tables were highly helpful because they speeded computation enormously, for the forecasters their use was somewhat cumbersome as the search for data through columns and rows was time-consuming.[160] These and other factors made graphical techniques the preferred procedure of calculation in weather forecasting.

Although the principles used in the preparation of synoptic charts were very far from uniform, there are also some common general patterns. A study by Alfred J. Henry from the Weather Bureau well illustrates the essence of these patterns. As was described in the introduction of Chapter 6, Henry published an article on 'Weather forecasting from synoptic charts' in 1906 in which he discerned five steps in the daily practice of forecasting.[161] Let us recall again those steps. First, a team of telegraphers decoded the messages containing reports from weather stations. Next other clerks recorded the data in statistical registers and delivered the information to a group of cartographers. Then the cartographers transcribed the numerical expressions and symbols of the various meteorological elements on separate charts—one principal and three auxiliary in the case of Henry; and two principal and several auxiliary, according to Harrington's 1895 survey. These data usually corresponded to each station's air temperature, atmospheric pressure, wind speed and direction, amount of precipitation, and degree of cloudiness, and were plotted on a blank outline map presenting rivers, state boundaries, and other features. The next step was to draw isobars and isotherms, so that 'the eye quickly perceives the salient features

---

159 This need was specially felt in places like Britain. Gold (1934, p. 124): 'Much preliminary work was being done to permit of pressures being expressed in centibars or millibars and rainfall in millimetres. Instruments had to be regraduated or replaced and tables of reduction and conversion prepared. The *Daily Weather Report* was remodeled and these changes were incorporated in it at the beginning of May, 1914'.

160 This point—the use of numerical or graphical procedures for calculation in weather forecasting—has been stressed by Nebeker (1995, p. 43). Alfred Henry was a professor of meteorology at the Weather Bureau from 1900 to 1917.

161 Henry (1906, pp. 297–316). Other accounts describing the construction of synoptic charts at the Weather Bureau include Abbe (1899, pp. 228–34), Milham (1936, pp. 360–68), Prindle (1898, pp. 307–33), Smith (1891, pp. 49–52), and Whitnah (1961, p. 63). On weather mapping in general, Monmonier (1999, pp. 7–9).

of the chart', as well as the auxiliary charts.[162] Thus, a clerk outlined the changes in air pressure occurred in twelve and twenty-four hours; another compared the maximum and minimum temperatures for each twenty-four hours; and a third prepared two maps, one showing regional variations in humidity and the other describing the amount, kind, and direction of clouds. Finally, all these maps were handed over to forecaster, who, after identifying centres of low and high pressure, made a forecast for the next twenty-four or thirty-six hours.

The main pattern that emerges from this and other testimonies of the forecasters at that time is unequivocal. For reasons of immediate application and operational convenience the diverse meteorological variables were plotted on separate and independent charts rather than on one single chart. Various considerations related to the preparation of weather forecasts and storm warnings led meteorologists to draw charts showing either only the temperature field and wind, like Buys Ballot's charts, or isobars and wind data on one hand and isotherms on the other hand, as Harrington noted in his 1895 survey. Some had to do with work distribution, others with the increasing specialization of cartographers, and all with image-production.[163] However, by showing data separately and partially they all in part hid information that could be valuable for researchers.[164]

To an influential meteorologist like Jerome Namias, who was to pioneer long-range weather forecasts in the United States some decades later, the view of the Weather Bureau on the production and utility of maps in its early years was 'rather provincial and narrow':

> For many decades the procedure used in the Weather Bureau had been to construct a series of charts each showing separate elements. For example, one map showed pressure, another pressure changes, another cloud and cloud movements, another precipitation, etc. The entire series of maps had to be studied by the chief forecaster, since there was

---

162 Henry (1906, p. 300).

163 The process of image-production not only required a synchronized team of cartographers, but also was subject to the huge demands of the printing process. In 1909, thirty-three large weather stations outside Washington and seventy-nine smaller stations printed 10.5 million copies of the daily weather map in the United States. While stations like New York, Boston, and Chicago daily printed over a thousand maps, many smaller stations produced around a dozen maps daily. See U.S. Weather Bureau, *Report of the Chief of the Weather Bureau, 1908–1909*. Washington, D.C., 1910, 31—quoted by Monmonier (1988, p. 18).

164 While not analyzing its causes, Kutzbach (1979, p. 71) describes the prevalence and implications of this practice: 'It became customary at most weather services, when preparing weather reports and forecasts, to plot the various meteorological elements on separate charts rather than one single chart. This procedure saved time, but it had the disadvantage that a certain amount of information was lost to the viewer. Synoptic charts prepared for research purposes, however, generally incorporated the maximum amount of information available'.

no central map containing all the information [...] It must have been frustrating for the forecasters to have to walk around to inspect several maps on different tables to obtain a synoptic view.[165]

Within the forecast-oriented weather chart production, one of the few cases that seemed to vie as an exception was Alexander Buchan. From 1860, he was secretary of the Scottish Meteorological Society and editor of its journal until his death in 1907.[166] Buchan played an active role in synoptic research. In the early 1860s he constructed a series of synoptic charts which were comparable to those of Paris or even those of the Weather Bureau in the 1870s, but only in some aspects. Like his colleagues in Paris and Washington, he contributed to the progress of weather forecasting. However, his synoptic charts were conceived and intended to follow the movement of weather systems across Europe, particularly the midlatitude storms. Weather forecasting was an objective in his aspirations, but the scientific target was to ascertain the general laws governing atmospheric changes, the finding of which would lead him to a calculation of the coming weather. Thus, from an outsider position, with no direct link to a national weather service, Buchan would devise charts which not only allowed him to extend Buys Ballot's principle of the relationship between the pressure and wind fields but also became exemplary for thorough synoptic analysis.[167]

In 1865, Buchan published a series of eighteen synoptic charts constructed from data from 135 stations scattered over Europe. He reduced pressure observations to sea level and opted for temperature departures from the mean instead of absolute values to correct the anomalies derived by the use of different instruments.[168] His mapping technique involved incorporating isobars and isotherms, as well as data on winds, cloud cover, and precipitation for a twenty-four-hour period into one single chart (see Figure 5.2 in Chapter 5). By this means he identified thirty-two areas of low pressure or depressions. From a careful perusal of his charts Buchan concluded that winds blew in a counterclockwise direction around a depression in the Northern Hemisphere, their speed being proportional to the closeness of the isobars.[169] In short, Buchan saw in the synoptic chart and its isoline technique something entirely different than what forecasters saw: a means to investigate the general laws regulating atmospheric changes.

---

165 Namias (1981, p. 492).
166 Shaw (1907, pp. 83–84); Mitchell et al. (1908, pp. 101–18).
167 Buchan was inspector of Scottish stations for the British Meteorological Office from 1877 to 1906. Yet, in the 1860s he conducted research through the Scottish Meteorological Society. Shaw (1907, p. 84).
168 Buchan (1865, pp. 191–205).
169 The closeness of the isobars was defined as the barometric gradient by Thomas Stevenson in 1868. For more details, see Chapter 5.

However original we may find Buchan and his subtle distinction between investigation and forecast, we must recognize that his synoptic chart was a *rara avis*—the exception that proved the general rule at that time.[170] The testimony of another outsider, Henri Peslin, a mining engineer at Tarbes, France, bears witness to this.[171] In 1868, Peslin published the theoretical study entitled 'Sur les mouvements généraux de l'atmosphère', in which he supported the thermal theory of cyclones.[172] In this work, he suggested that the thermal energy and the gravitational potential energy should be considered as the source of the kinetic energy in storms, an energy that was continuously dissipated by friction. Although he deduced his results with no assistance of daily synoptic charts, only drawing on the mathematico-physical reasoning, he resorted to the synoptic charts to obtain empirical evidence for his theory.[173]

Peslin expressed the insufficiencies inherent to the Paris Observatory's synoptic charts more clearly and forcefully than any other meteorologist of that era:

> We have examined the synoptic charts published by the Imperial Observatory in the last years, but we have found no evidence that clearly confirms or refutes the theory that we have just developed. The synoptic charts provide the barometric pressure, and wind direction and force; they show the tracks followed by storms over the earth's surface, but they do not afford temperature or hygrometrical degree, or especially the law of temperature decrease in the vertical component; in brief, none of the data that would lead us to verify our ideas.[174]

To a theorist like Peslin, atmospheric processes should be elucidated from physico-mathematical formulations rather than merely empirical evidence. A year later he submitted a paper to the Academy of Sciences in Paris, but

---

170 Kutzbach (1979, p. 87) states that Buchan's synoptic charts were virtually 'the only ones that contained information on temperatures'. Although Henrik Mohn published a storm-atlas in 1870 containing daily temperature data, he opted for separate charts instead of seeking combination of those data with the pressure and wind fields on one single chart. Mohn (1870).

171 Peslin's work and biographical details have been unnoticed in France. See Rochas (2005). See also the brief note in *Quarterly Journal of the Royal Meteorological Society*, 1873, 1, 127; Kutzbach (1979, esp. pp. 45–49, 52–57, 64, 84–90, 101); and Cox (2002, p. 90).

172 Peslin (1868). For a thorough discussion on this work and Peslin's contribution to the convective or thermal theory of cyclones, see Kutzbach (1979, pp. 84–87). See also Middleton (1966, pp. 163–65).

173 Those days it was commonplace to see this type of approach. Hermann von Helmholtz and Theodor Reye carried out theoretical investigations on the energetics of cyclones without using synoptic charts. Their approach was in sharp contrast to that of Buchan and Mohn, who afforded empirical evidence on the convective theory of cyclones on the basis of examining collections of data capsulized in synoptic charts. For these studies, see Kutzbach (1979, pp. 71–99).

174 Peslin (1868, p. 319).

a commission including Le Verrier rejected it for publication.[175] In this paper he mathematically treated the relationship between pressure gradient, centrifugal force, and the earth's rotation. Furthermore, his results could well have been applied to weather forecasting for he formulated an equation whereby the velocity of winds could be easily calculated from the variations of barometric pressure.[176] Yet, there is every indication that Peslin's formula was never incorporated into the Paris Observatory's forecasting method.[177]

Similar conclusions can be drawn from the case of William Ferrel and his early studies on the mechanics of the atmosphere. As shown in Chapter 1, Ferrel formulated in 1859 the equations of motion for a body moving on a rotating earth, and showed how a particle (air, water, or the like) in motion was deflected by the Coriolis effect. By examining the influence of this effect on atmospheric motions, he sought to find common elements in both cyclones and the general circulation of the atmosphere.[178] When, in 1870, the work of storm predictions was imposed upon the US Signal Office, Abbe found a desolate landscape. Two decades later, he will write: 'Copies of Ferrel's [1859] memoir had indeed been distributed at the time of its first publication, but he was too far in advance of the ordinary student of meteorology to be fully appreciated at that time'.[179] Although he was imbued with Ferrel's mechanical principals and Espy's thermodynamic ideas when he began the work of daily weather predictions, Abbe always distinguished these early 'deductive predictions' from the subsequent empirical forecasts made at the Signal Office. It is 'fair to say', he admitted, that 'my predictions of the first few years were of a deductive character based upon my confidence in the truth of the principles developed by Espy and Ferrel'.[180]

---

175 For Rochas (2005, p. 45), this incident reveals not only the Academy's inability to recognize the value of Peslin's work but also how the Academy's ostracism aborted the meteorological career of amateurs like Peslin and others.

176 Peslin (1878, pp. 472–75). Its title, 'Relation between Barometric Variations and the General Atmospheric Currents', suggests such a possibility of application.

177 Cox (2002, p. 90). The vicissitudes of this paper reflect the obstacles encountered by Peslin. After its rejection by the Academy, the paper came to Marié-Davy's hands, then director of the Observatory at Montsouris, who was at enmity with Le Verrier, and who, contrary to his rival's view, helped the work be published in the *Bulletin International de l'Observatoire de Paris* in 1872. Five years later, Cleveland Abbe translated it into English, publishing together with other ten little known memoirs in the *Annual report of the Board of Regents of the Smithsonian Institution*.

178 For Ferrel's contribution to meteorology and the thermal theory of cyclones in particular, see Davis (1891), Abbe (1895a), Kutzbach (1979, pp. 35–41, 110–117), and Cox (2002, pp. 65–74)

179 Abbe (1891, p. 343). Abbe (1891, p. 347): 'In [Ferrel's] opinion the low state of meteorology throughout the world as compared with other exact sciences, and even as compared with climatology, arose from the fact that trained physicists, mechanicians and mathematicians had not yet been induced to take up the study of the phenomena of the atmosphere and there could be no progress until they did so'.

180 Abbe (1891, p. 345). Abbe introduced his American colleagues to Ferrel's theories, especially through Abbe's popular work *Practical Use of Meteorological Reports and*

In seeking the reasons for the flagrant inconsiderateness—not to say hostility, as has been suggested—towards Peslin's theoretical work circa 1870, and the tardy (albeit in practice modest) welcome of Ferrel's ideas, one cannot but point to the increasingly positive valuation of graphic methods, especially synoptic charts, among meteorologists at the time.[181] For meteorologists, in particular those from national services, were beginning to realize the sufficiency of the weather charts with isobars and isotherms for predictive purposes—despite the limitations evinced by Buchan, Mohn, and others—and to distrust other techniques not geared to image-production. In addition to this, there was the opposition to the thermal theory of cyclones in France and Germany, where early mechanical theories still dominated the debate, but its influence on meteorological services' staffs was little.

All the foregoing allusions to the limitations of synoptic charts are an indication of the increasing importance of isoline maps for forecast and storm warning in national weather services.

National expertise, voluntarist internationalism, and global knowledge

Meteorology has often been defined as a global science of massed observation, a science that symbolizes the relationship between local observations and global forces. Recent historiography of environmental and geophysical sciences has analyzed this double dimension, placing the focus on the changing relations and balance between the forces at play.[182] Thus, where the idiosyncrasy of diverse international ventures in meteorological sciences is studied, the aim pursued is often to reflect the relation and balance between the local and the global. However, international, intercontinental, and inter-regional enterprises and scales are often confused with global ones. In their search for this relation and balance, the global—and by extension, global knowledge—is used as a basic and unproblematic category.

Although it is undeniable that knowledge of the weather did not steadily progress from the local to the global, and that the movement towards the global was halted or even reverted, a careful perusal of the international

---

Weather-Maps. Washington: Government Printing Office, 1871. Although they assimilated the idea that atmospheric motions were deflected by the earth's rotation, it is far from clear that his fellow meteorologists applied his thermal theory of cyclonic storms to daily weather forecast practice.

181 According to Kutzbach (1979, p. 88), Peslin's work 'was met not merely with indifference but with outspoken hostility from the established French authorities on the subject of meteorology'.

182 Case studies dealing with the relation between local phenomena and global perspectives in these sciences include Anderson (2006), Coen (2006), and Westermann (2011). See also the introduction to the work edited by Jeremy Vetter, *Knowing Global Environments: New Historical Perspectives on the Field Sciences* (2011).

synoptic charts at the time shows a much more complex reality.[183] Before proceeding any further, however, it should be noted that in this section we understand global knowledge in two senses.[184] The first refers to a form of knowledge associated with informational structures (such as telegraph systems, observational networks, and postal services) and organizational and procedural realities, which moves between and beyond various territories, connecting institutions and individuals scattered throughout the world. This construal, applied to the case of meteorological cartography, will raise questions about how operational knowledge was extended to an international realm, in what way and in whose interests. The second sense concerns the investigation of phenomena, patterns, or trends considered by scientists as worldwide in scope, extent, and nature. This definition will attract attention to physical laws and regularities rather than international contacts, transcontinental communications, and institutional relations.

A discussion of meteorological cartography in the second half of the nineteenth century immediately calls to mind a host of scientific-cultural initiatives whose common feature is their tacit or unequivocal adhesion to an emerging ideology at the time: scientific internationalism.[185] Yet, when one observes that international meteorological relations were often characterized by national rivalries and enmities, by boycotts and counter-boycotts, then one is led to think that the expansion of global knowledge entailed obstacles and tensions. Was not the expansion of global knowledge in meteorological sciences only possible because the ideology of scientific internationalism triumphed and because the assertion of the universality of science prevailed over more spurious and mundane interests? That was not, however, the case—at least not for the vast majority of meteorologists. As I will show below, the expansion of global knowledge through synoptic weather charts is only explainable by reference to a struggle of forces expressed at three levels (local, national, and global), rather than to a supposed balance between the local and the global. In this struggle, the global was marked by an internationalism that was based on voluntarism and on its subordination to national interests.

The origins of the international synoptic charts can be traced back to the Paris Observatory and its director Le Verrier, who in 1858 was the first to publish a daily sheet of observations including a lithographed map of the

---

183 Some of the works collected by J.R. Fleming, V. Jankovic, and D.R. Coen in the work *Intimate Universality: Local and Global Themes in the History of Weather and Climate* clearly question the progressivist model that has been traditionally maintained in the history of meteorology.

184 Finnegan and Wright (2016, p. 5) suggest the need to differentiate two senses of 'global knowledge', though their proposal does not distinguish the operational aspects from the theoretical ones.

185 For an overview of the conceptions of scientific internationalism from the Enlightenment to the Cold War, and their relations to nationalism, cosmopolitanism, and 'the West', see Somsen (2008).

weather over Europe—the *Bulletin Meteorological International*. But it was not until 1868 that the first of a vast series of oceanic synoptic charts were printed. The series of five books, entitled *Atlas des mouvements généraux de l'atmosphère*, contained maps of isobars and winds over the Atlantic, and intended to follow storm tracks across Newfoundland, Iceland, and the Azores before arriving in Europe.[186] These oceanic maps did not corroborate the thesis of the tropical origin of storms held by Marié-Davy. Insofar as they displayed the formation of atmospheric perturbations in the Atlantic, the *Atlas* of synoptic maps was of paramount importance for the study of storms.[187]

It would be tempting to suggest that the oceanic synoptic charts helped to elucidate the Atlantic origin of European storms, allowing the Paris Observatory to take the lead. In a letter addressed to Quetelet in 1865, Marié-Davy admitted the non-tropical origin of midlatitude European storms.[188] But even if it seems clear that oceanic synoptic maps contributed to a better understanding of global atmospheric movements, it is not true that the only or even the most important driving factor behind this international initiative was the scientific one. On 24 July 1865, Le Verrier explained to the Paris Academy of Sciences the importance of these maps, constructed from seventy telegraph dispatches from different parts of Europe. In his allocution, Paris emerged as the core of a Pan-European telegraphic network:

> A central and European bureau for this part of international service seemed indispensable. Looking at the charts of pressure, one can immediately see that the curves covering the overall surface of Europe would have not their precise meaning if they were reduced by half of the size.

So, it became necessary to extend scale. Here, 'France fulfils the role of first occupant; her geographical situation would even give this leadership in a natural way'.[189] The correlation between meteorology and scientific leadership was manifest in this case: when we move out from the focus of storm origins and look at the telegraphic data exchange as a whole, then it is control and domination rather than cooperation that emerges as an outstanding feature of this international initiative.

This readiness to give precedence to national and geopolitical considerations in the issue of international meteorological relations can be shown

---

186 Valuable accounts of the Paris Observatory's maps can be found in Hildebrandsson and Teisserenc de Bort (1898, pp. 65–67, 134–35), Grandidier (1882, pp. 536, 541– 42), Hellmann (1897, p. 8). See also Cox (2002, pp. 295–99).

187 Locher (2008, p. 136) shows how these maps changed Marié-Davy and his assistants' view on the origin of storms.

188 Marié-Davy to Quetelet, 30 April 1865. Archives de l'Académie Royale des Sciences de Belgique, Brussels, M1720—quoted by Locher (2008, p. 136).

189 Le Verrier (1868, p. 56).

in another striking example, led by the US Chief Signal Officer, General Albert Myer, a few years later. By 1870, Myer had foreseen the need of international exchanges with Canada, West Indies, and Europe. From 1871 to 1872, Abbe often expressed to Myer the importance of a daily map of the atmosphere as a whole, especially the Northern Hemisphere. In his account of the discussions with Myer, he underlined the confluence of various interests: 'The work seemed likely to be for the benefit of American commerce in whatever port of the world its ships might be'. However, Abbe added,

> it was a work in which all nations would be likely to co-operate (provided only there could be found the proper central nation—the one friendly power which would be recognized by all as the one with which they could co-operate).

Leading this work was a formidable challenge for the post-Civil War geopolitical rebalancing: 'the United States was, fortunately, in an independent position that enabled it to command the hearty co-operation of all'.[190] After weighing the matter, Myer submitted his plan to the International Meteorological Congress in Vienna in 1873 in which he proposed that the nations of the world prepare a series of simultaneous observations for the study of world climatology and weather patterns.[191] A striking example of the lack of global aspirations, at least those related to standard procedures, is the strategy followed by Myer in this congress, according to his own confession to Abbe:

> I have followed your advice quite closely; I strictly avoided all discussions and controversies as to theories and methods, and stated that we did not care how the observations were made and reduced, provided only they were strictly simultaneous and promptly forwarded. [192]

Most national representatives in Vienna recognized Myer and the Signal Office as the legitimate voice of this international enterprise. This explains why seventeen national services contributed to the *Bulletin of International Simultaneous Observations*, a publication edited by the Signal Office from 1875 to 1889, which included a daily international weather map from 1878.[193] For Abbe, it represented 'the finest piece of international co-operation in scientific work' even seen.[194] Certainly, in its funding, organization, and

190 Abbe (1895b, pp. 256–57).
191 *Bericht über die verhandlungen des Internationalen meteorologen-congresses zu Wien : 2.-16. september 1873. Protokolle und Beilagen.* Wien, 1873, pp. 27–29, 58.
192 Abbe (1895b, p. 257).
193 'International Meteorology'. *Symons' Monthly Meteorological Magazine*, 8 (1873–74), 181; Fleming (2000, p. 327); Fuller (1990, p. 5); Raines (2011, p. 54).
194 Abbe (1895b, p. 257).

scale, it was an enterprise without precedent in the European systems. However, in its interests, infrastructures, and aspirations, it was an international and intercontinental, rather than fully global, enterprise. While the synoptic maps followed an already familiar global pattern (centred on the North Pole, with black heavy lines as isobars and red dotted lines as isotherms, see Figure 8.5), organizational and procedural issues related to observation and mapping, such as calibration and standardization, were far from being global.[195] Even the isoline technique, whose application entailed a global perspective, was inappropriately used in vast areas of the Atlantic where the few data collected came from scattered ships.[196]

Notwithstanding the foregoing, the pursuit of exclusively scientific ends could sometimes serve as justification for promoting international meteorological collaboration. This was the main reason adduced by Niels Hoffmeyer, the founder of the National Meteorological Institute in Denmark, for the preparation of the international maps (or *Nordatlantik-Karten*) published in Copenhagen from September 1873 to November 1876.[197] In 1873, the participants at the International Meteorological Congress in Vienna were asked to give their opinion on the need to prepare a form of publication suited for international purposes.[198] For Hoffmeyer, this was a pressing need, especially for the study of dynamical meteorology. 'While the observations of more or less local phenomena are not placed in relation with the general meteorological conditions of a vast surface, their causes shall not be sufficiently recognized'. And he added:

> that is why I have vigorously insisted [in Vienna] on the centralization of the service of meteorological telegraphy, for it is impossible for the national systems in Europe to have sufficient extension to determine the position of the high-pressure masses.[199]

The maps mentioned earlier, which were mostly financed by his own resources, responded to that need: their isobaric curves encompassing the north of the Atlantic and a part of the surrounding continents symbolized centralization and voluntary collaboration (Figure 8.6).[200] This was,

---

195 This fact seems to be corroborated from Abbe's own words: 'Although the international bulletins and charts were widely distributed throughout the world, [...] this data has been as yet used by very few persons outside of the Weather Bureau as a means of investigating problems in meteorology'. Abbe (1895b, p. 258).

196 Although describing these limitations, Anderson (2006, p. 85) adopts a 'globalist' stance on the simultaneous international weather maps, stating that they showed 'how the global organization of meteorologists matched the global forces of the atmosphere'.

197 Hoffmeyer (1874–1880). These maps were jointly published by Hoffmeyer and Georg Neumayer from the Deutsche Seewarte in Hamburg in 1880, and later by Neumayer himself, to 1914. See Schröder and Wiederkehr (1992, p. 52).

198 'The Meteorological Congress at Vienna'. *Nature*, May 7, 1874, p. 18.

199 Quoted by Zurcher (1875, p. 36).

200 Hoffmeyer recommended to use the intermediary of meteorological central stations to all those who wished to receive his maps.

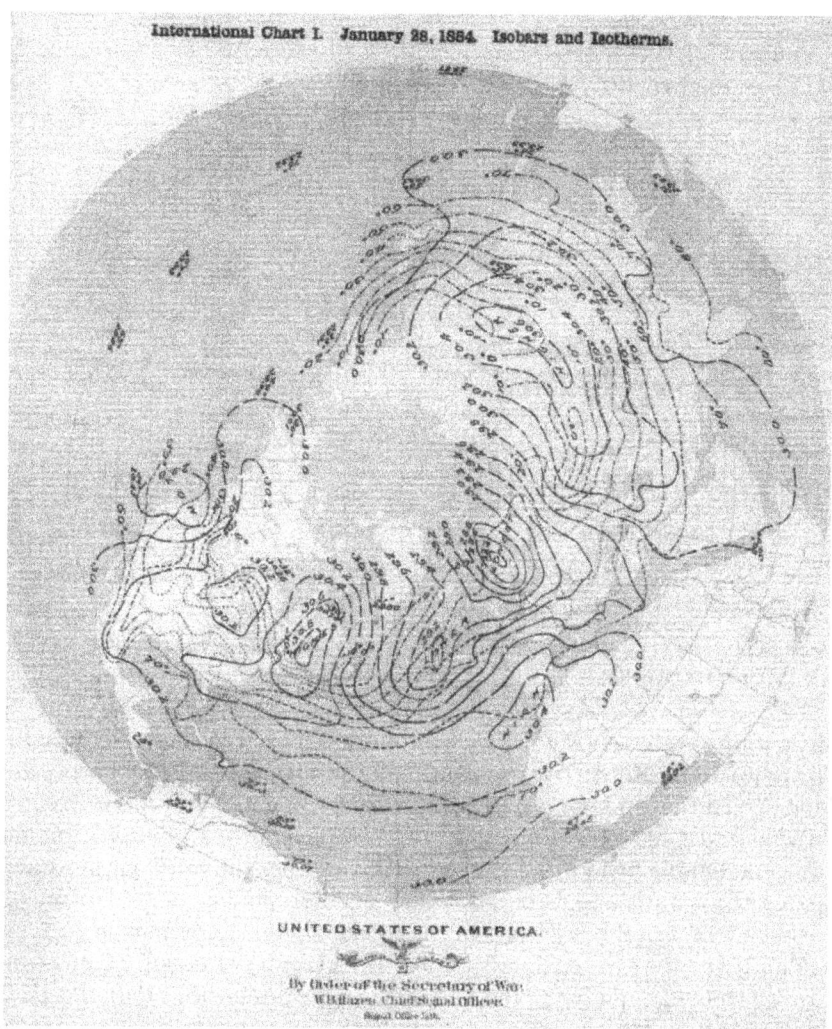

Figure 8.5 An international weather map illustrating pressure distribution by isobars (heavy lines) and temperature by isotherms (dotted lines) in the Northern Hemisphere.

Source: Bulletin of International Meteorology for the Month of January (Washington, 1885), Chart II.

therefore, the result of an individual action that aimed to pursue purely scientific ends, and was materialized through governmental services.[201]

201 These maps should not be confused with the daily charts prepared by Hoffmeyer for research purposes from 1868 onwards—four maps in total, one for the isobars for the month, another for the isothermals, the third for the minimum temperature at each station, and the fourth for isohyetal lines. 'Niels Henrik Cordulus Hoffmeyer'. *Nature,*

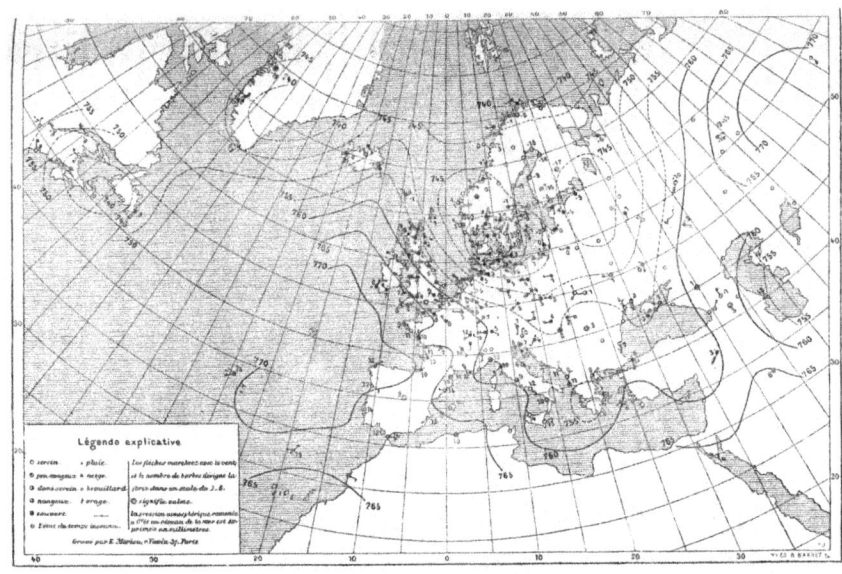

Fac simile d'une carte synoptique de l'Institut météorologique danois.

Figure 8.6 Hoffmeyer's international synoptic map, 20 March 1874.
Source: Zurcher (1875, p. 37).

More multifaceted, however, was the project that Hoffmeyer devised in 1880 on the establishment of telegraphic communication with the Faroes and Iceland.[202] To this end, he obtained a private audience with King Oscar II of Sweden and Norway, who agreed with him that this communication network would enable meteorologists to appreciate the state of the atmosphere in the North Atlantic and forecast weather further in advance.[203] Hoffmeyer had established weather stations in the former Danish colonies in Iceland, Greenland, and the Faroe Islands.[204] Yet, the viability of laying a cable from Europe to Iceland and Greenland had many more dimensions than the meteorological one; in fact, some French chambers of commerce did point out the services that such a cable would render to fishery.[205] Perhaps for this reason, Hoffmeyer proposed to the Danish Ministry of Naval Affairs to distribute expenses among the countries involved. But still, the project did not prosper.

March 6, 1884, 434–35. See also Hildebrandsson and Teisserenc de Bort (1898, pp. 135–36, 168–69).

202 Hoffmeyer (1880).

203 Hoffmeyer's project received support in international forums such as the Wien conference on agricultural weather services and the Bern meeting of the Permanent Meteorological Committee. See Smed (2011, pp. 10–11).

204 For Danish provincial and island weather stations, see Brandt (1994, pp. 10–12).

205 E. Mascart to the Minister of Public Instruction, 5 February 1882. National Archives of France, Public Instruction, F/17/3822.

Also in the 1880s one can find some examples of this same propensity to promote a voluntarist meteorological internationalism and even to subordinate the global to the nation's interests. The British Meteorological Council prepared daily weather charts of the North Atlantic for thirteen months, from 1 August 1882 to 31 August 1883.[206] As the Council asserted in an official statement issued in London, '[we] believe that any systematic information' from this part of the Ocean would 'be of immediate benefit to seamen' and 'would tend directly to the improvement of the forecasts and storm warnings issued to the British coasts'. And next they added: 'the object now proposed can only be achieved by the voluntary cooperation of an increased number of observers', especially 'owners, captains, and officers of ships', as well as 'the great companies whose steamers ply between this country and America'.[207] In mentioning the precedents of Le Verrier and Hoffmeyer, the Council asserted that the control and possession of such maritime information were essential to the process of weather forecasts at the service of Britain's seafaring community. In fact, the telegraphic predictions transmitted by the *New York Herald* to England and founded, according to the Council, on the reports of the vessels arriving in America could not 'be utilised in a scientific investigation of weather'.[208]

The meteorologists who headed the weather services of the British colonies developed international relations somewhat differently in other circumstances and at the service of other interests. These relations took shape in the standard weather chart of Australasia, a daily isobaric weather chart that was issued by government meteorologist in Queensland, Clement L. Wragge, from 1887 to 1892.[209] One year before its launch Wragge established the Meteorology Society of Australasia in Adelaide. Addressing an audience of educated authorities and meteorology enthusiasts at the inaugural sessions of this Society, he emphasized that 'as in politics, so in science we desire federation' which, for the case of meteorology at least, 'meant the continuance of the efforts towards uniformity in equipment, data-collecting, publishing and the training of observers'.[210] Although this Society did not survive much longer, the issue of federation took form in the Australasian weather chart, as a result of the daily exchange of telegraphic weather data between Australia and New Zealand.[211]

---

206 *Synchronous weather charts of the North Atlantic and the adjacent continents for every day from 1st August 1882 to 31 August 1883*. London: Her Majesty's Stationery Office, 1886.
207 Quoted in 'Daily Weather Charts in the North Atlantic'. *Nature*, April 27, 1882, p. 606.
208 Quoted in 'Daily Weather Charts in the North Atlantic'. *Nature*, April 27, 1882, p. 606.
209 Wragge (1898). He used to name the low-pressure systems and tropical cyclones on the weather charts. He usually followed an alphabetical sequence: the Greek alphabet for tropical cyclones over the Coral Sea, feminine Christian names for other series, etc. See Newman and Deacon (1956, p. 3).
210 Quoted by Hoare (1975, p. 12). The first quotation is from Wragge (quoted in *Meteorological Society of Australasia: History, Rules, Regulations and List of Members*. Adelaide, 1886, pp. 1–5); the second is from Hoare himself.
211 This exchange was secured at the third Australasian Meteorological Conference held in Melbourne in September 1888.

In view of the foregoing, it must be admitted that in considerable measure the bearers of international relations were professed 'globalists' who advocated the universality of the results of scientific research and the need to understand the global forces of atmospheric circulation. The widespread use of isoline technique is evidence of this concern. However, for the purpose of our present inquiry the most important circumstance is that in the pursuit of this global knowledge, meteorologists not only refrained from demanding the uniformity of procedural and organizational systems but also subordinated their scientific internationalism to specific interests of their states. Their strategy to undertake international weather charts was to promote an internationalism based on voluntary adhesion and at the service of their own nations. So should we ask which of the five initiatives examined of international weather maps responded to a genuinely internationalist desire, surely the answer would be almost none.

Conclusion

On 16 September 1863, Marié-Davy published the first weather map of isobars from data collected telegraphically. Though meteorological prediction was not born with telegraph, the synoptic map quickly became the central tool for generating and disseminating of weather forecasts in Europe and America, and as such it occupies a prominent place in the modern landscape. But if its importance in the history of meteorology has been studied, less known is its role in the history of nation-state building. The evolution of the synoptic weather map, especially in its first decades, illuminates the manner in which national weather services became unquestioned parts of the scientific and social structures of the modern states. But not only that, this evolution also shows the way of generating knowledge as an instrument of national control and surveillance.

The interconnection between meteorological science and the interests of state-building markedly shaped the way in which the weather data were collected and mapped. Until the mid-1850s, renowned scientists driven by climatologic concerns had produced synoptic maps of storms and weather (indeed, *retrospective* maps with reference to past events) in which the objective pursued was to ascertain the laws of nature and its physical causes—i.e. maps as tools for weather *diagnosis*. From then on, in the conviction—which originates from Le Verrier's yearning for order and security after the 1854 Black Sea storm—that an effective method of preventing such threats was the centralization of state-controlled observational systems, the French meteorologists constructed maps as tools for weather *prognosis*, representing what they called isobaric charts of *bourrasques*. Like their Italian predecessors, they shifted the focus from observation points to isolines. In doing so they pioneered and extended to other countries a method of weather forecasting based on an Eulerian approach and the idea that the atmospheric pressure distribution basically determined weather changes. They sowed the seeds for a method destined to become very entrenched in national weather services.

The developments shown in this chapter indicate the power of isolines, and isobars in particular, to create 'things which hold' (paraphrasing sociologist Alain Desrosières), that is, symbols and forms of representation which acquire the status of objective realities. Once the *Highs* and *Lows* were given official existence and their motion was traceable and trackable by the instrument of cartography, they entered the routine wheel of image-production. In contrast to the view of the Lagrangian approach's standard bearers, who held that these charts must represent the different air masses and their relationship with meteorological elements, the isobaric maps gained the scientific legitimacy from government weather services, attaining great popularity among the general public. They indeed became the basis for weather forecasting in most Western countries from the 1870s onwards—as the first international meteorological meetings enabled specific agreements regarding the drawing of isobars for forecasting purposes to be signed. Endorsed by these international agreements and reproduced in innumerable synoptic maps of national weather services, the isobaric *Highs* and *Lows* became endowed with a status of 'scientific fact' which masked their nature of artefacts and the context of their true conception. The air-mass or air currents concept, in contrast, did not become the basis for the production of weather forecasts that its promoters such as FitzRoy had envisaged. To that end, other methods were required, methods that had to do with physical research and that the authorities, however, were not willing, nor gifted, to promote.

The weather forecasts generated by national weather services were predicated on an idea of the 'highs' and 'lows' as an object to be known, traced, measured, and predicted. Not in vain did the 'lows' embody the true carriers of bad weather, and the 'highs' the announcers of good weather. They indeed formed part of the official picture of a national meteorological space that, until then, had not existed. They made up a space of which the state was the tutor and patron, its paternalistic promoter and the subsidizer of all meteorological bureaucratization. This role fitted well with the policies of modern nation-states whose interest in security, order, and surveillance for the sake of commerce and agriculture could be satisfied through what they considered valid methods to prevent uncontrollable natural threats.

There is, however, a paradox in this picture of the role of the nation-state in the shaping of weather forecasting. Meteorologists may well have regarded national weather spaces as the result of global knowledge and universal laws, but statesmen often conceived nations as monolithic and authoritative entities. In accounting for the continuing demand for international agreements in meteorology in this critical period of time, the dominant factor was the experience and needs of the states themselves. Procedural uniformity, standardization, and the rise of statistical cartography all played a part, but states were behind scientific internationalism just as scientific internationalism strengthened states. Nation-states, in turn, reformulated the old links of global ventures by promoting an internationalism based on voluntary adhesion and at the service of their own interests. Even when the first international meteorological meetings in Leipzig and Vienna claimed to represent universal principles and pursue global knowledge, they were indeed structured by the participation of nation-states. After two decades of international meetings, in which agreements on notation and procedures were the order of the day, the national defence of measurement units and graphic conventions became more and more apparent as the century drew to an end. While isolines, isobars, and depressions became more widespread, nation-states sought to handle and used them to their own ends.

Hence the weather forecast map—the most emblematic expression of any national meteorological space—did not spontaneously emerge. As I have shown, it had an unmistakable development: it shifted from a tool for diagnosis to a tool for prognosis. Once destined to prediction, it witnessed a struggle between approaches, a struggle in which the Eulerian line of thought prevailed over the Lagrangian approach. However, this dominant though not fully successful shift towards the integration of civil state-controlled weather data collection and image-production did mark something new in the history of state administration. If security and surveillance are two of the hallmarks of modern states, then the generation of weather forecast maps must be seen as one of the seminal facts in nation-state building.

References

Abbe, Cleveland, 1871. *Practical Use of Meteorological Reports and Weather-Maps*. Washington, DC: Government Printing Office.

Abbe, Cleveland, 1891. 'Ferrel's Influence in the Signal Office'. *American Meteorological Journal*, 8, 342–48.

Abbe, Cleveland, 1895a. 'Memoir of William Ferrel, 1817–1891'. *In Biographical Memoirs of the National Academy of Sciences*. Washington, DC: National Academy of Sciences, 265–309.

Abbe, Cleveland, 1895b. 'The Meteorological Work of the U.S. Signal Service, 1870 to 1891'. [U.S. Weather Bureau] *Bulletin*, 11, 232–85.

Abbe, Cleveland, 1897. 'Weather Telegraphy in England and America'. *Monthly Weather Review*, 25, 205–06.

Abbe, Cleveland, 1899. *The Aims and Methods of Meteorological Work*. Baltimore, MD: The Johns Hopkins Press.

Achbari, Azadeh, 2017. *Rulers of the Winds. How Academics Came to Dominate the Science of the Weather, 1830–1870*. Amsterdam: Vrije Universiteit.

Achbari, Azadeh and Lunteren, Frans van, 2016. 'Dutch Skies, Global Laws: The British Creation of "Buys Ballot's Law"'. *Historical Studies in the Natural Sciences*, 46 (1), 1–43.

Agnew, Duncan Carr, 2004. 'Robert Fitzroy and the Myth of the "Marsden Square": Transatlantic Rivalries in Early Marine Meteorology'. *Notes and Records of the Royal Society of London*, 58, 21–46.

Anderson, Katharine, 2005. *Predicting the Weather: Victorians and the Science of Meteorology*. Chicago, IL: The University of Chicago Press.

Anderson, Katharine, 2006. 'Mapping Meteorology'. In James Rodger Fleming, Vladimir Jankovic, and Deborah R. Coen eds., *Intimate Universality. Local and Global Themes in the History of Weather and Climate*. Sagamore Beach, MA: Watson Publishing International, 69–91.

Annual Report of the Board of Regents of the Smithsonian Institution, for the Year 1857. Washington, DC, 1858.

Annual Report of the Board of Regents of the Smithsonian Institution, for the Year 1858. Washington, DC, 1859.

Annual Report of the Board of Regents of the Smithsonian Institution, Showing the Operations, Expenditures, and Condition of the Institution for the Year 1865. Washington, DC: Government Printing Office, 1866.

Barboza, Christina Helena da Motta, 2012. *As viagens do tempo. Uma história da meteorologia em meados do século XIX*. Rio de Janeiro: E-papers Serviços Editoriais Ltda.

Beniger, James R., and Robyn, Dorothy L., 1978. 'Quantitative Graphics in Statistics: A Brief History'. *The American Statistician*, 32 (1), 1–11.

Benjamin, Marcus, 1897. 'Meteorology'. In George Brown Goode ed., *The Smithsonian Institution 1846–1896. The History of its First Half Century*. Washington, DC: [s.n.], 647–78.

Bergeron, Tor O., Devik, Olaf, and Godske, Carl Ludvig, 1962. 'Vilhelm Bjerknes, March 14, 1862–April 9, 1951'. *Geofysiske Publikasjoner*, 24, 7–62 [*In Memory of Vilhelm Bjerknes on the 100th Anniversary of his Birth Oslo*: Det Norske Videnskaps-Akademi I Oslo, 1962].

Bericht über die verhandlungen des Internationalen meteorologen-congresses zu Wien : 2.-16. september 1873. Protokolle und Beilagen. Wien, 1873.

Blondheim, Menahem, 1994. *News over the Wires: The Telegraph and the Flow of Public Information in America, 1844–1897*. Cambridge, MA: Harvard University Press.

Brandes, Heinrich Wilhelm, 1817. 'Aus einem Schreiben des Professor Brandes, meteorologischen Inhalts'. *Annalen der Physik*, 55, 112–14.

Brandes, Heinrich Wilhelm, 1819. 'Einige Resultate aus der Witterungs-Geschichte des Jahres 1783, und Bitte um Nachrichten aus jener Zeit; aus einem Schreiben des Professor Brandes an Gilbert'. *Annalen der Physik*, 61, 421–26.

Brandes, Heinrich Wilhelm, 1820. *Beiträge zur Witterungskunde*. Leipzig: Johann Ambrosius Barth.

Brandt, Marie Louise, 1994. *Summary of Meta Data from NACD-Stations in Denmark, Greenland and the Faroe Islands, 1872–1994*. København: Danish Meteorological Institute.

Brunt, David, 1951. 'A Hundred Years of Meteorology'. *The Advancement of Science*, 30, 114–24.

Buchan, Alexander, 1865. 'Examination of the Storms of Wind which Occurred in Europe during October, November and December 1863'. *Transactions of the Royal Society of Edinburgh*, 24, 191–205.

Burstyn, Harold L., 1966. 'Early Explanations of the Role of the Earth's Rotation in the Circulation of the Atmosphere and the Ocean'. *Isis*, 57 (2), 167–87.

Burton, James, 1986. 'Robert FitzRoy and the Early History of the Meteorological Office'. *British Journal for the History of Science*, 19, 147–76.

Buys Ballot, Christophorus H.D., 1854. 'Erläuterungen einer graphischen Methode zur gleichzeitigen Darstellung der Witterungserscheinungen an vielen

Orten, und Aufforderung der Beobachter das Sammeln der Beobachtungen an vielen Orten zu erleichtern'. *Annalen der Physik und Chemie*, Ergänzungsband, 4, 559–76.

Buys Ballot, Christophorus H.D., 1857. '[Note]'. *Verslagen en Mededeelingen der Koninklijke Akademie van Wetenschappen: Afdeeling Natuurkunde*, 7, 75–77.

Buys Ballot, Christophorus H.D., 1860. *Eenige regelen voor aanstaande weersveranderingen in Nederland*. Utrecht: Kemink & Zoon.

Buys Ballot, Christophorus H.D., 1868. *De Invoering en verklaring van den aeroklinoskoop. Regelen naar welke hij ons de aanstaande windveranderingen met eenige waarschijnlijkheid doet vermoeden*. Utrecht: Kemink & Zoon.

Buys Ballot, Christophorus H.D., 1882. *Beredeneerd register op de werken van het Koninklijk Nederlandsch Meteorologisch Instituut tot 1882*. Utrecht: Kemink & Zoon.

Camerini, Jane R., 1993. 'The Physical Atlas of Heinrich Berghaus: Distribution Maps as Scientific Knowledge'. In Renato G. Mazzolini ed., *Non-Verbal Communication in Science Prior to 1900*. Florence: Leo S. Olschki, 479–512.

Cannegieter, Hendrik Gerrit, 1963. 'The History of the International Meteorological Organization, 1872–1951'. *Annalen der Meteorologie*, 1, 7–280.

Chapman, Sidney, 1941. 'Edmond Halley As Physical Geographer, and the Story of His Charts'. *Occasional Notes, Royal Astronomical Society*, 1, 122–34.

Cheysson, Émile, 1878. 'Les méthodes de statistique graphique à l'Exposition universelle de 1878'. *Journal de la Société de Statistique de Paris*, 19, 323–33.

Coen, Deborah R., 2006. 'Scaling Down: Mapping the Austrian Climate between Empire and Republic'. In James Rodger Fleming, Vladimir Jankovic, and Deborah R. Coen eds., *Intimate Universality: Local and Global Themes in the History of Weather and Climate*. Sagamore Beach, MA: Science History Publications, 115–40.

Congrès international de statistique à la Haye. *Compte rendu des travaux de la septième session*. La Haye, 1870.

Cox, John D. 2002, *Storm Watchers: The Turbulent History of Weather Prediction from Franklin's Kite to El Niño*. Hoboken, NJ: John Wiley & Sons, Inc.

'Daily Weather Charts in the North Atlantic'. *Nature*, April 27, 1882, 605–06.

Davis, John L., 1984. 'Weather Forecasting and the Development of Meteorological Theory at the Paris Observatory, 1853–1878'. *Annals of Science*, 41 (4), 359–82.

Davis, William Morris, 1891. 'Ferrel's Contributions to Meteorology'. *American Meteorological Journal*, 8, 348–59.

De Young, G., 1985. 'The Storm Controversy (1830–1860) and its Impact on American Science'. *EOS*, 66 (38), 657–60.

Dettelbach, Michael, 1993. 'Romanticism and Administration: Mining, Galvanism and Oversight in Alexander von Humboldt's Global Physics'. Ph.D. diss., University of Cambridge.

Dettelbach, Michael, 1999. 'The Face of Nature: Precise Measurement, Mapping, and Sensibility in the Work of Alexander von Humboldt'. *Studies in the History and Philosophy of Biology and the Biomedical Sciences*, 30, 473–504.

Dettwiller, Jacques, 1982. 'La loi de Buys Ballot'. *La Météorologie*, 28, 61–70.

Dove, Heinrich Wilhelm, 1852. *Die Verbreitung der Wärme auf der Oberfläche der Erde: erläutert durch Isothermen, thermische Isanomalen und Temperaturcurven*. Berlin: D. Reimer.

Dry, Sarah, 2009. 'Safety Networks: Fishery Barometers and the Outsourcing of Judgement at the Early Meteorological Department'. *The British Journal for the History of Science*, 42 (1), 35–56.

Edwards, Paul N., 2006. 'Meteorology as Infrastructural Globalism'. *Osiris*, 21, 229–50.

Engelmann, Gerhard, 1964. 'Der Physikalische Atlas des Heinrich Berghaus und Alexander Keith Johnstons Physical Atlas'. *Petermanns Geographische Mitteilungen*, 108, 133–49.

Farr, William ed., 1861. *Report of the Proceedings of the Fourth Session of the International Statistical Congress: Held in London, July 16th, 1860 and the Five Following Days.* London: G.E. Eyre and W. Spottiswoode, for H.M. Stationery Off.

Feldman, Theodore, 1990. 'Late Enlightenment Meteorology'. In Tore Frängsmyr, John L. Heilbron, and Robin E. Rider eds., *The Quantifying Spirit in the 18th Century.* Berkeley: University of California Press, 164–77.

Ficker, Adolf ed., 1857. *Rechenschafts-Bericht über die dritte Versammlung des Internationalen Congresses für Statistik, abgehalten zu Wien vom 31. August bis 5. September 1857.* Vienna: K.K. Holf -und Staatsdruckerei.

Finnegan, Diarmid A., and Wright, Jonathan Jeffrey eds., 2016. *Spaces of Global Knowledge: Exhibition, Encounter and Exchange in an Age of Empire.* London: Routledge.

FitzRoy, Robert, 1855. 'Wind Charts of the Atlantic, Compiled from Maury's Pilot Charts'. *Report of the British Association for the Advancement of Science.* London, 2, 39–40.

FitzRoy, Robert, 1861. 'On British Storms, Illustrated with Diagrams and Charts'. *Report of the British Association for the Advancement of Science 1860: Transactions of the Section.* London, 39–44.

FitzRoy, Robert, 1863. *The Weather Book: A Manual of Practical Meteorology.* London: Longman and Green.

FitzRoy, Robert, 1864. *Report of the Meteorological Department of the Board of Trade.* London: H.M.S.O.

Fleming, James Rodger, 1988. 'Meteorology in America, 1814–1874: Theoretical, Observational, and Instrumental Horizons'. Ph.D. diss., Princeton University.

Fleming, James Rodger, 1990. *Meteorology in America, 1800–1870.* Baltimore, MD: Johns Hopkins University Press.

Fleming, James Rodger, 1997. 'Meteorological Observing Systems before 1870 in England, France, Germany, Russia, and the United States: A Review and Comparison'. *World Meteorological Organization Bulletin*, 46, 249–58.

Fleming, James Rodger, 2000. 'Storms, Strikes, and Surveillance: The U.S. Army Signal Office, 1861–1891'. *Historical Studies in the Physical and Biological Sciences*, 30, 315–32.

Friendly, Michael, 2008. 'The Golden Age of Statistical Graphics'. *Statistical Science*, 29 (4), 502–35.

Friis, Herman R., 1974. 'Statistical Cartography in the United States Prior to 1870 and the Role of Joseph C. G. Kennedy and the U.S. Census Office'. *The American Cartographer*, 1 (2), 131–57.

Fuller, John F., 1990. *Thor's Legions. Weather Support to the U.S. Air Force and Army, 1937–1987.* Boston, MA: American Meteorological Society.

Funkhouser, H. Gray, 1937. 'Historical Development of the Graphical Representation of Statistical Data'. *Osiris*, 3, 269–404.

Galton, Francis, 1863. *Meteorographica, or, Methods of Mapping the Weather: Illustrated by Upwards of 600 Printed and Lithographed Diagrams Referring to the Weather of a Large Part of Europe, During the Month of December 1861.* Cambridge: Macmillan.

Gillham, Nicholas Wright, 2001. *A Life of Sir Francis Galton. From African Exploration to the Birth of Eugenics.* Oxford: Oxford University Press.

Gold, Ernest, 1934. 'Incidents in the "March," 1906–1914'. *Quarterly Journal of the Royal Meteorological Society*, 60, 121–25.

Grahame, Nick, 2000. 'The Development of Meteorology over the Last 150 Years As Illustrated by Historical Weather Charts'. *Weather*, 55, 108–17.

Grandidier, Alfred, 1882. *Exposition universelle international de 1878 à Paris. Rapport sur les cartes et les appareils de géographie et de cosmographie, sur les cartes géologiques et sur les ouvrages de météorologie et de statistique.* Paris: Imprimerie Nationale.

Halford, Pauline, 2004. *Storm Warning. The Origins of the Weather Forecast.* Stroud: Sutton Publishing.

Hall, Granville Stanley, 1879. 'The Graphic Method'. *The Nation*, 29, 238–39.

Halley, Edmond, 1686. 'An Historical Account of the Trade Winds, and Monsoons, Observable in the Seas between and near the Tropicks, with an Attempt to Assign the Physical Cause of the Said Winds'. *Philosophical Transactions*, 16, 153–68.

Hankins, Thomas L., 1999. 'Blood, Dirt, and Nomograms: A Particular History of Graphs'. *Isis*, 90, 50–80.

Harrington, Mark, 1895. 'History of the Weather Map'. In Oliver L. Fassig ed., *Report of the International Meteorological Congress, Chicago, Ill., August 21–24, 1893.* Washington, DC: Weather Bureau, Part 2, 327–35.

Headrick, Daniel R., 1991. *The Invisible Weapon: Telecommunications and International Politics, 1851–1945.* New York: Oxford University Press.

Hellmann, Gustav, 1897. *Neudrucke von Schriften und Karten über Meteorologie und Erdmagnetismus. N. 8. Meteorologische Karten.* Berlin: A. Asher & Co.

Henry, Alfred J., 1906. 'Weather Forecasting from Synoptic Charts'. *Journal of the Franklin Institute*, 162, 297–316.

Henry, Joseph, 1854. 'Report of the Secretary of the Smithsonian Institution to the Board of Regents, December 8, 1847'. Reprinted in *Eight Annual Report of the Board of Regents of the Smithsonian Institution, up to January 1, 1854, and the Proceedings of the Board up to July 8, 1854.* Washington, DC: A.O.P. Nicholson, 119–39.

Hildebrandsson, Hugo H., and Teisserenc de Bort, Léon Philippe, 1898. *Les bases de la météorologie dynamique.* Paris: Gauthier-Villars.

Hoare, Michael E., 1975. 'The Intercolonial Science Movement in Australasia 1870–1890'. *Records of Australian Academy of Science*, 3 (2), 7–28.

Hoffmeyer, Niels Henrik Cordelus, 1874–1880. *Cartes synoptiques journalières embrassant l'Europe et le nord de l'Atlantique.* Copenhagen: [s.n.].

Hoffmeyer, Niels Henrik Cordelus, 1880. *Étude sur les tempêtes de l'Atlantique septentrional et projet d'un service télégraphique international relatif à cet océan.* Copenhagen: M.P. Hauberg.

Hughes, Patrick, 1994. 'The Great Leap Forward'. *Weatherwise*, 47 (5), 22–27.

Humboldt, Alexander, 1811. *Political Essay on the Kingdom of New Spain.* London: Longman.

Humboldt, Alexandre de, 1816. 'Sur les lois que l'on observe dans la distribution des formes végétales'. *Annales de chimie et de physique*, 1, 225–39.

Humboldt, Alexandre de, 1817a. 'Des lignes isothermes et de la distribution de la chaleur sur le globe'. *Mémoires de physique et de chimie, de la Société d'Arcueil*, 3, 462–602.

Humboldt, Alexandre de, 1817b. 'Sur les lignes isothermes'. *Annales de chimie et de physique*, 5, 102–11.

Hunt, Bruce, 1997. 'Doing Science in a Global Empire: Cable Telegraphy and Electrical Physics in Victorian Britain'. In Bernard Lightman ed., *Victorian Science in Context*. Chicago, IL: University of Chicago Press, 312–33.

'International Meteorology'. *Symons' Monthly Meteorological Magazine*, 8 (1873–74), 181.

Javelle, Michel Rochas, Pastre, Claude, Beaurepaire, Michel, and Jacomy, Bruno, 2000. *La météorologie, du baromètre au satellite. Mesurer l'atmosphère et prévoir le temps*. Lausanne: Éditions Delachaux et Niestlé.

Jordanova, Ludmilla J., 1979. 'Earth Science and Environmental Medicine: The Synthesis of the Late Enlightenment'. In Ludmilla J. Jordanova and Roy Porter eds., *Images of the Earth: Essays in the History of the Environmental Sciences*. Chalfont St. Giles: British Society for the History of Science, 119–46.

Kämtz, Ludwig Friedrich, 1832. *Lehrbuch der Meteorologie. Bd. 2*. Halle: Gebauer.

Khrgian, Aleksandr Khristoforovich, 1970. *Meteorology: A Historical Survey*, 2nd ed. Jerusalem: Israel Program for Scientific Translations.

Kington, John A., 1980. 'Daily Weather Mapping From 1781: A Detailed Synoptic Examination of Weather and Climate during the Decade Leading Up to the French Revolution'. *Climatic Change*, 3, 7–36.

Kington, John A., 1988. *The Weather of the 1780s over Europe*. Cambridge: Cambridge University Press.

Kington, John A., 1980. 'Daily Weather Mapping From 1781: A Detailed Synoptic Examination of Weather and Climate during the Decade Leading Up to the French Revolution'. *Climatic Change*, 3, 7–36.

Kington, John A., 1988. *The Weather of the 1780s over Europe*. Cambridge: Cambridge University Press.

Konvitz, Joseph W., 1987. *Cartography in France, 1660–1848. Science, Engineering and Statecraft*. Chicago, IL: Chicago University Press.

Köppen, Wladimir, 1921. 'Dove und wir'. *Meteorologische Zeitschrift*, 38, 289–92.

Kutzbach, Gisela, 1979. *The Thermal Theory of Cyclones: A History of Meteorological Thought in the Nineteenth Century*. Boston, MA: American Meteorological Society.

Landsberg, Helmut, 1954. 'Storm of Balaklava and the Daily Weather Forecast'. *The Scientific Monthly*, 79 (6), 347–52.

Langley, Samuel Pierpont. 1895. 'The Meteorological Work of the Smithsonian Institution'. In Oliver L. Fassig ed., *Report of the International Meteorological Congress Held at Chicago, Ill., August 21–24, 1893*. Washington, DC: Weather Bureau, 2, 216–20.

Le Verrier, Urbain, 1855. 'Travail fait à l'Observatoire Impérial, par M. Liais, sur la Tempête de la Mer Noire, en novembre 1854'. *CR*, 41, 1197–204.

Le Verrier, Urbain Jean Joseph, 1868. *Historique des entreprises météorologiques: 1864–1867, Observatoire impérial de Paris*. Paris: Gauthier-Villars.

Lempfert, R.G.K., 1913. 'British Weather Forecasts: Past and Present'. *Quarterly Journal of the Royal Meteorological Society*, 39, 173–84.

Lequeux, James, 2013. *Le Verrier–Magnificent and Detestable Astronomer.* New York: Springer. Edited and with an introduction by William Sheehan. Translated By Bernard Sheehan.

Lewis, Charles Lee, 1927. *Matthew Fontaine Maury: Pathfinder of Seas.* Annapolis, MD: U.S. Naval Institute.

Locher, Fabien, 2008. *Le savant et la tempête. Étudier l'atmosphère et prévoir le temps au XIX siècle.* Rennes: Presses Universitaires de Rennes.

Locher, Fabien, 2009. 'Les météores de la modernité: la dépression, le télégraphe et la prévision savante du temps (1850–1914)'. *Revue d'histoire moderne et contemporaine,* 56 (4), 77–103. English translation as 'Atmosphere of Globalisation. Depressions, the Astronomer and the Telegraph (1850–1914)'. *Revue d'histoire moderne et contemporaine,* 56 (4) (2009), 77–103.

Loomis, Elias, 1841. 'On the Storm Which Was Experienced Throughout the United States about the 20th of December, 1836'. *Transactions of the American Philosophical Society,* 7, 125–63.

Loomis, Elias, 1846. 'On Two Storms Which Were Experienced Throughout the United States, in the Month of February, 1842'. *Transactions of the American Philosophical Society,* 9 (2), 161–84.

Lüdecke, Cornelia, 2008. 'Gründung der Deutschen Meteorologischen Gesellschaft (Ära Neumayer 1883–1889)'. *Annalen der Meteorologie,* 43, 41–47.

Lunteren, Frans van, 1998. 'De oprichting van het Koninklijk Nederlands Meteorologisch Instituut: Humboldtiaanse wetenschap, internationale samenwerking en praktisch nut'. *Gewina,* 21 (4), 216–43.

MacMunn, George, 1935. *The Crimea in Perspective.* London: G. Bell & Sons.

Marié-Davy, Edme Hippolyte, 1863a. 'Sur la météorologie'. *Revue des sociétés savantes,* 3, 232–35.

Marié-Davy, Edme Hippolyte, 1863b. 'Sur l'état de l'atmosphère pendant la première quinzaine d'août, d'après les renseignements recueillis à l'Observatoire impérial de Paris'. *CR,* 57, 384–86.

Marié-Davy, Edme Hippolyte, 1866. *Les mouvements de l'atmosphère et des mers, considérés du point de vue de la prévision du temps.* Paris: Victor Masson et fils.

Marriott, William, 1903. 'The Earliest Telegraphic Daily Meteorological Reports and Weather Maps'. *Quarterly Journal of the Royal Meteorological Society,* 29, 123–31.

Marvin, Carolyn, 1988. *When Old Technologies Were New: Thinking about Electric Communication in the Late Nineteenth Century.* New York: Oxford University Press.

Meinardus, Wilhelm, 1899. 'Die Entwicklung des Karten der Jahres-Isothermen von Alexander von Humboldt bis auf Heinrich Wilhelm Dove'. In Gesellschaft für Erdkunde zu Berlin ed., *Wissenschaftliche Beiträge zum Gedächtnis der hundertjährigen Wiederkehr des Antritts von Alexander von Humboldt's Reise nach Amerika am 5. Juni 1799: aus Anlaß des Siebenten Internationalen Geographen-Kongresses.* Berlin: Kühl, 1–32.

Meteorological Society of Australasia: History, Rules, Regulations and List of Members. Adelaide, 1886.

Middleton, William Edgar Knowles, 1966. *A History of the Theories of Rain and Other Forms of Precipitation.* New York: Franklin Watts.

Milham, Willis Isbister, 1936. *Meteorology: A Text-Book on the Weather, the Causes of Its Changes, and Weather Forecasting.* New York: Macmillan.

Miller, Eric R., 1931. 'The Evolution of Meteorological Institutions in the United States'. *Monthly Weather Review,* 59, 1–6.

Minard, Charles Joseph, 1862. *Des tableaux graphiques et des cartes figuratives.* Paris: Thunot.

Mitchell, A. et al., 1908. 'Contributions towards a Memorial Notice of Alexander Buchan, M.A., L.L.D., F.R.S.'. *Journal of the Scottish Meteorological Society,* 14, 101–18.

Mohn, Henrik, 1870. *Det Norske meteorologiske instituts storm-atlas = Atlas des tempêtes de l'Institut météorologique de Norvége.* Christiania: Bentzen.

Monmonier, Mark, 1984. 'The Rise of the National Atlas'. In John Amadeus Wolter and John Young Cole eds., *Images of the World: The Atlas through History.* Washington, DC: Library of Congress. Reprinted in *Cartographica: The International Journal for Geographic Information and Geovisualization,* 1994, 31 (1), 1–15.

Monmonier, Mark, 1988. 'Telegraphy, Iconography, and the Weather Map: Cartographic Weather Reports by the United States Weather Bureau, 1870–1935'. *Imago Mundi,* 40 (1), 15–31.

Monmonier, Mark, 1999. *Air Apparent. How Meteorologists Learned to Map, Predict, and Dramatize Weather.* Chicago, IL: The University of Chicago Press.

Mood, Fulmer, 1946. 'The Rise of Official Statistical Cartography in Austria, Prussia, and the United States, 1855–1872'. *Agricultural History,* 20, 209–25.

Moore, Willis Luther, 1898–1899. *Weather Forecasting: Some Facts Historical, Practical, and Theoretical.* Washington, DC: U.S. Department of Agriculture, *Bulletin* No. 25.

Moore, Willis Luther, 1922. *The New Air World: The Science of Meteorology Simplified.* Boston, MA: Little, Brown.

Munzar, Jan, 1967. 'Alexander von Humboldt and His Isotherms: On the Occasion of the 150th Anniversary of the First Map of Isotherms'. *Weather,* 22, 360–63.

Nalbach, Alex, 2003. '"The Software of Empire": Telegraphic News Agencies and Imperial Publicity, 1865–1914'. In Julie F. Codell ed., *Imperial Co-Histories: National Identities and the British and Colonial Press.* Madison, NJ: Fairleigh Dickinson University Press, 46–67.

Namias, Jerome, 1981. 'The Early Influence of the Bergen School on Synoptic Meteorology in the United States'. In Gösta H. Liljequist ed., *Weather and Weather Maps: A Vol. Dedicated to the Memory of Tor Bergeron (15.8.1891–13.6.1977).* Basel: Birkhäuser Basel, 491–99.

Naylor, Simon, 2015. 'Weather Instruments All at Sea: Meteorology and the Royal Navy in the Nineteenth Century'. In Fraser MacDonald and Charles W.J. Withers eds., *Geography, Technology and Instruments of Exploration.* London: Routledge, 77–96.

Nebeker, Frederik, 1995. *Calculating the Weather. Meteorology in the 20th Century.* San Diego, CA: Academic Press.

Newman, B. W., and Deacon, E. L., 1956. 'A "Dynamic" Meteorologist—Clement Wragge, 1852–1922'. *Weather,* 11, 3–7.

Newton, H. A., 1895. 'Memoir of Elias Loomis, 1811–1889'. *Biographical Memoirs* (United States, National Academy of Science), 3, 213–52.

'Niels Henrik Cordulus Hoffmeyer'. *Nature,* March 6, 1884, 434–35.

Obreen, J. M. 1869, 'La méthode graphique'. In *Congrès international de statistique à la Haye, septième session, 1869, Programme.* La Haye, 29–31.

Palsky, Gilles, 1996. *Des chiffres et des cartes: Naissance et développement de la cartographie quantitative française au XIXe siècle.* Paris: Comité des Travaux Historiques et Scientifiques.

Palsky, Gilles, 1998. 'Origines et évolution de la Cartographie Thématique (XVIIe–XIXe siècles)'. *Revista da Faculdade de Letras–Geografia* (Porto), 14, 39–60.

Patriarca, Silvana, 1996. *Numbers and Nationhood. Writing Statistics in Nineteenth-Century Italy.* Cambridge: Cambridge University Press.

Pearson, Karl, 1914–30. *Life, Letters and Labours of Francis Galton*, 3 vols. Cambridge: Cambridge University Press.

Peslin, Henri, 1868. 'Sur les mouvements généraux de l'atmosphère'. *Bulletin Hebdomadaire de l'Association Scientifique de France*, 3, 299–319. Also reproduced in the *Atlas Météorologique de l'Observatoire Impérial. Année 1867.* Paris: Charles Chauvin, 1868, D24–D28.

Peslin, H., 1872. 'Sur la relation entre les variations du baromètre et les grands courants atmosphériques'. *Bulletin International de l'Observatoire de Paris*, 1872, 29 May–7 July (11pp).

Peslin, H., 1878. 'Relation between Barometric Variations and the General Atmospheric Currents'. *Annual Report of the Board of Regents of the Smithsonian Institution..., for the Year 1877*, 465–78. [Translated from the French original by Cleveland Abbe].

Pinsel, Marc, 1981. 'The Wind and Current Chart Series Produced by Matthew Fontaine Maury'. *Navigation*, 28, 123–36.

Prindle, E.J., 1898. 'Weather Forecasts: The Manner of Making Them and Their Practical Value'. *Popular Science Monthly*, 53, 307–33.

Raines, Rebecca Robbins, 2011. *Getting the Message through a Branch History of the U.S. Army Signal Corps.* Washington, DC: Center of Military History, United States Army.

Redfield, William C., 1846. 'On Three Several Hurricanes of the American Seas, and Their Relation to the Northers So Called of the Gulf of Mexico and the Bay of Honduras, with Charts Illustrating the Same'. *The American Journal of Science* (new series), 1, 1–16, 153–69, 333–69; 2, 162–87, 311–34.

Reingold, Nathan, 1972. 'Joseph Henry'. In Charles Coulston Gillispie ed., *Dictionary of Scientific Biography.* New York: Charles Scribner's Sons, 6, 277–81.

Report of the Meteorological Department of the Board of Trade. London, 1857–1864.

Report of the Proceedings of the Meteorological Congress at Vienna. Protocols and Appendices. London: G.E. Eyre & W. Spottiswoode for H.M. Stationery Off., 1874.

Report of the Proceedings of the Second International Meteorological Congress at Rome. London: Her Majesty's Stationery Office, 1879.

Report on Weather Telegraphy and Storm Warnings, Presented to the Meteorological Congress at Vienna, by a Committee Appointed at the Leipzig Conference. London: Her Majesty's Stationery Office, 1874.

Robinson, Arthur H., 1967. 'The Thematic Maps of Charles Joseph Minard'. *Imago Mundi*, 21, 95–108.

Robinson, Arthur H., 1971. 'The Genealogy of the Isopleth'. *Cartographical Innovations*, 8, 49–53.

Robinson, Arthur H., 1982. *Early Thematic Mapping in the History of Cartography.* Chicago, IL: The University of Chicago Press.

Robinson, Arthur H., Pentchenik, Barbara Bartz, and Wallis, Helen, 1976. *The Nature of Maps: Essays towards Understanding Maps and Mapping.* Chicago, IL: The University of Chicago Press.

Robinson, Arthur H., and Wallis, Helen, 1967. 'Humboldt's Map of Isothermal Lines: A Milestone in the History of Thematic Cartography'. *Cartographic Journal*, 2, 119–23.

Robinson, Arthur H., and Wallis, Helen, 1987. *Cartographical Innovations: An International Handbook of Mapping Terms to 1900*. Tring: Map Collector Publications.

Rochas, Michel, 2005. 'H. Peslin, ingénieur des Mines à Tarbes'. *La Météorologie*, 49, 42–45.

Schneider-Carius, Karl, 1975. *Weather Science, Weather Research: History of Their Problems and Findings from Documents during Three Thousand Years*. New Delhi: Indian National Scientific Documentation Centre; originally published as *Wetterkunde, Wetterforschung: Geschichte ihrer Probleme und Erkenntnisse in Dokumenten aus drei Jahrtausenden*. Freiburg: Verlag Karl Albert, 1955.

Schoder, Hugo von, 1874. 'Appendix 1. To the Protocol of the Fourth Meeting'. *Report of the Proceedings of the Meteorological Congress at Vienna. Protocols and Appendices*. London: G.E. Eyre & W. Spottiswoode for Her Majesty's Stationery Office, 48–49.

Schröder, Wilfried, and Wiederkehr, Karl Heinrich, 1992. 'Georg von Neumayer (1826–1909) und die internationale Entwicklung der Geophysik: I. Teil: Meteorologie'. *Gesnerus: Swiss Journal of the History of Medicine and Sciences*, 49, 45–62.

Scultetus, H.R., 1943. 'Dove und Loomis als Wegbereiter der Synopsis'. *Meteorologische Zeitschrift*, 60, 419–22.

Shaw, William Napier, 1907. 'Dr. Alexander Buchan, F.R.S.'. *Nature*, 76 (1960), 83–84.

Shaw, William Napier, 1926. *Manual of Meteorology. Vol. I: Meteorology in History*. Cambridge: Cambridge University Press.

Smed, Jens, 2011. 'Early Plans for Telegraphic Communication with the Faroes and Iceland in the Interests of Meteorology and Fishery'. *History of Oceanography*, 22, 10–13.

Smith, J. Warren, 1891. 'The U.S. Signal Service Weather Forecasts'. *American Meteorological Journal*, 8, 49–52.

Stigler, Stephen M., 1986. *The History of Statistics. The Measurement of Uncertainty before 1900*. Cambridge, MA: The Belknap Press of Harvard University Press.

Synchronous Weather Charts of the North Atlantic and the Adjacent Continents for Every Day from 1st August 1882 to 31 August 1883. London: Her Majesty's Stationery Office, 1886.

'The Meteorological Congress at Rome'. *Nature*, May 15, 1879, 57–59.

'The Meteorological Congress at Vienna'. *Nature*, May 7, 1874, 17–18.

Thorburn, Thomas, 1855. *Diagrams, Illustrative of Facts, Principles, & Theories, Relating to Agriculture, Annuity, Army, Banking, &c. &c.* London.

Thrower, Norman J.W., 1969. 'Edmond Halley as a Thematic Geo-Cartographer'. *Annals of the Association of American Geographers*, 59 (4), 652–76.

Tickell, Sir Crispin, 1993. 'Meteorographica and Weather'. In Milo Keynes ed., *Sir Francis Galton, FRS. The Legacy of His Ideas*. New York: Palgrave Macmillan, 140–51.

Trabert, Wilhelm, 1905. *Meteorologie und Klimatologie*. Leipzig: Franz Deuticke.

True, Webster Prentiss, 1929. *The Smithsonian Institution*. Washington, DC: Smithsonian Institution series.

U.S. Weather Bureau, 1910. *Report of the Chief of the Weather Bureau, 1908–1909*. Washington, DC.

Velkar, Aashish, 2012. *Markets and Measurements in Nineteenth-Century Britain.* Cambridge: Cambridge University Press.

Vetter, Jeremy ed., 2011. *Knowing Global Environments: New Historical Perspectives on the Field Sciences.* London: Rutgers University Press.

Walker, Francis A., 1874. *Statistical Atlas of the United States, Based on the Results of the Ninth Census, 1870, with Contributions from Many Eminent Men of Science and Several Departments of the Government.* Washington, DC.

Walker, J. Malcolm, 2012. *History of the Meteorological Office.* Cambridge: Cambridge University Press.

Westermann, Andrea, 2011. 'Disciplining the Earth: Earthquake Observation in Switzerland and Germany at the Turn of the Nineteenth Century'. *Environment and History,* 17, 53–77.

Whitnah, Donald R., 1961. *A History of the United States Weather Bureau.* Urbana: University of the Illinois Press.

Wragge, Clement L., 1898. *Meteorology of Australia .... Standard Weather Chart[s] of Australasia, Including the Malay Archipelago and the Western Pacific. [For] January, 1898.* Brisbane: Chief Weather Bureau.

Zurcher, F., 1875. 'Cartes synoptiques de l'Institut Météorologique Danois'. *La Nature: revue des sciences et de leurs applications aux arts,* 3, 36–38.

Index

Note: **Bold** page numbers refer to tables and *italic* page numbers refer to figures.